BIOINFORMATICS

SECOND EDITION

METHODS OF
BIOCHEMICAL ANALYSIS

Volume 43

BIOINFORMATICS

A Practical Guide to the Analysis of Genes and Proteins

SECOND EDITION

Andreas D. Baxevanis
Genome Technology Branch
National Human Genome Research Institute
National Institutes of Health
Bethesda, Maryland
USA

B. F. Francis Ouellette
Centre for Molecular Medicine and Therapeutics
Children's and Women's Health Centre of British Columbia
University of British Columbia
Vancouver, British Columbia
Canada

A JOHN WILEY & SONS, INC., PUBLICATION

New York • Chichester • Weinheim • Brisbane • Singapore • Toronto

This book is printed on acid-free paper. ♾

Published simultaneously in Canada.

For ordering and customer service, call 1-800-CALL-WILEY.

Library of Congress Cataloging-in-Publication Data:

Bioinformatics : a practical guide to the analysis of genes and proteins / edited by Andreas D. Baxevanis, B. F. Francis Ouellette.—2nd ed.
p. cm.
Includes index.
ISBN 0-471-38390-2 (cloth : alk. paper) ISBN 0-471-38391-0 (paper)
1. Nucleotide sequence—Databases. 2. Amino acid sequence—Databases. 3. Genes—Analysis—Data processing. 4. Proteins—Analysis—Data processing. I. Baxevanis, Andreas D. II. Ouellette, B. F. Francis.

QP620 .B565 2001
572.8′633′0285—dc21

00-068610

Printed in the United States of America.

10 9 8 7 6 5 4

ADB dedicates this book to his Goddaughter, Anne Terzian, for her constant kindness, good humor, and love—and for always making me smile.

BFFO dedicates this book to his daughter, Maya. Her sheer joy and delight in the simplest of things lights up my world everyday.

CONTENTS

13 SEQUENCE ASSEMBLY AND FINISHING METHODS 303

Rodger Staden, David P. Judge, and James K. Bonfield

14 PHYLOGENETIC ANALYSIS 323

Fiona S. L. Brinkman and Detlef D. Leipe

15 COMPARATIVE GENOME ANALYSIS 359

Michael Y. Galperin and Eugene V. Koonin

FOREWORD

I am writing these words on a watershed day in molecular biology. This morning, a paper was officially published in the journal *Nature* reporting an initial sequence and analysis of the human genome. One of the fruits of the Human Genome Project, the paper describes the broad landscape of the nearly 3 billion bases of the euchromatic portion of the human chromosomes.

In the most narrow sense, the paper was the product of a remarkable international collaboration involving six countries, twenty genome centers, and more than a thousand scientists (myself included) to produce the information and to make it available to the world freely and without restriction.

In a broader sense, though, the paper is the product of a century-long scientific program to understand genetic information. The program began with the rediscovery of Mendel's laws at the beginning of the 20th century, showing that information was somehow transmitted from generation to generation in discrete form. During the first quarter-century, biologists found that the cellular basis of the information was the chromosomes. During the second quarter-century, they discovered that the molecular basis of the information was DNA. During the third quarter-century, they unraveled the mechanisms by which cells read this information and developed the recombinant DNA tools by which scientists can do the same. During the last quarter-century, biologists have been trying voraciously to gather genetic information-first from genes, then entire genomes.

The result is that biology in the 21st century is being transformed from a purely laboratory-based science to an information science as well. The information includes comprehensive global views of DNA sequence, RNA expression, protein interactions or molecular conformations. Increasingly, biological studies begin with the study of huge databases to help formulate specific hypotheses or design large-scale experiments. In turn, laboratory work ends with the accumulation of massive collections of data that must be sifted. These changes represent a dramatic shift in the biological sciences.

One of the crucial steps in this transformation will be training a new generation of biologists who are both computational scientists and laboratory scientists. This major challenge requires both vision and hard work: vision to set an appropriate agenda for the computational biologist of the future and hard work to develop a curriculum and textbook.

James Watson changed the world with his co-discovery of the double-helical structure of DNA in 1953. But, he also helped train a new generation to inhabit that new world in the 1960s and beyond through his textbook, *The Molecular Biology of the Gene*. Discovery and teaching go hand-in-hand in changing the world.

In this book, Andy Baxevanis and Francis Ouellette have taken on the tremendously important challenge of training the 21st century computational biologist. Toward this end, they have undertaken the difficult task of organizing the knowledge in this field in a logical progression and presenting it in a digestible form. And, they have done an excellent job. This fine text will make a major impact on biological research and, in turn, on progress in biomedicine. We are all in their debt.

Eric S. Lander

February 15, 2001
Cambridge, Massachusetts

PREFACE

With the advent of the new millenium, the scientific community marked a significant milestone in the study of biology—the completion of the "working draft" of the human genome. This work, which was chronicled in special editions of *Nature* and *Science* in early 2001, signals a new beginning for modern biology, one in which the majority of biological and biomedical research would be conducted in a "sequence-based" fashion. This new approach, long-awaited and much-debated, promises to quickly lead to advances not only in the understanding of basic biological processes, but in the prevention, diagnosis, and treatment of many genetic and genomic disorders. While the fruits of sequencing the human genome may not be known or appreciated for another hundred years or more, the implications to the basic way in which science and medicine will be practiced in the future are staggering. The availability of this flood of raw information has had a significant effect on the field of bioinformatics as well, with a significant amount of effort being spent on how to effectively and efficiently warehouse and access these data, as well as on new methods aimed at mining this warehoused data in order to make novel biological discoveries.

This new edition of *Bioinformatics* attempts to keep up with the quick pace of change in this field, reinforcing concepts that have stood the test of time while making the reader aware of new approaches and algorithms that have emerged since the publication of the first edition. Based on our experience both as scientists and as teachers, we have tried to improve upon the first edition by introducing a number of new features in the current version. Five chapters have been added on topics that have emerged as being important enough in their own right to warrant distinct and separate discussion: expressed sequence tags, sequence assembly, comparative genomics, large-scale genome analysis, and BioPerl. We have also included problem sets at the end of most of the chapters with the hopes that the readers will work through these examples, thereby reinforcing their command of the concepts presented therein. The solutions to these problems are available through the book's Web site, at **www.wiley.com/bioinformatics**. We have been heartened by the large number of instructors who have adopted the first edition as their book of choice, and hope that these new features will continue to make the book useful both in the classroom and at the bench.

There are many individuals we both thank, without whose efforts this volume would not have become a reality. First and foremost, our thanks go to all of the authors whose individual contributions make up this book. The expertise and professional viewpoints that these individuals bring to bear go a long way in making this book's contents as strong as it is. That, coupled with their general good-naturedness under tight time constraints, has made working with these men and women an absolute pleasure.

Since the databases and tools discussed in this book are unique in that they are freely shared amongst fellow academics, we would be remiss if we did not thank all of the people who, on a daily basis, devote their efforts to curating and maintaining the public databases, as well as those who have developed the now-indispensible tools for mining the data contained in those databases. As we pointed out in the preface to the first edition, the bioinformatics community is truly unique in that the *esprit de corps* characterizing this group is one of openness, and this underlying philosophy is one that has enabled the field of bioinformatics to make the substantial strides that it has in such a short period of time.

We also thank our editor, Luna Han, for her steadfast patience and support throughout the entire process of making this new edition a reality. Through our extended discussions both on the phone and in person, and in going from deadline to deadline, we've developed a wonderful relationship with Luna, and look forward to working with her again on related projects in the future. We also would like to thank Camille Carter and Danielle Lacourciere at Wiley for making the entire copy-editing process a quick and (relatively) painless one, as well as Eloise Nelson for all of her hard work in making sure all of the loose ends came together on schedule.

BFFO would like to acknowledge the continued support of Nancy Ryder. Nancy is not only a friend, spouse, and mother to our daughter Maya, but a continuous source of inspiration to do better, and to challenge; this is something that I try to do every day, and her love and support enables this. BFFO also wants to acknowledge the continued friendship and support from ADB throughout both of these editions. It has been an honor and a privilege to be a co-editor with him. Little did we know seven years ago, in the *second* basement of the Lister Hill Building at NIH where we shared an office, that so many words would be shared between our respective computers.

ADB would also like to specifically thank Debbie Wilson for all of her help throughout the editing process, whose help and moral support went a long way in making sure that this project got done the right way the first time around. I would also like to extend special thanks to Jeff Trent, who I have had the pleasure of working with for the past several years and with whom I've developed a special bond, both professionally and personally. Jeff has enthusiastically provided me the latitude to work on projects like these and has been a wonderful colleague and friend, and I look forward to our continued associations in the future.

Andreas D. Baxevanis
B. F. Francis Ouellette

CONTRIBUTORS

Sharmila Banerjee-Basu, Genome Technology Branch, National Human Genome Research Institute, National Institutes of Health, Bethesda, Maryland

Geoffrey J. Barton, European Molecular Biology Laboratory, European Bioinformatics Institute, Wellcome Trust Genome Campus, Hinxton, Cambridge, United Kingdom

Andreas D. Baxevanis, Genome Technology Branch, National Human Genome Research Institute, National Institutes of Health, Bethesda, Maryland

James K. Bonfield, Medical Research Council, Laboratory of Molecular Biology, Cambridge, United Kingdom

Fiona S. L. Brinkman, Department of Microbiology and Immunology, University of British Columbia, Vancouver, British Columbia, Canada

Michael Y. Galperin, National Center for Biotechnology Information, National Library of Medicine, National Institutes of Health, Bethesda, Maryland

Christopher W. V. Hogue, Samuel Lunenfeld Research Institute, Mount Sinai Hospital, Toronto, Ontario, Canada

David P. Judge, Department of Biochemistry, University of Cambridge, Cambridge, United Kingdom

Jonathan A. Kans, National Center for Biotechnology Information, National Library of Medicine, National Institutes of Health, Bethesda, Maryland

Ilene Karsch-Mizrachi, National Center for Biotechnology Information, National Library of Medicine, National Institutes of Health, Bethesda, Maryland

Eugene V. Koonin, National Center for Biotechnology Information, National Library of Medicine, National Institutes of Health, Bethesda, Maryland

David Landsman, Computational Biology Branch, National Center for Biotechnology Information, National Library of Medicine, National Institutes of Health, Bethesda, Maryland

Detlef D. Leipe, National Center for Biotechnology Information, National Library of Medicine, National Institutes of Health, Bethesda, Maryland

Tara C. Matise, Department of Genetics, Rutgers University, New Brunswick, New Jersey

Paul S. Meltzer, Cancer Genetics Branch, National Human Genome Research Institute, National Institutes of Health, Bethesda, Maryland

James M. Ostell, National Center for Biotechnology Information, National Library of Medicine, National Institutes of Health, Bethesda, Maryland

B. F. Francis Ouellette, Centre for Molecular Medicine and Therapeutics, Children's and Women's Health Centre of British Columbia, The University of British Columbia, Vancouver, British Columbia, Canada

Gregory D. Schuler, National Center for Biotechnology Information, National Library of Medicine, National Institutes of Health, Bethesda, Maryland

Rodger Staden, Medical Research Council, Laboratory of Molecular Biology, Cambridge, United Kingdom

Lincoln D. Stein, The Cold Spring Harbor Laboratory, Cold Spring Harbor, New York

Sarah J. Wheelan, National Center for Biotechnology Information, National Library of Medicine, National Institutes of Health, Bethesda, Maryland and Department of Molecular Biology and Genetics, The Johns Hopkins School of Medicine, Baltimore, Maryland

Peter S. White, Department of Pediatrics, University of Pennsylvania, Philadelphia, Pennsylvania

Tyra G. Wolfsberg, Genome Technology Branch, National Human Genome Research Institute, National Institutes of Health, Bethesda, Maryland

1

BIOINFORMATICS AND THE INTERNET

Andreas D. Baxevanis

Genome Technology Branch
National Human Genome Research Institute
National Institutes of Health
Bethesda, Maryland

Bioinformatics represents a new, growing area of science that uses computational approaches to answer biological questions. Answering these questions requires that investigators take advantage of large, complex data sets (both public and private) in a rigorous fashion to reach valid, biological conclusions. The potential of such an approach is beginning to change the fundamental way in which basic science is done, helping to more efficiently guide experimental design in the laboratory.

With the explosion of sequence and structural information available to researchers, the field of bioinformatics is playing an increasingly large role in the study of fundamental biomedical problems. The challenge facing computational biologists will be to aid in gene discovery and in the design of molecular modeling, site-directed mutagenesis, and experiments of other types that can potentially reveal previously unknown relationships with respect to the structure and function of genes and proteins. This challenge becomes particularly daunting in light of the vast amount of data that has been produced by the Human Genome Project and other systematic sequencing efforts to date.

Before embarking on any practical discussion of computational methods in solving biological problems, it is necessary to lay the common groundwork that will enable users to both access and implement the algorithms and tools discussed in this book. We begin with a review of the Internet and its terminology, discussing major Internet protocol classes as well, without becoming overly engaged in the engineering

Bioinformatics: A Practical Guide to the Analysis of Genes and Proteins
Edited by A. D. Baxevanis and B. F. F. Ouellette
ISBN 0-471-38390-2 (cloth), ISBN 0-471-383910 (paper)

minutiae underlying these protocols. A more in-depth treatment on the inner workings of these protocols may be found in a number of well-written reference books intended for the lay audience (Rankin, 1996; Conner-Sax and Krol, 1999; Kennedy, 1999). This chapter will also discuss matters of connectivity, ranging from simple modem connections to digital subscriber lines (DSL). Finally, we will address one of the most common problems that has arisen with the proliferation of Web pages throughout the world—finding useful information on the World Wide Web.

INTERNET BASICS

Despite the impression that it is a single entity, the Internet is actually a network of networks, composed of interconnected local and regional networks in over 100 countries. Although work on remote communications began in the early 1960s, the true origins of the Internet lie with a research project on networking at the Advanced Research Projects Agency (ARPA) of the US Department of Defense in 1969 named ARPANET. The original ARPANET connected four nodes on the West Coast, with the immediate goal of being able to transmit information on defense-related research between laboratories. A number of different network projects subsequently surfaced, with the next landmark developments coming over 10 years later. In 1981, BITNET ("Because It's Time") was introduced, providing point-to-point connections between universities for the transfer of electronic mail and files. In 1982, ARPA introduced the Transmission Control Protocol (TCP) and the Internet Protocol (IP); TCP/IP allowed different networks to be connected to and communicate with one another, creating the system in place today. A number of references chronicle the development of the Internet and communications protocols in detail (Quarterman, 1990; Froehlich and Kent, 1991; Conner-Sax and Krol, 1999). Most users, however, are content to leave the details of *how* the Internet works to their systems administrators; the relevant fact to most is that it *does* work.

Once the machines on a network have been connected to one another, there needs to be an unambiguous way to specify a single computer so that messages and files actually find their intended recipient. To accomplish this, all machines directly connected to the Internet have an *IP number*. IP addresses are unique, identifying one and only one machine. The IP address is made up of four numbers separated by periods; for example, the IP address for the main file server at the National Center for Biotechnology Information (NCBI) at the National Institutes of Health (NIH) is 130.14.25.1. The numbers themselves represent, from left to right, the domain (130.14 for NIH), the subnet (.25 for the National Library of Medicine at NIH), and the machine itself (.1). The use of IP numbers aids the computers in directing data; however, it is obviously very difficult for users to remember these strings, so IP addresses often have associated with them a *fully qualified domain name* (FQDN) that is dynamically translated in the background by *domain name servers*. Going back to the NCBI example, rather than use `130.14.25.1` to access the NCBI computer, a user could instead use `ncbi.nlm.nih.gov` and achieve the same result. Reading from left to right, notice that the IP address goes from least to most specific, whereas the FQDN equivalent goes from most specific to least. The name of any given computer can then be thought of as taking the general form *computer.domain*, with the top-level domain (the portion coming after the last period in the FQDN) falling into one of the broad categories shown in Table 1.1. Outside the

TABLE 1.1. Top-Level Doman Names

Top-level domain names	
.com	Commercial site
.edu	Educational site
.gov	Government site
.mil	Military site
.net	Gateway or network host
.org	Private (usually not-for-profit) organizations
Examples of top-level domain names used outside the United States	
.ca	Canadian site
.ac.uk	Academic site in the United Kingdom
.co.uk	Commercial site in the United Kingdom
Generic top-level domains proposed by IAHC	
.firm	Firms or businesses
.shop	Businesses offering goods to purchase (stores)
.web	Entities emphasizing activities relating to the World Wide Web
.arts	Cultural and entertainment organizations
.rec	Recreational organizations
.info	Information sources
.nom	Personal names (e.g., *yourlastname.nom*)

A complete listing of domain suffixes, including country codes, can be found at http://www.currents.net/resources/directory/noframes/nf.domains.html.

United States, the top-level domain names *may* be replaced with a two-letter code specifying the country in which the machine is located (e.g., .ca for Canada and .uk for the United Kingdom). In an effort to anticipate the needs of Internet users in the future, as well as to try to erase the arbitrary line between top-level domain names based on country, the now-dissolved International Ad Hoc Committee (IAHC) was charged with developing a new framework of generic top-level domains (gTLD). The new, recommended gTLDs were set forth in a document entitled *The Generic Top Level Domain Memorandum of Understanding* (gTLD-MOU); these gTLDs are overseen by a number of governing bodies and are also shown in Table 1.1.

The most concrete measure of the size of the Internet lies in actually counting the number of machines physically connected to it. The Internet Software Consortium (ISC) conducts an Internet Domain Survey twice each year to count these machines, otherwise known as *hosts*. In performing this survey, ISC considers not only how many hostnames have been assigned, but how many of those are actually in use; a hostname might be issued, but the requestor may be holding the name in abeyance for future use. To test for this, a representative sample of host machines are sent a probe (a "ping"), with a signal being sent back to the originating machine if the host was indeed found. The rate of growth of the number of hosts has been phenomenal; from a paltry 213 hosts in August 1981, the Internet now has more than 60 million "live" hosts. The doubling time for the number of hosts is on the order of 18 months. At this time, most of this growth has come from the commercial sector, capitalizing on the growing popularity of multimedia platforms for advertising and communications such as the World Wide Web.

CONNECTING TO THE INTERNET

Of course, before being able to use all the resources that the Internet has to offer, one needs to actually make a physical connection between one's own computer and "the information superhighway." For purposes of this discussion, the elements of this connection have been separated into two discrete parts: the actual, physical connection (meaning the "wire" running from one's computer to the Internet backbone) and the service provider, who handles issues of routing and content once connected. Keep in mind that, in practice, these are not necessarily treated as two separate parts—for instance, one's service provider may also be the same company that will run cables or fibers right into one's home or office.

Copper Wires, Coaxial Cables, and Fiber Optics

Traditionally, users attempting to connect to the Internet away from the office had one and only one option—a modem, which uses the existing copper twisted-pair cables carrying telephone signals to transmit data. Data transfer rates using modems are relatively slow, allowing for data transmission in the range of 28.8 to 56 kilobits per second (kbps). The problem with using conventional copper wire to transmit data lies not in the copper wire itself but in the switches that are found along the way that route information to their intended destinations. These switches were designed for the efficient and effective transfer of voice data but were never intended to handle the high-speed transmission of data. Although most people still use modems from their home, a number of new technologies are already in place and will become more and more prevalent for accessing the Internet away from hardwired Ethernet networks. The maximum speeds at which each of the services that are discussed below can operate are shown in Figure 1.1.

The first of these "new solutions" is the integrated services digital network or ISDN. The advent of ISDN was originally heralded as the way to bring the Internet into the home in a speed-efficient manner; however, it required that special wiring be brought into the home. It also required that users be within a fixed distance from a central office, on the order of 20,000 feet or less. The cost of running this special, dedicated wiring, along with a per-minute pricing structure, effectively placed ISDN out of reach for most individuals. Although ISDN is still available in many areas, this type of service is quickly being supplanted by more cost-effective alternatives.

In looking at alternatives that did not require new wiring, cable television providers began to look at ways in which the coaxial cable already running into a substantial number of households could be used to also transmit data. Cable companies are able to use bandwidth that is not being used to transmit television signals (effectively, unused channels) to push data into the home at very high speeds, up to 4.0 megabits per second (Mbps). The actual computer is connected to this network through a cable modem, which uses an Ethernet connection to the computer and a coaxial cable to the wall. Homes in a given area all share a single cable, in a wiring scheme very similar to how individual computers are connected via the Ethernet in an office or laboratory setting. Although this branching arrangement can serve to connect a large number of locations, there is one major disadvantage: as more and more homes connect through their cable modems, service effectively slows down as more signals attempt to pass through any given node. One way of circumventing

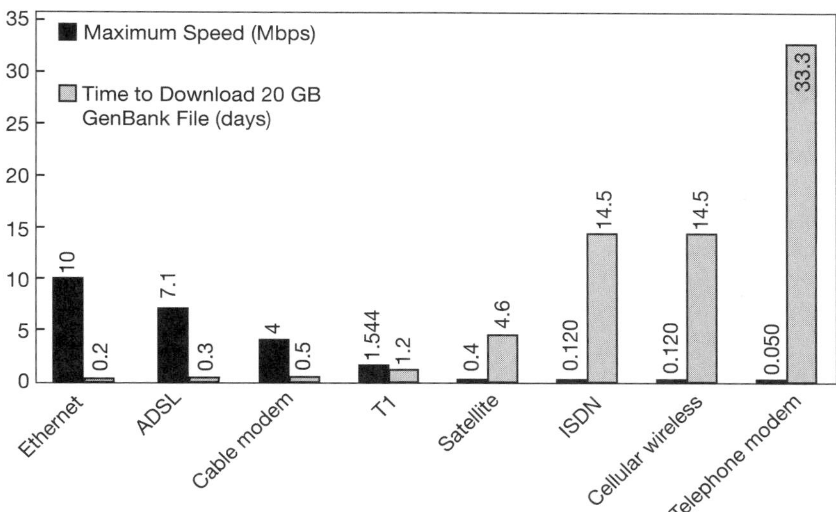

Figure 1.1. Performance of various types of Internet connections, by maximum throughput. The numbers indicated in the graph refer to peak performance; often times, the actual performance of any given method may be on the order of one-half slower, depending on configurations and system conditions.

this problem is the installation of more switching equipment and reducing the size of a given "neighborhood."

Because the local telephone companies were the primary ISDN providers, they quickly turned their attention to ways that the existing, conventional copper wire already in the home could be used to transmit data at high speed. The solution here is the digital subscriber line or DSL. By using new, dedicated switches that are designed for rapid data transfer, DSL providers can circumvent the old voice switches that slowed down transfer speeds. Depending on the user's distance from the central office and whether a particular neighborhood has been wired for DSL service, speeds are on the order of 0.8 to 7.1 Mbps. The data transfers do not interfere with voice signals, and users can use the telephone while connected to the Internet; the signals are "split" by a special modem that passes the data signals to the computer and a microfilter that passes voice signals to the handset. There is a special type of DSL called *asynchronous* DSL or ADSL. This is the variety of DSL service that is becoming more and more prevalent. Most home users download much more information than they send out; therefore, systems are engineered to provide super-fast transmission in the "in" direction, with transmissions in the "out" direction being 5–10 times slower. Using this approach maximizes the amount of bandwidth that can be used without necessitating new wiring. One of the advantages of ADSL over cable is that ADSL subscribers effectively have a direct line to the central office, meaning that they do not have to compete with their neighbors for bandwidth. This, of course, comes at a price; at the time of this writing, ADSL connectivity options were on the order of twice as expensive as cable Internet, but this will vary from region to region.

Some of the newer technologies involve wireless connections to the Internet. These include using one's own cell phone or a special cell phone service (such as

Ricochet) to upload and download information. These cellular providers can provide speeds on the order of 28.8–128 kbps, depending on the density of cellular towers in the service area. Fixed-point wireless services can be substantially faster because the cellular phone does not have to "find" the closest tower at any given time. Along these same lines, satellite providers are also coming on-line. These providers allow for data download directly to a satellite dish with a southern exposure, with uploads occuring through traditional telephone lines. Along the satellite option has the potential to be among the fastest of the options discussed, current operating speeds are only on the order of 400 kbps.

Content Providers vs. ISPs

Once an appropriately fast and price-effective connectivity solution is found, users will then need to actually connect to some sort of service that will enable them to traverse the Internet space. The two major categories in this respect are *online services* and *Internet service providers* (ISPs). Online services, such as America Online (AOL) and CompuServe, offer a large number of interactive digital services, including information retrieval, electronic mail (E-mail; see below), bulletin boards, and "chat rooms," where users who are online at the same time can converse about any number of subjects. Although the online services now provide access to the World Wide Web, most of the specialized features and services available through these systems reside in a proprietary, closed network. Once a connection has been made between the user's computer and the online service, one can access the special features, or content, of these systems without ever leaving the online system's host computer. Specialized content can range from access to online travel reservation systems to encyclopedias that are constantly being updated—items that are not available to nonsubscribers to the particular online service.

Internet service providers take the opposite tack. Instead of focusing on providing content, the ISPs provide the tools necessary for users to send and receive E-mail, upload and download files, and navigate around the World Wide Web, finding information at remote locations. The major advantage of ISPs is connection speed; often the smaller providers offer faster connection speeds than can be had from the online services. Most ISPs charge a monthly fee for unlimited use.

The line between online services and ISPs has already begun to blur. For instance, AOL's now monthly flat-fee pricing structure in the United States allows users to obtain all the proprietary content found on AOL as well as all the Internet tools available through ISPs, often at the same cost as a simple ISP connection. The extensive AOL network puts access to AOL as close as a local phone call in most of the United States, providing access to E-mail no matter where the user is located, a feature small, local ISPs cannot match. Not to be outdone, many of the major national ISP providers now also provide content through the concept of *portals*. Portals are Web pages that can be customized to the needs of the individual user and that serve as a jumping-off point to other sources of news or entertainment on the Net. In addition, many national firms such as Mindspring are able to match AOL's ease of connectivity on the road, and both ISPs and online providers are becoming more and more generous in providing users the capacity to publish their own Web pages. Developments such as this, coupled with the move of local telephone and cable companies into providing Internet access through new, faster fiber optic networks, foretell major changes in how people will access the Net in the future, changes that should favor the end user in both price and performance.

ELECTRONIC MAIL

Most people are introduced to the Internet through the use of electronic mail or *E-mail*. The use of E-mail has become practically indispensable in many settings because of its convenience as a medium for sending, receiving, and replying to messages. Its advantages are many:

- It is much quicker than the postal service or "snail mail."
- Messages tend to be much clearer and more to the point than is the case for typical telephone or face-to-face conversations.
- Recipients have more flexibility in deciding whether a response needs to be sent immediately, relatively soon, or at all, giving individuals more control over workflow.
- It provides a convenient method by which messages can be filed or stored.
- There is little or no cost involved in sending an E-mail message.

These and other advantages have pushed E-mail to the forefront of interpersonal communication in both industry and the academic community; however, users should be aware of several major disadvantages. First is the issue of security. As mail travels toward its recipient, it may pass through a number of remote nodes, at any one of which the message may be intercepted and read by someone with high-level access, such as a systems administrator. Second is the issue of privacy. In industrial settings, E-mail is often considered to be an asset of the company for use in official communication only and, as such, is subject to monitoring by supervisors. The opposite is often true in academic, quasi-academic, or research settings; for example, the National Institutes of Health's policy encourages personal use of E-mail within the bounds of certain published guidelines. The key words here are "published guidelines"; no matter what the setting, users of E-mail systems should always find out their organization's policy regarding appropriate use and confidentiality so that they may use the tool properly and effectively. An excellent, basic guide to the effective use of E-mail (Rankin, 1996) is recommended.

Sending E-Mail. E-mail addresses take the general form *user@computer.domain*, where *user* is the name of the individual user and *computer.domain* specifies the actual computer that the E-mail account is located on. Like a postal letter, an E-mail message is comprised of an *envelope* or *header*, showing the E-mail addresses of sender and recipient, a line indicating the subject of the E-mail, and information about how the E-mail message actually traveled from the sender to the recipient. The header is followed by the actual message, or *body*, analogous to what would go inside a postal envelope. Figure 1.2 illustrates all the components of an E-mail message.

E-mail programs vary widely, depending on both the platform and the needs of the users. Most often, the characteristics of the local area network (LAN) dictate what types of mail programs can be used, and the decision is often left to systems

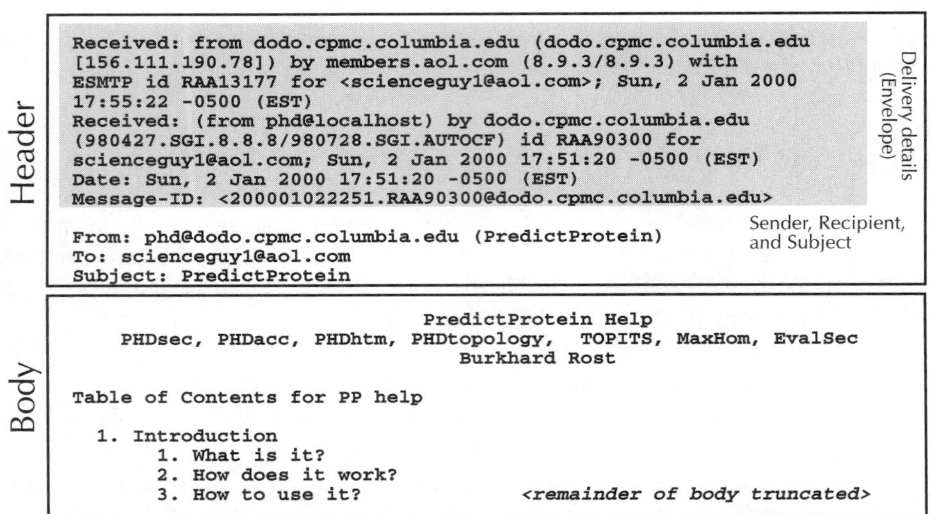

```
Received: from dodo.cpmc.columbia.edu (dodo.cpmc.columbia.edu
[156.111.190.78]) by members.aol.com (8.9.3/8.9.3) with
ESMTP id RAA13177 for <scienceguy1@aol.com>; Sun, 2 Jan 2000
17:55:22 -0500 (EST)
Received: (from phd@localhost) by dodo.cpmc.columbia.edu
(980427.SGI.8.8.8/980728.SGI.AUTOCF) id RAA90300 for
scienceguy1@aol.com; Sun, 2 Jan 2000 17:51:20 -0500 (EST)
Date: Sun, 2 Jan 2000 17:51:20 -0500 (EST)
Message-ID: <200001022251.RAA90300@dodo.cpmc.columbia.edu>

From: phd@dodo.cpmc.columbia.edu (PredictProtein)
To: scienceguy1@aol.com
Subject: PredictProtein

                        PredictProtein Help
    PHDsec, PHDacc, PHDhtm, PHDtopology,  TOPITS, MaxHom, EvalSec
                           Burkhard Rost

Table of Contents for PP help

  1. Introduction
      1. What is it?
      2. How does it work?
      3. How to use it?               <remainder of body truncated>
```


Figure 1.2. Anatomy of an E-mail message, with relevant components indicated. This message is an automated reply to a request for help file for the PredictProtein E-mail server.

administrators rather than individual users. Among the most widely used E-mail packages with a graphical user interface are Eudora for the Macintosh and both Netscape Messenger and Microsoft Exchange for the Mac, Windows, and UNIX platforms. Text-based E-mail programs, which are accessed by logging in to a UNIX-based account, include Elm and Pine.

Bulk E-Mail. As with postal mail, there has been an upsurge in "spam" or "junk E-mail," where companies compile bulk lists of E-mail addresses for use in commercial promotions. Because most of these lists are compiled from online registration forms and similar sources, the best defense for remaining off these bulk E-mail lists is to be selective as to whom E-mail addresses are provided. Most newsgroups keep their mailing lists confidential; if in doubt and if this is a concern, one should ask.

E-Mail Servers. Most often, E-mail is thought of a way to simply send messages, whether it be to one recipient or many. It is also possible to use E-mail as a mechanism for making predictions or retrieving records from biological databases. Users can send E-mail messages in a format defining the action to be performed to remote computers known as *servers*; the servers will then perform the desired operation and E-mail back the results. Although this method is not interactive (in that the user cannot adjust parameters or have control over the execution of the method in real time), it does place the responsibility for hardware maintenance and software upgrades on the individuals maintaining the server, allowing users to concentrate on their results instead of on programming. The use of a number of E-mail servers is discussed in greater detail in context in later chapters. For most of these servers, sending the message `help` to the server E-mail address will result in a detailed set of instructions for using that server being returned, including ways in which queries need to be formatted.

Aliases and Newsgroups. In the example in Figure 1.2, the E-mail message is being sent to a single recipient. One of the strengths of E-mail is that a single piece of E-mail can be sent to a large number of people. The primary mechanism for doing this is through *aliases*; a user can define a group of people within their mail program and give the group a special name or alias. Instead of using individual E-mail addresses for all of the people in the group, the user can just send the E-mail to the alias name, and the mail program will handle broadcasting the message to each person in that group. Setting up alias names is a tremendous time-saver even for small groups; it also ensures that all members of a given group actually receive all E-mail messages intended for the group.

The second mechanism for broadcasting messages is through *newsgroups*. This model works slightly differently in that the list of E-mail addresses is compiled and maintained on a remote computer through subscriptions, much like magazine subscriptions. To participate in a newsgroup discussions, one first would have to subscribe to the newsgroup of interest. Depending on the newsgroup, this is done either by sending an E-mail to the host server or by visiting the host's Web site and using a form to subscribe. For example, the BIOSCI newsgroups are among the most highly trafficked, offering a forum for discussion or the exchange of ideas in a wide variety of biological subject areas. Information on how to subscribe to one of the constituent BIOSCI newsgroups is posted on the BIOSCI Web site. To actually participate in the discussion, one would simply send an E-mail to the address corresponding to the group that you wish to reach. For example, to post messages to the computational biology newsgroup, mail would simply be addressed to `comp-bio@net.bio.net`, and, once that mail is sent, *everyone* subscribing to that newsgroup would receive (and have the opportunity to respond to) that message. The ease of reaching a large audience in such a simple fashion is both a blessing and a curse, so many newsgroups require that postings be reviewed by a moderator before they get disseminated to the individual subscribers to assure that the contents of the message are actually of interest to the readers.

It is also possible to participate in newsgroups without having each and every piece of E-mail flood into one's private mailbox. Instead, interested participants can use news-reading software, such as NewsWatcher for the Macintosh, which provides access to the individual messages making up a discussion. The major advantage is that the user can pick and choose which messages to read by scanning the subject lines; the remainder can be discarded by a single operation. NewsWatcher is an example of what is known as a *client-server application*; the client software (here, NewsWatcher) runs on a client computer (a Macintosh), which in turn interacts with a machine at a remote location (the server). Client-server architecture is interactive in nature, with a direct connection being made between the client and server machines.

Once NewsWatcher is started, the user is presented with a list of newsgroups available to them (Fig. 1.3). This list will vary, depending on the user's location, as system administrators have the discretion to allow or to block certain groups at a given site. From the rear-most window in the figure, the user double-clicks on the newsgroup of interest (here, *bionet.genome.arabidopsis*), which spawns the window shown in the center. At the top of the center window is the current unread message count, and any message within the list can be read by double-clicking on that particular line. This, in turn, spawns the last window (in the foreground), which shows the actual message. If a user decides not to read any of the messages, or is done

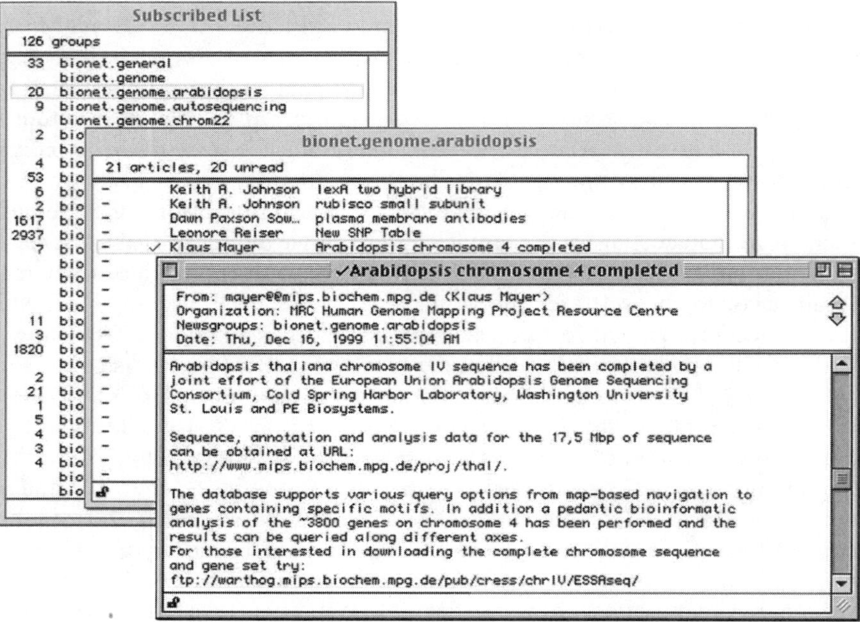

Figure 1.3. Using NewsWatcher to read postings to newsgroups. The list of newsgroups that the user has subscribed to is shown in the Subscribed List window (*left*). The list of new postings for the highlighted newsgroup (*bionet.genome.arabidopsis*) is shown in the center window. The window in the foreground shows the contents of the posting selected from the center window.

reading individual messages, the balance of the messages within the newsgroup (center) window can be deleted by first choosing Select All from the File menu and then selecting Mark Read from the News menu. Once the newsgroup window is closed, the unread message count is reset to zero. Every time NewsWatcher is restarted, it will automatically poll the news server for new messages that have been created since the last session. As with most of the tools that will be discussed in this chapter, news-reading capability is built into Web browsers such as Netscape Navigator and Microsoft Internet Explorer.

FILE TRANSFER PROTOCOL

Despite the many advantages afforded by E-mail in transmitting messages, many users have no doubt experienced frustration in trying to transmit files, or *attachments*, along with an E-mail message. The mere fact that a file can be attached to an E-mail message and sent does not mean that the recipient will be able to detach, decode, and actually use the attached file. Although more cross-platform E-mail packages such as Microsoft Exchange are being developed, the use of different E-mail packages by people at different locations means that sending files via E-mail is not an effective, foolproof method, at least in the short term. One solution to this problem is through the use of a *file transfer protocol* or FTP. The workings of FTP are quite simple: a connection is made between a user's computer (the *client*) and a remote server, and that connection remains in place for the duration of the FTP session. File transfers are very fast, at rates on the order of 5–10 kilobytes per second, with speeds varying with the time of day, the distance between the client and server machines, and the overall traffic on the network.

In the ordinary case, making an FTP connection and transferring files requires that a user have an account on the remote server. However, there are many files and programs that are made freely available, and access to those files does not require having an account on each and every machine where these programs are stored. Instead, connections are made using a system called *anonymous FTP*. Under this system, the user connects to the remote machine and, instead of entering a username/password pair, types `anonymous` as the username and enters their E-mail address in place of a password. Providing one's E-mail address allows the server's system administrators to compile access statistics that may, in turn, be of use to those actually providing the public files or programs. An example of an anonymous FTP session using UNIX is shown in Figure 1.4.

Although FTP actually occurs within the UNIX environment, Macintosh and PC users can use programs that rely on graphical user interfaces (GUI, pronounced

```
$ ftp ftp.bio.indiana.edu
Connected to magpie.bio.indiana.edu.
220 iubio.bio.indiana.edu FTP server ready.
Name: anonymous
331 Guest login ok, send your complete e-mail address as password.
Password: ********
230-                       Welcome to IUBio archive!
230-
230-  This is a user-supported archive for biology software and data.
230-
230-       See the file Archive.Doc for details of this archive.
230-
230-          See IUBio Bio-Mirror archive of large data sets at
230-     ftp to iubio.bio.indiana.edu, user: iubio, password: iubio
230-   This includes GenBank, EMBL and DDBJ and other biosequence data.
230-
230-     Report problems, uploads and other matters via e-mail to
230-                      archive@bio.indiana.edu.
230-
230 Guest login ok, access restrictions apply.
Remote system type is UNIX.
Using binary mode to transfer files.
ftp> cd /molbio/align/clustal
250 CWD command successful.
ftp> get clustalw1.75.unix.tar.Z
local: clustalw1.75.unix.tar.Z remote: clustalw1.75.unix.tar.Z
200 PORT command successful.
150 Opening BINARY mode data connection for clustalw1.75.unix.tar.Z (230379 bytes).
226 Transfer complete.
230379 bytes received in 0.45 seconds (500.75 Kbytes/s)
ftp> quit
221-You have transferred 230379 bytes in 1 files.
221-Total traffic for this session was 231859 bytes in 1 transfers.
221-Thank you for using the FTP service on iubio.bio.indiana.edu.
221 Goodbye.
```

Figure 1.4. Using UNIX FTP to download a file. An anonymous FTP session is established with the molecular biology FTP server at the University of Indiana to download the CLUSTAL W alignment program. The user inputs are shown in boldface.

"gooey") to navigate through the UNIX directories on the FTP server. Users need not have any knowledge of UNIX commands to download files; instead, they select from pop-up menus and point and click their way through the UNIX file structure. The most popular FTP program on the Macintosh platform for FTP sessions is Fetch. A sample Fetch window is shown in Figure 1.5 to illustrate the difference between using a GUI-based FTP program and the equivalent UNIX FTP in Figure 1.4. In the figure, notice that the Automatic radio button (near the bottom of the second window under the Get File button) is selected, meaning that Fetch will determine the appropriate type of file transfer to perform. This may be manually overridden by selecting either Text or Binary, depending on the nature of the file being transferred. As a rule, text files should be transferred as Text, programs or executables as Binary, and graphic format files such as PICT and TIFF files as Raw Data.

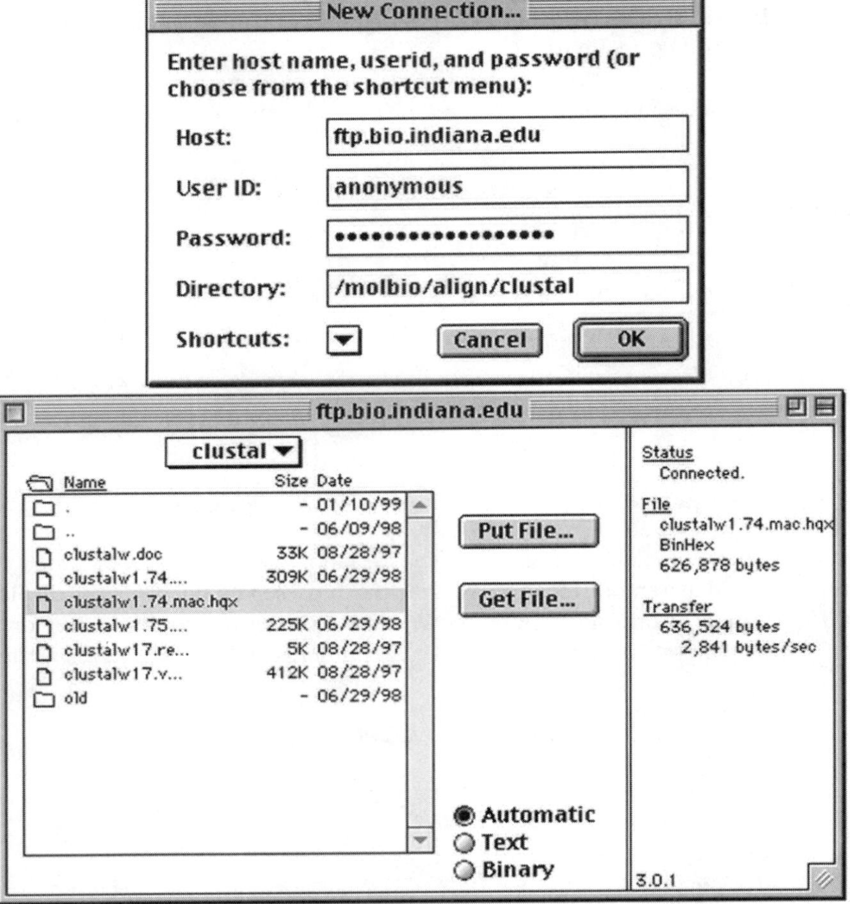

Figure 1.5. Using Fetch to download a file. An anonymous FTP session is established with the molecular biology FTP server at the University of Indiana (*top*) to download the CLUSTAL W alignment program (*bottom*). Notice the difference between this GUI-based program and the UNIX equivalent illustrated in Figure 1.4.

THE WORLD WIDE WEB

Although FTP is of tremendous use in the transfer of files from one computer to another, it does suffer from some limitations. When working with FTP, once a user enters a particular directory, they can only see the names of the directories or files. To actually view what is within the files, it is necessary to physically download the files onto one's own computer. This inherent drawback led to the development of a number of *distributed document delivery systems* (DDDS), interactive client-server applications that allowed information to be viewed without having to perform a download. The first generation of DDDS development led to programs like Gopher, which allowed plain text to be viewed directly through a client-server application. From this evolved the most widely known and widely used DDDS, namely, the World Wide Web. The Web is an outgrowth of research performed at the European Nuclear Research Council (CERN) in 1989 that was aimed at sharing research data between several locations. That work led to a medium through which text, images, sounds, and videos could be delivered to users on demand, anywhere in the world.

Navigation on the World Wide Web

Navigation on the Web does not require advance knowledge of the location of the information being sought. Instead, users can navigate by clicking on specific text, buttons, or pictures. These clickable items are collectively known as *hyperlinks*. Once one of these hyperlinks is clicked, the user is taken to another Web location, which could be at the same site or halfway around the world. Each document displayed on the Web is called a *Web page*, and all of the related Web pages on a particular server are collectively called a *Web site*. Navigation strictly through the use of hyperlinks has been nicknamed "Web surfing."

Users can take a more direct approach to finding information by entering a specific address. One of the strengths of the Web is that the programs used to view Web pages (appropriately termed *browsers*) can be used to visit FTP and Gopher sites as well, somewhat obviating the need for separate Gopher or FTP applications. As such, a unified naming convention was introduced to indicate to the browser program both the location of the remote site and, more importantly, the type of information at that remote location so that the browser could properly display the data. This standard-form address is known as a *uniform resource locator*, or URL, and takes the general form *protocol://computer.domain*, where *protocol* specifies the type of site and *computer.domain* specifies the location (Table 1.2). The *http* used for the protocol in World Wide Web URLs stands for *hypertext transfer protocol*, the method used in transferring Web files from the host computer to the client.

TABLE 1.2. Uniform Resource Locator (URL) Format for Each Type of Transfer Protocol

General form	*protocol://computer.domain*
FTP site	*ftp://ftp.ncbi.nlm.nih.gov*
Gopher site	*gopher://gopher.iubio.indiana.edu*
Web site	*http://www.nhgri.nih.gov*

Browsers

Browsers, which are used to look at Web pages, are client-server applications that connect to a remote site, download the requested information at that site, and display the information on a user's monitor, then disconnecting from the remote host. The information retrieved from the remote host is in a platform-independent format named *hypertext markup language* (HTML). HTML code is strictly text-based, and any associated graphics or sounds for that document exist as separate files in a common format. For example, images may be stored and transferred in GIF format, a proprietary format developed by CompuServe for the quick and efficient transfer of graphics; other formats, such as JPEG and BMP, may also be used. Because of this, a browser can display any Web page on any type of computer, whether it be a Macintosh, IBM compatible, or UNIX machine. The text is usually displayed first, with the remaining elements being placed on the page as they are downloaded. With minor exception, a given Web page will look the same when the same browser is used on any of the above platforms. The two major players in the area of browser software are Netscape, with their Communicator product, and Microsoft, with Internet Explorer. As with many other areas where multiple software products are available, the choice between Netscape and Internet Explorer comes down to one of personal preference. Whereas the computer literati will debate the fine points of difference between these two packages, for the average user, both packages perform equally well and offer the same types of features, adequately addressing the Web-browser needs of most users.

It is worth mentioning that, although the Web is by definition a visually-based medium, it is also possible to travel through Web space and view documents without the associated graphics. For users limited to line-by-line terminals, a browser called Lynx is available. Developed at the University of Kansas, Lynx allows users to use their keyboard arrow keys to highlight and select hyperlinks, using their return key the same way that Netscape and Internet Explorer users would click their mouse.

Internet vs. Intranet

The Web is normally thought of as a way to communicate with people at a distance, but the same infrastructure can be used to connect people within an organization. Such *intranets* provide an easily accessible repository of relevant information, capitalizing on the simplicity of the Web interface. They also provide another channel for broadcast or confidential communication within the organization. Having an intranet is of particular value when members of an organization are physically separated, whether in different buildings or different cities. Intranets are protected: that is, people who are not on the organization's network are prohibited from accessing the internal Web pages; additional protections through the use of passwords are also common.

Finding Information on the World Wide Web

Most people find information on the Web the old-fashioned way: by word of mouth, either using lists such as those preceding the References in the chapters of this book or by simply following hyperlinks put in place by Web authors. Continuously clicking from page to page can be a highly ineffective way of finding information, though, especially when the information sought is of a very focused nature. One way of finding interesting and relevant Web sites is to consult *virtual libraries*, which are curated lists of Web resources arranged by subject. Virtual libraries of special interest to biologists include the WWW Virtual Library, maintained by Keith Robison at Harvard, and the EBI BioCatalog, based at the European Bioinformatics Institute. The URLs for these sites can be found in the list at the end of this chapter.

It is also possible to directly search the Web by using *search engines*. A search engine is simply a specialized program that can perform full-text or keyword searches on databases that catalog Web content. The result of a search is a hyperlinked list of Web sites fitting the search criteria from which the user can visit any or all of the found sites. However, the search engines use slightly different methods in compiling their databases. One variation is the attempt to capture most or all of the text of every Web page that the search engine is able to find and catalog ("Web crawling"). Another technique is to catalog only the title of each Web page rather than its entire text. A third is to consider words that must appear next to each other or only relatively close to one another. Because of these differences in search-engine algorithms, the results returned by issuing the same query to a number of different search engines can produce wildly different results (Table 1.3). The other important feature of Table 1.3 is that most of the numbers are exceedingly large, reflecting the overall size of the World Wide Web. Unless a particular search engine ranks its results by relevance (e.g., by scoring words in a title higher than words in the body of the Web page), the results obtained may not be particularly useful. Also keep in mind that, depending on the indexing scheme that the search engine is using, the found pages may actually no longer exist, leading the user to the dreaded "404 Not Found" error.

Compounding this problem is the issue of *coverage*—the number of Web pages that any given search engine is actually able to survey and analyze. A comprehensive study by Lawrence and Giles (1998) indicates that the coverage provided by any of the search engines studied is both small and highly variable. For example, the HotBot engine produced 57.5% coverage of what was estimated to be the size of the "indexable Web," whereas Lycos had only 4.41% coverage, a full order of magnitude less than HotBot. The most important conclusion from this study was that the extent of coverage increased as the number of search engines was increased and the results from those individual searches were combined. Combining the results obtained from the six search engines examined in this study produced coverage approaching 100%.

To address this point, a new class of search engines called *meta-search engines* have been developed. These programs will take the user's query and poll anywhere from 5–10 of the "traditional" search engines. The meta-search engine will then

TABLE 1.3. Number of Hits Returned for Four Defined Search Queries on Some of the More Popular Search and Meta-Search Engines

	Search Engine				Meta-Search Engine		
Search Term	HotBot	Excite	Infoseek	Lycos	Google	MetaCrawler	SavvySearch
Genetic mapping	478	1,040	4,326	9,395	7,043	62	58
Human genome	13,213	34,760	15,980	19,536	19,797	42	54
Positional cloning	279	735	1,143	666	3,987	40	52
Prostate cancer	14,044	53,940	24,376	33,538	23,100	0	57

collect the results, filter out duplicates, and return a single, annotated list to the user. One big advantage is that the meta-search engines take relevance statistics into account, returning much smaller lists of results. Although the hit list is substantially smaller, it is much more likely to contain sites that directly address the original query. Because the programs must poll a number of different search engines, searches conducted this way obviously take longer to perform, but the higher degree of confidence in the compiled results for a given query outweighs the extra few minutes (and sometimes only seconds) of search time. Reliable and easy-to-use meta-search engines include MetaCrawler and Savvy Search.

INTERNET RESOURCES FOR TOPICS PRESENTED IN CHAPTER 1

DOMAIN NAMES

gTLD-MOU	*http://www.gtld-mou.org*
Internet Software Consortium	*http://www.isc.org*

ELECTRONIC MAIL AND NEWSGROUPS

BIOSCI Newsgroups	*http://www.bio.net/docs/biosci.FAQ.html*
Eudora	*http://www.eudora.com*
Microsoft Exchange	*http://www.microsoft.com/exchange/*
NewsWatcher	*ftp://ftp.acns.nwu.edu/pub/newswatcher/*

FILE TRANSFER PROTOCOL

Fetch 3.0/Mac	*http://www.dartmouth.edu/pages/softdev/fetch.html*
LeechFTP/PC	*http://stud.fh-heilbronn.de/jdebis/leechftp/*

INTERNET ACCESS

America Online	*http://www.aol.com*
AT&T	*http://www.att.com/worldnet*
Bell Atlantic	*http://www.verizon.net*
Bell Canada	*http://www.bell.ca*
CompuServe	*http://www.compuserve.com*
Ricochet	*http://www.ricochet.net*
Telus	*http://www.telus.net*
Worldcom	*http://www. worldcom.com*

VIRTUAL LIBRARIES

EBI BioCatalog	*http://www.ebi.ac.uk/biocat/biocat.html*
Amos' WWW Links Page	*http://www.expasy.ch/alinks.html*
NAR Database Collection	*http://www.nar.oupjournals.org*
WWW Virtual Library	*http://mcb.harvard.edu/BioLinks.html*

WORLD WIDE WEB BROWSERS

Internet Explorer	*http://explorer.msn.com/home.htm*
Lynx	*ftp://ftp2.cc.ukans.edu/pub/lynx*
Netscape Navigator	*http://home.netscape.com*

WORLD WIDE WEB SEARCH ENGINES

AltaVista	*http://www.altavista.com*
Excite	*http://www.excite.com*
Google	*http://www.google.com*

HotBot	*http://hotbot.lycos.com*
Infoseek	*http://infoseek.go.com*
Lycos	*http://www.lycos.com*
Northern Light	*http://www.northernlight.com*

World Wide Web Meta-Search Engines

MetaCrawler	*http://www.metacrawler.com*
Savvy Search	*http://www.savvysearch.com*

REFERENCES

Conner-Sax, K., and Krol, E. (1999). *The Whole Internet: The Next Generation* (Sebastopol, CA: O'Reilly and Associates).

Froehlich, F., and Kent, A. (1991). ARPANET, the Defense Data Network, and Internet. In *Encyclopedia of Communications* (New York: Marcel Dekker).

Kennedy, A. J. (1999). *The Internet: Rough Guide 2000* (London: Rough Guides).

Lawrence, S., and Giles, C. L. (1998). Searching the World Wide Web. *Science* 280, 98–100.

Quarterman, J. (1990). *The Matrix: Computer Networks and Conferencing Systems Worldwide* (Bedford, MA: Digital Press).

Rankin, B. (1996). *Dr. Bob's Painless Guide to the Internet and Amazing Things You Can Do With E-mail* (San Francisco: No Starch Press).

2

THE NCBI DATA MODEL

James M. Ostell

National Center for Biotechnology Information
National Library of Medicine
National Institutes of Health
Bethesda, Maryland

Sarah J. Wheelan

Department of Molecular Biology and Genetics
The Johns Hopkins School of Medicine
Baltimore, Maryland

Jonathan A. Kans

National Center for Biotechnology Information
National Library of Medicine
National Institutes of Health
Bethesda, Maryland

INTRODUCTION

Why Use a Data Model?

Most biologists are familiar with the use of animal models to study human diseases. Although a disease that occurs in humans may not be found in exactly the same form in animals, often an animal disease shares enough attributes with a human counterpart to allow data gathered on the animal disease to be used to make inferences about the process in humans. Mathematical models describing the forces involved in musculoskeletal motions can be built by imagining that muscles are combinations of springs and hydraulic pistons and bones are lever arms, and, often times,

Bioinformatics: A Practical Guide to the Analysis of Genes and Proteins
Edited by A. D. Baxevanis and B. F. F. Ouellette
ISBN 0-471-38390-2 (cloth), ISBN 0-471-383910 (paper)

such models allow meaningful predictions to be made and tested about the obviously much more complex biological system under consideration. The more closely and elegantly a model follows a real phenomenon, the more useful it is in predicting or understanding the natural phenomenon it is intended to mimic.

In this same vein, some 12 years ago, the National Center for Biotechnology Information (NCBI) introduced a new model for sequence-related information. This new and more powerful model made possible the rapid development of software and the integration of databases that underlie the popular Entrez retrieval system and on which the GenBank database is now built (cf. Chapter 7 for more information on Entrez). The advantages of the model (e.g., the ability to move effortlessly from the published literature to DNA sequences to the proteins they encode, to chromosome maps of the genes, and to the three-dimensional structures of the proteins) have been apparent for years to biologists using Entrez, but very few biologists understand the foundation on which this model is built. As genome information becomes richer and more complex, more of the real, underlying data model is appearing in common representations such as GenBank files. Without going into great detail, this chapter attempts to present a practical guide to the principles of the NCBI data model and its importance to biologists at the bench.

Some Examples of the Model

The GenBank flatfile is a "DNA-centered" report, meaning that a region of DNA coding for a protein is represented by a "CDS feature," or "coding region," on the DNA. A *qualifier* (`/translation="MLLYY"`) describes a sequence of amino acids produced by translating the CDS. A limited set of additional *features* of the DNA, such as `mat_peptide`, are occasionally used in GenBank flatfiles to describe cleavage products of the (possibly unnamed) protein that is described by a `/translation`, but clearly this is not a satisfactory solution. Conversely, most protein sequence databases present a "protein-centered" view in which the connection to the encoding gene may be completely lost or may be only indirectly referenced by an accession number. Often times, these connections do not provide the exact codon-to-amino acid correspondences that are important in performing mutation analysis.

The NCBI data model deals directly with the two sequences involved: a DNA sequence and a protein sequence. The translation process is represented as a link between the two sequences rather than an annotation on one with respect to the other. Protein-related annotations, such as peptide cleavage products, are represented as features annotated directly on the protein sequence. In this way, it becomes very natural to analyze the protein sequences derived from translations of CDS features by BLAST or any other sequence search tool without losing the precise linkage back to the gene. A collection of a DNA sequence and its translation products is called a *Nuc-prot set*, and this is how such data is represented by NCBI. The GenBank flatfile format that many readers are already accustomed to is simply a particular style of report, one that is more "human-readable" and that ultimately flattens the connected collection of sequences back into the familiar one-sequence, DNA-centered view. The navigation provided by tools such as Entrez much more directly reflects the underlying structure of such data. The protein sequences derived from GenBank translations that are returned by BLAST searches are, in fact, the protein sequences from the Nuc-prot sets described above.

The standard GenBank format can also hide the multiple-sequence nature of some DNA sequences. For example, three genomic exons of a particular gene are sequenced, and partial flanking, noncoding regions around the exons may also be available, but the full-length sequences of these intronic sequences may not yet be available. Because the exons are not in their complete genomic context, there would be three GenBank flatfiles in this case, one for each exon. There is no explicit representation of the complete set of sequences over that genomic region; these three exons come in genomic order and are separated by a certain length of unsequenced DNA. In GenBank format there would be a Segment line of the form `SEGMENT 1 of 3` in the first record, `SEGMENT 2 of 3` in the second, and `SEGMENT 3 of 3` in the third, but this only tells the user that the lines are part of some undefined, ordered series (Fig. 2.1*A*). Out of the whole GenBank release, one locates the correct Segment records to place together by an algorithm involving the `LOCUS` name. All segments that go together use the same first combination of letters, ending with the numbers appropriate to the segment, e.g., `HSDDT1`, `HSDDT2`, and `HSDDT3`. Obviously, this complicated arrangement can result in problems when `LOCUS` names include numbers that inadvertently interfere with such series. In addition, there is no one sequence record that describes the whole assembled series, and there is no way to describe the distance between the individual pieces. There is no segmenting convention in the EMBL sequence database at all, so records derived from that source or distributed in that format lack even this imperfect information.

The NCBI data model defines a sequence type that directly represents such a segmented series, called a "segmented sequence." Rather than containing the letters A, G, C, and T, the segmented sequence contains instructions on how it can be built from other sequences. Considering again the example above, the segmented sequence would contain the instructions "take all of `HSDDT1`, then a gap of unknown length, then all of `HSDDT2`, then a gap of unknown length, then all of `HSDDT3`." The segmented sequence itself can have a name (e.g., `HSDDT`), an accession number, features, citations, and comments, like any other GenBank record. Data of this type are commonly stored in a so-called "Seg-set" containing the sequences `HSDDT`, `HSDDT1`, `HSDDT2`, `HSDDT3` and all of their connections and features. When the GenBank release is made, as in the case of Nuc-prot sets, the Seg-sets are broken up into multiple records, and the segmented sequence itself is not visible. However, GenBank, EMBL, and DDBJ have recently agreed on a way to represent these constructed assemblies, and they will be placed in a new CON division, with CON standing for "contig" (Fig. 2.1*B*). In the Entrez graphical view of segmented sequences, the segmented sequence is shown as a line connecting all of its component sequences (Fig. 2.1*C*).

An NCBI segmented sequence does not require that there be gaps between the individual pieces. In fact the pieces can overlap, unlike the case of a segmented series in GenBank format. This makes the segmented sequence ideal for representing large sequences such as bacterial genomes, which may be many megabases in length. This is what currently is done within the Entrez Genomes division for bacterial genomes, as well as other complete chromosomes such as yeast. The NCBI Software Toolkit (Ostell, 1996) contains functions that can gather the data that a segmented sequence refers to "on the fly," including constituent sequence and features, and this information can automatically be remapped from the coordinates of a small, individual record to that of a complete chromosome. This makes it possible to provide graphical views, GenBank flatfile views, or FASTA views or to perform analyses on

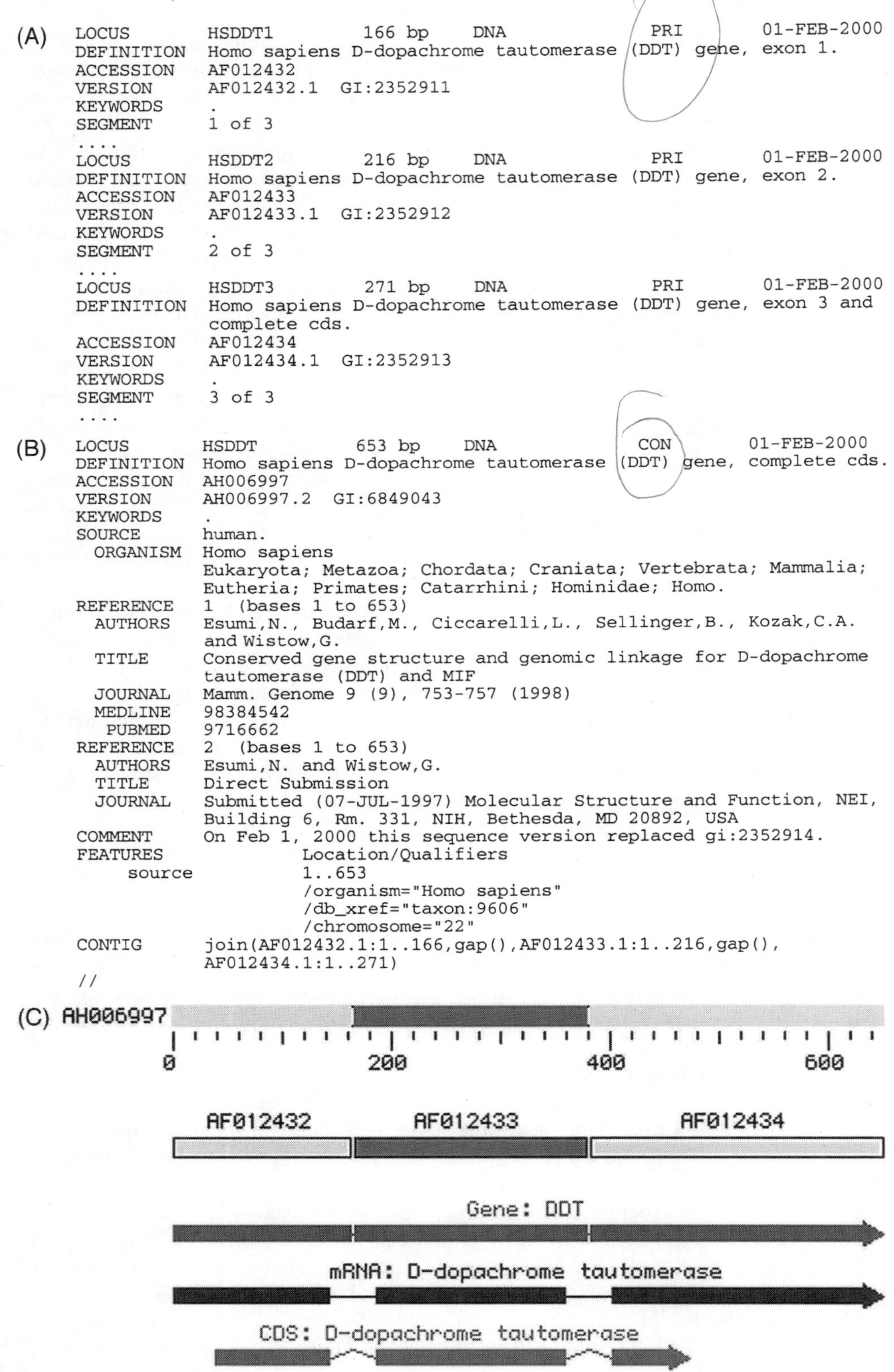

(A)

```
LOCUS       HSDDT1        166 bp    DNA              PRI        01-FEB-2000
DEFINITION  Homo sapiens D-dopachrome tautomerase (DDT) gene, exon 1.
ACCESSION   AF012432
VERSION     AF012432.1  GI:2352911
KEYWORDS    .
SEGMENT     1 of 3
....
LOCUS       HSDDT2        216 bp    DNA              PRI        01-FEB-2000
DEFINITION  Homo sapiens D-dopachrome tautomerase (DDT) gene, exon 2.
ACCESSION   AF012433
VERSION     AF012433.1  GI:2352912
KEYWORDS    .
SEGMENT     2 of 3
....
LOCUS       HSDDT3        271 bp    DNA              PRI        01-FEB-2000
DEFINITION  Homo sapiens D-dopachrome tautomerase (DDT) gene, exon 3 and
            complete cds.
ACCESSION   AF012434
VERSION     AF012434.1  GI:2352913
KEYWORDS    .
SEGMENT     3 of 3
....
```

(B)

```
LOCUS       HSDDT         653 bp    DNA              CON        01-FEB-2000
DEFINITION  Homo sapiens D-dopachrome tautomerase (DDT) gene, complete cds.
ACCESSION   AH006997
VERSION     AH006997.2  GI:6849043
KEYWORDS    .
SOURCE      human.
  ORGANISM  Homo sapiens
            Eukaryota; Metazoa; Chordata; Craniata; Vertebrata; Mammalia;
            Eutheria; Primates; Catarrhini; Hominidae; Homo.
REFERENCE   1  (bases 1 to 653)
  AUTHORS   Esumi,N., Budarf,M., Ciccarelli,L., Sellinger,B., Kozak,C.A.
            and Wistow,G.
  TITLE     Conserved gene structure and genomic linkage for D-dopachrome
            tautomerase (DDT) and MIF
  JOURNAL   Mamm. Genome 9 (9), 753-757 (1998)
  MEDLINE   98384542
   PUBMED   9716662
REFERENCE   2  (bases 1 to 653)
  AUTHORS   Esumi,N. and Wistow,G.
  TITLE     Direct Submission
  JOURNAL   Submitted (07-JUL-1997) Molecular Structure and Function, NEI,
            Building 6, Rm. 331, NIH, Bethesda, MD 20892, USA
COMMENT     On Feb 1, 2000 this sequence version replaced gi:2352914.
FEATURES             Location/Qualifiers
     source          1..653
                     /organism="Homo sapiens"
                     /db_xref="taxon:9606"
                     /chromosome="22"
CONTIG      join(AF012432.1:1..166,gap(),AF012433.1:1..216,gap(),
            AF012434.1:1..271)
//
```

(C)

whole chromosomes quite easily, even though data exist only in small, individual pieces. This ability to readily assemble a set of related sequences on demand for any region of a very large chromosome has already proven to be valuable for bacterial genomes. Assembly on demand will become more and more important as larger and larger regions are sequenced, perhaps by many different groups, and the notion that an investigator will be working on one huge sequence record becomes completely impractical.

What Does ASN.1 Have to Do With It?

The NCBI data model is often referred to as, and confused with, the "NCBI ASN.1" or "ASN.1 Data Model." *Abstract Syntax Notation 1* (ASN.1) is an International Standards Organization (ISO) standard for describing structured data that reliably encodes data in a way that permits computers and software systems of all types to reliably exchange both the structure and the content of the entries. Saying that a data model is written in ASN.1 is like saying a computer program is written in C or FORTRAN. The statement identifies the *language*; it does not say what the program *does*. The familiar GenBank flatfile was really designed for humans to read, from a DNA-centered viewpoint. ASN.1 is designed for a *computer* to read and is amenable to describing complicated data relationships in a very specific way. NCBI describes and processes data using the ASN.1 format. Based on that single, common format, a number of human-readable formats and tools are produced, such as Entrez, GenBank, and the BLAST databases. Without the existence of a common format such as this, the neighboring and hard-link relationships that Entrez depends on would not be possible. This chapter deals with the structure and content of the NCBI data model and its implications for biomedical databases and tools. Detailed discussions about the choice of ASN.1 for this task and its overall form can be found elsewhere (Ostell, 1995).

What to Define?

We have alluded to how the NCBI data model defines sequences in a way that supports a richer and more explicit description of the experimental data than can be

Figure 2.1. (*A*) Selected parts of GenBank-formatted records in a segmented sequence. GenBank format historically indicates merely that records are part of some ordered series; it offers no information on what the other components are or how they are connected. To see the complete view of these records, see *http://www.ncbi.nlm.nih.gov/htbin-post/Entrez/query?uid=6849043&form=6&db=n&Dopt=g*. (*B*) Representation of segmented sequences in the new CON (contig) division. A new extension of GenBank format allows the details of the construction of segmented records to be presented. The `CONTIG` line can include individual accessions, gaps of known length, and gaps of unknown length. The individual components can still be displayed in the traditional form, although no features or sequences are present in this format. (*C*) Graphical representation of a segmented sequence. This view displays features mapped to the coordinates of the segmented sequence. The segments include all exonic and untranslated regions plus 20 base pairs of sequence at the ends of each intron. The segment gaps cover the remaining intronic sequence.

obtained with the GenBank format. The details of the model are important, and will be expanded on in the ensuing discussion. At this point, we need to pause and briefly describe the reasoning and general principles behind the model as a whole.

There are two main reasons for putting data on a computer: retrieval and discovery. Retrieval is basically being able to get back out what was put in. Amassing sequence information without providing a way to retrieve it makes the sequence information, in essence, useless. Although this is important, it is even *more* valuable to be able to get back from the system *more* knowledge than was put in to begin with—that is, to be able to use the information to make biological discoveries. Scientists can make these kinds of discoveries by discerning connections between two pieces of information that were not known when the pieces were entered separately into the database or by performing computations on the data that offer new insight into the records. In the NCBI data model, the emphasis is on facilitating discovery; that means the data must be defined in a way that is amenable to both linkage and computation.

A second, general consideration for the model is stability. NCBI is a US Government agency, not a group supported year-to-year by competitive grants. Thus, the NCBI staff takes a very long-term view of its role in supporting bioinformatics efforts. NCBI provides large-scale information systems that will support scientific inquiry well into the future. As anyone who is involved in biomedical research knows, many major conceptual and technical revolutions can happen when dealing with such a long time span. Somehow, NCBI must address these changing views and needs with software and data that may have been created years (or decades) earlier. For that reason, basic observations have been chosen as the central data elements, with interpretations and nomenclature (elements more subject to change) being placed outside the basic, core representation of the data.

Taking all factors into account, NCBI uses four core data elements: bibliographic citations, DNA sequences, protein sequences, and three-dimensional structures. In addition, two projects (taxonomy and genome maps) are more interpretive but nonetheless are so important as organizing and linking resources that NCBI has built a considerable base in these areas as well.

PUBs: PUBLICATIONS OR PERISH

Publication is at the core of every scientific endeavor. It is the common process whereby scientific information is reviewed, evaluated, distributed, and entered into the permanent record of scientific progress. Publications serve as vital links between factual databases of different structures or content domains (e.g., a record in a sequence database and a record in a genetic database may cite the same article). They serve as valuable entry points into factual databases ("I have read an article about this, now I want to see the primary data").

Publications also act as essential annotation of function and context to records in factual databases. One reason for this is that factual databases have a structure that is essential for efficient use of the database but may not have the representational capacity to set forward the full biological, experimental, or historical context of a particular record. In contrast, the published paper is limited only by language and contains much fuller and more detailed explanatory information than will ever be in a record in a factual database. Perhaps more importantly, authors are evaluated by their scientific peers based on the content of their published papers, not by the content of the associated database records. Despite the best of intentions, scientists move on and database records become static, even though the knowledge about them has expanded, and there is very little incentive for busy scientists to learn a database system and keep records based on their own laboratory studies up to date.

Generally, the form and content of citations have not been thought about carefully by those designing factual databases, and the quality, form, and content of citations can vary widely from one database to the next. Awareness of the importance of having a link to the published literature and the realization that bibliographic citations are much less volatile than scientific knowledge led to a decision that a careful and complete job of defining citations was a worthwhile endeavor. Some components of the publication specification described below may be of particular interest to scientists or users of the NCBI databases, but a full discussion of all the issues leading to the decisions governing the specifications themselves would require another chapter in itself.

Authors

Author names are represented in many formats by various databases: last name only, last name and initials, last name-comma-initials, last name and first name, all authors with initials and the last with a full first name, with or without honorifics (Ph.D.) or suffixes (Jr., III), to name only a few. Some bibliographic databases (such as MEDLINE) might represent only a fixed number of authors. Although this inconsistency is merely ugly to a human reader, it poses *severe* problems for database systems incorporating names from many sources and providing functions as simple as looking up citations by author last name, such as Entrez does. For this reason, the specification provides two alternative forms of author name representation: one a simple string and the other a structured form with fields for last name, first name, and so on. When data are submitted directly to NCBI or in cases when there is a consistent format of author names from a particular source (such as MEDLINE), the structured form is used. When the form cannot be deciphered, the author name remains as a string. This limits its use for retrieval but at least allows data to be viewed when the record is retrieved by other means.

Even the structured form of author names must support diversity, since some sources give only initials whereas others provide a first and middle name. This is mentioned to specifically emphasize two points. First, the NCBI data model is designed both to direct our view of the data into a more useful form and to accommodate the available existing data. (This pair of functions can be confusing to people reading the specification and seeing alternative forms of the same data defined.) Second, software developers must be aware of this range of representations and accommodate whatever form had to be used when a particular source was being converted. In general, NCBI tries to get as much of the data into a uniform, structured form as possible but carries the rest in a less optimal way rather than losing it altogether.

Author affiliations (i.e., authors' institutional addresses) are even more complicated. As with author names, there is the problem of supporting both structured forms and unparsed strings. However, even sources with reasonably consistent author name conventions often produce affiliation information that cannot be parsed from text into a structured format. In addition, there may be an affiliation associated with the whole author list, or there may be different affiliations associated with each author. The NCBI data model allows for both scenarios. At the time of this writing only the first form is supported in either MEDLINE or GenBank, both types may appear in published articles.

Articles

The most commonly cited bibliographic entity in biological science is an article in a journal; therefore, the citation formats of most biological databases are defined with that type in mind. However, "articles" can also appear in books, manuscripts, theses, and now in electronic journals as well. The data model defines the fields necessary to cite a book, a journal, or a manuscript. An article citation occupies one field; other fields display additional information necessary to uniquely identify the article in the book, journal, or manuscript—the author(s) of the article (as opposed to the author or editor of the book), the title of the article, page numbers, and so on.

There is an important distinction between the fields necessary to uniquely identify a published article from a citation and those necessary to describe the same article meaningfully to a database user. The NCBI Citation Matching Service takes fields from a citation and attempts to locate the article to which they refer. In this process, a successful match would involve only correctly matching the journal title, the year, the first page of the article, and the last name of an author of the article. Other information (e.g., article title, volume, issue, full pages, author list) is useful to look at but very often is either not available or outright incorrect. Once again, the data model must allow the minimum information set to come in as a citation, be matched against MEDLINE, and then be replaced by a citation having the full set of desired fields obtained from MEDLINE to produce accurate, useful data for consumption by the scientific public.

Patents

With the advent of patented sequences it became necessary to cite a patent as a bibliographic entity instead of an article. The data model supports a very complete patent citation, a format developed in cooperation with the US Patent Office. In practice, however, patented sequences tend to have limited value to the scientific public. Because a patent is a *legal* document, not a scientific one, its purpose is to present and support the claims of the patent, *not* to fully describe the biology of the sequence itself. It is often prepared in a lawyer's office, not by the scientist who did the research. The sequences presented in the patent may function only to illustrate some discreet aspect of the patent, rather than being the focus of the document. Organism information, location of biological features, and so on may not appear at all if they are not germane to the patent. Thus far, the vast majority of sequences appearing in patents also appear in a more useful form (to scientists) in the public databases.

In NCBI's view, the main purpose of listing patented sequences in GenBank is to be able to retrieve sequences by similarity searches that may serve to locate patents related to a given sequence. To make a legal determination in the case, however, one would still have to examine the full text of the patent. To evaluate the biology of the sequence, one generally must locate information other than that contained in the patent. Thus, the critical linkage is between the sequence and its patent number.

Additional fields in the patent citation itself may be of some interest, such as the title of the patent and the names of the inventors.

Citing Electronic Data Submission

A relatively new class of citations comprises the act of data submission to a database, such as GenBank. This is an act of publication, similar but not identical to the publication of an article in a journal. In some cases, data submission precedes article publication by a considerable period of time, or a publication regarding a particular sequence may never appear in press. Because of this, there is a separate citation designed for deposited sequence data. The submission citation, because it is indeed an act of publication, may have an author list, showing the names of scientists who worked on the record. This may or may not be the same as the author list on a subsequently published paper also cited in the same record. In most cases, the scientist who submitted the data to the database is also an author on the submission citation. (In the case of large sequencing centers, this may not always be the case.) Finally, NCBI has begun the practice of citing the update of a record with a submission citation as well. A comment can be included with the update, briefly describing the changes made in the record. All the submission citations can be retained in the record, providing a history of the record over time.

MEDLINE and PubMed Identifiers

Once an article citation has been matched to MEDLINE, the simplest and most reliable key to point to the article is the MEDLINE unique identifier (MUID). This is simply an integer number. NCBI provides many services that use MUID to retrieve the citation and abstract from MEDLINE, to link together data citing the same article, or to provide Web hyperlinks.

Recently, in concert with MEDLINE and a large number of publishers, NCBI has introduced *PubMed*. PubMed contains *all* of MEDLINE, as well as citations provided directly by the publishers. As such, PubMed contains more recent articles than MEDLINE, as well as articles that may never appear in MEDLINE because of their subject matter. This development led NCBI to introduce a new article identifier, called a PubMed identifier (PMID). Articles appearing in MEDLINE will have *both* a PMID and an MUID. Articles appearing only in PubMed will have only a PMID. PMID serves the same purpose as MUID in providing a simple, reliable link to the citation, a means of linking records together, and a means of setting up hyperlinks.

Publishers have also started to send information on ahead-of-print articles to PubMed, so this information may now appear before the printed journal. A new project, *PubMed Central*, is meant to allow electronic publication to occur in lieu of or ahead of publication in a traditional, printed journal. PubMed Central records contain the full text of the article, not just the abstract, and include all figures and references.

The NCBI data model stores most citations as a collection called a Pub-equiv, a set of equivalent citations that includes a reliable identifier (PMID or MUID) and the citation itself. The presence of the citation form allows a useful display without an extra retrieval from the database, whereas the identifier provides a reliable key for linking or indexing the same citation in the record.

SEQ-IDs: WHAT'S IN A NAME?

The NCBI data model defines a whole class of objects called Sequence Identifiers (Seq-id). There has to be a whole class of such objects because NCBI integrates sequence data from many sources that name sequence records in different ways and where, of course, the individual names have different meanings. In one simple case, PIR, SWISS-PROT, and the nucleotide sequence databases all use a string called an "accession number," all having a similar format. Just saying "A10234" is not enough to uniquely identify a sequence record from the collection of all these databases. One must distinguish "A10234" in SWISS-PROT from "A10234" in PIR. (The DDBJ/EMBL/GenBank nucleotide databases share a common set of accession numbers; therefore, "A12345" in EMBL is the same as "A12345" in GenBank or DDBJ.) To further complicate matters, although the sequence databases define their records as containing a single sequence, PDB records contain a single *structure*, which may contain more than one sequence. Because of this, a PDB Seq-id contains a molecule name and a chain ID to identify a single unique sequence. The subsections that follow describe the form and use of a few commonly used types of Seq-ids.

Locus Name

The *locus* appears on the `LOCUS` line in GenBank and DDBJ records and in the `ID` line in EMBL records. These originally were the only identifier of a discrete GenBank record. Like a genetic locus name, it was intended to act both as a unique identifier for the record and as a mnemonic for the function and source organism of the sequence. Because the `LOCUS` line is in a fixed format, the locus name is restricted to ten or fewer numbers and uppercase letters. For many years in GenBank, the first three letters of the name were an organism code and the remaining letters a code for the gene (e.g., `HUMHBB` was used for "human β-globin region"). However, as with genetic locus names, locus names were changed when the function of a region was discovered to be different from what was originally thought. This instability in locus names is obviously a problem for an identifier for retrieval. In addition, as the number of sequences and organisms represented in GenBank increased geometrically over the years, it became impossible to invent and update such mnemonic names in an efficient and timely manner. At this point, the locus name is dying out as a useful name in GenBank, although it continues to appear prominently on the first line of the flatfile to avoid breaking the established format.

Accession Number

Because of the difficulties in using the locus/ID name as the unique identifier for a nucleotide sequence record, the International Nucleotide Sequence Database Collaborators (DDBJ/EMBL/GenBank) introduced the accession number. It intentionally carries no biological meaning, to ensure that it will remain (relatively) stable. It originally consisted of one uppercase letter followed by five digits. New accessions consist of two uppercase letters followed by six digits. The first letters were allocated to the individual collaborating databases so that accession numbers would be unique across the Collaboration (e.g., an entry beginning with a "U" was from GenBank).

The accession number was an improvement over the locus/ID name, but, with use, problems and deficiencies became apparent. For example, although the accession is stable over time, many users noticed that the sequence retrieved by a particular accession was not always the same. This is because the accession identifies the *whole database record*. If the sequence in a record was updated (say by the insertion of 1000 bp at the beginning), the accession number did not change, as it was an updated version of the same record. If one had analyzed the original sequence and recorded that at position 100 of accession U00001 there was a putative protein-binding site, after the update a completely different sequence would be found at position 100!

The accession number appears on the `ACCESSION` line of the GenBank record. The first accession on the line, called the "primary" accession, is the key for retrieving this record. Most records have only this type of accession number. However, other accessions may follow the primary accession on the `ACCESSION` line. These "secondary" accessions are intended to give some notion of the history of the record. For example, if U00001 and U00002 were merged into a single updated record, then U00001 would be the primary accession on the new record and U00002 would appear as a secondary accession. In standard practice, the U00002 record would be removed from GenBank, since the older record had become obsolete, and the secondary accessions would allow users to retrieve whatever records superseded the old one. It should also be noted that, historically, secondary accession numbers do not always mean the same thing; therefore, users should exercise care in their interpretations. (Policies at individual databases differed, and even shifted over time in a given database.) The use of secondary accession numbers also caused problems in that there was still not enough information to determine exactly what happened and why. Nonetheless, the accession number remains the most controlled and reliable way to point to a record in DDBJ/EMBL/GenBank.

gi Number

In 1992, NCBI began assigning GenInfo Identifiers (gi) to all sequences processed into Entrez, including nucleotide sequences from DDBJ/EMBL/GenBank, the protein sequences from the translated CDS features, protein sequences from SWISS-PROT, PIR, PRF, PDB, patents, and others. The gi is assigned in addition to the accession number provided by the source database. Although the form and meaning of the accession Seq-id varied depending on the source, the meaning and form of the gi is the same for all sequences regardless of the source.

The gi is simply an integer number, sometimes referred to as a *GI number*. It is an identifier *for a particular sequence only*. Suppose a sequence enters GenBank and is given an accession number U00001. When the sequence is processed internally at NCBI, it enters a database called ID. ID determines that it has not seen U00001 before and assigns it a gi number—for example, 54. Later, the submitter might update the record by changing the citation, so U00001 enters ID again. ID, recognizing the record, retrieves the first U00001 and compares its sequence with the new one. If the two are completely identical, ID reassigns gi 54 to the record. If the sequence differs in any way, even by a single base pair, it is given a new gi number, say 88. However, the new sequence retains accession number U00001 because of the semantics of the source database. At this time, ID marks the old record (gi 54) with the date it was replaced and adds a "history" indicating that it was replaced by gi 88. ID also adds a history to gi 88 indicating that it replaced gi 54.

The gi number serves three major purposes:

- It provides a single identifier across sequences from many sources.
- It provides an identifier that specifies an exact sequence. Anyone who analyzes gi 54 and stores the analysis can be sure that it will be valid as long as U00001 has gi 54 attached to it.
- It is stable and retrievable. NCBI keeps the last version of every gi number. Because the history is included in the record, anyone who discovers that gi 54 is no longer part of the GenBank release can still retrieve it from ID through NCBI and examine the history to see that it was replaced by gi 88. Upon aligning gi 54 to gi 88 to determine their relationship, a researcher may decide to remap the former analysis to gi 88 or perhaps to reanalyze the data. This can be done at any time, not just at GenBank release time, because gi 54 will always be available from ID.

For these reasons, all internal processing of sequences at NCBI, from computing Entrez sequence neighbors to determining when new sequence should be processed or producing the BLAST databases, is based on gi numbers.

Accession.Version Combined Identifier

Recently, the members of the International Nucleotide Sequence Database Collaboration (GenBank, EMBL, and DDBJ) introduced a "better" sequence identifier, one that combines an accession (which identifies a particular sequence record) with a version number (which tracks changes to the sequence itself). It is expected that this kind of Seq-id will become the preferred method of citing sequences.

Users will still be able to retrieve a record based on the accession number alone, without having to specify a particular version. In that case, the latest version of the record will be obtained by default, which is the current behavior for queries using Entrez and other retrieval programs.

Scientists who are analyzing sequences in the database (e.g., aligning all alcohol dehydrogenase sequences from a particular taxonomic group) and wish to have their conclusions remain valid over time will want to reference sequences by accession and the given version number. Subsequent modification of one of the sequences by its owner (e.g., 5′ extension during a study of the gene's regulation) will result in the version number being incremented appropriately. The analysis that cited accession and version remains valid because a query using both the accession and version will return the desired record.

Combining accession and version makes it clear to the casual user that a sequence has changed since an analysis was done. Also, determining how many times a sequence has changed becomes trivial with a version number. The accession.version number appears on the `VERSION` line of the GenBank flatfile. For sequence retrieval, the accession.version is simply mapped to the appropriate gi number, which remains the underlying tracking identifier at NCBI.

Accession Numbers on Protein Sequences

The International Sequence Database Collaborators also started assigning accession.version numbers to *protein* sequences within the records. Previously, it was difficult to reliably cite the translated product of a given coding region feature, except by its gi number. This limited the usefulness of translated products found in BLAST results, for example. These sequences will now have the same status as protein sequences submitted directly to the protein databases, and they have the benefit of direct linkage to the nucleotide sequence in which they are encoded, showing up as a CDS feature's `/protein_id` qualifier in the flatfile view. Protein accessions in these records consist of three uppercase letters followed by five digits and an integer indicating the version.

Reference Seq-id

The NCBI RefSeq project provides a curated, nonredundant set of reference sequence standards for naturally occurring biological molecules, ranging from chromosomes to transcripts to proteins. RefSeq identifiers are in accession.version form but are prefixed with `NC_` (chromosomes), `NM_` (mRNAs), `NP_` (proteins), or `NT_` (constructed genomic contigs). The `NG_` prefix will be used for genomic regions or gene clusters (e.g., immunoglobulin region) in the future. RefSeq records are a stable reference point for functional annotation, point mutation analysis, gene expression studies, and polymorphism discovery.

General Seq-id

The General Seq-id is meant to be used by genome centers and other groups as a way of identifying their sequences. Some of these sequences may never appear in public databases, and others may be preliminary data that eventually will be submitted. For example, records of human chromosomes in the Entrez Genomes division contain multiple physical and genetic maps, in addition to sequence components. The physical maps are generated by various groups, and they use General Seq-ids to identify the proper group.

Local Seq-id

The Local sequence identifier is most prominently used in the data submission tool Sequin (see Chapter 4). Each sequence will eventually get an `accession.version` identifier and a gi number, but only when the completed submission has been processed by one of the public databases. During the submission process, Sequin assigns a local identifier to each sequence. Because many of the software tools made by NCBI require a sequence identifier, having a local Seq-id allows the use of these tools without having to first submit data to a public database.

BIOSEQs: SEQUENCES

The Bioseq, or biological sequence, is a central element in the NCBI data model. It comprises a single, continuous molecule of either nucleic acid or protein, thereby defining a linear, integer coordinate system for the sequence. A Bioseq must have at least one sequence identifier (Seq-id). It has information on the physical type of molecule (DNA, RNA, or protein). It may also have annotations, such as biological features referring to specific locations on specific Bioseqs, as well as descriptors.

Descriptors provide additional information, such as the organism from which the molecule was obtained. Information in the descriptors describe the entire Bioseq.

However, the Bioseq isn't necessarily a fully sequenced molecule. It may be a segmented sequence in which, for example, the exons have been sequenced but not all of the intronic sequences have been determined. It could also be a genetic or physical map, where only a few landmarks have been positioned.

Sequences are the Same

All Bioseqs have an integer coordinate system, with an integer length value, even if the actual sequence has not been completely determined. Thus, for physical maps, or for exons in highly spliced genes, the spacing between markers or exons may be known only from a band on a gel. Although the coordinates of a fully sequenced chromosome are known exactly, those in a genetic or physical map are a best guess, with the possibility of significant error from the "real" coordinates.

Nevertheless, any Bioseq can be annotated with the same kinds of information. For example, a gene feature can be placed on a region of sequenced DNA or at a discrete location on a physical map. The map and the sequence can then be aligned on the basis of their common gene features. This greatly simplifies the task of writing software that can display these seemingly disparate kinds of data.

Sequences are Different

Despite the benefits derived from having a common coordinate system, the different Bioseq classes do differ in the way they are represented. The most common classes (Fig. 2.2) are described briefly below.

Virtual Bioseq. In the virtual Bioseq, the molecule type is known, and its length and topology (e.g., linear, circular) may also be known, but the actual sequence is not known. A virtual Bioseq can represent an intron in a genomic molecule in which only the exon sequences have been determined. The length of the putative sequence may be known only by the size of a band on an agarose gel.

Figure 2.2. Classes of Bioseqs. All Bioseqs represent a single, continuous molecule of nucleic acid or protein, although the complete sequence may not be known. In a virtual Bioseq, the type of molecule is known, but the sequence is not known, and the precise length may not be known (e.g., from the size of a band on an electrophoresis gel). A raw Bioseq contains a single contiguous string of bases or residues. A segmented Bioseq points to its components, which are other raw or virtual Bioseqs (e.g., sequenced exons and undetermined introns). A constructed sequence takes its original components and subsumes them, resulting in a Bioseq that contains the string of bases or residues and a "history" of how it was built. A map Bioseq places genes or physical markers, rather than sequence, on its coordinates. A delta Bioseq can represent a segmented sequence but without the requirement of assigning identifiers to each component (including gaps of known length), although separate raw sequences can still be referenced as components. The delta sequence is used for unfinished high-throughput genome sequences (HTGS) from genome centers and for genomic contigs.

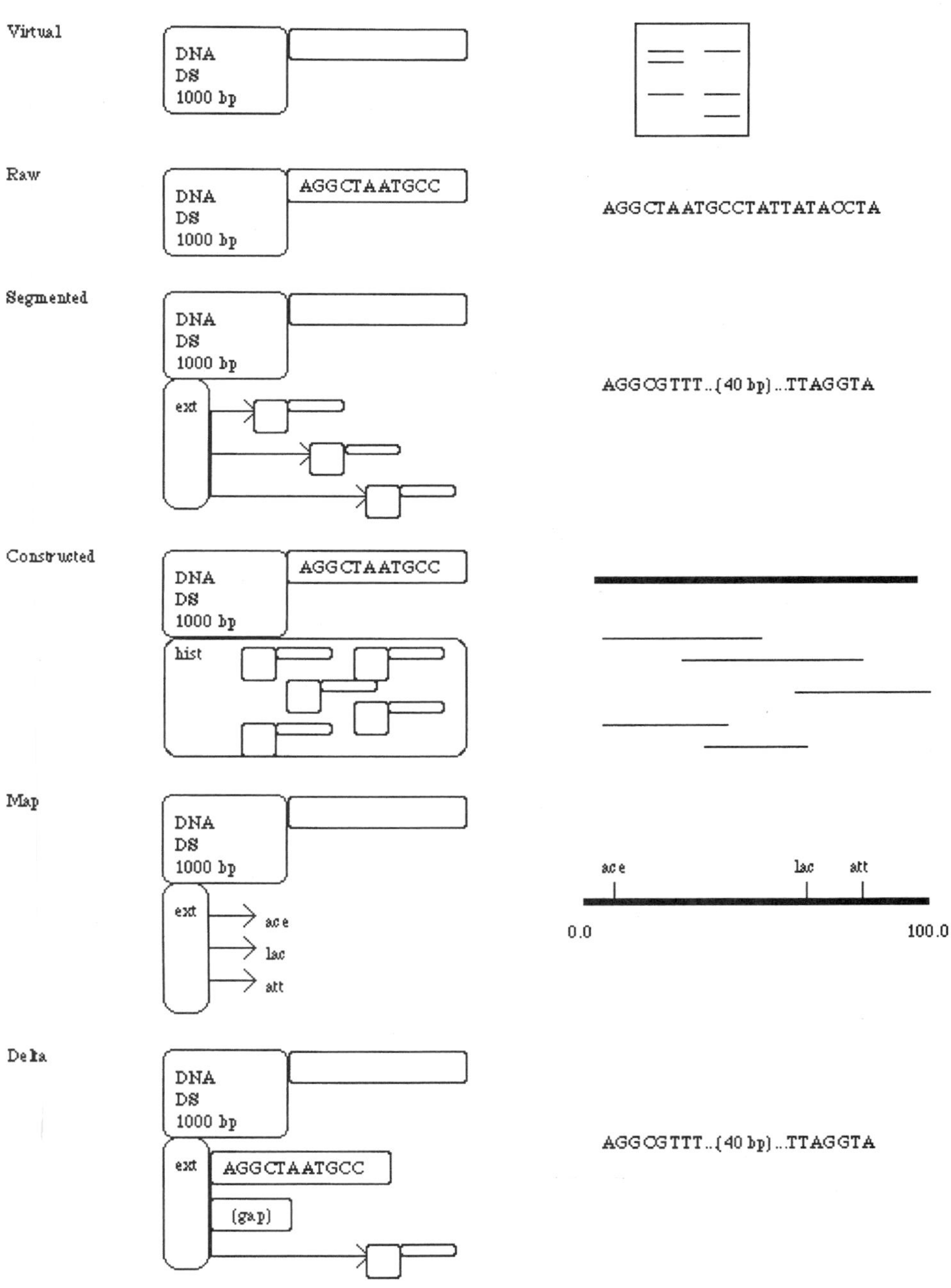
Virtual
DNA
DS
1000 bp
Raw
DNA
DS
1000 bp
AGGCTAATGCC
AGGCTAATGCCTATTATACCTA
Segmented
DNA
DS
1000 bp
ext
AGGCGTTT...(40 bp)...TTAGGTA
Constructed
DNA
DS
1000 bp
AGGCTAATGCC
hist
Map
DNA
DS
1000 bp
ext
ace
lac
att
ace
lac
att
0.0
100.0
Delta
DNA
DS
1000 bp
ext
AGGCTAATGCC
(gap)
AGGCGTTT...(40 bp)...TTAGGTA

Raw Bioseq. This is what most people would think of as a sequence, a single contiguous string of bases or residues, in which the actual sequence is known. The length is obviously known in this case, matching the number of bases or residues in the sequence.

Segmented Bioseq. A segmented Bioseq does not contain raw sequences but instead contains the identifiers of other Bioseqs from which it is made. This type of Bioseq can be used to represent a genomic sequence in which only the exons are known. The "parts" in the segmented Bioseq would be the individual, raw Bioseqs representing the exons and the virtual Bioseqs representing the introns.

Delta Bioseq. Delta Bioseqs are used to represent the unfinished high-throughput genome sequences (HTGS) derived at the various genome sequencing centers. Using delta Bioseqs instead of segmented Bioseqs means that only one Seq-id is needed for the entire sequence, even though subregions of the Bioseq are not known at the sequence level. Implicitly, then, even at the early stages of their presence in the databases, delta Bioseqs maintain the same accession number.

Map Bioseq. Used to represent genetic and physical maps, a map Bioseq is similar to a virtual Bioseq in that it has a molecule type, perhaps a topology, and a length that may be a very rough estimate of the molecule's actual length. This information merely supplies the coordinate system, a property of every Bioseq. Given this coordinate system for a genetic map, we estimate the positions of genes on it based on genetic evidence. The table of the resulting gene features is the essential data of the map Bioseq, just as bases or residues constitute the raw Bioseq's data.

BIOSEQ-SETs: COLLECTIONS OF SEQUENCES

A biological sequence is often most appropriately stored in the context of other, related sequences. For example, a nucleotide sequence and the sequences of the protein products it encodes naturally belong in a set. The NCBI data model provides the Bioseq-set for this purpose.

A Bioseq-set can have a list of *descriptors*. When packaged on a Bioseq, a descriptor applies to all of that Bioseq. When packaged on a Bioseq-set, the descriptor applies to every Bioseq in the set. This arrangement is convenient for attaching publications and biological source information, which are expected on all sequences but frequently are identical within sets of sequences. For example, both the DNA and protein sequences are obviously from the same organism, so this descriptor information can be applied to the set. The same logic may apply to a publication.

The most common Bioseq-sets are described in the sections that follow.

Nucleotide/Protein Sets

The Nuc-prot set, containing a nucleotide and one or more protein products, is the type of set most frequently produced by a Sequin data submission. The component Bioseqs are connected by coding sequence region (CDS) features that describe how translation from nucleotide to protein sequence is to proceed. In a traditional nucleotide or protein sequence database, these records might have cross-references to each other to indicate this relationship. The Nuc-prot set makes this explicit by packaging them together. It also allows descriptive information that applies to all sequences (e.g., the organism or publication citation) to be entered once (see *Seq-descr: Describing the Sequence*, below).

Population and Phylogenetic Studies

A major class of sequence submissions represent the results of population or phylogenetic studies. Such research involves sequencing the same gene from a number of individuals in the same species (population study) or in different species (phylogenetic study). An alignment of the individual sequences may also be submitted (see *Seq-align: Alignments*, below). If the gene encodes a protein, the components of the Population or Phylogenetic Bioseq-set may themselves be Nuc-prot sets.

Other Bioseq-sets

A Seg set contains a segmented Bioseq and a Parts Bioseq-set, which in turn contains the raw Bioseqs that are referenced by the segmented Bioseq. This may constitute the nucleotide component of a Nuc-prot set.

An Equiv Bioseq-set is used in the Entrez Genomes division to hold multiple equivalent Bioseqs. For example, human chromosomes have one or more genetic maps, physical maps derived by different methods and a segmented Bioseq on which "islands" of sequenced regions are placed. An alignment between the various Bioseqs is made based on references to any available common markers.

SEQ-ANNOT: ANNOTATING THE SEQUENCE

A Seq-annot is a self-contained package of sequence annotations or information that refers to specific locations on specific Bioseqs. It may contain a feature table, a set of sequence alignments, or a set of graphs of attributes along the sequence.

Multiple Seq-annots can be placed on a Bioseq or on a Bioseq-set. Each Seq-annot can have specific attribution. For example, PowerBLAST (Zhang and Madden, 1997) produces a Seq-annot containing sequence alignments, and each Seq-annot is named based on the BLAST program used (e.g., BLASTN, BLASTX, etc.). The individual blocks of alignments are visible in the Entrez and Sequin viewers.

Because the components of a Seq-annot have specific references to locations on Bioseqs, the Seq-annot can stand alone or be exchanged with other scientists, and it need not reside in a sequence record. The scope of descriptors, on the other hand, does depend on where they are packaged. Thus, information *about* Bioseqs can be created, exchanged, and compared independently of the Bioseq itself. This is an important attribute of the Seq-annot and of the NCBI data model.

Seq-feat: Features

A sequence feature (Seq-feat) is a block of structured data explicitly attached to a region of a Bioseq through one or two sequence locations (Seq-locs). The Seq-feat itself can carry information common to all features. For example, there are flags to indicate whether a feature is partial (i.e., goes beyond the end of the sequence of the Bioseq), whether there is a biological exception (e.g., RNA editing that explains why a codon on the genomic sequence does not translate to the expected amino acid), and whether the feature was experimentally determined (e.g., an mRNA was isolated from a proposed coding region).

A feature must always have a location. This is the Seq-loc that states where on the sequence the feature resides. A coding region's location usually starts at the ATG and ends at the terminator codon. The location can have more than one interval if it is on a genomic sequence and mRNA splicing occurs. In cases of alternative splicing, separate coding region features are created, with one multi-interval Seq-loc for each isolated molecular species.

Optionally, a feature may have a product. For a coding region, the product Seq-loc points to the resulting protein sequence. This is the link that allows the data model to separately maintain the nucleotide and protein sequences, with annotation on each sequence appropriate to that molecule. An mRNA feature on a genomic sequence could have as its product an mRNA Bioseq whose sequence reflects the results of posttranscriptional RNA editing. Features also have information unique to the kind of feature. For example, the CDS feature has fields for the genetic code and reading frame, whereas the tRNA feature has information on the amino acid transferred.

This design completely modularizes the components required by each feature type. If a particular feature type calls for a new field, no other field is affected. A new feature type, even a very complex one, can be added without changing the existing features. This means that software used to display feature locations on a sequence need consider only the location field common to all features.

Although the DDBJ/EMBL/GenBank feature table allows numerous kinds of features to be included (see Chapter 3), the NCBI data model treats some features as "more equal" than others. Specifically, certain features directly model the central dogma of molecular biology and are most likely to be used in making connections between records and in discovering new information by computation. These features are discussed next.

Genes. A gene is a feature in its own right. In the past, it was merely a qualifier on other features. The Gene feature indicates the location of a gene, a heritable region of nucleic acid sequence that confers a measurable phenotype. That phenotype may be achieved by many components of the gene being studied, including, but not limited to, coding regions, promoters, enhancers, and terminators. The Gene feature is meant to approximately cover the region of nucleic acid considered by workers in the field to be the gene. This admittedly fuzzy concept has an appealing simplicity, and it fits in well with higher-level views of genes such as genetic maps. It has practical utility in the era of large genomic sequencing when a biologist may wish to see just the "xyz gene" and not a whole chromosome. The Gene feature may also contain cross-references to genetic databases, where more detailed information on the gene may be found.

RNAs. An RNA feature can describe both coding intermediates (e.g., mRNAs) and structural RNAs (e.g., tRNAs, rRNAs). The locations of an mRNA and the corresponding coding region (CDS) completely determine the locations of 5′ and 3′ untranslated regions (UTRs), exons, and introns.

Coding Regions. A Coding Region (CDS) feature in the NCBI data model can be thought of as "instructions to translate" a nucleic acid into its protein product, via a genetic code (Fig. 2.3). A coding region serves as a link between the nucleotide and protein. It is important to note that several situations can provide exceptions to the classical colinearity of gene and protein. Translational stuttering (ribosomal slippage), for example, merely results in the presence of overlapping intervals in the feature's location Seq-loc.

The genetic code is assumed to be universal unless explicitly given in the Coding Region feature. When the genetic code is not followed at specific positions in the sequence—for example, when alternative initiation codons are used in the first position, when suppressor tRNAs bypass a terminator, or when selenocysteine is added—the Coding Region feature allows these anomalies to be indicated.

Proteins. A Protein feature names (or at least describes) a protein or proteolytic product of a protein. A single protein Bioseq may have many Protein features on it. It may have one over its full length describing a pro-peptide, the primary product of translation. (The name in this feature is used for the `/product` qualifier in the CDS feature that produces the protein.) It may have a shorter protein feature describing the mature peptide or, in the case of viral polyproteins, several mature peptide features. Signal peptides that guide a protein through a membrane may also be indicated.

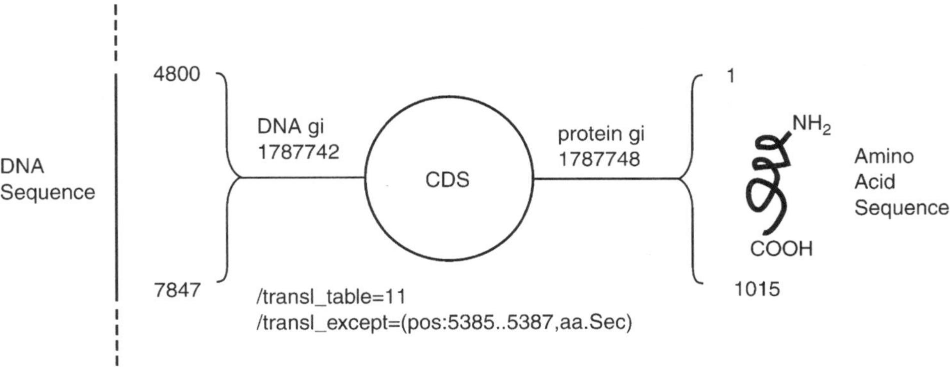

Figure 2.3. The Coding Region (CDS) feature links specific regions on a nucleotide sequence with its encoded protein product. All features in the NCBI data model have a "location" field, which is usually one or more intervals on a sequence. (Multiple intervals on a CDS feature would correspond to individual exons.) Features may optionally have a "product" field, which for a CDS feature is the entirety of the resulting protein sequence. The CDS feature also contains a field for the genetic code. This appears in the GenBank flat file as a `/transl_table` qualifier. In this example, the Bacterial genetic code (code 11) is indicated. A CDS may also have translation exceptions indicating that a particular residue is not what is expected, given the codon and the genetic code. In this example, residue 196 in the protein is selenocysteine, indicated by the `/transl_except` qualifier. NCBI software includes functions for converting between codon locations and residue locations, using the CDS as its guide. This capability is used to support the historical conventions of GenBank format, allowing a signal peptide, annotated on the protein sequence, to appear in the GenBank flat file with a location on the nucleotide sequence.

Others. Several other features are less commonly used. A Region feature provides a simple way to name a region of a chromosome (e.g., "major histocompatibility complex") or a domain on a polypeptide. A Bond feature annotates a bond between two residues in a protein (e.g., disulfide). A Site feature annotates a known site (e.g., active, binding, glycosylation, methylation, phosphorylation).

Finally, numerous features exist in the table of legal features, covering many aspects of biology. However, they are less likely than the above-mentioned features to be used for making connections between records or for making discoveries based on computation.

Seq-align: Alignments

Sequence alignments simply describe the relationships between biological sequences by designating portions of sequences that correspond to each other. This correspondence can reflect evolutionary conservation, structural similarity, functional similarity, or a random event. An alignment can be generated algorithmically by software (e.g., BLAST produces a Seq-annot containing one or more Seq-aligns) or directly by a scientist (e.g., one who is submitting an aligned population study using a favorite alignment tool and a submission program like Sequin; cf. Chapter 4). The Seq-align is designed to capture the final result of the process, not the process itself. Aligned regions can be given scores according to the probability that the alignment is a chance occurrence.

Regardless of how or why an alignment is generated or what its biological significance may be, the data model records, in a condensed format, which regions of which sequences are said to correspond. The fundamental unit of an alignment is a segment, which is defined as an unbroken region of the alignment. In these segments, each sequence is present either without gaps or is not present at all (completely gapped). The alignment below has four segments, delineated by vertical lines:

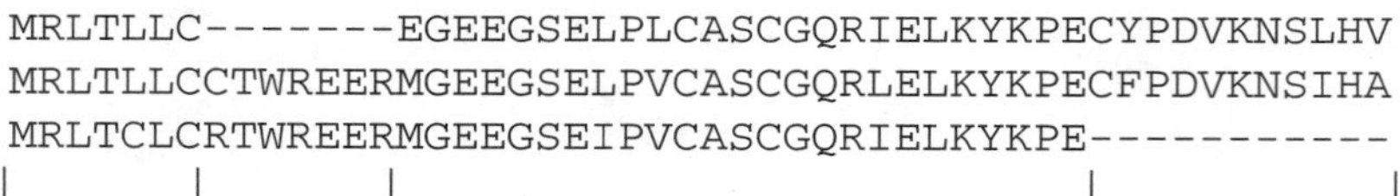

```
MRLTLLC-------EGEEGSELPLCASCGQRIELKYKPECYPDVKNSLHV
MRLTLLCCTWREERMGEEGSELPVCASCGQRLELKYKPECFPDVKNSIHA
MRLTCLCRTWREERMGEEGSEIPVCASCGQRIELKYKPE-----------
|      |      |                        |          |
```

Note that mismatches do not break a segment; only a gap opening or closing event will force the creation of a new segment.

By structuring the alignment in this fashion, it can be saved in condensed form. The data representation records the start position in sequence coordinates for each sequence in a segment and the length of the segment. If a sequence is gapped in a segment, its start position is -1. Note that this representation is independent of the actual sequence; that is, nucleotide and protein alignments are represented the same way, and only the score of an alignment gives a clue as to how many matches and mismatches are present in the data.

The Sequence Is Not the Alignment

Note that the gaps in the alignment are not actually represented in the Bioseqs as dashes. A fundamental property of the genetic code is that it is "commaless" (Crick et al., 1961). That is, there is no "punctuation" to distinguish one codon from the next or to keep translation in the right frame. The gene is a contiguous string of nucleotides. We remind the reader that sequences themselves are also "gapless." Gaps are shown only in the alignment report, generated from the alignment data; they are used only for comparison.

Classes of Alignments

Alignments can exist by themselves or in sets and can therefore represent quite complicated relationships between sequences. A single alignment can only represent a continuous and linear correspondence, but a set of alignments can denote a continuous, discontinuous, linear, or nonlinear relationship among sequences. Alignments can also be local, meaning that only portions of the sequences are included in the alignment, or they can be global, so that the alignment completely spans all the sequences involved.

A continuous alignment does not have regions that are unaligned; that is, for each sequence in the alignment, each residue between the lowest-numbered and highest-numbered residues of the alignment is also contained in the alignment. More simply put, there are no pieces missing. Because such alignments are necessarily linear, they can be displayed with one sequence on each line, with gaps representing deletions or insertions. To show the differences from a "master" sequence, one of the sequences can be displayed with no gaps and no insertions; the remaining sequences can have gaps or inserted segments (often displayed above or below the rest of the sequence), as needed. If pairwise, the alignment can be displayed in a square matrix as a squiggly line traversing the two sequences.

A discontinuous alignment contains regions that are unaligned. For example, the alignment below is a set of two local alignments between two protein sequences. The regions in between are simply not aligned at all:

```
12 MA-TLICCTWREGRMG 26 45 KPECFPDVKNSIHV 58
15 MRLTLLCCTWREERMG 30 35 KPECFPDAKNSLHV 48
```

This alignment could be between two proteins that have two matching (but not identical) structural domains linked by a divergent segment. There is simply no alignment for the regions that are not shown above. A discontinuous alignment can be linear, like the one in the current example, so that the sequences can still be shown one to a line without breaking the residue order. More complicated discontinuous alignments may have overlapping segments, alignments on opposite strands (for nucleotides), or repeated segments, so that they cannot be displayed in linear order. These nonlinear alignments are the norm and can be displayed in square matrices (if pairwise), in lists of aligned regions, or by complex shading schemes.

Data Representations of Alignments

A continuous alignment can be represented as a single list of coordinates, as described above. Depending on whether the alignment spans all of the sequences, it can be designated global or local.

Discontinuous alignments must be represented as sets of alignments, each of which is a single list of coordinates. The regions between discontinuous alignments are not represented at all in the data, and, to display these regions, the missing pieces must be calculated. If the alignment as a whole is linear, the missing pieces can be fairly simply calculated from the boundaries of the aligned regions. A discontinuous alignment is usually local, although if it consists of several overlapping pieces it may in fact represent a global correspondence between the sequences.

Seq-graph: Graphs

Graphs are the third kind of annotation that can go into Seq-annots. A Seq-graph defines some continuous set of values over a defined interval on a Bioseq. It can be used to show properties like G+C content, surface potential, hydrophobicity, or base accuracy over the length of the sequence.

SEQ-DESCR: DESCRIBING THE SEQUENCE

A Seq-descr is meant to describe a Bioseq (or Bioseq-set) and place it in its biological and/or bibliographic context. Seq-descrs apply to the whole Bioseq or to the whole of each Bioseq in the Bioseq-set to which the Seq-descr is attached.

Descriptors were introduced in the NCBI data model to reduce redundant information in records. For example, the protein products of a nucleotide sequence should always be from the same biological source (organism, tissue) as the nucleotide itself. And the publication that describes the sequencing of the DNA in many cases also discusses the translated proteins. By placement of these items as descriptors at the Nuc-prot set level, only one copy of each item is needed to properly describe all the sequences.

BioSource: The Biological Source

The BioSource includes information on the source organism (scientific name and common name), its lineage in the NCBI integrated taxonomy, and its nuclear and (if appropriate) mitochondrial genetic code. It also includes information on the location of the sequence in the cell (e.g., nuclear genome or mitochondrion) and additional modifiers (e.g., strain, clone, isolate, chromosomal map location).

A sequence record for a gene and its protein product will typically have a single BioSource descriptor at the Nuc-prot set level. A population or phylogenetic study, however, will have BioSource descriptors for each component. (The components can be nucleotide Bioseqs or they can themselves be Nuc-prot sets.) The BioSources in a population study will have the same organism name and usually will be distinguished from each other by modifier information, such as strain or clone name.

MolInfo: Molecule Information

The MolInfo descriptor indicates the type of molecule [e.g., genomic, mRNA (usually isolated as cDNA), rRNA, tRNA, or peptide], the technique with which it was sequenced (e.g., standard, EST, conceptual translation with partial peptide sequencing for confirmation), and the completeness of the sequence [e.g., complete, missing the left (5′ or amino) end, missing both ends]. Each nucleotide and each protein should get its own MolInfo descriptor. Normally, then, this descriptor will not appear attached at the Nuc-prot set level. (It may go on a Seg set, since all parts of a segmented Bioseq should be of the same type.)

USING THE MODEL

There are a number of consequences of using the NCBI data model for building databases and generating reports. Some of these are discussed in the remainder of this section.

GenBank Format

GenBank presents a "DNA-centered" view of a sequence record. (GenPept presents the equivalent "protein-centered" view.) To maintain compatibility with these historical views, some mappings are performed between features on different sequences or between overlapping features on the same sequence.

In GenBank format, the protein product of a coding region feature is displayed as a `/translation` qualifier, not as a sequence that can have its own features. The largest protein feature on the product Bioseq is used as the `/product` qualifier. Some of the features that are actually annotated on the protein Bioseq in the NCBI data model, such as mature peptide or signal peptide, are mapped onto the DNA coordinate system (through the CDS intervals) in GenBank format.

The Gene feature names a region on a sequence, typically covering anything known to affect that gene's phenotype. Other features contained in this region will pick up a `/gene` qualifier from the Gene feature. Thus, there is no need to separately annotate the `/gene` qualifier on the other features.

FASTA Format

FASTA format contains a definition line and sequence characters and may be used as input to a variety of analysis programs (see Chapter 3). The definition line starts with a right angle bracket (>) and is usually followed by the sequence identifiers in a parsable form, as in this example:

```
>gi|2352912|gb|AF012433.1|HSDDT2
```

The remainder of the definition line, which is usually a title for the sequence, can be generated by software from features and other information in a Nuc-prot set.

For a segmented Bioseq, each raw Bioseq part can be presented separately, with a dash separating the segments. (The regular BLAST search service uses this method for producing search databases, so that the resulting "hits" will map to individual GenBank records.) The segmented Bioseq can also be treated as a single sequence, in which case the raw components will be catenated. (This form is used for generating the BLAST neighbors in Entrez; see Chapter 7.)

BLAST

The Basic Local Alignment Search Tool (BLAST; Altschul et al., 1990) is a popular method of ascertaining sequence similarity. The BLAST program takes a query sequence supplied by the user and searches it against the entire database of sequences maintained at NCBI. The output for each "hit" is a Seq-align, and these are combined into a Seq-annot. (Details on performing BLAST searches can be found in Chapter 8.)

The resulting Seq-annot can be used to generate the traditional BLAST printed report, but it is much more useful when viewed with software tools such as Entrez and Sequin. The viewer in these programs is now designed to display alignment information in useful forms. For example, the Graphical view shows only insertions and deletions relative to the query sequence, whereas the Alignment view fetches the individual sequences and displays mismatches between bases or residues in aligned regions. The Sequence view shows the alignment details at the level of individual bases or residues. This ability to zoom in from an overview to fine details makes it much easier to see the relationships between sequences than with a single report.

Finally, the Seq-annot, or any of its Seq-aligns, can be passed to other tools (such as banded or gapped alignment programs) for refinement. The results may then be sent back into a display program.

Entrez

The Entrez sequence retrieval program (Schuler et al., 1996; cf. Chapter 7) was designed to take advantage of connections that are captured by the NCBI data model. For example, the publication in a sequence record may contain a MEDLINE UID or PubMed ID. These are direct links to the PubMed article, which Entrez can retrieve. In addition, the product Seq-loc of a Coding Region feature points to the protein product Bioseq, which Entrez can also retrieve. The links in the data model allow retrieval of linked records at the touch of a button. The Genomes division in Entrez takes further advantage of the data model by providing "on the fly" display of certain regions of large genomes, as is the case when one hits the ProtTable button in Web Entrez.

Sequin

Sequin is a submission tool that takes raw sequence data and other biological information and assembles a record (usually a Bioseq-set) for submission to one of the DDBJ/EMBL/GenBank databases (Chapter 4). It makes full use of the NCBI data model and takes advantage of redundant information to validate entries. For example, because the user supplies both the nucleotide and protein sequences, Sequin can determine the coding region location (one or more intervals on the nucleotide that, through the genetic code, produce the protein product). It compares the translation of the coding region to the supplied protein and reports any discrepancy. It also makes sure that each Bioseq has BioSource information applied to it. This requirement can be satisfied for a nucleotide and its protein products by placing a single BioSource descriptor on the Nuc-prot set.

Sequin's viewers are all interactive, in that double-clicking on an existing item (shown as a GenBank flatfile paragraph or a line in the graphical display of features on a sequence) will launch an editor for that item (e.g., feature, descriptor, or sequence data).

LocusLink

LocusLink is an NCBI project to link information applicable to specific genetic loci from several disparate databases. Information maintained by LocusLink includes official nomenclature, aliases, sequence accessions (particularly RefSeq accessions), phenotypes, Enzyme Commission numbers, map information, and Mendelian Inheritance in Man numbers. Each locus is assigned a unique identification number, which additional databases can then reference. LocusLink is described in greater detail in Chapter 7.

CONCLUSIONS

The NCBI data model is a natural mapping of how biologists think of sequence relationships and how they annotate these sequences. The data that results can be saved, passed to other analysis programs, modified, and then displayed, all without having to go through multiple format conversions. The model definition concentrates on fundamental data elements that can be measured in a laboratory, such as the sequence of an isolated molecule. As new biological concepts are defined and understood, the specification for data can be easily expanded without the need to change existing data. Software tools are stable over time, and only incremental changes are needed for a program to take advantage of new data fields. Separating the specification into domains (e.g., citations, sequences, structures, maps) reduces the complexity of the data model. Providing neighbors and links between individual records increases the richness of the data and enhances the likelihood of making discoveries from the databases.

REFERENCES

Altschul, S. F., Gish, W., Miller, W., Meyers, E. W., and Lipman, D. J. (1990). Basic Local Alignment Search Tool. *J. Mol. Biol.* 215, 403–410.

Crick, F. H. C., Barnett, L., Brenner, S., and Watts-Tobin, R. J. (1961). General nature of the genetic code for proteins. *Nature* 192, 1227–1232.

Ostell, J. M. (1995). Integrated access to heterogeneous biomedical data from NCBI. *IEEE Eng. Med. Biol.* 14, 730–736.

Ostell, J. M. (1996). The NCBI software tools. In *Nucleic Acid and Protein Analysis: A Practical Approach*, M. Bishop and C. Rawlings, Eds. (IRL Press, Oxford), p. 31–43.

Schuler, G. D., Epstein, J. A., Ohkawa, H., and Kans, J. A. (1996). Entrez: Molecular biology database and retrieval system. *Methods Enzymol.* 266, 141–162.

Zhang, J., and Madden, T. L. (1997). Power BLAST: A new network BLAST application for interactive or automated sequence analysis and annotation. *Genome Res.* 7, 649–656.

3

THE GENBANK SEQUENCE DATABASE

Ilene Karsch-Mizrachi

National Center for Biotechnology Information
National Library of Medicine
National Institutes of Health
Bethesda, Maryland

B. F. Francis Ouellette

Centre for Molecular Medicine and Therapeutics
Children's and Women's Health Centre of British Columbia
University of British Columbia
Vancouver, British Columbia

INTRODUCTION

Primary protein and nucleic acid sequence databases are so pervasive to our way of thinking in molecular biology that few of us stop to wonder how these ubiquitous tools are built. Understanding how the these databases are put together will allow us to move forward in our understanding of biology and in fully harvesting the abstracted information present in these records.

GenBank, the National Institutes of Health (NIH) genetic sequence database, is an annotated collection of all publicly available nucleotide and protein sequences. The records within GenBank represent, in most cases, single, contiguous stretches of DNA or RNA with annotations. GenBank files are grouped into divisions; some of these divisions are phylogenetically based, whereas others are based on the technical approach that was used to generate the sequence information. Presently, all records in GenBank are generated from direct submissions to the DNA sequence

Bioinformatics: A Practical Guide to the Analysis of Genes and Proteins
Edited by A. D. Baxevanis and B. F. F. Ouellette
ISBN 0-471-38390-2 (cloth), ISBN 0-471-383910 (paper)

databases from the original authors, who volunteer their records to make the data publicly available or do so as part of the publication process. GenBank, which is built by the National Center for Biotechnology Information (NCBI), is part of the International Nucleotide Sequence Database Collaboration, along with its two partners, the DNA Data Bank of Japan (DDBJ, Mishima, Japan) and the European Molecular Biology Laboratory (EMBL) nucleotide database from the European Bioinformatics Institute (EBI, Hinxton, UK). All three centers provide separate points of data submission, yet all three centers exchange this information daily, making the same database (albeit in slightly different format and with different information systems) available to the community at-large.

This chapter describes how the GenBank database is structured, how it fits into the realm of the protein databases, and how its various components are interpreted by database users. Numerous works have dealt with the topic of sequence databases (Bairoch and Apweiller, 2000; Baker et al., 2000; Barker et al., 2000; Benson et al., 2000; Mewes et al., 2000; Tateno et al., 1997). These publications emphasize the great rate at which the databases have grown, and they suggest various ways of utilizing such vast biological resources. From a practical scientific point of view, as well as from a historical perspective, the sequence data have been separated into protein and nucleotide databases. The nucleotides are the primary entry points to the databases for both protein and nucleotide sequences, and there appears to be a migration toward having the nucleotide databases also involved in "managing" the protein data sets, as will be illustrated below. This is not a surprising development, since submitters are encouraged to provide annotation for the coding sequence (CDS) feature, the feature that tells how a translation product is produced. This trend toward the comanagement of protein and nucleotide sequences is apparent from the nucleotide sequences available through Entrez (cf. Chapter 7) as well as with GenBank and the formatting of records in the GenPept format. It is also apparent at EBI, where SWISS-PROT and TREMBL are being comanaged along with EMBL nucleotide databases. Nonetheless, the beginnings of each database set are distinct. Also implicit in the discussion of this chapter is the underlying data model described in Chapter 2.

Historically, the protein databases preceded the nucleotide databases. In the early 1960s, Dayhoff and colleagues collected all of the protein sequences known at that time; these sequences were catalogued as the Atlas of Protein Sequences and Structures (Dayhoff et al., 1965). This *printed* book laid the foundation for the resources that the entire bioinformatics community now depends on for day-to-day work in computational biology. A data set, which in 1965 could easily reside on a single floppy disk (although these did not exist then), represented years of work from a small group of people. Today, this amount of data can be generated in a fraction of a day. The advent of the DNA sequence databases in 1982, initiated by EMBL, led to the next phase, that of the explosion in database sequence information. Joined shortly thereafter by GenBank (then managed by the Los Alamos National Laboratory), both centers were contributing to the input activity, which consisted mainly of transcribing what was published in the printed journals to an electronic format more appropriate for use with computers. The DNA Data Bank of Japan (DDBJ) joined the data-collecting collaboration a few years later. In 1988, following a meeting of these three groups (now referred to as the International Nucleotide Sequence Database Collaboration), there was an agreement to use a common format for data elements within a unit record and to have each database update only the records that were directly submitted to it. Now, all three centers are collecting direct submissions and distributing them so that each center has copies of all of the sequences, meaning that they can act as a primary distribution center for these sequences. However, each record is owned by the database that created it and can only be updated by that database, preventing "update clashes" that are bound to occur when any database can update any record.

PRIMARY AND SECONDARY DATABASES

Although this chapter is about the GenBank nucleotide database, GenBank is just one member of a community of databases that includes three important protein databases: SWISS-PROT, the Protein Information Resource (PIR), and the Protein DataBank (PDB). PDB, the database of nucleic acid and protein structures, is described in Chapter 5. SWISS-PROT and PIR can be considered secondary databases, curated databases that add value to what is already present in the primary databases. Both SWISS-PROT and PIR take the majority of their protein sequences from nucleotide databases. A small proportion of SWISS-PROT sequence data is submitted directly or enters through a journal-scanning effort, in which the sequence is (quite literally) taken directly from the published literature. This process, for both SWISS-PROT and PIR, has been described in detail elsewhere (Bairoch and Apweiller, 2000; Barker et al., 2000.)

As alluded to above, there is an important distinction between primary (archival) and secondary (curated) databases. The most important contribution that the sequence databases make to the scientific community is making the sequences themselves accessible. The primary databases represent experimental results (with some interpretation) but are not a curated review. Curated reviews are found in the secondary databases. GenBank nucleotide sequence records are derived from the sequencing of a biological molecule that exists in a test tube, somewhere in a lab. They do not represent sequences that are a consensus of a population, nor do they represent some other computer-generated string of letters. This framework has consequences in the interpretation of sequence analysis. In most cases, all a researcher will need is a given sequence. Each such DNA and RNA sequence will be annotated to describe the analysis from experimental results that indicate why that sequence was determined in the first place. One common type of annotation on a DNA sequence record is the coding sequence (CDS). A great majority of the protein sequences have not been experimentally determined, which may have downstream implications when analyses are performed. For example, the assignment of a product name or function qualifier based on a subjective interpretation of a similarity analysis can be very useful, but it can sometimes be misleading. Therefore, the DNA, RNA, or protein sequences are the "computable" items to be analyzed and represent the most valuable component of the primary databases.

FORMAT VS. CONTENT: COMPUTERS VS. HUMANS

Database records are used to hold raw sequence data as well as an array of ancillary annotations. In a survey of the various database formats, we can observe that, although different sets of rules are applied, it is still possible in many cases to interchange data among formats. The best format for a human to read may not be the most efficient for a computer program. The simplicity of the flat file, which does lend itself to simple tools that are available to all, is in great part responsible for the popularity of the EMBL and GenBank flatfile formats. In its simplest form, a DNA sequence record can be represented as a string of nucleotides with some tag or identifier. Here is a nucleotide file as represented in a FASTA (or Pearson format) file:

```
>L04459
GCAGCGCACGACAGCTGTGCTATCCCGGCGAGCCCGTGGCAGAGGACCTCGCTTGCGAAAGCATCGAGTACC
GCTACAGAGCCAACCCGGTGGACAAACTCGAAGTCATTGTGGACCGAATGAGGCTCAATAACGAGATTAGCG
ACCTCGAAGGCCTGCGCAAATATTTCCACTCCTTCCCGGGTGCTCCTGAGTTGAACCCGCTTAGAGACTCCG
AAATCAACGACGACTTCCACCAGTGGGCCCAGTGTGACCGCCACACTGGACCCCATACCACTTCTTTTTGTT
ATTCTTAAATATGTTGTAACGCTATGTAATTCCACCCTTCATTACTAATAATTAGCCATTCACGTGATCTCA
GCCAGTTGTGGCGCCACACTTTTTTTTCCATAAAAATCCTCGAGGAAAAGAAAAGAAAAAAATATTTCAGTT
ATTTAAAGCATAAGATGCCAGGTAGATGGAACTTGTGCCGTGCCAGATTGAATTTTGAAAGTACAATTGAGG
CCTATACACATAGACATTTGCACCTTATACATATAC
```

Similarly, for a protein record, the FASTA record would appear as follows:

```
>P31373
MTLQESDKFATKAIHAGEHVDVHGSVIEPISLSTTFKQSSPANPIGTYEYSRSQNPNRENLERAVAALENAQ
YGLAFSSGSATTATILQSLPQGSHAVSIGDVYGGTHRYFTKVANAHGVETSFTNDLLNDLPQLIKENTKLVW
IETPTNPTLKVTDIQKVADLIKKHAAGQDVILVVDNTFLSPYISNPLNFGADIVVHSATKYINGHSDVVLGV
LATNNKPLYERLQFLQNAIGAIPSPFDAWLTHRGLKTLHLRVRQAALSANKIAEFLAADKENVVAVNYPGLK
THPNYDVVLKQHRDALGGGMISFRIKGGAEAASKFASSTRLFTLAESLGGIESLLEVPAVMTHGGIPKEARA
SGVFDDLVRISVGIEDTDDLLEDIKQALKQATN
```

The FASTA format is used in a variety of molecular biology software suites. In its simplest incarnation (as shown above) the "greater than" character (>) designates the beginning of a new file. An identifier (L04459 in the first of the preceding examples) is followed by the DNA sequence in lowercase or uppercase letters, usually with 60 characters per line. Users and databases can then, if they wish, add a certain degree of complexity to this format. For example, without breaking any of the rules just outlined, one could add more information to the FASTA definition line, making the simple format a little more informative, as follows:

```
>gi|171361|gb|L04459|YSCCYS3A Saccharomyces cerevisiae cystathionine gamma-lyase
(CYS3) gene, complete cds.
GCAGCGCACGACAGCTGTGCTATCCCGGCGAGCCCGTGGCAGAGGACCTCGCTTGCGAAAGCATCGAGTACC
GCTACAGAGCCAACCCGGTGGACAAACTCGAAGTCATTGTGGACCGAATGAGGCTCAATAACGAGATTAGCG
ACCTCGAAGGCCTGCGCAAATATTTCCACTCCTTCCCGGGTGCTCCTGAGTTGAACCCGCTTAGAGACTCCG
AAATCAACGACGACTTCCACCAGTGGGCCCAGTGTGACCGCCACACTGGACCCCATACCACTTCTTTTTGTT
ATTCTTAAATATGTTGTAACGCTATGTAATTCCACCCTTCATTACTAATAATTAGCCATTCACGTGATCTCA
GCCAGTTGTGGCGCCACACTTTTTTTTCCATAAAAATCCTCGAGGAAAAGAAAAGAAAAAAATATTTCAGTT
ATTTAAAGCATAAGATGCCAGGTAGATGGAACTTGTGCCGTGCCAGATTGAATTTTGAAAGTACAATTGAGG
CCTATACACATAGACATTTGCACCTTATACATATAC
```

This modified FASTA file now has the gi number (see below and Chapter 2), the GenBank accession number, the LOCUS name, and the DEFINITION line from the GenBank record. The record was passed from the underlying ASN.1 record (see Appendix 3.2), which NCBI uses to actually store and maintain all its data.

Over the years, many file formats have come and gone. Tools exist to convert the sequence itself into the minimalist view of one format or another. NCBI's asn2ff (ASN.1 to flatfile) will convert an ASN.1 file into a variety of flatfiles. The asn2ff program will generate GenBank, EMBL, GenPept, SWISS-PROT, and FASTA formats and is available from the NCBI Toolkit. READSEQ is another tool that has been widely used and incorporated into many work environments. Users should be aware that the features from a GenBank or EMBL format may be lost when passed through such utilities. Programs that need only the sequence (e.g., BLAST; see Chapter 8) are best used with a FASTA format for the query sequence. Although less informative than other formats, the FASTA format offers a simple way of dealing with the primary data in a human- and computer-readable fashion.

THE DATABASE

A full release of GenBank occurs on a bimonthly schedule with incremental (and nonincremental) daily updates available by anonymous FTP. The International Nucleotide Sequence Database Collaboration also exchanges new and updated records daily. Therefore, all sequences present in GenBank are also present in DDBJ and EMBL, as described in the introduction to this chapter. The three databases rely on a common data format for information described in the feature table documentation (see below). This represents the *lingua franca* for nucleotide sequence database annotations. Together, the nucleotide sequence databases have developed defined submission procedures (see Chapter 4), a series of guidelines for the content and format of all records.

As mentioned above, nucleotide records are often the primary source of sequence and biological information from which protein sequences in the protein databases are derived. There are three important consequences of not having the correct or proper information on the nucleotide record:

- If a coding sequence is not indicated on a nucleic acid record, it will not be represented in the protein databases. Thus, because querying the protein databases is the most sensitive way of doing similarity discoveries (Chapter 8), failure to indicate the CDS intervals on an mRNA or genomic sequence of interest (when one should be present) may cause important discoveries to be missed.
- The set of features usable in the nucleotide feature table that are specific to protein sequences themselves is limited. Important information about the protein will not be entered in the records in a "parsable place." (The information may be present in a note, but it cannot reliably be found in the same place under all circumstances.)
- If a coding feature on a nucleotide record contains incorrect information about the protein, this could be propagated to other records in both the nucleotide and protein databases on the basis of sequence similarity.

THE GENBANK FLATFILE: A DISSECTION

The GenBank flatfile (GBFF) is the elementary unit of information in the GenBank database. It is one of the most commonly used formats in the representation of biological sequences. At the time of this writing, it is the format of exchange from GenBank to the DDBJ and EMBL databases and vice versa. The DDBJ flatfile format and the GBFF format are now nearly identical to the GenBank format (Appendix 3.1). Subtle differences exist in the formatting of the definition line and the use of the gene feature. EMBL uses line-type prefixes, which indicate the type of information present in each line of the record (Appendix 3.2). The feature section (see below), prefixed with `FT`, is identical in content to the other databases. All these formats are really reports from what is represented in a much more structured way in the underlying ASN.1 file.

The GBFF can be separated into three parts: the *header*, which contains the information (descriptors) that apply to the whole record; the *features*, which are the annotations on the record; and the nucleotide sequence itself. All major nucleotide database flat files end with // on the last line of the record.

The Header

The header is the most database-specific part of the record. The various databases are not obliged to carry the same information in this segment, and minor variations exist, but some effort is made to ensure that the same information is carried from one to the other. The first line of all GBFFs is the Locus line:

```
LOCUS       AF111785     5925 bp    mRNA            PRI        01-SEP-1999
```

The first element on this line is the locus name. This element was historically used to represent the locus that was the subject of the record, and submitters and database staff spent considerable time in devising it so that it would serve as a mnemonic. Characters after the first can be numerical or alphabetic, and all letters are uppercase. The locusname was most useful back when most DNA sequence records represented only one genetic locus, and it was simple to find in GenBank a unique name that could represent the biology of the organism in a few letters and numbers. Classic examples include HUMHBB for the human β-globin locus or SV40 for the Simian virus (one of the copies anyway; there are many now). To be usable, the locus name needs to be unique within the database; because virtually all the meaningful designators have been taken, the LOCUS name has passed its time as a useful format element. Nowadays, this element must begin with a letter, and its length cannot exceed 10 characters. Because so many software packages rely on the presence of a unique LOCUS name, the databases have been reluctant to remove it altogether. The preferred path has been to instead put a unique word, and the simplest way to do this has been to use an accession number of ensured uniqueness: AF111785 in the example above conforms to the locus name requirement.

The second item on the locus line is the length of the sequence. Sequences can range from 1 to 350,000 base pairs (bp) in a single record. In practice, GenBank and the other databases seldom accept sequences shorter than 50 bp; therefore, the inclusion of polymerase chain reaction (PCR) primers as sequences (i.e., submissions of 24 bp) is discouraged. The 350 kb limit is a practical one, and the various databases represent longer contigs in a variety of different and inventive ways (see Chapters 2 and 6 and Appendix 3.3). Records of greater than 350 kb are acceptable in the database if the sequence represents a single gene.

The third item on the locus line indicates the molecule type. The "mol type" usually is DNA or RNA, and it can also indicate the strandedness (single or double, as ss or ds, respectively); however, these attributes are rarely used these days (another historical leftover). The acceptable mol types are DNA, RNA, tRNA, rRNA, mRNA, and uRNA and are intended to represent the original biological molecule. For example, a cDNA that is sequenced really represents an mRNA, and mRNA is the indicated mol type for such a sequence. If the tRNA or rRNA has been sequenced directly or via some cDNA intermediate, then tRNA or rRNA is shown as the mol type. If the ribosomal RNA gene sequence was obtained via the PCR from genomic DNA, then DNA is the mol type, even if the sequence encodes a structural RNA.

The fourth item on the locus line is the GenBank division code: three letters, which have either taxonomic inferences or other classification purposes. Again, these codes exist for historical reasons, recalling the time when the various GenBank divisions were used to break up the database files into what was then a more manageable size. The GenBank divisions are slightly different from those of EMBL or DDBJ, as described elsewhere (Ouellette and Boguski, 1997). NCBI has not introduced additional organism-based divisions in quite a few years, but new, function-based divisions have been very useful because they represent functional and definable sequence types (Ouellette and Boguski, 1997). The Expressed Sequence Tags (EST) division was introduced in 1993 (Boguski et al., 1993) and was soon followed by a division for Sequence Tagged Sites (STS). These, along with the Genome Survey Sequences (GSS) and unfinished, High Throughput Genome sequences (HTG), represent functional categories that need to be dealt with by the users and the database staff in very different ways. For example, a user can query these data sets specifically (e.g., via a BLASTN search against the EST or HTG division). Knowing that the hit is derived from a specific technique-oriented database allows one to interpret the data accordingly. At this time, GenBank, EMBL, and DDBJ interpret the various functional divisions in the same way, and all data sets are represented in the same division from one database to the next. The CON division is a new division for constructed (or "contigged") records. This division contains segmented sets as well as all large assemblies, which may exceed (sometimes quite substantially) the 350,000-bp limit presently imposed on single records. Such records may take the form shown in Appendix 3.3. The record from the CON division shown in Appendix 3.3 gives the complete genomic sequence of *Mycoplasma pneumoniae*, which is more than 800,000 base pairs in length. This CON record does not include sequences or annotations; rather, it includes instructions on how to assemble pieces present in other divisions into larger or assembled pieces. Records within the CON division have accession and version numbers and are exchanged, like all other records within the collaboration.

The date on the locus line is the date the record was last made public. If the record has not been updated since being made public, the date would be the date that it was first made public. If any of the features or annotations were updated and the record was rereleased, then the date corresponds to the last date the entry was released. Another date contained in the record is the date the record was submitted (see below) to the database. It should be noted that none of these dates is legally binding on the promulgating organization. The databases make no claim that the dates are error-free; they are included as guides to users and should not be submitted in any arbitration dispute. To the authors' knowledge, they have never been used in establishing priority and publication dates for patent application.

```
DEFINITION Homo sapiens myosin heavy chain IIx/d mRNA, complete cds.
```

The definition line (also referred to as the "def line") is the line in the GenBank record that attempts to summarize the biology of the record. This is the line that appears in the FASTA files that NCBI generates and is what is seen in the summary line for BLAST hits generated from a BLAST similarity search (Chapter 8). Much care is taken in the generation of these lines, and, although many of them can be generated automatically from the features in the record, they are still reviewed by the database staff to make sure that consistency and richness of information are maintained. Nonetheless, it is not always possible to capture all the biology in a single line of text, and databases cope with this in a variety of ways. There are some agreements in force between the databases, and the databases are aware of each other's guidelines and try to conform to them.

The generalized syntax for an mRNA definition line is as follows:

```
Genus species product name (gene symbol) mRNA, complete cds.
```

The generalized syntax for a genomic record is

```
Genus species product name (gene symbol) gene, complete cds.
```

Of course, records of many other types of data are accounted for by the guidelines used by the various databases. The following set of rules, however, applies to organelle sequences, and these rules are used to ensure that the biology and source of the DNA are clear to the user and to the database staff (assuming they are clear to the submitter):

```
DEFINITION Genus species protein X(xxx) gene, complete cds;
    [one choice from below], OR
DEFINITION Genus species XXS ribosomal RNA gene, complete sequence;
    [one choice from below].
  nuclear gene(s) for mitochondrial product(s)
  nuclear gene(s) for chloroplast product(s)
  mitochondrial gene(s) for mitochondrial product(s)
  chloroplast gene(s) for chloroplast product(s)
```

In accordance with a recent agreement among the collaborative databases, the full genus-species names are given in the definition lines; common names (e.g., human) or abbreviated genus names (e.g., *H. sapiens* for Homo sapiens) are no longer used. The many records in the database that precede this agreement will eventually be updated. One organism has escaped this agreement: the human immunodeficiency virus is to be represented in the definition line as HIV1 and HIV2.

```
ACCESSION    AF111785
```

The accession number, on the third line of the record, represents the primary key to reference a given record in the database. This is the number that is cited in publications and is always associated with this record; that is, if the sequence is updated (e.g., by changing a single nucleotide), the accession number will not change. At this time, accession numbers exist in one of two formats: the "1 + 5" and "2 + 6" varieties, where 1 + 5 indicates one uppercase letter followed by five digits and 2 + 6 is two letters plus six digits. Most of the new records now entering the databases are of the latter variety. All GenBank records have only a single line with the word ACCESSION on it; however, there may be more than one accession number. The vast majority of records only have one accession number. This number is always referred to as the primary accession number; all others are secondary. In cases where more than one accession number is shown, the first accession number is the primary one.

Unfortunately, secondary accession numbers have meant a variety of things over the years, and no single definition applies. The secondary accession number may be related to the primary one, or the primary accession number may be a replacement for the secondary, which no longer exists. There is an ongoing effort within the Collaboration to make the latter the default for all cases, but, because secondary accession numbers have been used for more than 15 years (a period during which the management of GenBank changed), all data needed to elucidate all cases are not available.

```
ACCESSION    AF111785
VERSION      AF111785.1 GI:4808814
```

The version line contains the Accession.version and the gi (geninfo identifier). These identifiers are associated with a unique nucleotide sequence. Protein sequences also have accession numbers (protein_ids). These are also represented as Accession.version and gi numbers for unique sequences (see below). If the sequence changes, the version number in the Accession.version will be incremented by one and the gi will change (although not by one, but to the next available integer). The accession number stays the same. The example above shows version 1 of the sequence having accession number AF111785 and gi number 4808814.

```
KEYWORDS
```

The keywords line is another historical relic that is, in many cases, unfortunately misused. Adding keywords to an entry is often not very useful because over the years so many authors have selected words not on a list of controlled vocabulary and not uniformly applied to the whole database. NCBI, therefore, discourages the use of keywords but will include them on request, especially if the words are not present elsewhere in the record or are used in a controlled fashion (e.g., for EST, STS, GSS, and HTG records). At this time, the resistance to adding keywords is a matter of policy at NCBI/GenBank only.

```
SOURCE       human.
ORGANISM     Homo sapiens
 Eukaryota; Metazoa; Chordata; Craniata; Vertebrata;
 Mammalia; Eutheria; Primates; Catarrhini; Hominidae; Homo.
```

The source line will either have the common name for the organism or its scientific name. Older records may contain other source information (see below) in this field. A concerted effort is now under way to assure that all other information present in the source feature (as opposed to the source line) and all lines in the taxonomy block (source and organism lines) can be derived from what is in the source feature and the taxonomy server at NCBI. Those interested in the lineage and other aspects of the taxonomy are encouraged to visit the taxonomy home page at NCBI. This taxonomy database is used by all nucleotide sequence databases, as well as SWISS-PROT.

```
REFERENCE 1 (bases 1 to 5925)
AUTHORS   Weiss,A., McDonough,D., Wertman,B., Acakpo-Satchivi,L.,
          Montgomery,K., Kucherlapati,R., Leinwand,L. and Krauter,K.
TITLE     Organization of human and mouse skeletal myosin heavy chain
          gene clusters is highly conserved
JOURNAL   Proc. Natl. Acad. Sci. U.S.A. 96 (6), 2958-2963 (1999)
MEDLINE   99178997
PUBMED    10077619
```

Each GenBank record must have at least one reference or citation. It offers scientific credit and sets a context explaining why this particular sequence was determined. In many cases, the record will have two or more reference blocks, as shown in Appendix 3.1. The preceding sample indicates a published paper. There is a MEDLINE and PubMed identifier present that provides a link to the MEDLINE/PubMed databases (see Chapter 7). Other references may be annotated as unpublished (which could be "submitted) or as placeholders for a publication, as shown.

```
REFERENCE 1 (bases 1 to 3291)
AUTHORS   Morcillo, P., Rosen, C.,Baylies, M.K. and Dorsett, D.
TITLE     CHIP, a widely expressed chromosomal protein required for
          remote enhancer activity and segmentation in Drosophila
JOURNAL   Unpublished
REFERENCE 3 (bases 1 to 5925)
AUTHORS   Weiss,A. and Leinwand,L.A.
TITLE     Direct Submission
JOURNAL   Submitted (09-DEC-1998) MCDB, University of Colorado at
          Boulder, Campus Box 0347, Boulder, Colorado 80309-0347, USA
```

The last citation is present on most GenBank records and gives scientific credit to the people responsible for the work surrounding the submitted sequence. It usually includes the postal address of the first author or the lab where the work was done. The date represents the date the record was submitted to the database but not the date on which the data were first made public, which is the date on the locus line if the record was not updated. Additional submitter blocks may be added to the record each time the sequences are updated.

The last part of the header section in the GBFF is the comment. This section includes a great variety of notes and comments (also called "descriptors") that refer to the whole record. Genome centers like to include their contact information in this section as well as give acknowledgments. This section is optional and not found in most records in GenBank. The comment section may also include E-mail addresses or URLs, but this practice is discouraged at NCBI (although certain exceptions have been made for genome centers as mentioned above). The simple reason is that E-mail addresses tend to change more than the postal addresses of buildings. DDBJ has been including E-mail addresses for some years, again representing a subtle difference in policy. The comment section also contains information about the history of the sequence. If the sequence of a particular record is updated, the comment will contain a pointer to the previous version of the record.

```
COMMENT On Dec 23, 1999 this sequence version replaced gi:4454562.
```

Alternatively, if you retrieve an earlier version of the record, this comment will point forward to the newer version of the sequence and also backward if there was an earlier still version

```
COMMENT [WARNING] On Dec 23, 1999 this sequence was replaced by
        a newer version gi:6633795.
```

The Feature Table

The middle segment of the GBFF record, the feature table, is the most important direct representation of the biological information in the record. One could argue that the biology is best represented in the bibliographic reference, cited by the record. Nonetheless, a full set of annotations within the record facilitates quick extraction of the relevant biological features and allows the submitter to indicate why this record was submitted to the database. What becomes relevant here is the choice of annotations presented in this section. The GenBank feature table documentation describes in great detail the legal features (i.e., the ones that are allowed) and what qualifiers are permitted with them. This, unfortunately, has often invited an excess of invalid, speculative, or computed annotations. If an annotation is simply computed, its usefulness as a comment within the record is diminished.

Described below are some of the key GenBank features, with information on why they are important and what information can be extracted from them. The discussion here is limited to the biological underlyings of these features and guidelines applied to this segment by the NCBI staff. This material will also give the reader some insight into the NCBI data model (Chapter 2) and the important place the GBFF occupies in the analysis of sequences, serving also to introduce the concept of features and qualifiers in GenBank language. The features are slightly different from other features discussed in Chapter 2. In the GBFF report format, any component of this section designated as "feature." In the NCBI data model, "features" refer to annotations that are on a part of the sequences, whereas annotations that describe the whole sequence are called "descriptors." Thus, the source feature in the GenBank flatfile is really a descriptor in the data model view (the BioSource, which refers to the whole sequence), not a feature as used elsewhere. Because this is a chapter on the GenBank database, the "feature" will refer to all components of the feature table. The readers should be aware of this subtle difference, especially when referring to other parts of this book.

The Source Feature. The source feature is the only feature that must be present on all GenBank records. All features have a series of legal qualifiers, some of which are mandatory (e.g., `/organism` for source). All DNA sequence records have some origin, even if synthetic in the extreme case. In most cases, there will be a single source feature, and it will contain the `/organism`. Here is what we have in the example from Appendix 3.1:

```
source 1..5925
       /organism="Homo sapiens"
       /db_xref="taxon:9606"
       /chromosome="17"
       /map="17p13.1"
       /tissue_type="skeletal muscle"
```

The organism qualifier contains the scientific genus and species name. In some cases, "organisms" can be described at the subspecies level. For the source feature, the series of qualifiers will contain all matters relating to the BioSource, and these may include mapping, chromosome or tissue from which the molecule that was sequenced was obtained, clone identification, and other library information. For the source feature, as is true for all features in a GenBank record, care should be taken to avoid adding superfluous information to the record. For the reader of these records, anything that cannot be computationally validated should be taken with a grain of salt. Tissue source and library origin are only as good as the controls present in the associated publication (if any such publication exists) and only insofar as that type of information is applied uniformly across all records in GenBank. With sets of records in which the qualifiers are applied in a systematic way, as they are for many large EST sets, the taxonomy can be validated (i.e., the organism does exist in the database of all organisms that is maintained at the NCBI). If, in addition, the qualifier is applied uniformly across all records, it is of value to the researcher. Unfortunately, however, many qualifiers are derived without sufficient uniformity across the database and hence are of less value.

Implicit in the BioSource and the organism that is assigned to it is the genetic code used by the DNA/RNA, which will be used to translate the nucleic acid to represent the protein sequence (if one is present in the record). This information is shown on the CDS feature.

The CDS Feature. The CDS feature contains instructions to the reader on how to join two sequences together or on how to make an amino acid sequence from the indicated coordinates and the inferred genetic code. The GBFF view, being as DNA-centric as it is, maps all features through a DNA sequence coordinate system, not that of amino acid reference points, as in the following example from GenBank accession X59698 (contributed by a submission to EMBL).

```
sig peptide 160..231
CDS         160..>2301
            /codon_start=1
            /product="EGF-receptor"
            /protein_id="CAA42219.1"
            /db_xref="GI:50804"
            /db_xref="MGD:MGI:95294"
            /db_xref="SWISS-PROT:Q01279"
            /translation="MRPSGTARTTLLVLLTALCAAGGALEEKKVCQGTSNRLTQLGTF
            EDHFLSLQRMYNNCEVVLGNLEITYVQRNYDLSFLKTIQEVAGYVLIALNTVERIPLE
            NLQIIRGNALYENTYALAILSNYGTNRTGLRELPMRNLQEILIGAVRFSNNPILCNMD
            TIQWRDIVQNVFMSNMSMDLQSHPSSCPKCDPSCPNGSCWGGGEENCQKLTKIICAQQ
            CSHRCRGRSPSDCCHNQCAAGCTGPRESDCLVCQKFQDEATCKDTCPPLMLYNPTTYQ
            MDVNPEGKYSFGATCVKKCPRNYVVTDHGSCVRACGPDYYEVEEDGIRKCKKCDGPCR
```

```
               KVCNGIGIGEFKDTLSINATNIKHFKYCTAISGDLHILPVAFKGDSFTRTPPLDPREL
               EILKTVKEITGFLLIQAWPDNWTDLHAFENLEIIRGRTKQHGQFSLAVVGLNITSLGL
               RSLKEISDGDVIISGNRNLCYANTINWKKLFGTPNQKTKIMNNRAEKDCKAVNHVCNP
               LCSSEGCWGPEPRDCVSCQNVSRGRECVEKWNILEGEPREFVENSECIQCHPECLPQA
               MNITCTGRGPDNCIQCAHYIDGPHCVKTCPAGIMGENNTLVWKYADANNVCHLCHANC
               TYGCAGPGLQGCEVWPSGPKIPSIATGIVGGLLFIVVVALGIGLFMRRRHIVRKRTLR
               RLLQERELVEPLTPSGEAPNQAHLRILKETEF"
mat peptide    232..>2301
               /product="EGF-receptor"
```

This example also illustrates the use of the database cross-reference (db_xref). This controlled qualifier allows the databases to cross-reference the sequence in question to an external database (the first identifier) with an identifier used in that database. The list of allowed db_xref databases is maintained by the International Nucleotide Sequence Database Collaboration.

```
/protein_id="CAA42219.1"
/db_xref="GI:50804"
```

As mentioned above, NCBI assigns an accession number and a gi (geninfo) identifier to all sequences. This means that translation products, which are sequences in their own right (not simply attachments to a DNA record, as they are shown in a GenBank record), also get an accession number (/protein_id) and a gi number. These unique identifiers will change when the sequence changes. Each protein sequence is assigned a protein_id or protein accession number. The format of this accession number is "3 + 5," or three letters and five digits. Like the nucleotide sequence accession number, the protein accession number is represented as Accession.version. The protein gi numbers appear as a gi db_xref. When the protein sequence in the record changes, the version of the accession number is incremented by one and the gi is also changed.

Thus, the version number of the accession number presents the user with an easy way to look up the previous version of the record, if one is present. Because amino acid sequences represent one of the most important by-products of the nucleotide sequence database, much attention is devoted to making sure they are valid. (If a translation is present in a GenBank record, there are valid coordinates present that can direct the translation of nucleotide sequence.) These sequences are the starting material for the protein databases and offer the most sensitive way of making new gene discoveries (Chapter 8). Because these annotations can be validated, they have added value, and having the correct identifiers also becomes important. The correct product name, or protein name, can be subjective and often is assigned via weak similarities to other poorly annotated sequences, which themselves have poor annotations. Thus, users should be aware of potential circular amplification of paucity of information. A good rule is that more information is usually obtained from records describing single genes or full-length mRNA sequences with which a published paper is associated. These records usually describe the work from a group that has studied a gene of interest in some detail. Fortunately, quite a few records of these types are in the database, representing a foundation of knowledge used by many.

The Gene Feature. The gene feature, which has been explicitly included in the GenBank flatfile for only a few years, has nevertheless been implicitly in use since the beginning of the databases as a gene qualifier on a number of other features. By making this a separate feature, the de facto status has been made explicit, greatly facilitating the generation and validation of components now annotated with this feature. The new feature has also clearly shown in its short existence that biologists have very different definitions and uses for the gene feature in GenBank records. Although it is obvious that not all biologists will agree on a single definition of the gene feature, at its simplest interpretation, the gene feature represents a segment of DNA that can be identified with a name (e.g., the MyHC gene example from Appendix 3.1) or some arbitrary number, as is often used in genome sequencing project (e.g., T23J18.1 from GenBank accession number AC011661). The gene feature allows the user to see the gene area of interest and in some cases to select it.

The RNA Features. The various structural RNA features can be used to annotate RNA on genomic sequences (e.g., mRNA, rRNA, tRNA). Although these are presently not instantiated into separate records as protein sequences are, these sequences (especially the mRNA) are essential to our understanding of how higher genomes are organized. RNAs deserves special mention because they represent biological entities that can be measured in the lab and thus are pieces of information of great value for a genomic record and are often mRNA records on their own. This is in contrast to the promoter feature, which is poorly characterized, unevenly assigned in a great number of records, poorly defined from a biology point of view, and of lesser use in a GenBank record. The RNA feature on a genomic record should represent the experimental evidence of the presence of that biological molecule.

CONCLUDING REMARKS

The DDBJ/EMBL/GenBank database is the most commonly used nucleotide and protein sequence database. It represents a public repository of molecular biology information. Knowing what the various fields mean and how much biology can be obtained from these records greatly advances our understanding of this file format. Although the database was never meant to be read from computers, an army of computer-happy biologists have nevertheless parsed, converted, and extracted these records by means of entire suites of programs. THE DDBJ/EMBL/GenBank flatfile remains the format of exchange between the International Nucleotide Sequence Database Collaboration members, and this is unlikely to change for years to come, *despite* the availability of better, richer alternatives, such as the data described in ASN.1. However, therein lays the usefulness of the present arrangement: it is a readily available, simple format which can represent some abstraction of the biology it wishes to depict.

INTERNET RESOURCES FOR TOPICS PRESENTED IN CHAPTER 3

GenBank Release Notes	*ftp://ncbi.nlm.nih.gov/genbank/gbrel.txt*
READSEQ Sequence Conversion Tool	*http://magpie.bio.indiana.edu/MolecularBiology/Molbio_archive/readseq/*
Taxonomy Browser	*http://www.ncbi.nlm.nih.gov/Taxonomy/tax.html*

TREMBL and Swiss-Prot Release Notes — *http://www.ebi.ac.uk/ebi_docs/swissprot_db/documentation.html*

REFERENCES

Bairoch, A., and Apweiler, R. (2000). The SWISS-PROT protein sequence data bank and its supplement TrEMBL. *Nucl. Acids Res.* 28, 45–48.

Baker, W., van den Broek, A., Camon, E., Hingamp, P., Sterk, P., Stoesser, G., and Tuli, M. A. (2000). The EMBL Nucleotide Sequence Database. *Nucl. Acids Res.* 28, 19–23.

Barker, W. C., Garavelli, J. S., Huang, H., McGarvey, P. B., Orcutt, B. C., Srinivasarao, G. Y., Xiao, C., Yeh, L. S., Ledley, R. S., Janda, J. F., Pfeiffer, F., Mewes, H.-W., Tsugita, A., and Wu, C. (2000). The Protein Information Resource (PIR). *Nucl. Acids Res.* 28, 41–44.

Benson, D. A., Karsch-Mizrachi, I., Lipman, D. J., Ostell, J., Rapp, B. A. and Wheeler, D. L. (1997). GenBank. *Nucl. Acids Res.* 25, 1–6.

Boguski, M. S., Lowe, T. M., Tolstoshev, C. M. (1993). dbEST—database for "expressed sequence tags." *Nat. Genetics* 4: 332–333.

Cook-Deagan, R. (1993). *The Gene Wars. Science, Politics and the HumanGenome* (New York and London: W. W. Norton & Company).

Dayhoff, M. O., Eck, R. V., Chang, M. A., Sochard, M. R. (1965). *Atlas of Protein Sequence and Structure.* (National Biomedical Research Foundation, Silver Spring MD).

Mewes, H. W., Frischman, D., Gruber, C., Geier, B., Haase, D., Kaps, A., Lemcke, K., Mannhaupt, G., Pfeiffer, F., Schuller, C., Stocker, S., and Weil, B. (2000). MIPS: A database for genomes and protein sequences. *Nucl. Acids Res.* 28, 37–40.

Ouellette, B. F. F., and Boguski, M. S. (1997). Database divisions and homology search files: a guide for the perplexed. *Genome Res.* 7, 952–955.

Schuler, G. D., Epstein, J. A., Ohkawa, H., Kans, J. A. (1996). Entrez: Molecular biology database and retrieval system. *Methods Enzymol.* 266, 141–162.

Tateno, Y., Miyazaki, S., Ota, M., Sugawara, H., and Gojobori, T. (1997). DNA Data Bank of Japan (DDBJ) in collaboration with mass sequencing teams. *Nucl. Acids Res.* 28, 24–26.

APPENDICES

Appendix 3.1. Example of GenBank Flatfile Format

```
LOCUS       AF111785      5925 bp    mRNA          PRI         01-SEP-1999
DEFINITION  Homo sapiens myosin heavy chain IIx/d mRNA, complete cds.
ACCESSION   AF111785
VERSION     AF111785.1 GI:4808814
KEYWORDS    .
SOURCE      human.
ORGANISM    Homo sapiens
            Eukaryota; Metazoa; Chordata; Craniata; Vertebrata; Euteleos-
            tomi; Mammalia; Eutheria; Primates; Catarrhini; Hominidae; Homo.
REFERENCE   1 (bases 1 to 5925)
AUTHORS     Weiss,A., McDonough,D., Wertman,B., Acakpo-Satchivi,L., Mont-
            gomery,K., Kucherlapati,R., Leinwand,L. and Krauter,K.
TITLE       Organization of human and mouse skeletal myosin heavy chain gene
            clusters is highly conserved
JOURNAL     Proc. Natl. Acad. Sci. U.S.A. 96 (6), 2958-2963 (1999)
MEDLINE     99178997
PUBMED      10077619
```

```
REFERENCE   2 (bases 1 to 5925)
AUTHORS     Weiss,A., Schiaffino,S. and Leinwand,L.A.
TITLE       Comparative sequence analysis of the complete human sarcomeric
            myosin heavy chain family: implications for functional diversity
JOURNAL     J. Mol. Biol. 290 (1), 61-75 (1999)
MEDLINE     99318869
PUBMED      10388558
REFERENCE   3 (bases 1 to 5925)
AUTHORS     Weiss,A. and Leinwand,L.A.
TITLE       Direct Submission
JOURNAL     Submitted (09-DEC-1998) MCDB, University of Colorado at Boulder,
            Campus Box 0347, Boulder, Colorado 80309-0347, USA
FEATURES    Location/Qualifiers
source      1..5925
            /organism="Homo sapiens"
            /db_xref="taxon:9606"
            /chromosome="17"
            /map="17p13.1"
            /tissue_type="skeletal muscle"
CDS         1..5820
            /note="MyHC"
            /codon_start=1
            /product="myosin heavy chain IIx/d"
            /protein_id="AAD29951.1"
            /db_xref="GI:4808815"
            /translation="MSSDSEMAIFGEAAPFLRKSERERIEAQNKPFDAKTSVFVVDPK
            ESFVKATVQSREGGKVTAKTEAGATVTVKDDQVFPMNPPKYDKIEDMAMMTHLHEPAV
            LYNLKERYAAWMIYTYSGLFCVTVNPYKWLPVYNAEVVTAYRGKKRQEAPPHIFSISD
            NAYQFMLTDRENQSILITGESGAGKTVNTKRVIQYFATIAVTGEKKKEEVTSGKMQGT
            LEDQIISANPLLEAFGNAKTVRNDNSSRFGKFIRIHFGTTGKLASADIETYLLEKSRV
            TFQLKAERSYHIFYQIMSNKKPDLIEMLLITTNPYDYAFVSQGEITVPSIDDQEELMA
            TDSAIEILGFTSDERVSIYKLTGAVMHYGNMKFKQKQREEQAEPDGTEVADKAAYLQN
            LNSADLLKALCYPRVKVGNEYVTKGQTVQQVYNAVGALAKAVYDKMFLWMVTRINQQL
            DTKQPRQYFIGVLDIAGFEIFDFNSLEQLCINFTNEKLQQFFNHHMFVLEQEEYKKEG
            IEWTFIDFGMDLAACIELIEKPMGIFSILEEECMFPKATDTSFKNKLYEQHLGKSNNF
            QKPKPAKGKPEAHFSLIHYAGTVDYNIAGWLDKNKDPLNETVVGLYQKSAMKTLALLF
            VGATGAEAEAGGGKKGGKKKGSSFQTVSALFRENLNKLMTNLRSTHPHFVRCIIPNET
            KTPGAMEHELVLHQLRCNGVLEGIRICRKGFPSRILYADFKQRYKVLNASAIPEGQFI
            DSKKASEKLLGSIDIDHTQYKFGHTKVFFKAGLLGLLEEMRDEKLAQLITRTQAMCRG
            FLARVEYQKMVERRESIFCIQYNVRAFMNVKHWPWMKLYFKIKPLLKSAETEKEMANM
            KEEFEKTKEELAKTEAKRKELEEKMVTLMQEKNDLQLQVQAEADSLADAEERCDQLIK
            TKIQLEAKIKEVTERAEDEEEINAELTAKKRKLEDECSELKKDIDDLELTLAKVEKEK
            HATENKVKNLTEEMAGLDETIAKLTKEKKALQEAHQQTLDDLQAEEDKVNTLTKAKIK
            LEQQVDDLEGSLEQEKKIRMDLERAKRKLEGDLKLAQESAMDIENDKQQLDEKLKKKE
            FEMSGLQSKIEDEQALGMQLQKKIKELQARIEELEEEIEAERASRAKAEKQRSDLSRE
            LEEISERLEEAGGATSAQIEMNKKREAEFQKMRRDLEEATLQHEATAATLRKKHADSV
            AELGEQIDNLQRVKQKLEKEKSEMKMEIDDLASNMETVSKAKGNLEKMCRALEDQLSE
            IKTKEEEQQRLINDLTAQRARLQTESGEYSRQLDEKDTLVSQLSRGKQAFTQQIEELK
            RQLEEEIKAKSALAHALQSSRHDCDLLREQYEEEQEAKAELQRAMSKANSEVAQWRTK
            YETDAIQRTEELEEAKKKLAQRLQDAEEHVEAVNAKCASLEKTKQRLQNEVEDLMIDV
            ERTNAACAALDKKQRNFDKILAEWKQKCEETHAELEASQKESRSLSTELFKIKNAYEE
            SLDQLETLKRENKNLQQEISDLTEQIAEGGKRIHELEKIKKQVEQEKSELQAALEEAE
            ASLEHEEGKILRIQLELNQVKSEVDRKIAEKDEEIDQMKRNHIRIVESMQSTLDAEIR
            SRNDAIRLKKKMEGDLNEMEIQLNHANRMAAEALRNYRNTQAILKDTQLHLDDALRSQ
            EDLKEQLAMVERRANLLQAEIEELRATLEQTERSRKIAEQELLDASERVQLLHTQNTS
            LINTKKKLETDISQIQGEMEDIIQEARNAEEKAKKAITDAAMMAEELKKEQDTSAHLE
            RMKKNLEQTVKDLQHRLDEAEQLALKGGKKQIQKLEARVRELEGEVESEQKRNVEAVK
            GLRKHERKVKELTYQTEEDRKNILRLQDLVDKLQAKVKSYKRQAEEAEEQSNVNLSKF
            RRIQHELEEAEERADIAESQVNKLRVKSREVHTKIISEE"
```

```
BASE COUNT     1890 a 1300 c 1613 g 1122 t
ORIGIN
        1 atgagttctg actctgagat ggccattttt ggggaggctg ctcctttcct ccgaaagtct
       61 gaaagggagc gaattgaagc ccagaacaag ccttttgatg ccaagacatc agtctttgtg
      121 gtggacccta aggagtcctt tgtgaaagca acagtgcaga gcagggaagg ggggaaggtg
<< Sequence deleted to save space >>
     5701 cggaggatcc agcacgagct ggaggaggcc gaggaaaggg ctgacattgc tgagtcccag
     5761 gtcaacaagc tgagggtgaa gagcagggag gttcacacaa aaatcataag tgaagagtaa
     5821 tttatctaac tgctgaaagg tgaccaaaga aatgcacaaa atgtgaaaat ctttgtcact
     5881 ccattttgta cttatgactt ttggagataa aaaatttatc tgcca
//
```

Appendix 3.2. Example of EMBL Flatfile Format

```
ID AF111785 standard; RNA; HUM; 5925 BP.
XX
AC AF111785;
XX
SV AF111785.1
XX
DT 13-MAY-1999 (Rel. 59, Created)
DT 07-SEP-1999 (Rel. 61, Last updated, Version 3)
XX
DE Homo sapiens myosin heavy chain IIx/d mRNA, complete cds.
XX
KW .
XX
OS Homo sapiens (human)
OC Eukaryota; Metazoa; Chordata; Craniata; Vertebrata; Euteleostomi; Mammalia;
OC Eutheria; Primates; Catarrhini; Hominidae; Homo.
XX
RN [1]
RP 1-5925
RX MEDLINE; 99178997.
RA Weiss A., McDonough D., Wertman B., Acakpo-Satchivi L., Montgomery K.,
RA Kucherlapati R., Leinwand L., Krauter K.;
RT "Organization of human and mouse skeletal myosin heavy chain gene clusters
RT is highly conserved";
RL Proc. Natl. Acad. Sci. U.S.A. 96(6):2958-2963(1999).
XX
RN [2]
RP 1-5925
RX MEDLINE; 99318869.
RA Weiss A., Schiaffino S., Leinwand L.A.;
RT "Comparative sequence analysis of the complete human sarcomeric myosin
RT heavy chain family: implications for functional diversity";
RL J. Mol. Biol. 290(1):61-75(1999).
XX
RN [3]
RP 1-5925
RA Weiss A., Leinwand L.A.;
RT ;
RL Submitted (09-DEC-1998) to the EMBL/GenBank/DDBJ databases.
RL MCDB, University of Colorado at Boulder, Campus Box 0347, Boulder, Colorado
RL 80309-0347, USA
```

```
XX
DR SPTREMBL; Q9Y622; Q9Y622.
XX
FH Key Location/Qualifiers
FH
FT source     1..5925
FT            /chromosome="17"
FT            /db_xref="taxon:9606"
FT            /organism="Homo sapiens"
FT            /map="17p13.1"
FT            /tissue_type="skeletal muscle"
FT CDS        1..5820
FT            /codon_start=1
FT            /db_xref="SPTREMBL:Q9Y622"
FT            /note="MyHC"'
FT            /product="myosin heavy chain IIx/d"
FT            /protein_id="AAD29951.1"
FT            /translation="MSSDSEMAIFGEAAPFLRKSERERIEAQNKPFDAKTSVFVVDPKE
FT            SFVKATVQSREGGKVTAKTEAGATVTVKDDQVFPMNPPKYDKIEDMAMMTHLHEPAVLY
FT            NLKERYAAWMIYTYSGLFCVTVNPYKWLPVYNAEVVTAYRGKKRQEAPPHIFSISDNAY
FT            QFMLTDRENQSILITGESGAGKTVNTKRVIQYFATIAVTGEKKKEEVTSGKMQGTLEDQ
FT            IISANPLLEAFGNAKTVRNDNSSRFGKFIRIHFGTTGKLASADIETYLLEKSRVTFQLK
FT            AERSYHIFYQIMSNKKPDLIEMLLITTNPYDYAFVSQGEITVPSIDDQEELMATDSAIE
FT            ILGFTSDERVSIYKLTGAVMHYGNMKFKQKQREEQAEPDGTEVADKAAYLQNLNSADLL
FT            KALCYPRVKVGNEYVTKGQTVQQVYNAVGALAKAVYDKMFLWMVTRINQQLDTKQPRQY
FT            FIGVLDIAGFEIFDFNSLEQLCINFTNEKLQQFFNHHMFVLEQEEYKKEGIEWTFIDFG
FT            MDLAACIELIEKPMGIFSILEEECMFPKATDTSFKNKLYEQHLGKSNNFQKPKPAKGKP
FT            EAHFSLIHYAGTVDYNIAGWLDKNKDPLNETVVGLYQKSAMKTLALLFVGATGAEAEAG
FT            GGKKGGKKKGSSFQTVSALFRENLNKLMTNLRSTHPHFVRCIIPNETKTPGAMEHELVL
FT            HQLRCNGVLEGIRICRKGFPSRILYADFKQRYKVLNASAIPEGQFIDSKKASEKLLGSI
FT            DIDHTQYKFGHTKVFFKAGLLGLLEEMRDEKLAQLITRTQAMCRGFLARVEYQKMVERR
FT            ESIFCIQYNVRAFMNVKHWPWMKLYFKIKPLLKSAETEKEMANMKEEFEKTKEELAKTE
FT            AKRKELEEKMVTLMQEKNDLQLQVQAEADSLADAEERCDQLIKTKIQLEAKIKEVTERA
FT            EDEEEINAELTAKKRKLEDECSELKKDIDDLELTLAKVEKEKHATENKVKNLTEEMAGL
FT            DETIAKLTKEKKALQEAHQQTLDDLQAEEDKVNTLTKAKIKLEQQVDDLEGSLEQEKKI
FT            RMDLERAKRKLEGDLKLAQESAMDIENDKQQLDEKLKKKEFEMSGLQSKIEDEQALGMQ
FT            LQKKIKELQARIEELEEEIEAERASRAKAEKQRSDLSRELEEISERLEEAGGATSAQIE
FT            MNKKREAEFQKMRRDLEEATLQHEATAATLRKKHADSVAELGEQIDNLQRVKQKLEKEK
FT            SEMKMEIDDLASNMETVSKAKGNLEKMCRALEDQLSEIKTKEEEQQRLINDLTAQRARL
FT            QTESGEYSRQLDEKDTLVSQLSRGKQAFTQQIEELKRQLEEEIKAKSALAHALQSSRHD
FT            CDLLREQYEEEQEAKAELQRAMSKANSEVAQWRTKYETDAIQRTEELEEAKKKLAQRLQ
FT            DAEEHVEAVNAKCASLEKTKQRLQNEVEDLMIDVERTNAACAALDKKQRNFDKILAEWK
FT            QKCEETHAELEASQKESRSLSTELFKIKNAYEESLDQLETLKRENKNLQQEISDLTEQI
FT            AEGGKRIHELEKIKKQVEQEKSELQAALEEAEASLEHEEGKILRIQLELNQVKSEVDRK
FT            IAEKDEEIDQMKRNHIRIVESMQSTLDAEIRSRNDAIRLKKKMEGDLNEMEIQLNHANR
FT            MAAEALRNYRNTQAILKDTQLHLDDALRSQEDLKEQLAMVERRANLLQAEIEELRATLE
FT            QTERSRKIAEQELLDASERVQLLHTQNTSLINTKKKLETDISQIQGEMEDIIQEARNAE
FT            EKAKKAITDAAMMAEELKKEQDTSAHLERMKKNLEQTVKDLQHRLDEAEQLALKGGKKQ
FT            IQKLEARVRELEGEVESEQKRNVEAVKGLRKHERKVKELTYQTEEDRKNILRLQDLVDK
FT            LQAKVKSYKRQAEEAEEQSNVNLSKFRRIQHELEEAEERADIAESQVNKLRVKSREVHT
FT            KIISEE"
XX
SQ Sequence 5925 BP; 1890 A; 1300 C; 1613 G; 1122 T; 0 other;
 atgagttctg actctgagat ggccattttt ggggaggctg ctcctttcct ccgaaagtct 60
 gaaagggagc gaattgaagc ccagaacaag ccttttgatg ccaagacatc agtctttgtg 120
  << Sequence deleted to save space >>
 cggaggatcc agcacgagct ggaggaggcc gaggaaaggg ctgacattgc tgagtcccag 5760
 gtcaacaagc tgagggtgaa gagcagggag gttcacacaa aaatcataag tgaagagtaa 5820
```

```
 tttatctaac tgctgaaagg tgaccaaaga aatgcacaaa atgtgaaaat ctttgtcact 5880
 ccattttgta cttatgactt ttggagataa aaaatttatc tgcca 5925
//
```

Appendix 3.3. Example of a Record in CON Division

```
LOCUS       U00089       816394 bp    DNA     circular   CON         10-MAY-1999
DEFINITION  Mycoplasma pneumoniae M129 complete genome.
ACCESSION   U00089
VERSION     U00089.1 GI:6626256
KEYWORDS    .
SOURCE      Mycoplasma pneumoniae.
ORGANISM    Mycoplasma pneumoniae
            Bacteria; Firmicutes; Bacillus/Clostridium group; Mollicutes;
            Mycoplasmataceae; Mycoplasma.
REFERENCE   1 (bases 1 to 816394)
AUTHORS     Himmelreich,R., Hilbert,H., Plagens,H., Pirkl,E., Li,B.C. and
            Herrmann,R.
TITLE       Complete sequence analysis of the genome of the bacterium Mycoplasma
            pneumoniae
JOURNAL     Nucleic Acids Res. 24 (22), 4420-4449 (1996)
MEDLINE     97105885
REFERENCE   2 (bases 1 to 816394)
AUTHORS     Himmelreich,R., Hilbert,H. and Li,B.-C.
TITLE       Direct Submission
JOURNAL     Submitted (15-NOV-1996) Zentrun fuer Molekulare Biologie Heidelberg,
            University Heidelberg, 69120 Heidelberg, Germany
FEATURES    Location/Qualifiers
source      1..816394
            /organism="Mycoplasma pneumoniae"
            /strain="M129"
            /db_xref="taxon:2104"
            /note="ATCC 29342"
CONTIG      join(AE000001.1:1..9255,AE000002.1:59..16876,AE000003.1:59..10078,
            AE000004.1:59..17393,AE000005.1:59..10859,AE000006.1:59..11441,
            AE000007.1:59..10275,AE000008.1:59..9752,AE000009.1:59..14075,
            AE000010.1:59..11203,AE000011.1:59..15501,AE000012.1:59..10228,
            AE000013.1:59..10328,AE000014.1:59..12581,AE000015.1:59..17518,
            AE000016.1:59..16518,AE000017.1:59..18813,AE000018.1:59..11147,
            AE000019.1:59..10270,AE000020.1:59..16613,AE000021.1:59..10701,
            AE000022.1:59..12807,AE000023.1:59..13289,AE000024.1:59..9989,
            AE000025.1:59..10770,AE000026.1:59..11104,AE000027.1:59..33190,
            AE000028.1:59..10560,AE000029.1:59..10640,AE000030.1:59..11802,
            AE000031.1:59..11081,AE000032.1:59..12622,AE000033.1:59..12491,
            AE000034.1:59..11844,AE000035.1:59..10167,AE000036.1:59..11865,
            AE000037.1:59..11391,AE000038.1:59..11399,AE000039.1:59..14233,
            AE000040.1:59..13130,AE000041.1:59..11259,AE000042.1:59..12490,
            AE000043.1:59..11643,AE000044.1:59..15473,AE000045.1:59..10855,
            AE000046.1:59..11562,AE000047.1:59..20217,AE000048.1:59..10109,
            AE000049.1:59..12787,AE000050.1:59..12516,AE000051.1:59..16249,
            AE000052.1:59..12390,AE000053.1:59..10305,AE000054.1:59..10348,
            AE000055.1:59..9893,AE000056.1:59..16213,AE000057.1:59..11119,
            AE000058.1:59..28530,AE000059.1:59..12377,AE000060.1:59..11670,
            AE000061.1:59..24316,AE000062.1:59..10077,AE000063.1:59..1793)
//
```

4

SUBMITTING DNA SEQUENCES TO THE DATABASES

Jonathan A. Kans

National Center for Biotechnology Information
National Library of Medicine
National Institutes of Health
Bethesda, Maryland

B. F. Francis Ouellette

Centre for Molecular Medicine and Therapeutics
Children's and Women's Health Centre of British Columbia
University of British Columbia
Vancouver, British Columbia

INTRODUCTION

DNA sequence records from the public databases (DDBJ/EMBL/GenBank) are essential components of computational analysis in molecular biology. The sequence records are also reagents for improved curated resources like LocusLink (see Chapter 7) or many of the protein databases. Accurate and informative biological annotation of sequence records is critical in determining the function of a disease gene by sequence similarity search. The names or functions of the encoded protein products, the name of the genetic locus, and the link to the original publication of that sequence make a sequence record of immediate value to the scientist who retrieves it as the result of a BLAST or Entrez search. Effective interpretation of recently finished human genome sequence data is only possible by making use of *all* submitted data provided along with the actual sequence. These complete, annotated records capture the biology associated with DNA sequences.

Bioinformatics: A Practical Guide to the Analysis of Genes and Proteins
Edited by A. D. Baxevanis and B. F. F. Ouellette
ISBN 0-471-38390-2 (cloth), ISBN 0-471-383910 (paper)

Journals no longer print full sequence data, but instead print a database accession number, and require authors to submit sequences to a public database when an article describing a new sequence is submitted for publication. Many scientists release their sequences before the article detailing them is in press. This practice is now the rule for large genome centers, and, although some individual laboratories still wait for acceptance of publication before making their data available, others consider the release of a record to be publication in its own right.

The submission process is governed by an international, collaborative agreement. Sequences submitted to any one of the three databases participating in this collaboration will appear in the other two databases within a few days of their release to the public. Sequence records are then distributed worldwide by various user groups and centers, including those that reformat the records for use within their own suites of programs and databases. Thus, by submitting a sequence to only one of the three "major" databases, researchers can quickly disseminate their sequence data and avoid the possibility that redundant records will be archived.

As mentioned often in this book, the growth of sequence databases has been exponential. Most sequence records in the early years were submitted by individual scientists studying a gene of interest. A program suitable for this type of submission should allow for the manual annotation of arbitrary biological information. However, the databases recently have had to adapt not only to new classes of data but also to a substantially higher rate of submission. A significant fraction of submissions now represents phylogenetic and population studies, in which relationships between sequences need to be explicitly demonstrated. Completed genomes are also becoming available at a growing rate.

This chapter is devoted to the submission of DNA and protein sequences and their annotations into the public databases. Presented here are two different approaches for submitting sequences to the databases, one Web-based (using BankIt) and the other using Sequin, a multi-platform program that can use a direct network connection. Sequin is also an ASN.1 editing tool that takes full advantage of the NCBI data model (see Chapter 2) and has become a platform for many sequence analysis tools that NCBI has developed over the years. (A separate bulk-submission protocol used for EST records, which are submitted to the databases at the rate of thousands per day, is discussed briefly at the end of this chapter. Fortunately, EST records are fairly simple and uniform in content, making them amenable to automatic processing.)

WHY, WHERE, AND WHAT TO SUBMIT?

One should submit to whichever of the three public databases is most convenient. This may be the database that is closest geographically, it may be the repository one has always used in the past, or it may simply be the place one's submission is likely to receive the best attention. All three databases have knowledgeable staff able to help submitters throughout the process. Under normal circumstances, an accession number will be returned within one workday, and a finished record should be available within 5–10 working days, depending on the information provided by the submitter. Submitting data to the database is not the end of one's scientific obligation. Updating the record as more information becomes available will ensure that the information within the record will survive time and scientific rigor.

Presently, it is assumed that all submissions of sequences are done electronically: via the World Wide Web, by electronic mail, or (at the very least) on a computer disk sent via regular postal mail. The URLs and E-mail addresses for electronic submissions are shown in the list at the end of the chapter.

All three databases want the same end result: a richly annotated, biologically and computationally sound record, one that allows other scientists to be able to reap the benefits of the work already performed by the submitting biologist and that affords links to the protein, bibliographic, and genomic databases (see Chapter 7). There is a rich set of biological features and other annotations available, but the important components are the ones that lend themselves to analysis. These include the nucleotide and protein sequences, the CDS (coding sequence, also known as coding region), gene, and mRNA features (i.e., features representing the central dogma of molecular biology), the organism from which the sequences were determined, and the bibliographic citation that links them to the information sphere and will have all the experimental details that give this sequence its *raison d'être*.

DNA/RNA

The submission process is quite simple, but care must be taken to provide information that is accurate (free of errors and vector or mitochondrial contamination) and as biologically sound as possible, to ensure maximal usability by the scientific community. Here are a few matters to consider before starting a submission, regardless of its form.

Nature of the Sequence. Is it of genomic or mRNA origin? Users of the databases like to know the nature of the physical DNA that is the origin of the molecule being sequenced. For example, although cDNA sequencing is performed on DNA (and not RNA), the type of the molecule present in the cell is mRNA. The same is true for the genomic sequencing of rRNA genes, in which the sequenced molecule is almost always genomic DNA. Copying the rRNA into DNA, like direct sequencing of rRNA, although possible, is rarely done. Bear in mind also that, because the sequence being submitted should be of a unique molecular type, it must not represent (for example) a mixture of genomic and mRNA molecule types that cannot actually be isolated from a living cell.

Is the Sequence Synthetic, But Not Artificial? There is a special division in the nucleotide databases for synthetic molecules, sequences put together experimentally that do not occur naturally in the environment (e.g., protein expression vector sequences). The DNA sequence databases do not accept computer-generated sequences, such as consensus sequences, and all sequences in the databases are experimentally derived from the actual sequencing of the molecule in question. They can, however, be the compilation of a shotgun sequencing exercise.

How Accurate is the Sequence? This question is poorly documented in the database literature, but the assumption that the submitted sequence is as accurate as possible usually means at least two-pass coverage (in opposite orientations) on the whole submitted sequence. Equally important is the verification of the final submitted sequence. It should be free of vector contamination (this can be verified with a BLASTN search against the VecScreen database; see Chapter 8 and later in this chapter) and possibly checked with known restriction maps, to eliminate the possibility of sequence rearrangement and to confirm correct sequence assembly.

Organism

All DNA sequence records must show the organism from which the sequence was derived. Many inferences are made from the phylogenetic position of the records present in the databases. If these are wrongly placed, an incorrect genetic code may be used for translation, with the possible consequence of an incorrectly translated or prematurely truncated protein product sequence. Just knowing the genus and species is usually enough to permit the database staff to identify the organism and its lineage. NCBI offers an important taxonomy service, and the staff taxonomists maintain the taxonomy that is used by all the nucleotide databases and by SWISS-PROT, a curated protein database.

Citation

As good as the annotations can be, they will never surpass a published article in fully representing the state of biological knowledge with respect to the sequence in any given record. It is therefore imperative to ensure the proper link between the research publication and the primary data it will cite. For this reason, having a citation in the submission being prepared is of great importance, even if it consists of just a temporary list of authors and a working title. Updating these citations at publication time is also important to the value of the record. (This is done routinely by the database staff and will happen more promptly if the submitter notifies the staff on publication of the article.)

Coding Sequence(s)

A submission of nucleotide also means the inclusion of the protein sequences it encodes. This is important for two reasons:

- Protein databases (e.g., SWISS-PROT and PIR) are almost entirely populated by protein sequences present in DNA sequence database records.
- The inclusion of the protein sequence serves as an important, if not essential, validation step in the submission process.

Proteins include the enzyme molecules that carry out many of the biological reactions we study, and their sequences are an intrinsic part of the submission process. Their importance, which is discussed in Chapter 2, is also reflected in the submission process, and this information must be captured for representation in the various databases. Also important are the protein product and gene names, if these are known. There are a variety of resources (many present in the lists that conclude these chapters) that offer the correct gene nomenclature for many organisms (cf. Genetic nomenclature guide, *Trends in Genetics*, 1998).

The coding sequence features, or CDS, are the links between the DNA or RNA and the protein sequences, and their correct positioning is central in the validation, as is the correct genetic code. The nucleotide databases now use 17 different genetic codes that are maintained by the taxonomy and molecular biology staff at NCBI. Because protein sequences are so important, comprising one of the main pieces of molecular biology information on which biologists can compute, they receive much deserved attention from the staff at the various databases. It is usually simple to find the correct open-reading frame in an mRNA (see Chapter 10), and various tools are available for this (e.g., NCBI's ORF Finder). Getting the correct CDS intervals in a genomic sequence from a higher eukaryote is a little trickier: the different exon-coding sequences must be joined, and this involves a variety of approaches, also described in Chapter 10. (The Suggest Intervals function in Sequin will calculate CDS intervals if given the sequence of the protein and the proper genetic code.) A submitted record will be validated by the database staff but even more immediately by the submission tool used as well. Validation checks that the start and stop codons are included in the CDS intervals, that these intervals are using exon/intron-consensus boundaries, and that the provided amino acid sequence can be translated from the designated CDS intervals using the appropriate genetic code.

Other Features

There are a variety of other features available for the feature sections of a submitted sequence record. The complete set of these is represented in the feature table documentation. Although many features are available, there is much inconsistent usage in the databases, mainly due to a lack of consistent guidelines and poor agreement among biologists as to what they really mean. Getting the organism, bibliography, gene, CDS, and mRNA correct usually suffices and makes for a record that can be validated, is informative, and allows a biologist to grasp in a few lines of text an overview of the biology of the sequence. Nonetheless, the full renditions of the feature table documentation are available for use as appropriate but with care taken as to the intent of the annotations.

POPULATION, PHYLOGENETIC, AND MUTATION STUDIES

The nucleotide databases are now accepting population, phylogenetic, and mutational studies as submitted sequence sets, and, although this information is not adequately represented in the flatfile records, it is appearing in the various databases. This allows the submission of a group of related sequences together, with entry of shared information required only once. Sequin also allows the user to include the alignment generated with a favorite alignment tool and to submit this information with the DNA sequence. New ways to display this information (such as Entrez) should soon make this kind of data more visible to the general scientific community.

PROTEIN-ONLY SUBMISSIONS

In most cases, protein sequences come with a DNA sequence. There are some exceptions—people do sequence proteins directly—and such sequences must be submitted without a corresponding DNA sequence. SWISS-PROT presently is the best venue for these submissions.

HOW TO SUBMIT ON THE WORLD WIDE WEB

The World Wide Web is now the most common interface used to submit sequences to the three databases. The Web-based submission systems include Sakura ("cherry blossoms") at DDBJ, WebIn at EBI, and BankIt at the NCBI. The Web is the preferred submission path for simple submissions or for those that do not require complicated annotations or too much repetition (i.e., 30 similar sequences, as typically found in a population study, would best be done with Sequin, see below). The Web form is ideal for a research group that makes few sequence submissions and needs something simple, entailing a short learning curve. The Web forms are more than adequate for the majority of the submissions: some 75–80% of individual submissions to NCBI are done via the Web. The alternative addresses (or URLs) for submitting to the three databases are presented in the list at the end of the chapter.

On entering a BankIt submission, the user is asked about the length of the nucleotide sequence to be submitted. The next BankIt form is straightforward: it asks about the contact person (the individual to whom the database staff may address any questions), the citations (who gets the scientific credit), the organism (the top 100 organisms are on the form; all others must be typed in), the location (nuclear vs. organelle), some map information, and the nucleotide sequence itself. At the end of the form, there is a BankIt button, which calls up the next form. At this point, some validation is made, and, if any necessary fields were not filled in, the form is presented again. If all is well, the next form asks how many features are to be added and prompts the user to indicate their types. If no features were added, BankIt will issue a warning and ask for confirmation that not even one CDS is to be added to the submission. The user can say no (zero new CDSs) or take the opportunity to add one or more CDS. At this point, structural RNA information or any other legal DDBJ/EMBL/GenBank features can be added as well.

To begin to save a record, press the BankIt button again. The view that now appears must be approved before the submission is completed; that is, more changes may be made, or other features may be added. To finish, press BankIt one more time. The final screen will then appear; after the user toggles the Update/Finished set of buttons and hits BankIt one last time, the submission will go to NCBI for processing. A copy of the just-finished submission should arrive promptly via E-mail; if not, one should contact the database to confirm receipt of the submission and to make any correction that may be necessary.

HOW TO SUBMIT WITH SEQUIN

Sequin is designed for preparing new sequence records and updating existing records for submission to DDBJ, EMBL, and GenBank. It is a tool that works on most computer platforms and is suitable for a wide range of sequence lengths and complexities, including traditional (gene-sized) nucleotide sequences, segmented entries (e.g., genomic sequences of a spliced gene for which not all intronic sequences have been determined), long (genome-sized) sequences with many annotated features, and sets of related sequences (i.e., population, phylogenetic, or mutation studies of a particular gene, region, or viral genome). Many of these submissions could be performed via the Web, but Sequin is more practical for more complex cases. Certain types of submission (e.g., segmented sets) cannot be made via the Web unless explicit instructions to the database staff are inserted.

Sequin also accepts sequences of proteins encoded by the submitted nucleotide sequences and allows annotation of features on these proteins (e.g., signal peptides, transmembrane regions, or cysteine disulfide bonds). For sets of related or similar sequences (e.g., population or phylogenetic studies), Sequin accepts information from the submitter on how the multiple sequences are aligned to each other. Finally, Sequin can be used to edit and resubmit a record that already exists in GenBank, either by extending (or replacing) the sequence or by annotating additional features or alignments.

Submission Made Easy

Sequin has a number of attributes that greatly simplify the process of building and annotating a record. The most profound aspect is automatic calculation of the intervals on a CDS feature given only the nucleotide sequence, the sequence of the protein product, and the genetic code (which is itself automatically obtained from the organism name). This "Suggest Intervals" process takes consensus splice sites into account in its calculations. Traditionally, these intervals were entered manually, a time-consuming and error-prone process, especially on a genomic sequence with many exons, in cases of alternative splicing, or on segmented sequences.

Another important attribute is the ability to enter relevant annotation in a simple format in the definition line of the sequence data file. Sequin recognizes and extracts this information when reading the sequences and then puts it in the proper places in the record. For nucleotide sequences, it is possible to enter the organism's scientific name, the strain or clone name, and several other source modifiers. For example

```
>eIF4E [organism=Drosophila melanogaster] [strain=Oregon R]
CGGTTGCTTGGGTTTTATAACATCAGTCAGTGACAGGCATTTCCAGAGTTGCCCTGTTCAACAATCGATA
GCTGCCTTTGGCCACCAAAATCCCAAACTTAATTAAAGAATTAAATAATTCGAATAATAATTAAGCCCAG
...
```

This is especially important for population and phylogenetic studies, where the source modifiers are necessary to distinguish one component from another.

For protein sequences, the gene and protein names can be entered. For example

```
>4E-I [gene=eIF4E] [protein=eukaryotic initiation factor 4E-I]
MQSDFHRMKNFANPKSMFKTSAPSTEQGRPEPPTSAAAPAEAKDVKPKEDPQETGEPAGNTATTTAPAGD
DAVRTEHLYKHPLMNVWTLWYLENDRSKSWEDMQNEITSFDTVEDFWSLYNHIKPPSEIKLGSDYSLFKK
...
```

If this information is not present in the sequence definition line, Sequin will prompt the user for it before proceeding. Annotations on the definition line can be very convenient, since the information stays with the sequence and cannot be forgotten or mixed-up later. In addition to building the proper CDS feature, Sequin will automatically make gene and protein features with this information.

Because the majority of submissions contain a single nucleotide sequence and one or more coding region features (and their associated protein sequences), the functionality just outlined can frequently result in a finished record, ready to submit without any further annotation. With gene and protein names properly recorded, the record becomes informative to other scientists who may retrieve it as a BLAST similarity result or from an Entrez search.

Starting a New Submission

Sequin begins with a window that allows the user to start a new submission or load a file containing a saved record. After the initial submission has been built, the record can be saved to a file and edited later, before finally being sent to the database. If Sequin has been configured to be network aware, this window also allows the downloading of existing database records that are to be updated.

A new submission is made by filling out several forms. The forms use folder tabs to subdivide a window into several pages, allowing all the requested data to be entered without the need for a huge computer screen. These entry forms have buttons for Prev(ious) Page and Next Page. When the user arrives at the last page on a form, the Next Page button changes to Next Form.

The Submitting Authors form requests a tentative title, information on the contact person, the authors of the sequence, and their institutional affiliations. This form is common to all submissions, and the contact, authors, and affiliation page data can be saved by means of the Export menu item. The resulting file can be read in when starting other submissions by choosing the Import menu item. However, because even population, phylogenetic, or mutation studies are submitted in one step as one record, there is less need to save the submitter information.

The Sequence Format form asks for the type of submission (single sequence, segmented sequence, or population, phylogenetic, or mutation study). For the last three types of submission, which involve comparative studies on related sequences, the format in which the data will be entered also can be indicated. The default is FASTA format (or raw sequence), but various contiguous and interleaved formats (e.g., PHYLIP, NEXUS, PAUP, and FASTA+GAP) are also supported. These latter formats contain alignment information, and this is stored in the sequence record.

The Organism and Sequences form asks for the biological data. On the Organism page, as the user starts to type the scientific name, the list of frequently used organisms scrolls automatically. (Sequin holds information on the top 800 organisms present in GenBank.) Thus, after typing a few letters, the user can fill in the rest of the organism name by clicking on the appropriate item in the list. Sequin now knows the scientific name, common name, GenBank division, taxonomic lineage, and, most importantly, the genetic code to use. (For mitochondrial genes, there is a control to indicate that the alternative genetic code should be used.) For organisms not on the list, it may be necessary to set the genetic code control manually. Sequin uses the standard code as the default. The remainder of the Organism and Sequences form differs depending on the type of submission.

Entering a Single Nucleotide Sequence and its Protein Products

For a single sequence or a segmented sequence, the rest of the Organism and Sequences form contains Nucleotide and Protein folder tabs. The Nucleotide page has controls for setting the molecule type (e.g., genomic DNA or mRNA) and topology (usually linear, occasionally circular) and for indicating whether the sequence is incomplete at the 5′ or 3′ ends. Similarly, the Protein page has controls for creating an initial mRNA feature and for indicating whether the sequence is incomplete at the amino or carboxyl ends.

For each protein sequence, Suggest Intervals is run against the nucleotide sequence (using the entered genetic code, which is usually deduced from the chosen organism), and a CDS feature is made with the resulting intervals. A Gene feature is generated, with a single interval spanning the CDS intervals. A protein product sequence is made, with a Protein feature to give it a name. The organism and publication are placed so as to apply to all nucleotide and protein sequences within the record. Appropriate molecule-type information is also placed on the sequences. In most cases, it is much easier to enter the protein sequence and let Sequin construct the record automatically than to manually add a CDS feature (and associated gene and protein features) later.

Entering an Aligned Set of Sequences

A growing class of submissions involves sets of related sequences: population, phylogenetic, or mutation studies. A large number of HIV sequences come in as population studies. A common phylogenetic study involves ribulose-1,5-bisphosphate carboxylase (RUBISCO), a major enzyme of photosynthesis and perhaps the most prevalent protein (by weight) on earth. Submitting such a set of sequences is not much more complex than submitting a single sequence. The same submission information form is used to enter author and contact information.

In the Sequence Format form, the user chooses the desired type of submission. Population studies are generally from different individuals in the same (crossbreeding) species. Phylogenetic studies are from different species. In the former case, it is best to embed in the definition lines strain, clone, isolate, or other source-identifying information. In the latter case, the organism's scientific name should be embedded. Multiple sequence studies can be submitted in FASTA format, in which case Sequin should later be called on to calculate an alignment. Better yet, alignment information can be indicated by encoding the data in one of several popular alignment formats.

The Organism and Sequences form is slightly different for sets of sequences. The Organism page for phylogenetic studies allows the setting of a default genetic code only for organisms not in Sequin's local list of popular species. The Nucleotide page has the same controls as for a single sequence submission. Instead of a Protein page, there is now an Annotation page. Many submissions are of rRNA sequence or no more than a complete CDS. (This means that the feature intervals span the full range of each sequence.) The Annotation page allows these to be created and named. A definition line (title) can be specified, and Sequin can prefix the individual organism name to the title. More complex situations, in which sequences have more than a single interval feature across the entire span, can be annotated by feature propagation after the initial record has been built and one of the sequences has been annotated.

As a final step, Sequin displays an editor that allows all organism and source modifiers on each sequence to be edited (or entered if the definition lines were not annotated). On confirmation of the modifiers, Sequin finishes assembling the record into the proper structure.

Viewing the Sequence Record

Sequin provides a number of different views of a sequence record. The traditional flatfile can be presented in FASTA, GenBank (Fig. 4.1), or EMBL format. (These can be exported to files on the user's computer, which can then be entered into other sequence analysis packages.) A graphical view (Fig. 4.2) shows feature intervals on a sequence. This is particularly useful for viewing alternatively spliced coding regions. (The style of the Graphical view can be customized, and these views can also be copied to the personal computer's clipboard for pasting into a word processor or drawing program that will be used in preparing a manuscript for publication.) There is a more detailed view that shows the features on the actual sequence. For records containing alignments (e.g., alignments between related sequences entered by a user, or the results of a BLAST search), one can request either a graphical

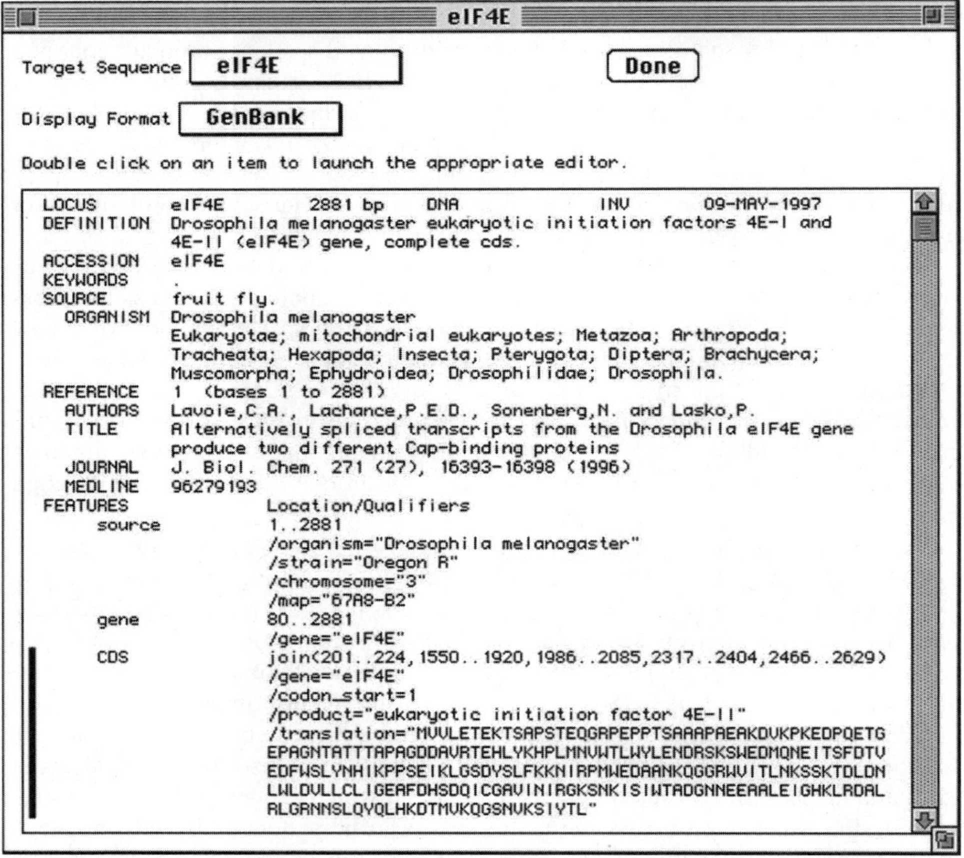

Figure 4.1. Viewing a sequence record with Sequin. The sequence record viewer uses GenBank format, by default. In this example, a CDS feature has been clicked, as indicated by the bar next to its paragraph. Double-clicking on a paragraph will launch an editor for the feature, descriptor, or sequence that was selected. The viewer can be duplicated, and multiple viewers can show the same record in different formats.

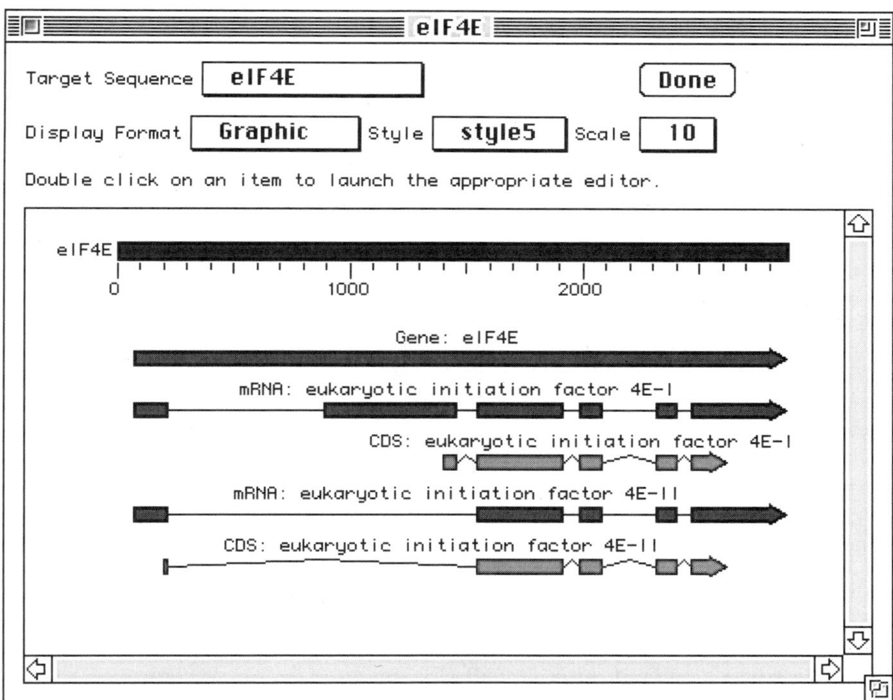

Figure 4.2. Sequin's graphical format can show segmented sequence construction and feature intervals. These can be compared with drawings in laboratory notebooks to see, at a glance, whether the features are annotated at the proper locations. Different styles can be used, and new styles can be created, to customize the appearance of the graphical view. The picture can be copied to a personal computer's clipboard for pasting into a word processor or drawing program.

overview showing insertions, deletions, and mismatches or a detailed view showing the alignment of sequence letters.

The above-mentioned viewers are interactive. Clicking on a feature, a sequence, or the graphical representation of an alignment between sequences will highlight that object. Double-clicking will launch the appropriate editor. Multiple viewers can be used on the same record, permitting different formats to be seen simultaneously. For example, it is quite convenient to have the graphical view and the GenBank (or EMBL) flatfile view present at the same time, especially on larger records containing more than one CDS. The graphical view can be compared to a scientist's lab notebook drawings, providing a quick reality check on the overall accuracy of the feature annotation.

Validation

To ensure the quality of data being submitted, Sequin has a built-in validator that searches for missing organism information, incorrect coding region lengths (compared to the submitted protein sequence), internal stop codons in coding regions, mismatched amino acids, and nonconsensus splice sites. Double-clicking on an item in the error report launches an editor on the "offending" feature.

The validator also checks for inconsistent use of "partial" indications, especially among coding regions, the protein product, and the protein feature on the product. For example, if the coding region is marked as incomplete at the 5′ end, the protein product and protein feature should be marked as incomplete at the amino end. (Unless told otherwise, the CDS editor will automatically synchronize these separate partial indicators, facilitating the correction of this kind of inconsistency.)

Advanced Annotation and Editing Functions

The sequence editor built into Sequin automatically adjusts feature intervals as the sequence is edited. This is particularly important if one is extending an existing record by adding new 5′ sequence. Prior to Sequin, this process entailed manually correcting the intervals on all biological features on the sequence or, more likely, redoing the entire submission from scratch. The sequence editor is used much like a text editor, with new sequence being pasted in or typed in at the position of a cursor.

For population or phylogenetic studies, Sequin allows annotation of one sequence, whereupon features from that sequence can be propagated to all other sequences through the supplied alignment. (In the case of a CDS feature, the feature intervals can be calculated automatically by reading in the sequence of its protein product rather than having to enter them by typing.) Feature propagation is accessed from the alignment editor. The result is the same as would have been achieved if features had been manually annotated on each sequence, but with feature propagation the entire process can be completed in minutes rather than hours.

The concepts behind feature propagation and the sequence editor combine to provide a simple and automatic method for updating an existing sequence. The Update Sequence functions allow the user to enter an overlapping sequence or a replacement sequence. Sequin makes an alignment, merges the sequences if necessary, propagates features onto the new sequence in their new positions, and uses these to replace the old sequence and features.

Genome centers frequently store feature coordinates in databases. Sequin can now annotate features by reading a simple tab-delimited file that specifies the location and type of each feature. The first line starts with `>Features`, a space, and the sequence identifier of the sequence. The table is composed of five columns: start, stop, feature key, qualifier key, and qualifier value. The columns are separated by tab characters. The first row for any given feature has start, stop, and feature key. Additional feature intervals just have start and stop. The qualifiers follow on lines starting with three tabs. An example of this format follows below.

```
>Features lcl|eIF4E
80      2881    gene
                        gene    eIF4E
1402    1458    CDS
1550    1920
1986    2085
2317    2404
2466    2629
                        product eukaryotic initiation factor 4E-I
```

Sending the Submission

A finished submission can be saved to disk and E-mailed to one of the databases. It is also a good practice to save frequently throughout the Sequin session, to make sure nothing is inadvertently lost. The list at the end of this chapter provides E-mail addresses and contact information for the three databases.

UPDATES

The database staffs at all three databases welcome all suggestions on making the update process as efficient and painless as possible. People who notice that records are published but not yet released are strongly encouraged to notify the databases as well. If errors are detected, these should also be forwarded to the updates addresses; the owner of the record is notified accordingly (by the database staff), and a correction usually results. This chain of events is to be distinguished from third-party annotations, which are presently not accepted by the databases. *The record belongs to the submitter(s)*; the database staff offers some curatorial, formatting guideline suggestions, but substantive changes come only from a listed submitter. Many scientists simply E-mail a newly extended sequence or feature update to the databases for updating.

CONSEQUENCES OF THE DATA MODEL

Sequin is, in reality, an ASN.1 editor. The NCBI data model, written in the ASN.1 data description language, is designed to keep associated information together in descriptors or features (see Chapter 2). Features are typically biological entities (e.g., genes, coding regions, RNAs, proteins) that always have a location (of one or more intervals) on a sequence. Descriptors were introduced to carry information that can apply to multiple sequences, eliminating the need to enter multiple copies of the same information.

For the simplest case, that of a single nucleotide sequence with one or more protein products, Sequin generally allows the user to work without needing to be aware of the data model's structural hierarchy. Navigation is necessary, as is at least a cursory understanding of the data model, if extensive annotation on protein product sequences is contemplated or for manual annotation of population and phylogenetic sets. Setting the Target control to a given sequence changes the viewer to show a graphical view or text report on that sequence. Any features or descriptors created with the Annotation submenus will be packaged on the currently targeted sequence.

Although Sequin does provide full navigation among all sequences within a structured record, building the original structure from the raw sequence data is a job best left to Sequin's "create new submission" functions described above. Sequin asks up front for information (e.g., organism and source modifiers, gene and protein names) and knows how to correctly package everything into the appropriate place. This was, in fact, one of the main design goals of Sequin. Manual annotation requires a more detailed understanding of the data model and expertise with the more esoteric functions of Sequin.

Using Sequin as a Workbench

Sequin also provides a number of sequence analysis functions. For example, one function will reverse-complement the sequence and the intervals of its features. New functions can easily be added. These functions appear in a window called the NCBI Desktop (Fig. 4.3), which directly displays the internal structure of the records currently loaded in memory. This window can be understood as a Venn diagram, with descriptors on a set (such as a population study) applying to all sequences in that set. The Desktop allows drag-and-drop of items within a record. For example, the

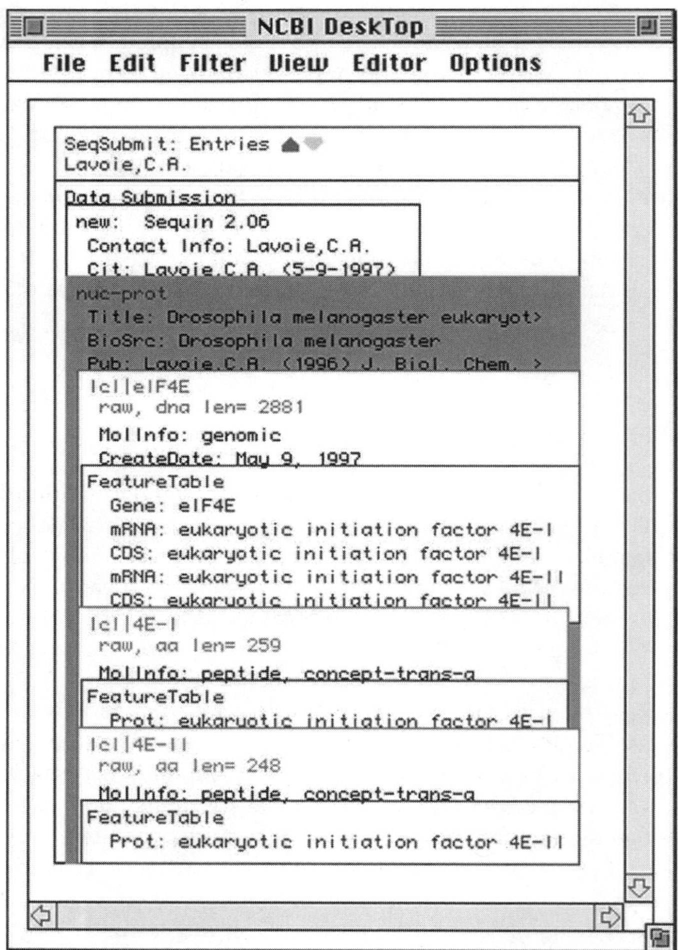

Figure 4.3. The NCBI Desktop displays a graphical overview of how the record is structured in memory, based on the NCBI data model (see Chapter 2). This view is most useful to a software developer or database sequence annotator. In this example, the submission contains a single Nuc-prot set, which in turn contains a nucleotide and two proteins. Each sequence has features associated with it. BioSource and publication descriptors on the Nuc-prot set apply the same organism (*Drosophila melanogaster*) and the same publication, respectively, to all sequences.

user can read the results of a BLAST analysis and then drag-and-drop this information onto a sequence record, thus adding the alignment data to the record, or a newly calculated feature can be dragged into the record. (A separate Seq-loc on the Desktop can be dragged onto a feature, in which case it changes the feature location.) The modifications are immediately displayed on any active viewers. Note, however, that not all annotations are visible in all viewers. The flatfile view does have its limitations; for example, it does not display alignments and does not even indicate that alignments are present. Understanding the Desktop is not necessary for the casual user submitting a simple sequence; however, for the advanced user, it can immediately take the mystery out of the data.

EST/STS/GSS/HTG/SNP AND GENOME CENTERS

Genome centers have now established a number of relationships with DNA sequence databases and have streamlined the submission process for a great number of record types. Not all genome centers deal with all sequence types, but all databases do. The databases have educated their more sophisticated users on this, and, conversely, some of the genomes centers have also encouraged certain database managers to learn their own data model as well (e.g., the use of AceDB to submit sequences at Stanford, Washington University at St. Louis, and the Sanger Centre or the use of XML at Celera).

CONCLUDING REMARKS

The act of depositing records into a database and seeing these records made public has always been an exercise of pride on the part of submitters, a segment of the scientific activity from their laboratory that they present to the scientific community. It is also a mandatory step that has been imposed by publishers as part of the publication process. In this process, submitters always hope to provide information in the most complete and useful fashion, allowing maximum use of their data by the scientific community.

Very few users are aware of the complete array of intricacies present in the databases, but they do know the biology they want these entries to represent. It is incumbent on the databases to provide tools that will facilitate this process. The database staff also provides expertise through their indexing staff (some databases also call them *curators* or *annotators*), who have extensive training in biology and are very familiar with the databases; they ensure that nothing is lost in the submission process. The submission exercise itself has not always been easy and was not even encouraged at the beginning of the sequencing era, simply because databases did not know how to handle this information. Now, however, the databases strongly encourage the submission of sequence data and of all appropriate updates. Many tools are available to facilitate this task, and together the databases support Sequin as the tool to use for new submissions, in addition to their respective Web submissions tools. Submitting data to the databases has now become a manageable (and sometimes enjoyable) task, with scientists no longer having good excuses for neglecting it.

CONTACT POINTS FOR SUBMISSION OF SEQUENCE DATA TO DDBJ/EMBL/GenBank

DDBJ (Center for Information Biology, NIG)

Address	DDBJ, 1111 Yata, Mishima, Shiznoka 411, Japan
Fax	81-559-81-6849
E-mail	
Submissions	*ddbjsub@ddbj.nig.ac.jp*
Updates	*ddbjupdt@ddbj.nig.ac.jp*
Information	*ddbj@ddbj.nig.ac.jp*
World Wide Web	
Home page	*http://www.ddbj.nig.ac.jp/*
Submissions	*http://sakura.ddbj.nig.ac.jp/*

EMBL (European Bioinformatics Institutes, EMBL Outstation)

Address	EMBL Outstation, EBI, Wellcome Trust Genome Campus, Hinxton Cambridge, CB10 1SD, United Kingdom
Voice	01.22.349.44.44
Fax	01.22.349.44.68
E-mail	
Submissions	*datasubs@ebi.ac.uk*
Updates	*update@ebi.ac.uk*
Information	*datalib@ebi.ac.uk*
World Wide Web	
Home page	*http://www.ebi.ac.uk/*
Submissions	*http://www.ebi.ac.uk/subs/allsubs.html*
WebIn	*http://www.ebi.ac.uk/submission/webin.html*

GenBank (National Center for Biotechnology Information, NIH)

Address	GenBank, National Center for Biotechnology Information, National Library of Medicine, National Institutes of Health, Building 38A, Room 8N805, Bethesda MD 20894
Telephone	301-496-2475
Fax	301-480-9241
E-mail	
Submissions	*gb-sub@ncbi.nlm.nih.gov*
EST/GSS/STS	*batch-sub@ncbi.nlm.nih.gov*
Updates	*update@ncbi.nlm.nih.gov*
Information	*info@ncbi.nlm.nih.gov*
World Wide Web	
Home page	*http://www.ncbi.nlm.nih.gov/*
Submissions	*http://www.ncbi.nlm.nih.gov/Web/GenBank/submit.html*
BankIt	*http://www.ncbi.nlm.nih.gov/BankIt/*

INTERNET RESOURCES FOR TOPICS PRESENTED IN CHAPTER 4

dbEST	*http://www.ncbi.nlm.nih.gov/dbEST/*
dbSTS	*http://www.ncbi.nlm.nih.gov/dbSTS/*
dbGSS	*http://www.ncbi.nlm.nih.gov/dbGSS/*

DDBJ/EMBL/GenBank Feature Table Documentation	*http://www.ncbi.nlm.nih.gov/collab/FT/*
EMBL Release Notes	*ftp://ftp.ebi.ac.uk/pub/databases/embl/release/relnotes.doc*
GenBank Release Notes	*ftp://ncbi.nlm.nih.gov/genbank/gbrel.txt*
Genetic codes used in DNA sequence databases	*http://www.ncbi.nlm.nih.gov/htbinpost/Taxonomy/wprintgc?mode_c*
HTGS	*http://www.ncbi.nlm.nih.gov/HTGS/*
ORF Finder	*http://www.ncbi.nlm.nih.gov/gorf/gorf.html*
Sequin	*http://www.ncbi.nlm.nih.gov/Sequin/*
Taxonomy browser	*http://www.ncbi.nlm.nih.gov/Taxonomy/tax.html*

REFERENCES

Boguski, M. S., Lowe, T. M., Tolstoshev, C. M. (1993). dbEST—database for "expressed sequence tags. *Nat. Genet.* 4, 332–333.

Ouellette, B. F. F., and Boguski, M. S. 1997. Database divisions and homology search files: a guide for the perplexed. *Genome Res.* 7, 952–955.

5

STRUCTURE DATABASES

Christopher W. V. Hogue

Samuel Lunenfeld Research Institute
Mount Sinai Hospital
Toronto, Ontario, Canada

INTRODUCTION TO STRUCTURES

This chapter introduces biomolecular structures from a bioinformatics perspective, with special emphasis on the sequences that are contained in three-dimensional structures. The major goal of this chapter is to inform the reader about the contents of structure database records and how they are treated, and sometimes mistreated, by popular software programs. This chapter does not cover the computational processes used by structural scientists to obtain three-dimensional structures, nor does it discuss the finer points of comparative protein architecture. Several excellent monographs regarding protein architecture and protein structure determination methods are already widely available and often found in campus bookstores (e.g., Branden and Tooze, 1999).

The imagery of protein and nucleic acid structures has become a common feature of biochemistry textbooks and research articles. This imagery can be beautiful and intriguing enough to blind us to the experimental details an image represents—the underlying biophysical methods and the effort of hard-working X-ray crystallographers and *nuclear magnetic resonance* (NMR) spectroscopists. The data stored in structure database records represents a practical summary of the experimental data. It is, however, not the data gathered directly by instruments, nor is it a simple mathematical transformation of that data. Each structure database record carries assumptions and biases that change as the state of the art in structure determination advances. Nevertheless, each biomolecular structure is a hard-won piece of crucial information and provides potentially critical information regarding the function of any given protein sequence.

Bioinformatics: A Practical Guide to the Analysis of Genes and Proteins
Edited by A. D. Baxevanis and B. F. F. Ouellette
ISBN 0-471-38390-2 (cloth), ISBN 0-471-383910 (paper)

Since the first edition of this book was released, the software for viewing and manipulating three-dimensional structures has improved dramatically. Another major change has come in the form of an organizational transition, with the Protein Data Bank (PDB) moving from the Brookhaven National Laboratories to the Research Collaboratory for Structural Biology. The result has been a complete change in the organization of the PDB web site. The impact of these changes for biologists will be discussed herein.

The Notion of Three-Dimensional Molecular Structure Data

Let us begin with a mental exercise in recording the three-dimensional data of a biopolymer. Consider how we might record, on paper, all the details and dimensions of a three-dimensional ball-and-stick model of a protein like myoglobin. One way to begin is with the sequence, which can be obtained by tracing out the backbone of the three-dimensional model. Beginning from the NH_2-terminus, we identify each amino acid side chain by comparing the atomic structure of each residue with the chemical structure of the 20 common amino acids, possibly guided by an illustration of amino acid structures from a textbook.

Once the sequence has been written down, we proceed with making a two-dimensional sketch of the biopolymer with all its atoms, element symbols, and bonds, possibly taking up several pieces of paper. The same must be done for the heme ligand, which is an important functional part of the myoglobin molecule. After drawing its chemical structure on paper, we might record the three-dimensional data by measuring the distance of each atom in the model starting from some origin point, along some orthogonal axis system. This would provide the x-, y-, and z-axis distances to each atomic "ball" in the ball-and-stick structure.

The next step is to come up with a bookkeeping scheme to keep all the (x, y, z) coordinate information connected to the identity of each atom. The easiest approach may be to write the (x, y, z) value as a coordinate triple on the same pieces of paper used for the two-dimensional sketch of the biopolymer, right next to each atom. This associates the (x, y, z) value with the atom it is attached to.

This mental exercise helps to conceptualize what a three-dimensional structure database record ought to contain. There are two things that have been recorded here: the chemical structure and the locations of the individual atoms in space. This is an adequate "human-readable" record of the structure, but one probably would not expect a computer to digest it easily. The computer needs clear encoding of the associations of atoms, bonds, coordinates, residues, and molecules, so that one may construct software that can read the data in an unambiguous manner. Here is where the real exercise in structural bioinformatics begins.

Coordinates, Sequences, and Chemical Graphs

The most obvious data in a typical three-dimensional structure record, regardless of the file format in use, is the *coordinate data*, the locations in space of the atoms of a molecule. These data are represented by (x, y, z) triples, distances along each axis to some arbitrary origin in space. The coordinate data for each atom is attached to a list of labeling information in the structure record: which element, residue, and molecule each point in space belongs to. For the standard biopolymers (DNA, RNA, and proteins), this labeling information can be derived starting with the raw sequence.

Implicit in each sequence is considerable chemical data. We can infer the complete chemical connectivity of the biopolymer molecule directly from a sequence, including all its atoms and bonds, and we could make a sketch, just like the one described earlier, from sequence information alone. We refer to this "sketch" of the molecule as the *chemical graph* component of a three-dimensional structure. Every time a sequence is presented in this book or elsewhere, remember that it can encode a fairly complete description of the chemistry of that molecule.

When we sketch all the underlying atoms and bonds representing a sequence, we may defer to a textbook showing the chemical structures of each residue, lest we forget a methyl group or two. Likewise, computers could build up a sketch like a representation of the chemical graph of a structure in memory using a *residue dictionary*, which contains a table of the atom types and bond information for each of the common amino acid and nucleic acid building blocks. What sequence is unable to encode is information about posttranslational modifications. For example, in the structure databases, a phosphorylated tyrosine residue is indicated as "X" in the one letter code—essentially an unknown! Any residue that has had an alteration to its standard chemical graph will, unfortunately, be indicated as X in the one-letter encoding of sequence.

Atoms, Bonds, and Completeness

Molecular graphics visualization software performs an elaborate "connect-the-dots" process to make the wonderful pictures of protein structure we see in textbooks of biomolecular structure, like the structure for insulin (3INS; Isaccs and Agarwa, 1978) shown in Figure 5.1. The connections used are, of course, the chemical bonds between all the atoms. In current use, three-dimensional molecular structure database records employ two different "minimalist" approaches regarding the storage of bond data.

The original approach to recording atoms and bonds is something we shall call the *chemistry rules* approach. The rules are the observable physical rules of chemistry, such as, "the average length of a stable C—C bond is about 1.5 angstroms." Applying these rules to derive the bonds means that any two coordinate locations in space that are 1.5 Å apart and are tagged as carbon atoms always form a single bond. With the chemistry rules approach, we can simply disregard the bonds. A perfect and complete structure can be recorded without any bond information, provided it does not break any of the rules of chemistry in atomic locations. Obviously, this is not always the case, and specific examples of this will be presented later in this chapter.

The chemistry rules approach ended up being the basis for the original three-dimensional biomolecular structure file format, the PDB format from the Protein Data Bank at Brookhaven (Bernstein et al., 1977). These records, in general, lack complete bond information for biopolymers. The working assumption is that no residue dictionary is required for interpretation of data encoded by this approach, just a table of bond lengths and bond types for every conceivable pair of bonded atoms is required.

Every software package that reads in PDB data files must reconstruct the bonds based on these rules. However, the rules we are describing have never been explicitly codified for programmers. This means that interpreting the bonding in PDB files is left for the *programmer* to decide, and, as a result, software can be inconsistent in

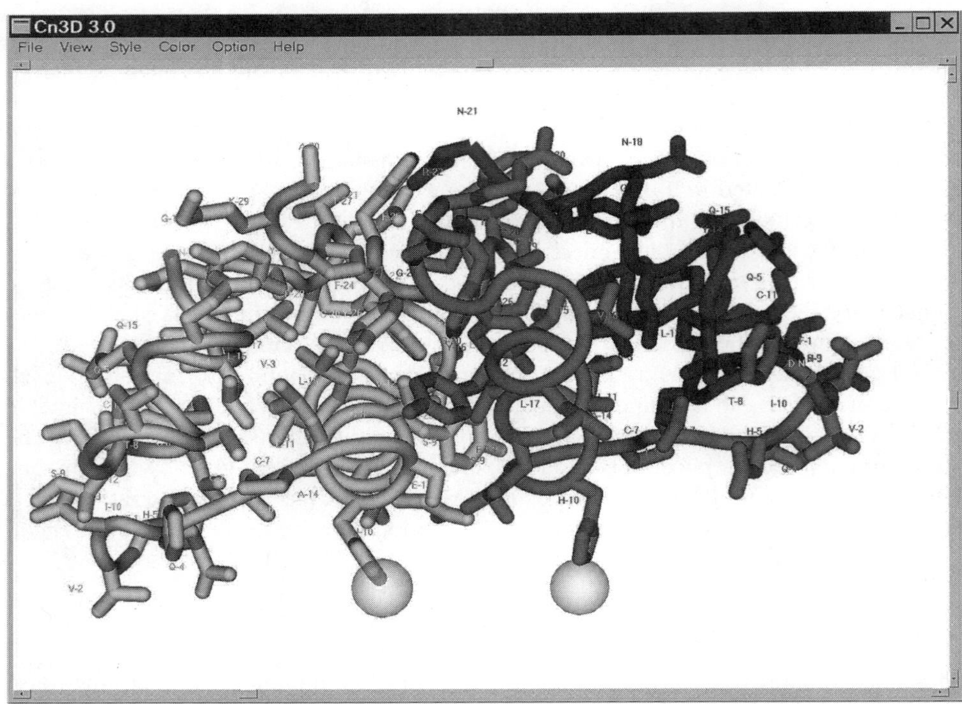

Figure 5.1. The insulin structure 3INS illustrated using Cn3D with OpenGL. Four chains are depicted in the crystallographic unit. This structure illustrates two of many bioinformatics bridges that must be spanned between sequence and structure databases, the lack of encoding of the active biological unit, and the lack of encoding of the relationship of the observed structure to the parent gene. (See color plate.)

the way it draws bonds, especially when different algorithms and distance tolerances are used. The PDB file approach is minimalist in terms of the data stored in a record, and deciphering it often requires much more sophisticated logic than would be needed if the bonding information and chemical graph were explicitly specified in the record. Rarely is this logic properly implemented, and it may in fact be impossible to deal with all the exceptions in the PDB file format. Each exception to the bonding rules needs to be captured by complicated logic statements programmed on a case-by-case basis.

The second approach to describing a molecule is what we call the *explicit bonding approach*, the method that is used in the database records of the Molecular Modeling Database (MMDB), which is, in turn, derived from the data in PDB. In the MMDB system, the data file contains all of its own explicit bonding information. MMDB uses a standard residue dictionary, a record of all the atoms and bonds in the polymer forms of amino acid and nucleic acid residues, plus end-terminal variants. Such data dictionaries are common in the specialized software used by scientists to solve X-ray or NMR structures. The software that reads in MMDB data can use the bonding information supplied in the dictionary to connect atoms together, without trying to enforce (or force) the rules of chemistry. As a result, the three-dimensional coordinate data are consistently interpreted by visualization software, regardless of type. This approach also lends itself to inherently simpler software, because exceptions to bonding rules are recorded within the database file itself and read in without the need for another layer of exception-handling codes.

Scientists that are unfamiliar with structure data often expect all structures in the public databases to be of "textbook" quality. They are often surprised when parts of a structure are missing. The availability of a three-dimensional database record for a particular molecule does not ever imply its completeness. Structural completeness is strictly defined as follows: *At least one coordinate value for each and every atom in the chemical graph is present.*

Structural completeness is quite rare in structure database records. Most X-ray structures lack coordinates for hydrogen atoms because the locations of hydrogens in space are not resolved by the experimental methods currently available. However, some modeling software can be used to predict the locations of these hydrogen atoms and reconstruct a structure record with the now-modeled hydrogens added. It is easy to identify the products of molecular modeling in structure databases. These often have overly complete coordinate data, usually with all possible hydrogen atoms present that could not have been found using an experimental method.

PDB: PROTEIN DATA BANK AT THE RESEARCH COLLABORATORY FOR STRUCTURAL BIOINFORMATICS (RCSB)

Overview

The use of computers in biology has its origins in biophysical methods, such as X-ray crystallography. Thus, it is not surprising that the first "bioinformatics" database was built to store complex three-dimensional data. The Protein Data Bank, originally developed and housed at the Brookhaven National Laboratories, is now managed and maintained by the Research Collaboratory for Structural Bioinformatics (RCSB). RCSB is a collaborative effort involving scientists at the San Diego Supercomputing Center, Rutgers University, and the National Institute of Standards and Technology. The collection contains all publicly available three-dimensional structures of proteins, nucleic acids, carbohydrates, and a variety of other complexes experimentally determined by X-ray crystallographers and NMR spectroscopists. This section focuses briefly on the database and bioinformatics services offered through RCSB.

RCSB Database Services

The World Wide Web site of the Protein Data Bank at the RCSB offers a number of services for submitting and retrieving three-dimensional structure data. The home page of the RCSB site provides links to services for depositing three-dimensional structures, information on how to obtain the status of structures undergoing processing for submission, ways to download the PDB database, and links to other relevant sites and software.

PDB Query and Reporting

Starting at the RCSB home page, one can retrieve three-dimensional structures using two different query engines. The SearchLite system is the one most often used, providing text searching across the database. The SearchFields interface provides the additional ability to search specific fields within the database. Both of these systems report structure matches to the query in the form of Structure Summary pages, an example of which is shown in Figure 5.2. The RCSB Structure Summary page links are to other Web pages that themselves provide a large number of links, and it may be confusing to a newcomer to not only sift through all this information but to decide which information sources are the most relevant ones for biological discovery.

Submitting Structures. For those who wish to submit three-dimensional structure information to PDB, the RCSB offers its ADIT service over the Web. This

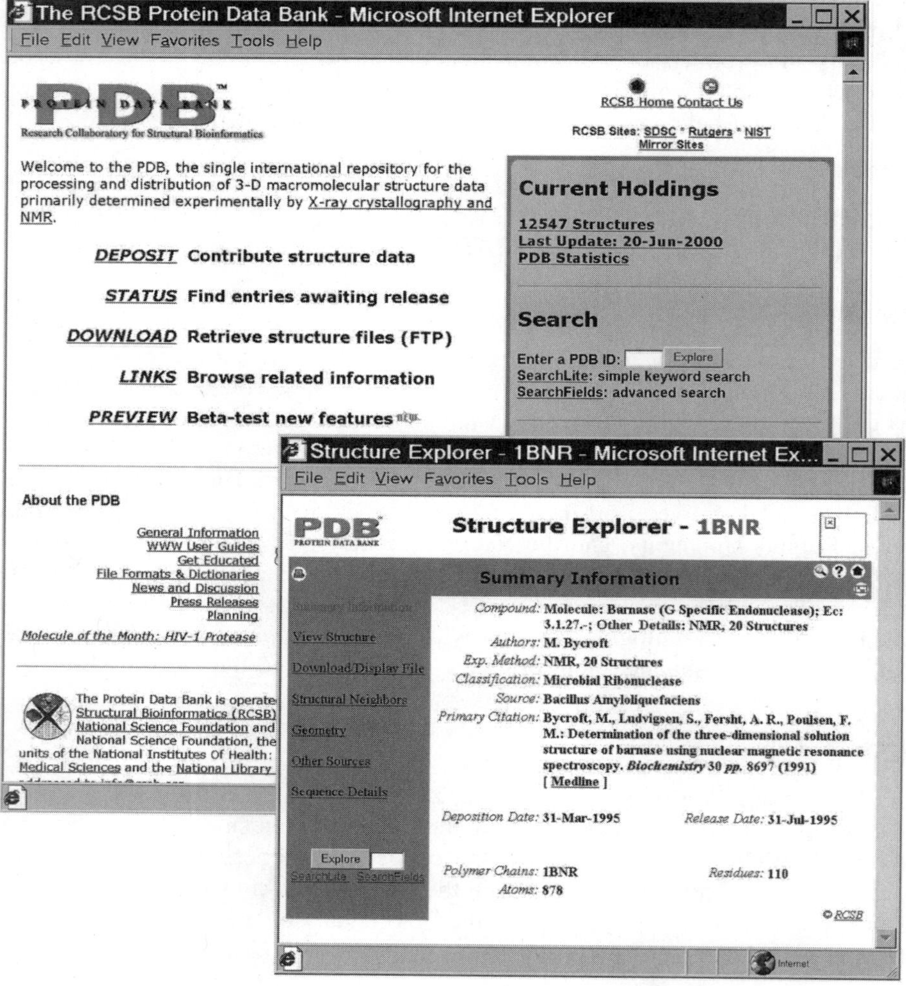

Figure 5.2. Structure query from RCSB with the structure 1BNR (Bycroft et al., 1991). The Structure Explorer can link the user to a variety of other pages with information about this structure including sequence, visualization tools, structure similarity (neighbors), and structure quality information, which are listed on subsequent Web pages.

service provides a data format check and can create automatic validation reports that provide diagnostics as to the quality of the structure, including bond distances and angles, torsion angles, nucleic acid comparisons, and crystal packing. Nucleic acid structures are accepted for deposition at NDB, the Nucleic Acids Database.

It has been the apparent working policy of PDB to reject three-dimensional structures that result from computational three-dimensional modeling procedures rather than from an actual physical experiment; submitting data to the PDB from a nonexperimental computational modeling exercise is strongly discouraged.

PDB-ID Codes. The structure record accessioning scheme of the Protein Data Bank is a unique four-character alphanumeric code, called a PDB-ID or PDB code. This scheme uses the digits 0 to 9 and the uppercase letters A to Z. This allows for over 1.3 million possible combinations and entries. Many older records have mnemonic names that make the structures easier to remember, such as 3INS, the record for insulin shown earlier. A different method is now being used to assign PDB-IDs, with the use of mnemonics apparently being abandoned.

Database Searching, PDB File Retrieval, mmCIF File Retrieval, and Links. PDB's search engine, the Structure Explorer, can be used to retrieve PDB records, as shown in Figure 5.2. The Structure Explorer is also the primary database of links to third-party annotation of PDB structure data. There are a number of links maintained in the Structure Explorer to Internet-based three-dimensional structure services on other Web sites. Figure 5.2 shows the Structure Summary for the protein barnase (1BNR; Bycroft et al., 1991). The Structure Explorer also provides links to special project databases maintained by researchers interested in related topics, such as structural evolution (FSSP; Holm and Sander, 1993), structure-structure similarity (DALI; Holm and Sander, 1996), and protein motions (Gerstein et al., 1994). Links to visualization tool-ready versions of the structure are provided, as well as authored two-dimensional images that can be very helpful to see how to orient a three-dimensional structure for best viewing of certain features such as binding sites.

Sequences from Structure Records

PDB file-encoded sequences are notoriously troublesome for programmers to work with. Because completeness of a structure is not always guaranteed, PDB records contain two copies of the sequence information: an *explicit sequence* and an *implicit sequence*. Both are required to reconstruct the chemical graph of a biopolymer.

Explicit sequences in a PDB file are provided in lines starting with the keyword SEQRES. Unlike other sequence databases, PDB records use the three-letter amino acid code, and nonstandard amino acids are found in many PDB record sequence entries with arbitrarily chosen three-letter names. Unfortunately, PDB records seem to lack sensible, consistent rules. In the past, some double-helical nucleic acid sequence entries in PDB were specified in a 3′-to-5′ order in an entry above the complementary strand, given in 5′-to-3′ order. Although the sequences may be obvious to a user as a representation of a double helix, the 3′-to-5′ explicit sequences are nonsense to a computer. Fortunately, the NDB project has fixed many of these types of problems, but the PDB data format is still open to ambiguity disasters from the standpoint of computer readability. As an aside, the most troubling glitch is the inability to encode element type separately from the atom name. Examples of where this becomes problematic include cases where atoms in structures having FAD or NAD cofactors are notorious for being interpreted as the wrong elements, such as neptunium (NP to Np), actinium (AC to Ac), and other nonsense elements.

Because three-dimensional structures can have multiple biopolymer chains, to specify a discrete sequence, the user must provide the *PDB chain identifier*. SEQRES entries in PDB files have a chain identifier, a single uppercase letter or blank space, identifying each individual biopolymer chain in an entry. For the structure 3INS shown in Figure 5.1, there are two insulin molecules in the record. The 3INS record contains sequences labeled A, B, C, and D. Knowledge of the biochemistry of insulin is required to understand that protein chains A and B are in fact derived from the same gene and that a posttranslational modification cuts the proinsulin sequence into the A and B chains observed in the PDB record. This information is not recorded in a three-dimensional structure record, nor in the sequence record for that matter. A place for such critical biological information is now being made within the BIND database (Bader and Hogue, 2000). The one-letter chain-naming scheme has difficulties with the enumeration of large oligomeric three-dimensional structures, such as viral capsids, as one quickly runs out of single-letter chain identifiers.

The *implicit* sequences in PDB records are contained in the embedded stereochemistry of the (x, y, z) data and names of each ATOM record in the PDB file. The implicit sequences are useful in resolving explicit sequence ambiguities such as the backward encoding of nucleic acid sequences or in verifying nonstandard amino acids. In practice, many PDB file viewers (such as RasMol) reconstruct the chemical graph of a protein in a PDB record using only the *implicit* sequence, ignoring the *explicit* SEQRES information. If this software then is asked to print the sequence of certain incomplete molecules, it will produce a nonphysiological and biologically irrelevant sequence. The implicit sequence, therefore, is not sufficient to reconstruct the complete chemical graph.

Consider an example in which the sequence `ELVISISALIVES` is represented in the SEQRES entry of a hypothetical PDB file, but the coordinate information is missing all (x, y, z) locations for the subsequence `ISA`. Software that reads the implicit sequence will often report the PDB sequence incorrectly from the chemical graph as `ELVISLIVES`. A test structure to determine whether software looks only at the implicit sequence is 3TS1 (Brick et al., 1989) as shown in the Java three-dimensional structure viewer WebMol in Figure 5.3. Here, both the implicit and explicit sequences in the PDB file to the last residue with coordinates are correctly displayed.

Validating PDB Sequences

To properly validate a sequence from a PDB record, one must first derive the *implicit* sequence in the ATOM records. This is a nontrivial processing step. If the structure has gaps because of lack of completeness, there may only be a set of *implicit sequence fragments* for a given chain. Each of these fragments must be aligned to the *explicit* sequence of the same chain provided within the SEQRES entry. This treatment will produce the complete chemical graph, including the parts of the biological sequence that may be missing coordinate data. This kind of validation is done on creation of records for the MMDB and mmCIF databases.

The best source of validated protein and nucleic acid sequences in single-letter code derived from PDB structure records is NCBI's MMDB service, which is part

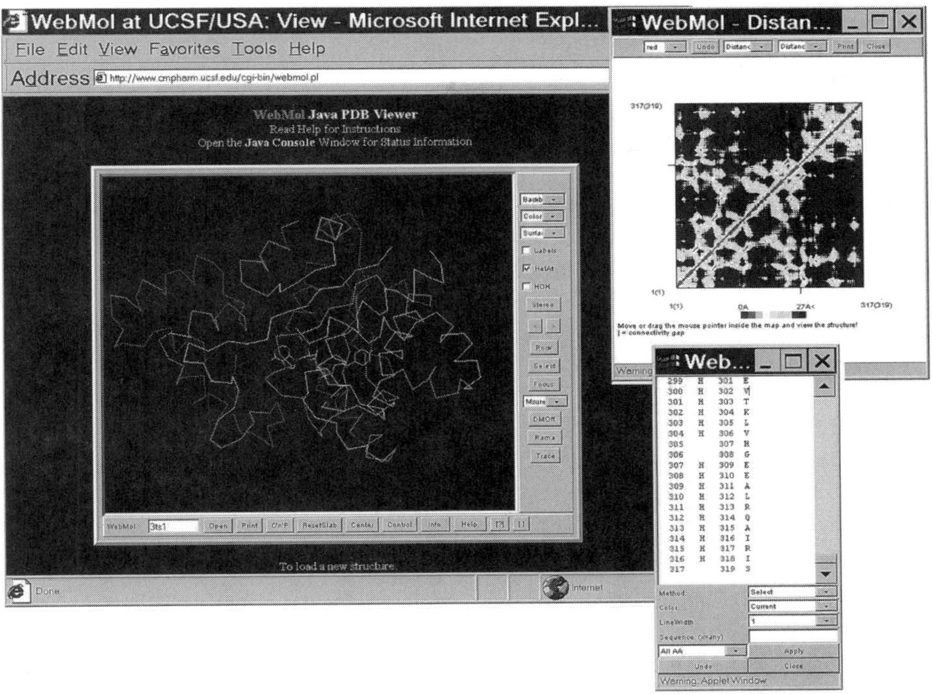

Figure 5.3. Testing a three-dimensional viewer for sequence numbering artifacts with the structure 3TS1 (Brick et al., 1989). WebMol, a Java applet, correctly indicates both the explicit and implicit sequences of the structure. Note the off-by-two difference in the numbering in the two columns of numbers in the inset window on the lower right. The actual sequence embedded in the PDB file is 419 residues long, but the COOH-terminal portion of the protein is lacking coordinates; it also has two missing residues. (See color plate.)

of the Entrez system. The sequence records from our insulin example have database accessions constructed systematically and can be retrieved from the protein sequence division of Entrez using the accessions `pdb|3INS|A`, `pdb|3INS|B`, `pdb|3INS|C`, and `pdb|3INS|D`. PDB files also have references in `db_xref` records to sequences in the SWISS-PROT protein database. Note that the SWISS-PROT sequences will not necessarily correspond to the structure, since the validation process described here is not carried out when these links are made! Also, note that many PDB files currently have ambiguously indicated taxonomy, reflecting the presence in some of three-dimensional structures of complexes of molecules that come from different species. The PDBeast project at NCBI has incorporated the correct taxonomic information for each biopolymer found within a given structure.

MMDB: MOLECULAR MODELING DATABASE AT NCBI

NCBI's Molecular Modeling Database (MMDB; Hogue et al., 1996) is an integral part of NCBI's Entrez information retrieval system (cf. Chapter 7). It is a compilation of all the Brookhaven Protein Data Bank (Bernstein et al., 1977) three-dimensional structures of biomolecules from crystallographic and NMR studies. MMDB records are in ASN.1 format (Rose, 1990) rather than in PDB format. Despite this, PDB-formatted files can also be obtained from MMDB. By representing the data in ASN.1 format, MMDB records have value-added information compared with the original PDB entries. Additional information includes explicit chemical graph information resulting from an extensive suite of validation procedures, the addition of uniformly derived secondary structure definitions, structure domain information, citation matching to MEDLINE, and the molecule-based assignment of taxonomy to each biologically derived protein or nucleic acid chain.

Free Text Query of Structure Records

The MMDB database can be searched from the NCBI home page using Entrez. (MMDB is also referred to as the NCBI Structure division.) Search fields in MMDB include PDB and MMDB ID codes, free text from the original PDB REMARK records, author name, and other bibliographic fields. For more specific, fielded queries, the RCSB site is recommended.

MMDB Structure Summary

MMDB's Web interface provides a Structure Summary page for each MMDB structure record, as shown in Figure 5.4. MMDB Structure Summary pages provide the FASTA-formatted sequences for each chain in the structure, links to MEDLINE references, links to the 3DBAtlas record and the Brookhaven PDB site, links to protein or nucleic acid sequence neighbors for each chain in the structure, and links to VAST structure-structure comparisons for each domain on each chain in the structure.

BLAST Against PDB Sequences: New Sequence Similarities

When a researcher wishes to find a structure related to a new sequence, NCBI's BLAST (Altschul et al., 1990) can be used because the BLAST databases contain a copy of all the validated sequences from MMDB. The BLAST Web interface can be used to perform the query by pasting a sequence in FASTA format into the sequence entry box and then selecting the "pdb" sequence database. This will yield a search against all the validated sequences in the current public structure database. More information on performing BLAST runs can be found in Chapter 8.

Entrez Neighboring: Known Sequence Similarities

If one is starting with a sequence that is already in Entrez, BLAST has, in essence, already been performed. Structures that are similar in sequence to a given protein sequence can be found by means of Entrez's neighboring facilities. Details on how to perform such searches are presented in Chapter 7.

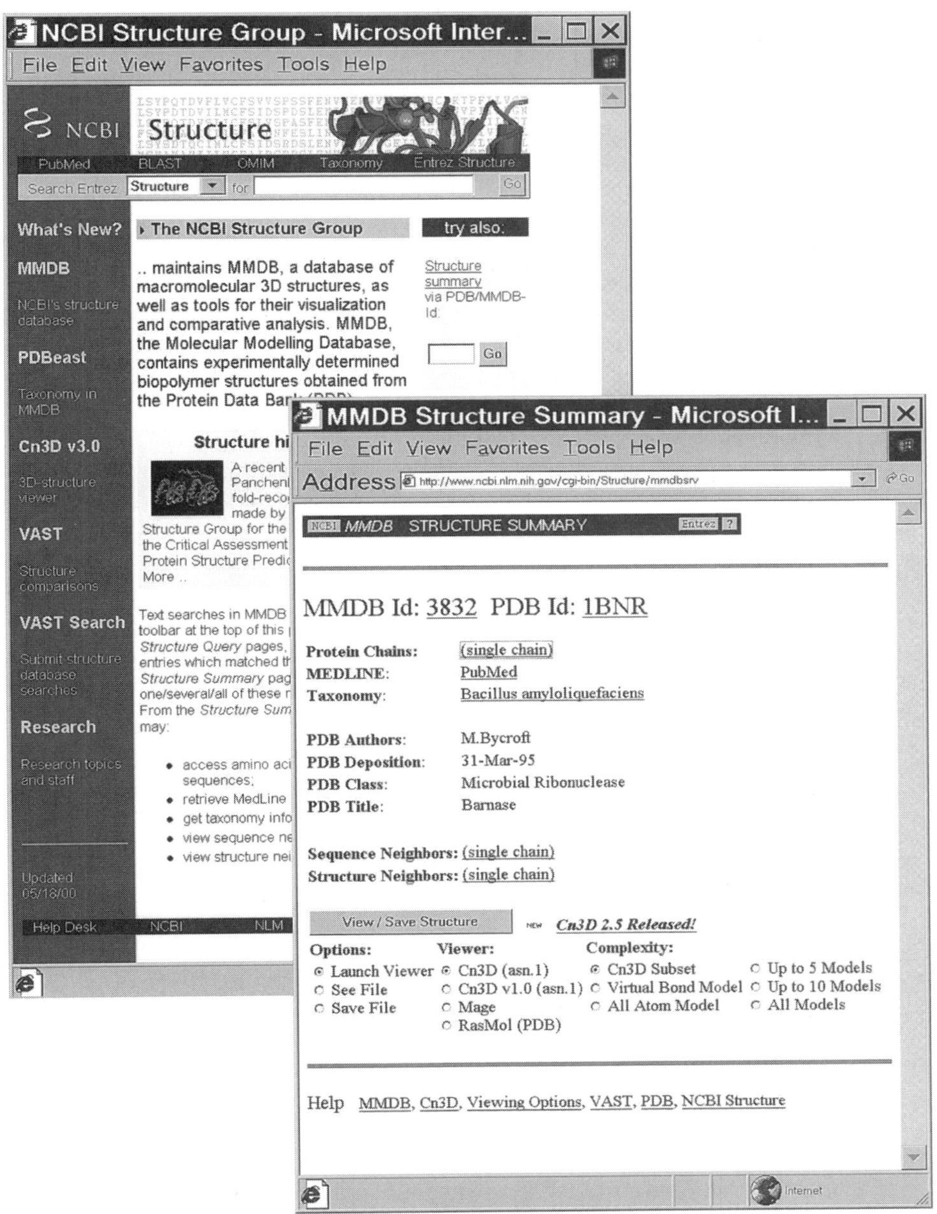

Figure 5.4. Structure query from NCBI with the structure 1BNR (Bycroft et al., 1991). The Structure Summary links the user to RCSB through the PDB ID link, as well as to validated sequence files for each biopolymer, sequence, and three-dimensional structure neighbors through the VAST system. This system is more efficient than the RCSB system (Fig. 5.2) for retrieval because visualization, sequence, and structure neighbor links are made directly on the structure summary page and do not require fetching more Web pages.

STRUCTURE FILE FORMATS

PDB

The PDB file format is column oriented, like that of the punched cards used by early FORTRAN programmers. The exact file format specification is available through the PDB Web site. Most software developed by structural scientists is written in FORTRAN, whereas the rest of the bioinformatics world has adopted other languages, such as those based on C. PDB files are often a paradox: they look rather easy to parse, but they have a few nasty surprises, as already alluded to in this chapter. To the uninitiated, the most obvious problem is that the information about biopolymer bonds is missing, obliging one to program in the rules of chemistry, clues to the identity of each atom given by the naming conventions of PDB, and robust exception handling. PDB parsing software often needs lists of synonyms and tables of exceptions to correctly interpret the information. However this chapter is not intended to be a manual of how to construct a PDB parser.

Two newer chemical-based formats have emerged: mmCIF (MacroMolecular Chemical Interchange Format) and MMDB (Molecular Modeling Database Format). Both of these file formats are attempts to modernize PDB information. Both start by using data description languages, which are consistently machine parsable. The data description languages use "tag value" pairs, which are like variable names and values used in a programming language. In both cases, the format specification is composed in a machine-readable form, and there is software that uses this format specification document to validate incoming streams of data. Both file formats are populated from PDB file data using the strategy of alignment-based reconstruction of the implicit ATOM and HETATM chemical graphs with the explicit SEQRES chemical graphs, together with extensive validation, which is recorded in the file. As a result, both of these file formats are superior for integrating with biomolecular sequence databases over PDB format data files, and their use in future software is encouraged.

mmCIF

The mmCIF (Bourne et al., 1995) file format was originally intended to be a biopolymer extension of the CIF (Chemical Interchange Format; Hall et al., 1991) familiar to small-molecule crystallographers and is based on a subset of the STAR syntax (Hall et al., 1991). CIF software for parsing and validating format specifications is not forward-compatible with mmCIF, since these have different implementations for the STAR syntax. The underlying data organization in an mmCIF record is a set of relational tables. The mmCIF project refers to their format specification as the mmCIF dictionary, kept on the Web at the Nucleic Acids Database site. The mmCIF dictionary is a large document containing specifications for holding the information stored in PDB files as well as many other data items derivable from the primary coordinate data, such as bond angles. The mmCIF data specification gives this data a consistent interface, which has been used to implement the NDB Protein Finder, a Web-based query format in a relational database style, and is also used as the basis for the new RCSB software systems.

Validating an incoming stream of data against the large mmCIF dictionary entails significant computational time; hence, mmCIF is probably destined to be an archival and advanced query format. Software libraries for reading mmCIF tables into relational tables into memory in FORTRAN and C are available.

MMDB

The MMDB file format is specified by means of the ASN.1 data description language (Rose, 1990), which is used in a variety of other settings, surprisingly enough including applications in telecommunications and automotive manufacturing. Because the US National Library of Medicine also uses ASN.1 data specifications for sequence and bibliographic information, the MMDB format borrows certain elements from other data specifications, such as the parts used in describing bibliographic references cited in the data record. ASN.1 files can appear as human-readable text files or as a variety of binary and packed binary files that can be decoded by any hardware platform. The MMDB standard residue dictionary is a lookup table of information about the chemical graphs of standard biopolymer residue types. The MMDB format specification is kept inside the NCBI toolkit distribution, but a browser is available over the Web for a quick look. The MMDB ASN.1 specification is much more compact and has fewer data items than the mmCIF dictionary, avoiding derivable data altogether.

In contrast to the relational table design of mmCIF, the MMDB data records are structured as hierarchical records. In terms of performance, ASN.1-formatted MMDB files provide for much faster input and output than do mmCIF or PDB records. Their nested hierarchy requires fewer validation steps at load time than the relational scheme in mmCIF or in the PDB file format; hence, ASN.1 files are ideal for three-dimensional structure database browsing.

A complete application programming interface is available for MMDB as part of the NCBI toolkit, containing a wide variety of C code libraries and applications. Both an ASN.1 input/output programming interface layer and a molecular computing layer (MMDB-API) are present in the NCBI toolkit. The NCBI toolkit supports x86 and alpha-based Windows' platforms, Macintosh 68K and PowerPC CPUs, and a wide variety of UNIX platforms. The three-dimensional structure database viewer (Cn3D) is an MMDB-API-based application with source code included in the NCBI toolkit.

VISUALIZING STRUCTURAL INFORMATION

Multiple Representation Styles

We often use multiple styles of graphical representation to see different aspects of molecular structure. Typical images of a protein structure are shown in Figure 5.5 (see also color plate). Here, the enzyme barnase 1BN1 (Buckle et al., 1993) appears both in wire-frame and space-filling model formats, as produced by RasMol (Sayle and Milner-White, 1995).

Because the protein structure record 1BN1 has three barnase molecules in the crystallographic unit, the PDB file has been hand-edited using a text editor to delete the superfluous chains. Editing data files is an accepted and widespread practice in three-dimensional molecular structure software, forcing the three-dimensional structure viewer to show what the user wants. In this case, the crystallographic data

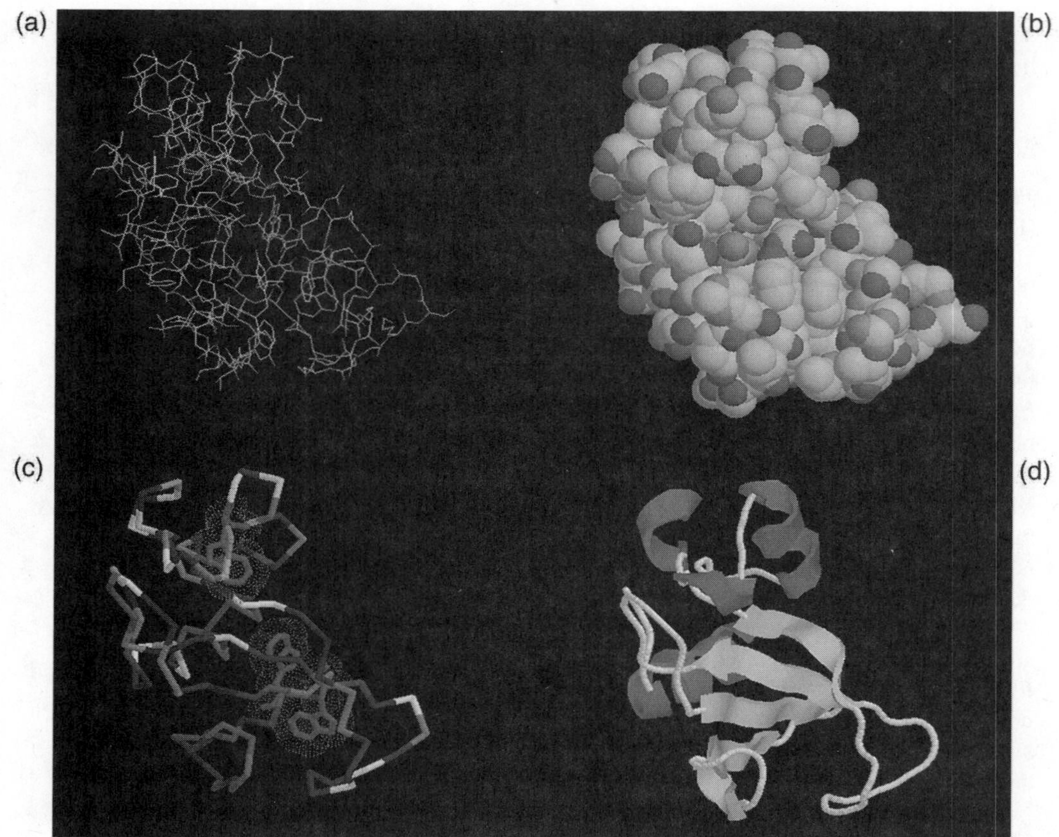

Figure 5.5. A constellation of viewing alternatives using RasMol with a portion of the barnase structure 1BN1 (Buckle et al., 1993). 1BN1 has three barnase molecules in the asymmetric unit. For this figure, the author edited the PDB file to remove two extra barnase molecules to make the images. Like most crystal structures, 1BN1 has no hydrogen locations. (a) Barnase in CPK coloring (element-based coloring) in a wire-frame representation. (b) Barnase in a space-filling representation. (c) Barnase in an α-carbon backbone representation, colored by residue type. The command line was used to select all the tryptophan residues, render them with "sticks," color them purple, and show a dot surface representation. (d) Barnase in a cartoon format showing secondary structure, α-helices in red; β-strands in yellow. Note that in all cases the default atom or residue coloring schemes used are at the discretion of the author of the software. (See color plate.)

recorded in the three-dimensional structure does not represent the functional *biological* unit. In our example, the molecule barnase is a monomer; however, we have three molecules in the crystallographic unit. In our other example, 3TS1 (Brick et al., 1989) (Fig. 5.3), the molecule is a dimer, but only one of the symmetric subunits is recorded in the PDB file.

The wire-frame image in Figure 5.5a clearly shows the chemistry of the barnase structure, and we can easily trace of the sequence of barnase on the image of its biopolymer in an interactive computer display. The space-filling model in Figure

5.5b gives a good indication of the size and surface of the biopolymer, yet it is difficult to follow the details of chemistry and bonding in this representation. The composite illustration in Figure 5.5c shows an α-carbon backbone in a typical pseudo-structure representation. The lines drawn are not actual chemical bonds, but they guide us along the path made by the α-carbons of the protein backbone. These are also called "virtual bonds." The purple tryptophan side chains have been selected and drawn together with a dot surface. This composite illustration highlights the volume taken up by the three tryptophan side chains in three hydrophobic core regions of barnase, while effectively hiding most of the structure's details.

The ribbon model in Figure 5.5d shows the organization of the structural path of the secondary structure elements of the protein chain (α-helix and β-sheet regions). This representation is very often used, with the arrowheads indicating the N-to-C-terminal direction of the secondary structure elements, and is most effective for identifying secondary structures within complex topologies.

The variety of information conveyed by the different views in Figure 5.5 illustrates the need to visualize three-dimensional biopolymer structure data in unique ways that are not common to other three-dimensional graphics applications. This requirement often precludes the effective use of software from the "macroscopic world," such as computer-aided design (CAD) or virtual reality modeling language (VRML) packages.

Picture the Data: Populations, Degeneracy, and Dynamics

Both X-ray and NMR techniques infer three-dimensional structure from a *synchronized* population of molecules—synchronized in space as an ordered crystal lattice or synchronized in behavior as nuclear spin states are organized by an external magnetic field. In both cases, information is gathered from the population as a whole. The coordinate (x, y, z) locations of atoms in a structure are derived using numerical methods. These fit the expected chemical graph of the sample into the three-dimensional data derived from the experimental data. The expected chemical graph can include a mixture of biopolymer sequence-derived information as well as the chemical graph of any other known small molecules present in the sample, such as substrates, prosthetic groups, and ions.

One somewhat unexpected result of the use of molecular populations is the assignment of degenerate coordinates in a database record, i.e., more than one coordinate location for a single atom in the chemical graph. This is recorded when the population of molecules has observable conformational heterogeneity.

NMR Models and Ensembles

Figure 5.6 (see also color plate) presents four three-dimensional structures (images on the left were determined using X-ray crystallography and the right using NMR). The NMR structures on the left appear "fuzzy." In fact, there are several different, complete structures piled one on top of another in these images. Each structure is referred to as a *model*, and the set of models is an *ensemble*. Each model in the ensemble is a chirally correct, plausible structure that fits the underlying NMR data as well as any other model in the ensemble.

The images from the ensemble of an NMR structure (Fig. 5.6, b and d) show the dynamic variation of a molecule in solution. This reflects the conditions of the

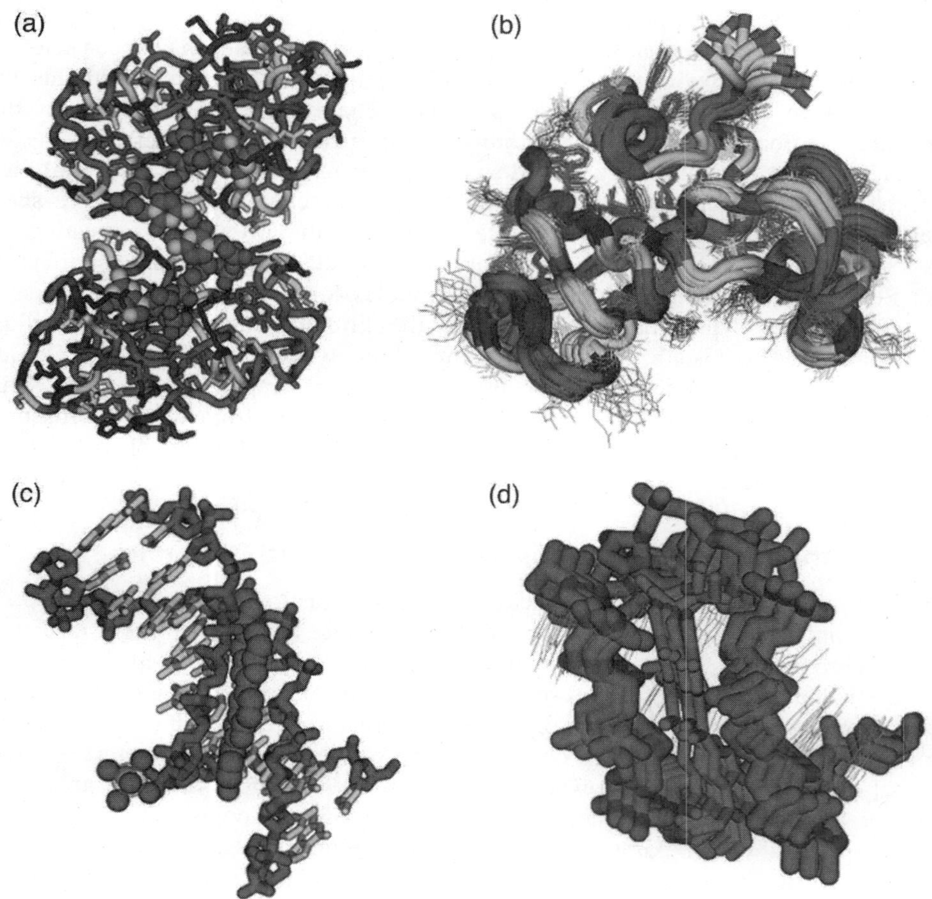

Figure 5.6. A comparison of three-dimensional structure data obtained by crystallography (left) and NMR methods (right), as seen in Cn3D. (a) The crystal structure 1BRN (Buckle and Fersht, 1994) has two barnase molecules in the asymmetric unit, although these are not dimers in solution. The image is rendered with an α-carbon backbone trace colored by secondary structure (green helices and yellow sheets), and the amino acid residues are shown with a wire-frame rendering, colored by residue type. (b) The NMR structure 1BNR (Bycroft et al., 1991) showing barnase in solution. Here, there are 20 different models in the ensemble of structures. The coloring and rendering are exactly as the crystal structure to its left. (c) The crystal structure 109D (Quintana et al., 1991) showing a complex between a minor-groove binding bis-benzimidazole drug and a DNA fragment. Note the phosphate ion in the lower left corner. (d) The NMR structure 107D showing four models of a complex between a different minor-groove binding compound (Duocarmycin A) and a different DNA fragment. It appears that the three-dimensional superposition of these ensembles is incorrectly shifted along the axis of the DNA, an error in PDB's processing of this particular file. (See color plate.)

experiment: molecules free in solution with freedom to pursue dynamic conformational changes. In contrast, the X-ray structures (Fig. 5.6, a and c) provide a very strong mental image of a static molecule. This also reflects the conditions of the experiment, an ordered crystal constrained in its freedom to explore its conformational dynamics. These mental images direct our interpretation of structure. If we measure distance between two atoms using an X-ray structure, we may get a single value. However, we can get a range of values for the same distance in each model looking at an ensemble of an NMR structure. Clearly, our interpretation of this distance can be dependent on the source of the three-dimensional structure data! It is prudent to steer clear of any software that ignores or fails to show the population degeneracy present in structure database records, since the absence of such information can further skew *biological* interpretations. Measuring the distance between two atoms in an NMR structure using software that hides the other members of the ensemble will give only one value and not the true range of distance observed by the experimentalist.

Correlated Disorder

Typically, X-ray structures have one and only one model. Some subsets of atoms, however, may have degenerate coordinates, which we will refer to as *correlated disorder* (Fig. 5.7a; see also color plate). Many X-ray structure database records show correlated disorder. Both correlated disorder and ensembles are often ignored by three-dimensional molecular graphics software. Some programs show only the first model in an ensemble, or the first location of each atom in a correlated disorder set, ignoring the rest of the degenerate coordinate values. Worse still, sometimes, erroneous bonds are drawn between the degenerate locations, making a mess of the structure, as seen in Figure 5.7b.

Local Dynamics

A single technique can be used to constrain the conformation of some atoms differently from others in the same structure. For example, an internal atom or a backbone atom that is locked in by a multitude of interactions may appear largely invariant in NMR or X-ray data, whereas an atom on the surface of the molecule may have much more conformational freedom (consider the size of the smears of different residues in Fig. 5.6b). Interior protein side chains typically show much less flexibility in ensembles, so it might be concluded that the interiors of proteins lack conformational dynamics altogether. However, a more sensitive, biophysical method, time-resolved fluorescence spectroscopy of single tryptophan residues, has a unique ability to detect heterogeneity (but not the actual coordinates) of the tryptophan side-chain conformation. Years of study using this method has shown that, time and time again, populations of interior tryptophans in pure proteins are more often in heterogeneous conformations than not (Beechem and Brand, 1985). This method was shown to be able to detect rotamers of tryptophan within single crystals of erabutoxin, where X-ray crystallography could not (Dahms and Szabo, 1995). When interpreting three-dimensional structure data, remember that heterogeneity does persist in the data, and that the NMR and X-ray methods can be blind to all but the most populated conformations in the sample.

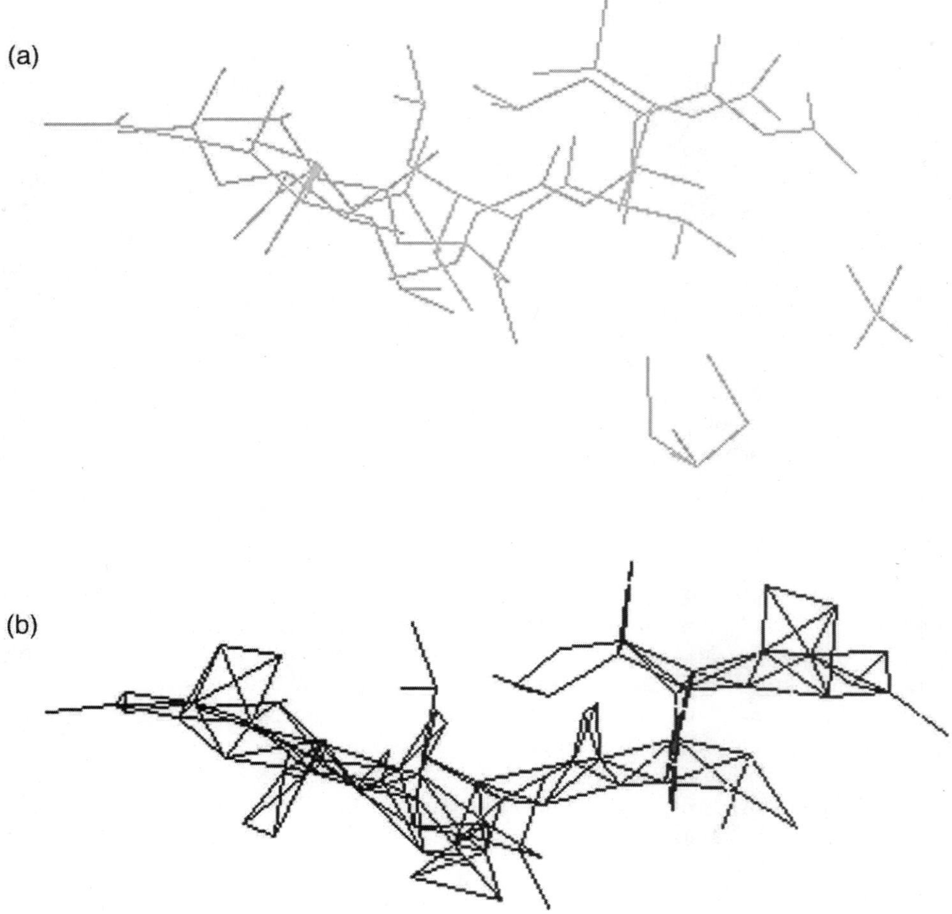

Figure 5.7. An example of crystallographic correlated disorder encoded in PDB files. This is chain C of the HIV protease structure 5HVP (Fitzgerald et al., 1990). This chain is in asymmetric binding site and can orient itself in two different directions. Therefore, it has a single chemical graph, but each atom can be in one of two different locations. (a) The correct bonding is shown with an MMDB-generated Kinemage file; magenta and red are the correlated disorder ensembles as originally recorded by the depositor, bonding calculated using standard-residue dictionary matching. (b) Bonding of the same chain in RasMol, wherein the disorder ensemble information is ignored, and all coordinates are displayed and all possible bonds are bonded together. (See color plate.)

DATABASE STRUCTURE VIEWERS

In the past several years, the software used to examine and display structure information has been greatly improved in terms of the quality of visualization and, more importantly, in terms of being able to relate sequence information to structure information.

Visualization Tools

Although the RCSB Web site provides a Java-based three-dimensional applet for visualizing PDB data, the applet does not currently support the display of nonprotein structures. For this and other reasons, the use of RasMol v2.7 is instead recommended for viewing structural data downloaded from RCSB; more information on RasMol appears in the following section. If a Java-based viewer is preferred, WebMol is recommended, and an example of WebMol output is shown in Figure 5.3. With the advent of many homemade visualization programs that can easily be downloaded from the Internet, the reader is strongly cautioned to *only* use mature, well-established visualization tools that have been thoroughly tested and have undergone critical peer review.

RasMol and RasMol-Based Viewers

As mentioned above, several viewers for examining PDB files are available (Sanchez-Ferrer et al., 1995). The most popular one is RasMol (Sayle and Milner-White, 1995). RasMol represents a breakthrough in software-driven three-dimensional graphics, and its source code is a recommended study material for anyone interested in high-performance three-dimensional graphics. RasMol treats PDB data with extreme caution and often recomputes information, making up for inconsistencies in the underlying database. It does not try to validate the chemical graph of sequences or structures encoded in PDB files. RasMol does not perform internally either dictionary-based standard residue validations or alignment of explicit and implicit sequences. RasMol 2.7.1 contains significant improvements that allow one to display information in correlated disorder ensembles and select different NMR models. It also is capable of reading mmCIF-formatted three-dimensional structure files and is thus the viewer of choice for such data. Other data elements encoded in PDB files, such as disulfide bonds, are recomputed based on rules of chemistry, rather than validated.

RasMol contains many excellent output formats and can be used with the Molscript program (Kraulis, 1991) to make wonderful PostScript™ ribbon diagrams for publication. To make optimal use of RasMol, however, one must master its command-line language, a familiar feature of many legacy three-dimensional structure programs.

Several new programs are becoming available and are free for academic users. Based on RasMol's software-driven three-dimensional-rendering algorithms and sparse PDB parser, these programs include Chime™, a Netscape™ plug-in. Another program, WebMol, is a Java-based three-dimensional structure viewer apparently based on RasMol-style rendering, as seen in Figure 5.3.

MMDB Viewer: Cn3D

Cn3D (for "see in 3-D") is a three-dimensional structure viewer used for viewing MMDB data records. Because the chemical graph ambiguities in data in PDB entries have been removed to make MMDB data records and because all the bonding information is explicit, Cn3D has the luxury of being able to display three-dimensional database structures consistently, without the parsing, validation, and exception-handling overhead required of programs that read PDB files. Cn3D's default image of a structure is more intelligently displayed because it works without fear of misrepresenting the data. However, Cn3D is dependent on the complete chemical graph information in the ASN.1 records of MMDB, and, as such, it does not read in PDB files.

Cn3D 3.0 has a much richer feature set than its predecessors, and it now allows selection of subsets of molecular structure and independent settings of rendering and coloring aspects of that feature. It has state-saving capabilities, making it possible to color and render a structure, and then save the information right into the ASN.1 structure record, a departure from the hand-editing of PDB files or writing scripts. This information can be shared with other Cn3D users on different platforms.

The images shown in Figures 5.1 and 5.6 are from Cn3D 3.0, now based on OpenGL three-dimensional graphics. This provides graphics for publication-quality images that are much better than previous versions, but the original Viewer3D version of Cn3D 3.0 is available for computers that are not capable of displaying OpenGL or that are too slow.

Also unique to Cn3D is a capacity to animate three-dimensional structures. Cn3D's animation controls resemble tape recorder controls and are used for displaying quickly the members of a multiple structure ensemble one after the other, making an animated three-dimensional movie. The GO button makes the images animated, and the user can rotate or zoom the structure while it is playing the animation. This is particularly useful for looking at NMR ensembles or a series of time steps of structures undergoing motions or protein folding. The animation feature also allows Cn3D to provide superior multiple structure alignment when used together with the VAST structure-structure comparison system, described later in this chapter.

Other 3D Viewers: Mage, CAD, and VRML

A variety of file formats have been used to present three-dimensional biomolecular structure data lacking in chemistry-specific data representations. These are viewed in generic three-dimensional data viewers such as those used for "macroscopic" data, like engineering software or virtual-reality browsers. File formats such as VRML contain three-dimensional graphical display information but little or no information about the underlying chemical graph of a molecule. Furthermore, it is difficult to encode the variety of rendering styles in such a file; one needs a separate VRML file for a space-filling model of a molecule, a wire-frame model, a ball-and-stick model, and so on, because each explicit list of graphics objects (cylinders, lines, spheres) must be contained in the file.

Biomolecular three-dimensional structure database records are currently not compatible with "macroscopic" software tools such as those based on CAD software. Computer-aided design software represents a mature, robust technology, generally superior to the available molecular structure software. However, CAD software and file formats in general are ill-suited to examine the molecular world, owing to the lack of certain "specialty" views and analytical functions built in for the examination of details of protein structures.

Making Presentation Graphics

To get the best possible publication-quality picture out of any molecular graphics software, first consider whether a bitmap or a vector-based graphic image is needed. Bitmaps are made by programs like RasMol and Cn3D—they reproduce exactly what you see on the screen, and are usually the source of trouble in terms of pixellation ("the jaggies"), as shown in Figure 5.7, a bitmap of 380–400 pixels. High-quality print resolution is usually at 300–600 dots per inch, but monitors have far less information in pixels per inch (normally 72 dpi), so a big image on a screen is quite tiny when printed at the same resolution on a printer. Expanding the image to fit a page causes exaggeration of pixel steps on diagonal lines.

The best advice for bitmaps is to use as big a monitor/desktop as possible, maximizing the number of pixels included in the image. This may mean borrowing a colleagues' 21-in monitor or using a graphics card that offers a "virtual desktop" that is larger than the monitor being used in pixel count. In any case, always fill the entire screen with the viewer window before saving a bitmap image for publication.

ADVANCED STRUCTURE MODELING

Tools that go beyond simple visualization are now emerging and are freely available. Biologists often want to display structures with information about charge distribution, surface accessibility, and molecular shape; they also want to be able to perform simple mutagenesis experiments and more complex structure modeling. SwissPDB Viewer, shown in Figure 5.8, also known as Deep View, is provided free of charge to academics and can address a good number of these needs. It is a multi platform (Mac, Win, and Linux) OpenGL-based tool that has the ability to generate molecular surfaces, align multiple proteins, use scoring functions, as well as do simple, fast modeling, including site-directed mutagenesis and more complex modeling such as loop rebuilding. An excellent tutorial for SwissPDB Viewer developed by Gale Rhodes is one of the best starting points for making the best use of this tool. It has the capability to dump formatted files for the free ray-tracing software POV-Ray, and it can be used to make stunning images of molecular structures, easily suitable for a journal cover.

STRUCTURE SIMILARITY SEARCHING

Although a sequence-sequence similarity program provides an alignment of two sequences, a structure-structure similarity program provides a three-dimensional structural superposition. This superposition results from a set of three-dimensional rotation-translation matrix operations that superimpose similar parts of the structure. A conventional sequence alignment can be derived from three-dimensional superposition by finding the α-carbons in the protein backbone that are superimposed in space. Structure similarity search services are based on the premise that some similarity metric can be computed between two structures and used to assess their similarity, much in the same way a BLAST alignment is scored. A structure similarity search service can take a three-dimensional protein structure, either already in PDB or a new one, and compare that structure, making three-dimensional superpositions with other structures in the database and reporting the best match without knowing anything about the sequence. If a match is made between two structures that are not related by any measurable sequence similarity, it is indeed a surprising discovery. For this type of data to be useful, the similarity metric *must* be meaningful. A large fraction of structures, for example, have β-sheets. Although a similar substructure may include a single β-hairpin turn with two strands, one can find an incredibly

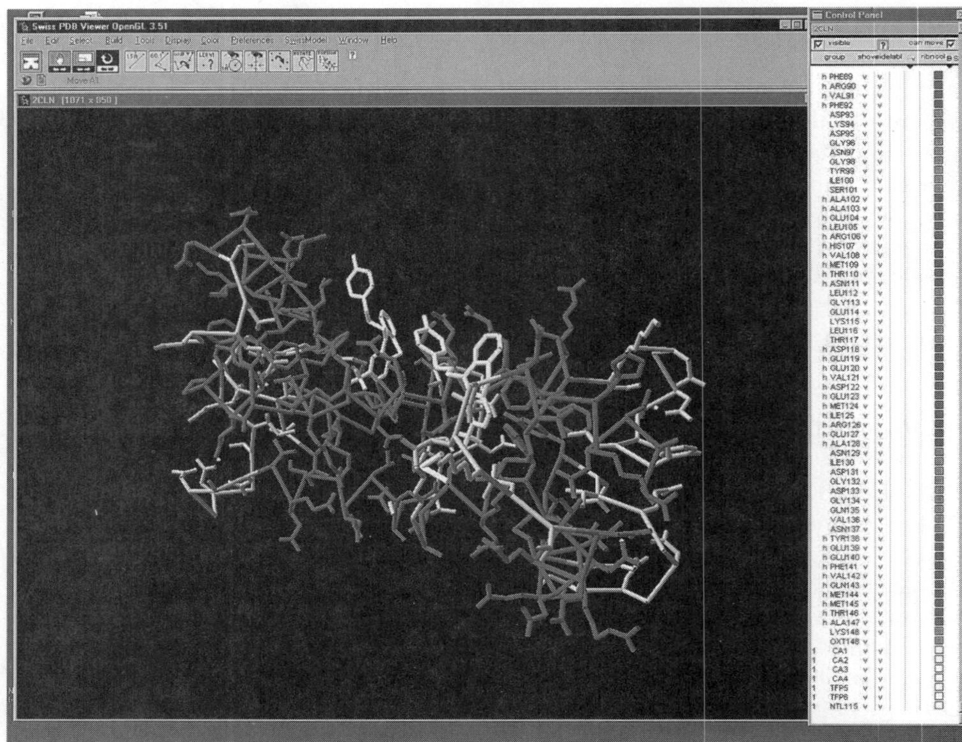

Figure 5.8. SwissPDB Viewer 3.51 with OpenGL, showing the calmodulin structure 2CLN. The binding of the inhibitor TFP is shown in yellow. The side panel allows great control over the rendering of the structure image, and menus provide a wealth of options and tools for structure superposition and modeling including mutagenesis and loop modeling, making it a complete structure modeling and analysis package. (See color plate.)

large number of such similarities in the PDB database, so these similarities are simply not surprising or informative. A number of structure similarity searching systems are now available on the Internet, and almost all of them can be found following links from the RCSB Structure Summary page. The process of similarity searching presents some interesting high-performance computational challenges, and this is addressed in different ways, ranging from human curation, as the SCOP system provides, to fully automated systems, such as DALI, SCOP, or the CE system provided by RCSB.

The Vector Alignment Search Tool (VAST; Gibrat et al., 1996) provides a similarity measure of three-dimensional structure. It uses vectors derived from secondary structure elements, with no sequence information being used in the search. VAST is capable of finding structural similarities when no sequence similarity is detected. VAST, like BLAST, is run on all entries in the database in an $N \times N$ manner, and the results are stored for fast retrieval using the Entrez interface. More than 20,000 domain substructures within the current three-dimensional structure database have been compared with one another using the VAST algorithm, the structure-structure (Fig. 5.9) superpositions recorded, and alignments of sequence derived from the

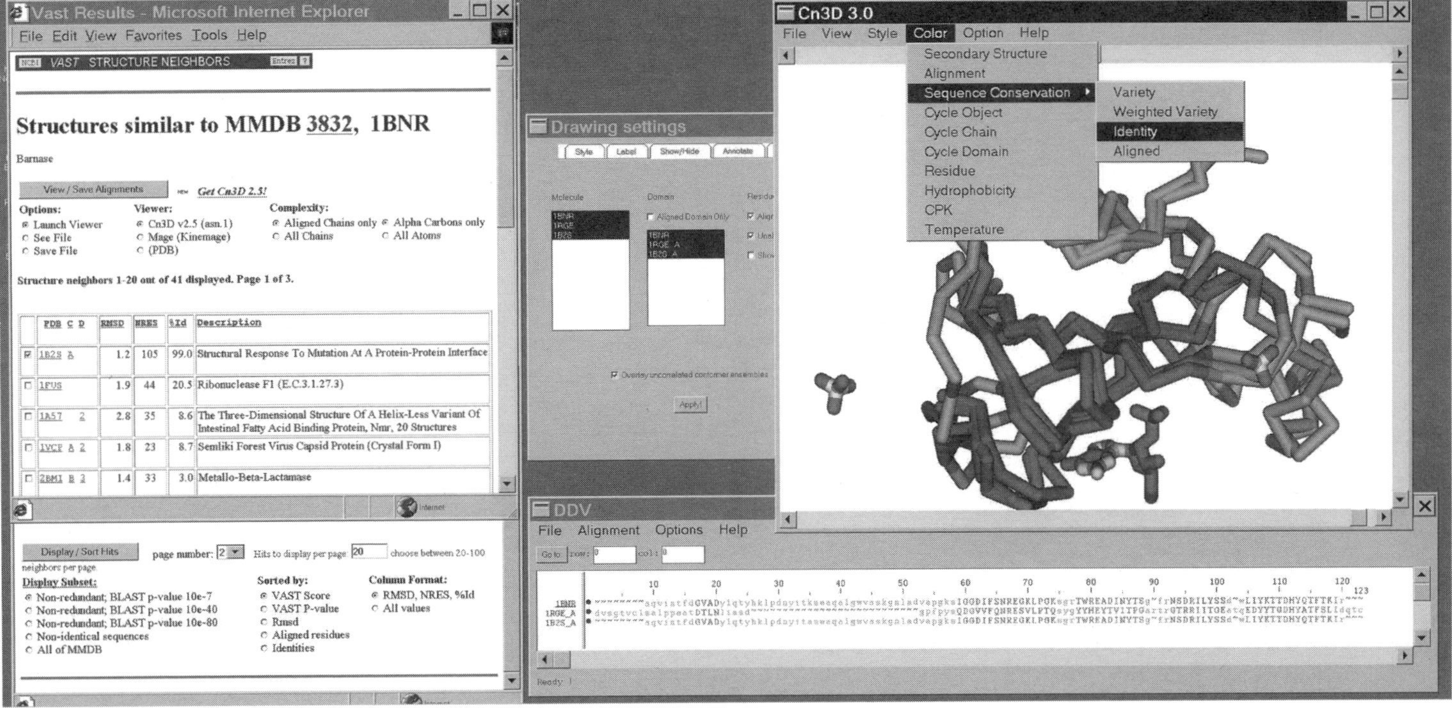

Figure 5.9. VAST structure neighbors of barnase. On the left is the query window obtained by clicking on the Structure Neighbors link from Figure 5.4. Structures to superposition are selected with the check boxes on the left, and Cn3D is launched from the top of the Web page. At the bottom left, controls that change the query are shown from the bottom of the VAST results page. The results shown here are selected as examples from a nonredundant set based on a BLAST probability of 10^{-7}, for the most concise display of hits that are not closely related to one another by sequence. The list may be sorted by a number of parameters, including RMSD from the query structure, number of identical residues, and the raw VAST score. More values can be displayed in the list as well. Cn3D is shown on the right, launched from the Web page with the structures 1RGE and 1B2S. Menu options show how Cn3D can highlight residues in the superposition (top right) and in the alignment (bottom right). The Cn3D drawing settings are shown in the top middle, where one can toggle structures on or off in the superposition window. (See color plate.)

superposition. The VAST algorithm focuses on similarities that are surprising in the *statistical* sense. One does not waste time examining many similarities of small substructures that occur by chance in protein structure comparison. For example, very many small segments of β-sheets have obvious, but not surprising, similarities. The similarities detected by VAST are often examples of remote homology, undetectable by sequence comparison. As such, they may provide a broader view of the structure, function, and evolution of a protein family.

The VAST system stands out amongst these comparative tools because (a) it has a clearly defined similarity metric leading to surprising relationships, (b) it has an adjustable interface that shows nonredundant hits for a quick first look at the most interesting relationships, without seeing the same relationships lots of times, (c) it provides a domain-based structure comparison rather than a whole protein comparison, and (d) it has the capability to integrate with Cn3D as a visualization tool for inspecting surprising structure relationships in detail. The interface between a VAST hit list search and the Cn3D structure superposition interface can be seen in Figure 5.9. In addition to a listing of similar structures, VAST-derived structure neighbors contain detailed residue-by-residue alignments and three-dimensional transformation matrices for structural superposition. In practice, refined alignments from VAST appear conservative, choosing a highly similar "core" substructure compared with DALI (Holm and Sander, 1996) superpositions. With the VAST superposition, one easily identifies regions in which protein evolution has modified the structure, whereas DALI superpositions may be more useful for comparisons involved in making structural models. Both VAST and DALI superpositions are excellent tools for investigating relationships in protein structure, especially when used together with the SCOP (Murzin et al., 1995) database of protein families.

INTERNET RESOURCES FOR TOPICS PRESENTED IN CHAPTER 5

BIND	*http://bioinfo.mshri.on.ca*
Imagemagick	*http://http://www.wizards.dupont.com/cristy/ImageMagick.html*
mmCif Project	*http://ndbserver.rutgers.edu/NDB/mmcif/index.html*
National Center for Biotechnology Information (NCBI)	*http://www.ncbi.nlm.nih.gov/*
NCBI Toolkit	*http://www.ncbi.nlm.nih.gov/Toolbox*
Nucleic Acids Database (NDB)	*http://ndbserver.rutgers.edu/*
POV-RayAY	*http://www.povray.org/*
Protein Data Bank at RCSB	*http://www.rcsb.org/*
RasMol	*http://www.bernstein-plus-sons.com/*
SwissPDB Viewer/Deep View	*http://http://www.expasy.ch/spdbv/mainpage.html*
WebMol	*http://www.cmpharm.ucsf.edu/~walther/webmol/*

PROBLEM SET

1. Calmodulin is a calcium-dependent protein that modulates other protein activity via protein interactions. The overall structure of calmodulin is variable, and is modulated by calcium. An NMR structure of calmodulin is found in PDB record 2BBN, complexed with a peptide. How many models are in this structure? Find other calmodulin structures from the PDB site, and inspect them using RasMol. How many "gross," unique conformations can this protein be found in? Where, in terms of secondary structure, is the site of the largest structural change?

2. The black beetle virus coat protein (pdb|2BBV) forms a very interesting triangular shape. Using VAST, examine the list of neighbors to the A chain. Examine some of the pairwise alignments in Cn3D. What is the extent of the similarity? What does this list of neighbors and the structure similarity shared by these proteins suggest about the origin and evolution of eukaryotic viruses?

3. Compare substrate binding in Rossman fold structures, between the tyrosinyl-5′-adenylate of tyrosyl-tRNA synthetase and the NADH of malate dehydrogenase. Describe the similarities and differences between the two substrates. Do you think these are homologous structures or are they related by convergent evolution?

4. Ribosomal protein synthesis or enzyme-based peptide synthesis—which came first?

   Repeat the analysis you did for question 3, examining the difference between substrates bound to tyrosyl-tRNA synthetase and D-Ala:D-Ala ligase (pdb|1IOV). Note the substrate is bound to domains 2 and 3 of 1IOV, but domain 1 is aligned with 3TS1. What does the superposition suggest about the activity of domain 1 of 1IOV? According to VAST, what is similar to domain 2 of 1IOV? How do you think D-Ala:D-Ala ligase arose in evolution? Speculate on whether enzyme-catalyzed protein synthesis such as that seen in 1IOV arose before or after ribosomal protein synthesis.

REFERENCES

Ahmed, F. R., Przybylska, M., Rose, D. R., Birnbaum, G. I., Pippy, M. E., and MacManus, J. P. (1990). Structure of oncomodulin refined at 1.85 angstroms resolution. An example of extensive molecular aggregation via Ca^2. *J. Mol. Biol.* 216, 127–140.

Altschul, S. F., Gish, W., Miller, W., Myers, E. W., and Lipman, D. J. (1990). Basic local alignment search tool. *J. Mol. Biol.* 215, 403–410.

Bader, G. D., and Hogue, C. W. V. (2000). BIND, a data specification for storing and describing biomolecular interactions, molecular complexes and pathways. *Bioinformatics* 16, 465–477.

Beechem, J. M., and Brand, L. (1985). Time-resolved fluorescence of proteins. *Annu. Rev. Biochem.* 54, 43–71.

Bernstein, F. C., Koetzle, T. F., Williams, G. J. B., Meyer, E. F., Jr., Brice, M. D., Rodgers, J. R., Kennard, O., Shimanouchi, T., and Tasumi, M. (1977). The Protein Data Bank. *J. Mol. Biol.* 112, 535–542.

Bourne, P. E., Berman, H. M., McMahon, B., Watenpaugh, K. D., Westbrook, J., and Fitzgerald, P. M. D. (1995). The macromolecular crystallographic information file (mmCIF). *Methods Enzymol.* 277.

Branden, C., and Tooze, J. (1999). *Introduction to Protein Structure* (New York: Garland).

Brick, P., Bhat, T. N., and Blow, D. M. (1989). Structure of tyrosyl-tRNA synthetase refined at 2.3 angstroms resolution. Interaction of the enzyme with the tyrosyl adenylate intermediate. *J. Mol. Biol.* 208, 83–98.

Buckle, A. M., and Fersht, A. R. (1994). Subsite binding in an RNase: Structure of a barnase-tetranucleotide complex at 1.76-angstroms resolution. *Biochemistry* 33, 1644–1653.

Buckle, A. M., Henrick, K., and Fersht, A. R. (1993). Crystal structural analysis of mutations in the hydrophobic cores of barnase. *J. Mol. Biol.* 23, 847–860.

Bycroft, M., Ludvigsen, S., Fersht, A. R., and Poulsen, F. M. (1991). Determination of the three-dimensional solution structure of barnase using nuclear magnetic resonance spectroscopy. *Biochemistry* 30, 8697–8701.

Dahms, T., and Szabo, A. G. (1995). Conformational heterogeneity of tryptophan in a protein crystal. *J. Am. Chem. Soc.* 117, 2321–2326.

Fitzgerald, P. M., McKeever, B. M., Van Middlesworth, J. F., Springer, J. P., Heimbach, J. C., Leu, C. T., Herber, W. K., Dixon, R. A., and Darke, P. L. (1990). Crystallographic analysis of a complex between human immunodeficiency virus type 1 protease and acetyl-pepstatin at 2.0-angstroms resolution. *J. Biol. Chem.* 265, 14209–14219.

Gerstein, M., Lesk, A., and Chothia, C. (1994). Structural mechanisms for domain movements in proteins. *Biochemistry* 33, 6739–6749.

Gibrat, J. F., Madej, T., and Bryant, S. H. (1996). Surprising similarities in structure comparison. *Curr. Opin. Struct. Biol.* 6, 377–385.

Hall, S. R. (1991). The STAR file: A new format for electronic data transfer and archiving. *J. Chem. Inf. Comput. Sci.* 31, 326–333.

Hall, S. R., Allen, A. H., and Brown, I. D. (1991). The crystallographic information file (CIF): A new standard archive file for crystallography. *Acta Crystallogr. Sect. A* 47, 655–685.

Hogue, C. W. V., Ohkawa, H., and Bryant, S. H. (1996). A dynamic look at structures: WWW-Entrez and the Molecular Modeling Database. *Trends Biochem. Sci.* 21, 226–229.

Hogue, C. W. V (1997) Cn3D: a new generation of three-dimensional molecular structure viewer. *Trends Biochem. Sci.* 22, 314–316.

Holm, L., and Sander, C. (1993). Protein structure comparison by alignment of distance matrices. *J. Mol. Biol.* 233, 123–138.

Holm, L., and Sander, C. (1996). The FSSP database: Fold classification based on structure-structure alignment of proteins. *Nucl. Acids Res.* 24, 206–210.

Issacs, N. W., and Agarwal, R. C. (1978). Experience with fast Fourier least squares in the refinement of the crystal structure of rhombohedral 2 zinc insulin at 1.5 angstroms resolution. *Acta Crystallogr. Sect. A* 34, 782.

Kraulis, P. J. (1991). MOLSCRIPT: A program to produce both detailed and schematic plots of protein structures. *J. Appl. Crystallogr.* 24, 946–950.

Murzin, A. G., Brenner, S. E., Hubbard, T., and Chothia, C. (1995). SCOP: A structural classification of proteins database for the investigation of sequences and structures. *J. Mol. Biol.* 24, 536–540.

Quintana, J. R., Lipanov, A. A., and Dickerson, R. E. (1991). Low-temperature crystallographic analyses of the binding of Hoechst 33258 to the double-helical DNA dodecamer C-G-C-G-A-A-T-T-C-G-C-G. *Biochemistry* 30, 10294–10306.

Richardson, D. C., and Richardson, J. S. (1992). The Kinemage: A tool for scientific communication. *Protein Sci.* 1, 3–9.

Richardson, D. C., and Richardson, J. S. (1994). KinemagesSimple macromolecular graphics for interactive teaching and publication. *Trends Biochem. Sci.* 19, 135–138.

Rose, M. T. (1990). *The Open Book, A Practical Perspective on OSI* (Englewood Cliffs, NJ: Prentice-Hall), p. 227–322.

Sanchez-Ferrer, A., Nunez-Delicado, E., and Bru, R. (1995). Software for viewing biomolecules in three dimensions on the Internet. *Trends Biochem. Sci.* 20, 286–288.

Sayle, R. A., and Milner-White, E. J. (1995). RASMOL: Biomolecular graphics for all. *Trends Biochem. Sci.* 20, 374–376.

Schuler, G. D., Epstein, J. A., Ohkawa, H., and Kans, J. A. (1996). Entrez: Molecular biology database and retrieval system. *Methods Enzymol.* 266, 141–162.

Walther D. (1997) WebMol—a Java based PDB viewer. *Trends Biochem. Sci.* 22, 274–275.

6

GENOMIC MAPPING AND MAPPING DATABASES

Peter S. White

Department of Pediatrics
University of Pennsylvania
Philadelphia, Pennsylvania

Tara C. Matise

Department of Genetics
Rutgers University
New Brunswick, New Jersey

A few years ago, only a handful of ready-made maps of the human genome existed, and these were low-resolution maps of small areas. Biomedical researchers wishing to localize and clone a disease gene were forced, by and large, to map their region of interest, a time-consuming and painstaking process. This situation has changed dramatically in recent years, and there are now high-quality genome-wide maps of several different types containing tens of thousands of DNA markers. With the pending availability of a finished human sequence, most efforts to construct genomic maps will come to a halt; however, integrated maps, genome catalogues, and comprehensive databases linking positional and functional genomic data will become even more valuable. Genome projects in other organisms are at various stages, ranging from having only a handful of available maps to having a complete sequence. By taking advantage of the available maps and DNA sequence, a researcher can, in many cases, focus in on a candidate region by searching public mapping databases in a matter of hours rather than by performing laboratory experiments over a course of months.

Bioinformatics: A Practical Guide to the Analysis of Genes and Proteins
Edited by A. D. Baxevanis and B. F. F. Ouellette
ISBN 0-471-38390-2 (cloth), ISBN 0-471-383910 (paper)

Subsequently, the researcher's burden has now shifted from mapping the genome to navigating a vast *terra incognita* of Web sites, FTP servers, and databases. There are large databases such as the National Center for Biotechnology Information (NCBI) Entrez Genomes Division, Genome Database (GDB), and Mouse Genome Database (MGD), smaller databases serving the primary maps published by genome centers, sites sponsored by individual chromosome committees, and sites used by smaller laboratories to publish highly detailed maps of specific regions. Each type of resource contains information that is valuable in its own right, even when it overlaps with the information found at others. Finding one's way around this information space is not easy. A recent search for the word "genome" using the AltaVista Web search engine turned up 400,000 potentially relevant documents.

This chapter is intended as a "map of the maps," a way to guide readers through the maze of publicly available genomic mapping resources. The different types of markers and methods used for genomic mapping will be reviewed and the inherent complexities in the construction and utilization of genome maps will be discussed. Several large community databases and method-specific mapping projects will be presented in detail. Finally, practical examples of how these tools and resources can be used to aid in specific types of mapping studies such as localizing a new gene or refining a region of interest will be provided. A complete description of the mapping resources available for all species would require an entire book. Therefore, this chapter focuses primarily on humans, with some references to resources for other organisms.

INTERPLAY OF MAPPING AND SEQUENCING

The recent advent of whole-genome sequencing projects for humans and select model organisms is dramatically impacting the use and utility of genomic map-based information and methodologies. Genomic maps and DNA sequence are often treated as separate entities, but large, uninterrupted DNA sequence tracts can be thought of and used as an ultra-high-resolution mapping technique. Traditional genomic maps that rely on genomic markers and either clone-based or statistical approaches for ordering are precursory to finished and completely annotated DNA sequences of whole chromosomes or genomes. However, such completed genome sequences are predicted to be publicly available only in 2002 for humans, 2005 for the mouse, and even later for other mammalian species, although complete sequences are now available for some individual human chromosomes and selected lower eukaryotes (see Chapter 15). Until these completed sequences are available, mapping and sequencing approaches to genomic analysis serve as complementary approaches for chromosome analysis.

Before determination of an entire chromosome's sequence, the types of sequences available can be roughly grouped into marker/gene-based tags [e.g., expressed sequence tags (ESTs) and sequence-tagged sites (STSs)], single gene sequences, prefinished DNA clone sequences, and completed, continuous genomic sequence tracts. The first two categories provide rich sources of the genomic markers used for mapping, but only the last two categories can reliably order genomic elements. The human genome draft sequence is an example of a prefinished sequence, in which >90% of the entire sequence is available, but most continuous sequence tracts are relatively short (usually <100 kb and often <10 kb), thus providing high local resolution but little long-range ordering information. Genomic maps can help provide a context for this sequence information. Thus, two or more sequences containing unique genomic markers can be oriented if these markers are ordered on a map. In this way, existing maps serve as a scaffold for orienting, directing, and troubleshooting sequencing projects. Similarly, users can first define a chromosomal region of interest using a traditional map approach and then can identify relevant DNA sequences to analyze by finding long sequences containing markers mapping within the defined region. NCBI tools such as BLAST and electronic PCR (e-PCR) are valuable for finding marker/sequence identities, and several of the resources discussed below provide marker/sequence integration.

As large sequence tracts emerge from the human and model organism projects, sequence-based ordering of genomic landmarks will eventually supplant map-based ordering methods. The evolution from a mapped chromosome to the determination of the chromosome's complete sequence is marked by increasing incorporation of partial genomic sequence tracts into the underlying map. Once complete, finished sequences can be used to confirm map-determined marker orders. Given the error rates inherent in both map and sequence-assembly methodology, it is good practice to use both map and sequence information simultaneously for independent verification of regional order.

GENOMIC MAP ELEMENTS

DNA Markers

A DNA *marker* is simply a uniquely identifiable segment of DNA. There are several different types of markers, usually ranging in size from one to 300–400 nucleotide bases in size. Markers can be thought of as landmarks, and a set of markers whose relative positions (or order) within a genome are known comprises a *map*. Markers can be categorized in several ways. Some markers are polymorphic, and others are not (monomorphic). Detection of markers may be either PCR based or hybridization based. Some markers lie in a sequence of DNA that is expressed; some do not, or their expression status may be unknown.

PCR-based markers are commonly referred to as sequence-tagged sites (STSs). An STS is defined as a segment of genomic DNA that can be uniquely PCR amplified by its primer sequences. STSs are commonly used in the construction of physical maps. STS markers may be developed from any genomic sequence of interest, such as from characterized and sequenced genes, or from expressed sequence tags (ESTs, Chapter 12). Alternatively, STSs may be randomly identified from total genomic DNA. The EST database (dbEST) at NCBI stores information on most STS markers.

Polymorphic Markers

Polymorphic markers are those that show sequence variation among individuals. Polymorphic markers are used to construct genetic linkage maps. The number of alleles observed in a population for a given polymorphism, which can vary from two to >30, determines the degree of polymorphism. For many studies, highly polymorphic markers (>5 alleles) are most useful.

Polymorphisms may arise from several types of sequence variations. One of the earlier types of polymorphic markers used for genomic mapping is a restriction fragment length polymorphism (RFLP). An RFLP arises from changes in the sequence of a restriction enzyme recognition site, which alters the digestion patterns observed during hybridization-based analysis. Another type of hybridization-based marker arises from a variable number of tandem repeat units (VNTR). A VNTR locus usually has several alleles, each containing a different number of copies of a common motif of at least 16 nucleotides tandemly oriented along a chromosome.

A third type of polymorphism is due to tandem repeats of short sequences that can be detected by PCR-based analysis. These are known variously as microsatellites, short tandem repeats (STRs), STR polymorphisms (STRPs), or short sequence length polymorphisms (SSLPs). These repeat sequences usually consist of two, three, or four nucleotides and are plentiful in most organisms. All PCR-converted STR markers (those for which a pair of oligonucleotides flanking the polymorphic site suitable for PCR amplification of the locus has been designed) are considered to be STSs. The advent of PCR-based analysis quickly made microsatellites the markers of choice for mapping.

Another polymorphic type of PCR-based marker is a single nucleotide polymorphism (SNP), which results from a base variation at a single nucleotide position. Most SNPs have only two alleles (biallelic). Because of their low heterozygosity, maps of SNPs require a much higher marker density than maps of microsatellites. SNPs occur frequently in most genomes, with one SNP occurring on average approximately once in every 100–300 bases in humans. SNPs lend themselves to highly automated fluidic or DNA chip-based analyses and have quickly become the focus of several large-scale development and mapping projects in humans and other organisms. Further details about all of these types of markers can be found elsewhere (Chakravarti and Lynn, 1999; Dietrich et al., 1999).

DNA Clones

The possibility of physically mapping eukaryotic genomes was largely realized with the advent of cloning vehicles that could efficiently and reproducibly propagate large DNA fragments. The first generation of large-insert cloning was made possible with yeast artificial chromosome (YAC) libraries (Burke et al., 1987). Because YACs can contain fragments up to 2 Mb, they are suitable for quickly making low-resolution maps of large chromosomal regions, and the first whole-genome physical maps of several eukaryotes were constructed with YACs. However, although YAC libraries work well for ordering STSs and for joining small physical maps, the high rate of chimerism and instability of these clones makes them unsuitable for DNA sequencing.

The second and current generation of large-insert clones consists of bacterial artificial chromosomes (BACs) and P1-artificial chromosomes, both of which act as episomes in bacterial cells rather than as eukaryotic artificial chromosomes. Bacterial propagation has several advantages, including higher DNA yields, ease-of-use for sequencing, and high integrity of the insert during propagation. As such, despite the relatively limited insert sizes (usually 100–300 kb), BACs and PACs have largely replaced YACs as the clones of choice for large-genome mapping and sequencing projects (Iaonnou et al., 1994; Shizuya et al., 1992). DNA fingerprinting has been applied to BACs and PACs to determine insert overlaps and to construct clone contigs. In this technique, clones are digested with a restriction enzyme, and the resulting fragment patterns are compared between clones to identify those sharing subsets of identically sized fragments. In addition, the ends of BAC and PAC inserts can be directly sequenced; clones whose insert-end sequences have been determined are referred to as sequence-tagged clones (STCs). Both DNA fingerprinting and STC generation now play instrumental roles in physical mapping strategies, as will be discussed below.

TYPES OF MAPS

Cytogenetic Maps

Cytogenetic maps are those in which the markers are localized to chromosomes in a manner that can be directly imaged. Traditional cytogenetic mapping hybridizes a radioactively or fluorescently labeled DNA probe to a chromosome preparation, usually in parallel with a chromosomal stain such as Giemsa, which produces a banded karyotype of each chromosome (Pinkel et al., 1986). This allows assignment of the probe to a specific chromosomal band or region. Assignment of cytogenetic positions in this manner is dependent on some subjective criteria (variability in technology, methodology, interpretation, reproducibility, and definition of band boundaries). Thus, inferred cytogenetic positions are often fairly large and occasionally overinterpreted, and some independent verification of cytogenetic position determinations is warranted for crucial genes, markers, or regions. Probes used for cytogenetic mapping are usually large-insert clones containing a gene or polymorphic marker of interest. Despite the subjective aspects of cytogenetic methodology, karyotype analysis is an important and relatively simple clinical genetic tool; thus, cytogenetic positioning remains an important parameter for defining genes, disease loci, and chromosomal rearrangements.

Newer cytogenetic techniques such as interphase fluorescence in situ hybridization (FISH) (Lawrence et al., 1990) and fiber FISH (Parra and Windle, 1993) instead examine chromosomal preparations in which the DNA is either naturally or mechanically extended. Studies of such extended chromatin have demonstrated a directly proportional relationship between the distances measured on the image and the actual physical distance for short stretches, so that a physical distance between two closely linked probes can be determined with some precision (van den Engh et al., 1992). However, these techniques have a limited ordering range ($\leq$1–2 Mb) and are not well-suited for high-throughput mapping.

Genetic Linkage Maps

Genetic linkage (GL) maps (also called meiotic maps) rely on the naturally occurring process of recombination for determination of the relative order of, and map distances between, polymorphic markers. Crossover and recombination events take place during meiosis and allow rearrangement of genetic material between homologous chromosomes. The likelihood of recombination between markers is evaluated using genotypes observed in multigenerational families. Markers between which only a few recombination occur are said to be linked, and such markers are usually located close to each other on the same chromosome. Markers between which many recombinations take place are unlinked and usually lie far apart, either at opposite ends of the same chromosome or on different chromosomes.

Because the recombination events cannot be easily quantified, a statistical method of maximum likelihood is usually applied in which the likelihood of two markers being linked is compared with the likelihood of being unlinked. This likelihood ratio is called a "lod" score (for "log of the odds"), and a lod score greater than 3 (corresponding to odds of 1,000:1 or greater) is usually taken as evidence that markers are linked. The lod score is computed at a range of recombination fraction values between markers (from 0 to 0.5), and the recombination fraction at which the lod score is maximized provides an estimate of the distance between markers. A map function (usually either Haldane or Kosambi) is then used to convert the recombination fraction into an additive unit of distance measured in centiMorgans (cM), with 1 cM representing a 1% probability that a recombination has occurred between two markers on a single chromosome. Because recombination events are not randomly distributed, map distances on linkage maps are not directly proportional to physical distances.

The majority of linkage maps are constructed using multipoint linkage analysis, although multiple pairwise linkage analysis and minimization of recombination are also valid approaches. Commonly used and publicly available computer programs for building linkage maps include LINKAGE (Lathrop et al., 1984), CRI-MAP (Lander and Green, 1987), MultiMap (Matise et al., 1994), MAPMAKER (Lander et al., 1987), and MAP (Collins et al., 1996). The MAP-O-MAT Web server is available for estimation of map distances and for evaluation of statistical support for order (Matise and Gitlin, 1999).

Because linkage mapping is a based on statistical methods, linkage maps are not guaranteed to show the correct order of markers. Therefore, it is important to be critical of the various available maps and to be aware of the statistical criteria that were used in map construction. Typically, only a subset of markers (framework or index markers) is mapped with high statistical support. The remainder are either placed into well-supported intervals or bins or placed into unique map positions but with low statistical support for order (see additional discussion below).

To facilitate global coordination of human linkage mapping, DNAs from a set of reference pedigrees collected for map construction were prepared and distributed by the Centre d'Etude du Polymorphism Humain (CEPH; Dausset et al., 1990). Nearly all human linkage maps are based on genotypes from the CEPH reference pedigrees, and genotypes for markers scored in the CEPH pedigrees are deposited in a public database maintained at CEPH. Most recent maps are composed almost entirely of highly polymorphic STR markers. These linkage maps have already exceeded the maximum map resolution possible given the subset of CEPH pedigrees that are commonly used for map construction, and no further large-scale efforts to place STR markers on human linkage maps are planned. Thousands of SNPs are currently being identified and characterized, and a subset are being placed on linkage maps (Wang et al., 1998).

Linkage mapping is also an important tool in experimental animals, with many maps already produced at high resolution and others still under development (see *Mapping Projects and Associated Resources*, below).

Radiation Hybrid Maps

Radiation hybrid (RH) mapping is very similar to linkage mapping. Both methods rely on the identification of chromosome breakage and reassortment. The primary difference is the mechanism of chromosome breakage. In the construction of radiation hybrids, breaks are induced by the application of lethal doses of radiation to a donor cell line, which is then rescued by fusion with a recipient cell line (typically mouse or hamster) and grown in a selective medium such that only fused cells survive. An RH panel is a library of fusion cells, each of which has a separate collection of donor fragments. The complete donor genome is represented multiple times across most RH panels. Each fusion cell, or radiation hybrid, is then scored by PCR to determine the presence or absence of each marker of interest. Markers that physically lie near each other will show similar patterns of retention or loss across a panel of RH cells and behave as if they are linked, whereas markers that physically lie far apart will show completely dissimilar patterns and behave as if they are unlinked. Because the breaks are largely randomly distributed, the break frequencies are roughly directly proportional to physical distances. The resulting data set is a series of positive and negative PCR scores for each marker across the hybrid panel.

These data can be used to statistically infer the position of chromosomal breaks, and, from that point on, the procedures for map construction are similar to those used in linkage mapping. A map function is used to convert estimates of breakage frequency to additive units of distance measured in centirays (cR), with 1 cR representing a 1% probability that a chromosomal break has occurred between two markers in a single hybrid. The resolution of a radiation hybrid map depends on the size of the chromosomal fragments contained in the hybrids, which in turn is proportional to the amount of irradiation to which the human cell line was exposed.

Most RH maps are built using multipoint linkage analysis, although multiple-pairwise linkage analysis and minimization of recombination are also valid approaches. Three genome-wide RH panels exist for humans and are commercially available, and RH panels are available for many other species as well. Widely used computer programs for RH mapping are RHMAP (Boehnke et al., 1991), RHMAPPER (Slonim et al., 1997), and MultiMap (Matise et al., 1994), and on-line servers that allow researchers to place their RH mapped markers on existing RH maps are available. The Radiation Hybrid Database (RHdb) is the central repository for RH data on panels available in all species. The Radiation Hybrid Information Web site also contains multi-species information about available RH panels, maps, ongoing projects, and available computer programs.

Transcript Maps

Of particular interest to researchers chasing disease genes are maps of transcribed sequences. Although the transcript sequences are mapped using one of the methods described in this section, and thus do not require a separate mapping technology, they are often set apart as a separate type of map. These maps consist of expressed sequences and sequences derived from known genes that have been converted into STSs and usually placed on conventional physical maps. Recent projects for creating large numbers of ESTs (Adams et al., 1991; Houlgatte et al., 1995; Hillier et al., 1996) have made tens of thousands of unique expressed sequences available to the mapping laboratories. Transcribed sequence maps can significantly speed the search for candidate genes once a disease locus has been identified. The largest human transcript map to date is the GeneMap '99, described below.

Physical Maps

Physical maps include maps that either are capable of directly measuring distances between genomic elements or that use cloned DNA fragments to directly order elements. Many techniques have been created to develop physical maps. The most widely adopted methodology, due largely to its relative simplicity, is STS content mapping (Green and Olson, 1990). This technique can resolve regions much larger than 1 Mb and has the advantage of using convenient PCR-based positional markers.

In STS content maps, STS markers are assayed by PCR against a library of large-insert clones. If two or more STSs are found to be contained in the same clone, chances are high that those markers are located close together. (The fact that they are not close 100% of the time is a reflection of various artifacts in the mapping procedure, such as the presence of chimeric clones.) The STS content mapping technique builds a series of contigs (i.e., overlapping clusters of clones joined together by shared STSs). The resolution and coverage of such a map are determined by a number of factors, including the density of STSs, the size of the clones, and the depth of the clone library. Maps that use cloning vectors with smaller insert sizes have a higher theoretical resolution but require more STSs to achieve coverage of the same area of the genome. Although it is generally possible to deduce the relative order of markers on STS content maps, the distances between adjacent markers cannot be measured with accuracy without further experimentation, such as by restriction mapping. However, STS content maps have the advantage of being associated with a clone resource that can be used for further studies, including subcloning, DNA sequencing, or transfection.

Several other techniques in addition to STS content and radiation hybrid mapping have also been used to produce physical maps. Clone maps rely on techniques other than STS content to determine the adjacency of clones. For example, the CEPH YAC map (see below) used a combination of fingerprinting, inter-Alu product hybridization, and STS content to create a map of overlapping YAC clones. Fingerprinting is commonly used by sequencing centers to assemble and/or verify BAC and PAC contigs before clones are chosen for sequencing, to select new clones for sequencing that can extend existing contigs, and to help order genomic sequence tracts generated in whole-genome sequencing projects (Chumakov et al., 1995). Sequencing of large-insert clone ends (STC generation), when applied to a whole-genome clone library of adequate coverage, is very effective for whole-genome mapping when used in combination with fingerprinting of the same library. Deletion and somatic cell hybrid maps relying on large genomic reorganizations (induced deliberately or naturally occurring) to place markers into bins defined by chromosomal breakpoints have been generated for some human chromosomes (Jensen et al., 1997; Lewis et al., 1995; Roberts et al., 1996; Vollrath et al., 1992). Optical mapping visualizes and measures the length of single DNA molecules extended and digested with restriction enzymes by high-resolution microscopy. This technique, although still in its infancy, has been successfully used to assemble whole chromosome maps of bacteria and lower eukaryotes and is now being applied to complex genomes (Aston et al., 1999; Jing et al., 1999; Schwartz et al., 1993).

Comparative Maps

Comparative mapping is the process of identifying conserved chromosome segments across different species. Because of the relatively small number of chromosomal breaks that have occurred during mammalian radiation, the order of genes usually is preserved over large chromosomal segments between related species. Orthologous genes (copies of the same genes from different species) can be identified through DNA sequence homology, and sets of orthologous genes sharing an identical linear order within a chromosomal region in two or more species are used to identify conserved segments and ancient chromosomal breakpoints.

Knowledge about which chromosomal segments are shared and how they have become rearranged over time greatly increases our understanding of the evolution of different plant and animal lineages. One of the most valuable applications of comparative maps is to use an established gene map of one species to predict positions of orthologous genes in another species. Many animal models exist for diseases observed in humans. In some cases, it is easier to identify the responsible genes in an animal model than in humans, and the availability of a good comparative map can simplify the process of identifying the responsible genes in humans. In other cases, more might be known about the gene(s) responsible in humans, and the same comparative map could be used to help identify the gene(s) responsible in the model species. There are several successful examples of comparative candidate gene mapping (O'Brien et al., 1999).

As mapping and sequencing efforts progress in many species, it is becoming possible to identify smaller homologous chromosome segments, and detailed comparative maps are being developed between many different species. Fairly dense gene-based comparative maps now exist between the human, mouse, and rat genomes and also between several agriculturally important mammalian species. Sequence- and protein-based comparative maps are also under development for several lower organisms for which complete sequence is available (Chapter 15). A comparative map is typically presented either graphically or in tabular format, with one species designated as the index species and one or more others as comparison species. Homologous regions are presented graphically with nonconsecutive segments from the comparison species shown aligned with their corresponding segments along the map of the index species.

Integrated Maps

Map integration provides interconnectivity between mapping data generated from two or more different experimental techniques. However, achieving accurate and useful integration is a difficult task. Most of the genomic maps and associated Web sites discussed in this section provide some measure of integration, ranging from the approximate cytogenetic coordinates provided in the Généthon GL map to the inter-associated GL, RH, and physical data provided by the Whitehead Institute (WICGR) Web site. Several integration projects have created truly integrated maps by placing genomic elements mapped by differing techniques relative to a single map scale. The most advanced sources of genomic information provide some level of genomic cataloguing, where considerable effort is made to collect, organize, and map all available positional information for a given genome.

COMPLEXITIES AND PITFALLS OF MAPPING

It is important to realize that the genomic mapping information currently available is a collection of a large number of individual data sets, each of which has unique characteristics. The experimental techniques, methods of data collection, annotation, presentation, and quality of the data differ considerably among these data sets. Although most mapping projects include procedures to detect and eliminate and/or correct errors, there are invariably some errors that occur, which often result in the incorrect ordering or labeling of individual markers. Although the error rate is usually very low (5% or less), a marker misplacement can obviously have a great impact on a study. A few mapping Web sites are beginning to flag and correct (or at least warn) users of potential errors, but most errors cannot be easily detected. Successful strategies for minimizing the effects of data error include (1) simultaneously assessing as many different maps as possible to maximize redundancy (note that ideally "different" maps use independently-derived data sets or different techniques); (2) increased emphasis on utilizing integrated maps and genomic catalogues that provide access to all available genomic information for the region of interest (while closely monitoring the map resolution and marker placement confidence of the integrated map); and (3) if possible, experimentally verifying the most critical marker positions or placements.

In addition to data errors, several other, more subtle complexities are notable. Foremost is the issue of nomenclature, or the naming of genomic markers and elements. Many markers have multiple names, and keeping track of all the names is a major bioinformatics challenge. For example, the polymorphic marker D1S243 has several assigned names: AFM214yg7, which is actually the name of the DNA clone from which this polymorphism was identified; SHGC-428 and stSG729, two examples of genome centers renaming a marker to fit their own nomenclature schemes; and both GDB:201358 and GDB:133491, which are database identifier numbers used to track the polymorphism and STS associated with this marker, respectively, in the Genome Database (GDB). Genomic mapping groups working with a particular marker often assign an additional name to simplify their own data management, but, too often, these alternate identifiers are subsequently used as a primary name. Furthermore, many genomic maps display only one or a few names, making comparisons of maps problematic. Mapping groups and Web sites are beginning to address these inherent problems, but the difficulty of precisely defining "markers," "genes," and "genomic elements" adds to the confusion. It is important to distinguish between groups of names defining different elements. A gene can have several names, and it can also be associated with one or more EST clusters, polymorphisms, and STSs. Genes spanning a large genomic stretch can even be represented by several markers that individually map to different positions. Web sites providing genomic cataloguing, such as LocusLink, UniGene, GDB, GeneCards, and eGenome, list most names associated with a given genomic element. Nevertheless, collecting, cross-referencing, and frequently updating one's own sets of names for markers of interest is also a good practice (see Chapter 4 for data management using Sequin), as even the genomic cataloguing sites do not always provide complete nomenclature collections.

Each mapping technique yields its own resolution limits. Cytogenetic banding potentially orders markers separated by $\geq$1–2 Mb, and genetic linkage (GL) and RH analyses yields long-range resolutions of $\geq$0.5–1 Mb, although localized ordering can achieve higher resolutions. The confidence level with which markers are ordered on statistically based maps is often overlooked, but this is crucial for assessing map quality. For genomes with abundant mapping data such as human or mouse, the number of markers used for mapping often far exceeds the ability of the technique to order all markers with high confidence (often, confidence levels of 1,000:1 or lod 3 are used as a cutoff, which usually means that a marker is $\geq$1,000: 1 times more likely to be in the given position than in any other). Mappers have taken two approaches to address this issue. The first is to order all markers in the best possible linear order, regardless of the confidence for map position of each marker [examples include GeneMap '99 (GM99) and the Genetic Location Database; Collins et al., 1996; Deloukas et al., 1998]. Alternatively, the high confidence linear order of a subset of markers is determined, and the remaining markers are then placed in high confidence "intervals," or regional positions (such as Généthon, SHGC, and eGenome; Dib et al., 1996; Stewart et al., 1997; White et al., 1999). The advantage of the first approach is that resolution is maximized, but it is important to pay attention to the odds for placement of individual markers, as alternative local orders are often almost equally likely. Thus, beyond the effective resolving power of a mapping technique, increased resolution often yields decreased accuracy, and researchers are cautioned to strike a healthy balance between the two.

Each mapping technique also yields very different measures of distance. Cytogenetic approaches, with the exception of high-resolution fiber FISH, provide only rough distance estimates, GL and STS content mapping provide marker orientation but only relative distances, and RH mapping yields distances roughly proportional to true physical distance. For GL analysis, unit measurements are in centMiorgans, with 1 cM equivalent to a 1% chance of recombination between two linked markers. The conversion factor of 1 cM $\simeq$ 1 Mb is often cited for the human genome but is overstated, as this is just the *average* ratio genome-wide, and many chromosomal regions have recombination hotspots and coldspots in which the cM-to-Mb ratio varies as much as 10-fold. In general, cytogenetic maps provide subband marker regionalization but limited localized ordering, GL and STS content maps provide excellent ordering and limited-to-moderate distance information, and RH maps provide the best combination of localized ordering and distance estimates.

Finally, there are various levels at which genomic information can be presented. *Single-resource maps* such as the Généthon GL maps use a single experimental technique and analyze a homogeneous set of markers. Strictly *comparative maps* make comparisons between two or more different single-dimension maps either within or between species but without combining data sets for integration. GDB's Mapview program can display multiple maps in this fashion (Letovsky et al., 1998). *Integrated maps* recalculate or completely integrate multiple data sets to display the map position of all genomic elements relative to a single scale; GDB's Comprehensive Maps are an example of such integration (Letovsky et al., 1998). Lastly, *genome cataloguing* is a relatively new way to display genomic information, in which many data sets and/or Web sites are integrated to provide a comprehensive listing and/or display of all identified genomic elements for a given chromosome or genome. Completely sequenced genomes such as *C. elegans* and *S. cerevisiae* have advanced cataloguing efforts (see Chapter 15), but catalogues for complex genome organisms are in the early stages. Examples include the interconnected NCBI databases, MGD, and eGenome (Blake et al., 2000; Wheeler et al., 2000). Catalogues provide a "one-stop shopping" solution to collecting and analyzing genomic data and are recommended as a maximum-impact means to begin a regional analysis. However, the individual data sets provide the highest quality positional information and are ultimately the most useful for region definition and refinement.

DATA REPOSITORIES

There are several valuable and well-developed data repositories that have greatly facilitated the dissemination of genome mapping resources for humans and other species. This section covers three of the most comprehensive resources for mapping in humans: the Genome Database (GDB), the National Center for Biotechnology Information (NCBI), and the Mouse Genome Database (MGD). More focused resources are mentioned in the *Mapping Projects and Associated Resources* section of this chapter.

GDB

The Genome Database (GDB) is the official central repository for genomic mapping data created by the Human Genome Project (Pearson, 1991). GDB's central node is located at the Hospital for Sick Children (Toronto, Ontario, Canada). Members of the scientific community as well as GDB staff curate data submitted to the GDB. Currently, GDB comprises descriptions of three types of objects from humans: Genomic Segments (genes, clones, amplimers, breakpoints, cytogenetic markers, fragile sites, ESTs, syndromic regions, contigs, and repeats), Maps (including cytogenetic, GL, RH, STS-content, and integrated), and Variations (primarily relating to polymorphisms). In addition, contributing investigator contact information and citations are also provided. The GDB holds a vast quantity of data submitted by hundreds of investigators. Therefore, like other large public databases, the data quality is variable. A more detailed description of the GDB can be found in Talbot and Cuticchia (1994).

GDB provides a full-featured query interface to its database with extensive online help. Several focused query interfaces and predefined reports, such as the Maps within a Region search and Lists of Genes by Chromosome report, present a more intuitive entry into GDB. In particular, GDB's Mapview program provides a graphical interface to the genetic and physical maps available at GDB.

A Simple Search is available on the home page of the GDB Web site. This query is used when searching for information on a specific genomic segment, such as a gene or STS (amplimer, in GDB terminology) and can be implemented by entering the segment name or GDB accession number. Depending on the type of segment queried and the available data, many different types of segment-specific information may be returned, such as alternate names (aliases), primer sequences, positions in various maps, related segments, polymorphism details, contributor contact information, citations, and relevant external links.

At the bottom of the GDB home page is a link to Other Search Options. From the Other Search Options page there are links to three customized search forms (Markers and Genes within a Region, Maps within a Region, and Genes by Name or Symbol), sequence-based searches, specific search forms for subclasses of GDB elements, and precompiled lists of data (Genetic Diseases by Chromosome, Lists of Genes by Chromosome, and Lists of Genes by Symbol Name).

A particularly useful query is the Maps within a Region search. This search allows retrieval of all maps stored in GDB that span a defined chromosomal region.

In a two-step process, the set of maps to be retrieved is first determined, and, from these, the specific set to be displayed is then selected.

Select the Maps within a Region link to display the search form. To view an entire chromosome, simply select it from the pop-up menu. However, entire chromosomes may take considerable time to download and display; therefore, it is usually best to choose a subchromosomal region. To view a chromosomal region, type the names of two cytogenetic bands or flanking genetic markers into the text fields labeled From and To. An example query is shown in Figure 6.1. If the flanking markers used in the query are stored in GDB as more than one type of object, the next form will request selection of the specific type of element for each marker. For the example shown in Figure 6.1, it is appropriate to select Amplimer.

The resulting form lists all maps stored in GDB that overlap the selected region. Given the flanking markers specified above, there are a total of 21 maps. The user selects which maps to display by marking the respective checkboxes. Note that GDB's Comprehensive Map is automatically selected. If a graphical display is requested, the size of the region and the number of maps to be displayed can significantly affect the time to fetch and display them. The resulting display will appear in a separate window showing the selected maps in side-by-side fashion.

While the Mapview display is loading, a new page is shown in the browser window. If your system is not configured to handle Java properly, a helpful message will be displayed in the browser window. (*Important*: Do not close the browser window behind Mapview. Because of an idiosyncrasy of Java's security specification, the applet cannot interact properly with GDB unless the browser window remains open.) To safely exit the Mapview display, select Exit from Mapview's File menu.

Mapview has many useful options, which are well described in the online help. Some maps have more than one *tier*, each displaying different types of markers, such as markers positioned with varying confidence thresholds on a linkage or radiation hybrid map. It is possible to zoom in and out, highlight markers across maps, color code different tiers, display markers using different aliases, change the relative position of the displayed maps, and search for specific markers. To retrieve additional information on a marker from any of the maps, double-click on its name to perform a *Simple Search* (as described above). A separate browser window will then display the GDB entry for the selected marker.

Two recently added GDB tools are GDB BLAST and e-PCR. These are available from the Other Search Options page and enable users to employ GDB's many data resources in their analysis of the emerging human genome sequence. GDB BLAST returns GDB objects associated with BLAST hits against the public human sequence. GDB's e-PCR finds which of its many amplimers are contained within queried DNA sequences and is thereby a quick means to determine or refine gene or marker localization. In addition, the GDB has many useful genome resource Web links on its Resources page.

NCBI

The NCBI has developed many useful resources and tools, several of which are described throughout this book. Of particular relevance to genome mapping is the Genomes Division of Entrez. Entrez provides integrated access to several different types of data for over 600 organisms, including nucleotide sequences, protein structures and sequences, PubMed/MEDLINE, and genomic mapping information. The

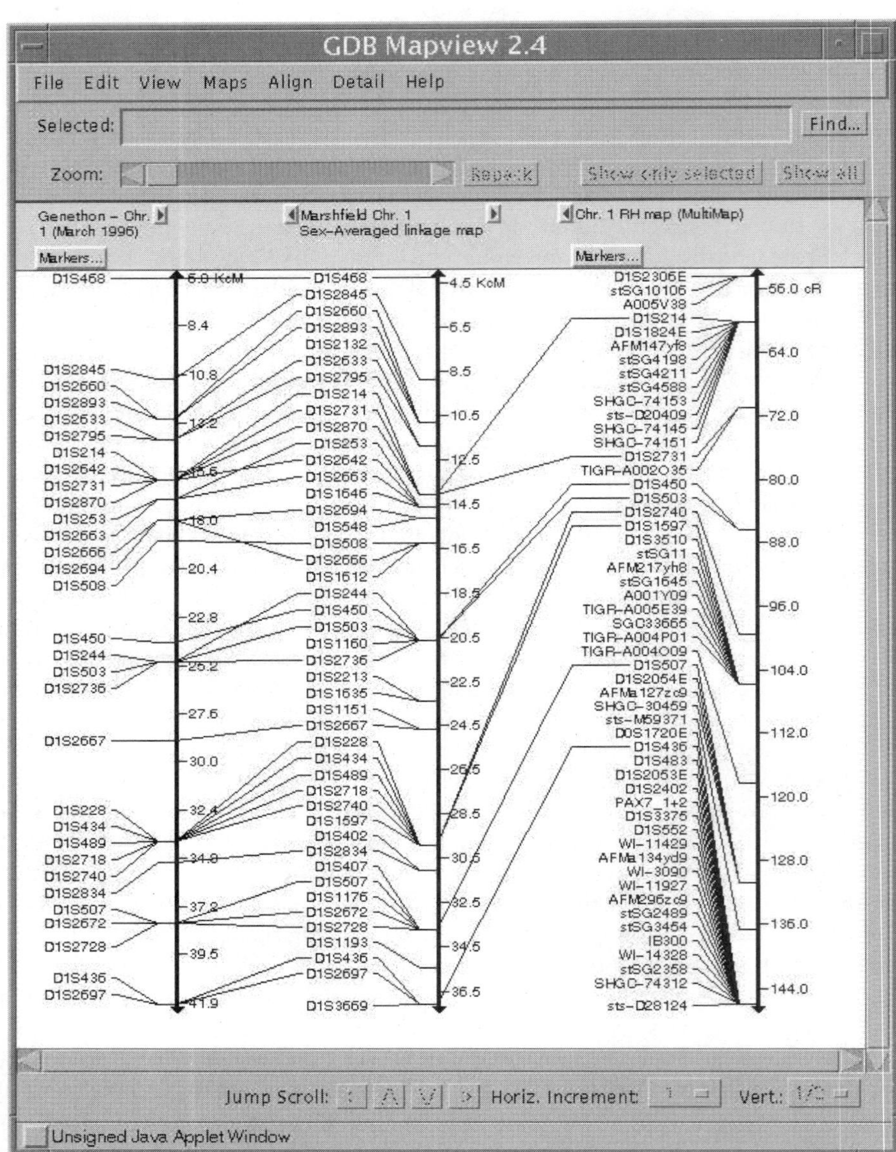

Figure 6.1. Results of a *Maps within a Region* GDB query for the region *D1S468–D1S214*, with no limits applied to the types of maps to be retrieved. Twenty-one maps were available for display. Only the Genethon and Marshfield linkage maps, as well as the Chromosome 1 RH map were selected for graphical display. Markers that are shared across maps are connected by lines.

NCBI Human Genome Map Viewer is a new tool that presents a graphical view of the available human genome sequence data as well as cytogenetic, genetic, physical, and radiation hybrid maps. Because the Map Viewer provides displays of the human genome sequence for the finished contigs, the BAC tiling path of finished and draft sequence, and the location of genes, STSs, and SNPs on finished and draft sequences, it is an especially useful tool for integrating maps and sequence. The only other organisms for which the Map Viewer is currently available is *M. musculus* and *D. melanogaster*.

The NCBI Map Viewer can simultaneously display up to seven maps that are selected from a set of 19, including cytogenetic, linkage, RH, physical, and sequence-based maps. Some of the maps have been previously published, and others are being computed at NCBI. An extensive set of help pages is available. There are many different paths to the Map Vieweron the NCBI Web site, as described in the help pages. The Viewer supports genome-wide or chromosome-specific searches.

A good starting point is the *Homo sapiens* Genome View page. This is reached from the NCBI home page by connecting to Human Genome Resources (listed on the right side), followed by the link to the Map Viewer (listed on the left side). From the Genome View page, a genome-wide search may be initiated using the search box at the top left, or a chromosome-specific search may be performed by entering a chromosome number(s) in the top right search box or by clicking on a chromosome idiogram. The searchable terms include gene symbol or name and marker name or alias. The search results include a list of hits for the search term on the available maps. Clicking on any of the resulting items will bring up a graphical view of the region surrounding the item on the specific map that was selected. For example, a genome-wide search for the term `CMT*` returns 33 hits, representing the loci for forms of Charcot-Marie-Tooth neuropathy on eight different chromosomes. Selecting the Genes_seq link for the PMP22 gene (the gene symbol for CMT1A, on chromosome 17) returns the view of the sequence map for the region surrounding this gene. The Display Settings window can then be used to select simultaneous display of additional maps (Fig. 6.2).

The second search box at the top right may be used to limit a genome-wide search to a single chromosome or range of chromosomes. Alternatively, to browse an entire chromosome, click on the link below each idiogram. Doing so will return a graphical representation of the chromosome using the default display settings. Currently, the default display settings select the STS map (shows placement of STSs using electronic PCR), the GenBank map (shows the BAC tiling path used for sequencing), and the contig map (shows the contig map assembled at NCBI from finished high-throughput genomic sequence) as additional maps to be displayed. To select a smaller region of interest from the view of the whole chromosome, either define the range (using base pairs, cytogenetic bands, gene symbols or marker names) in the main Map Viewer window or in the display settings or click on a region of interest from the thumbnail view graphic in the sidebar or the map view itself. As with the GDB map views, until all sequence is complete, alignment of multiple maps and inference of position from one map to another must be judged cautiously and should not be overinterpreted (see Complexities and Pitfalls of Mapping section above).

There are many other tools and databases at NCBI that are useful for gene mapping projects, including e-PCR, BLAST (Chapter 8), the GeneMap '99 (see Mapping Projects and Associated Resources), and the LocusLink, OMIM (Chapter 7), dbSTS, dbSNP, dbEST (Chapter 12), and UniGene (Chapter 12) databases. e-PCR and BLAST can be used to search DNA sequences for the presence of markers and to confirm and refine map localizations. In addition to EST alignment information and DNA sequence, UniGene reports include cytogenetic and RH map locations. The GeneMap '99 is a good starting point for finding approximate map

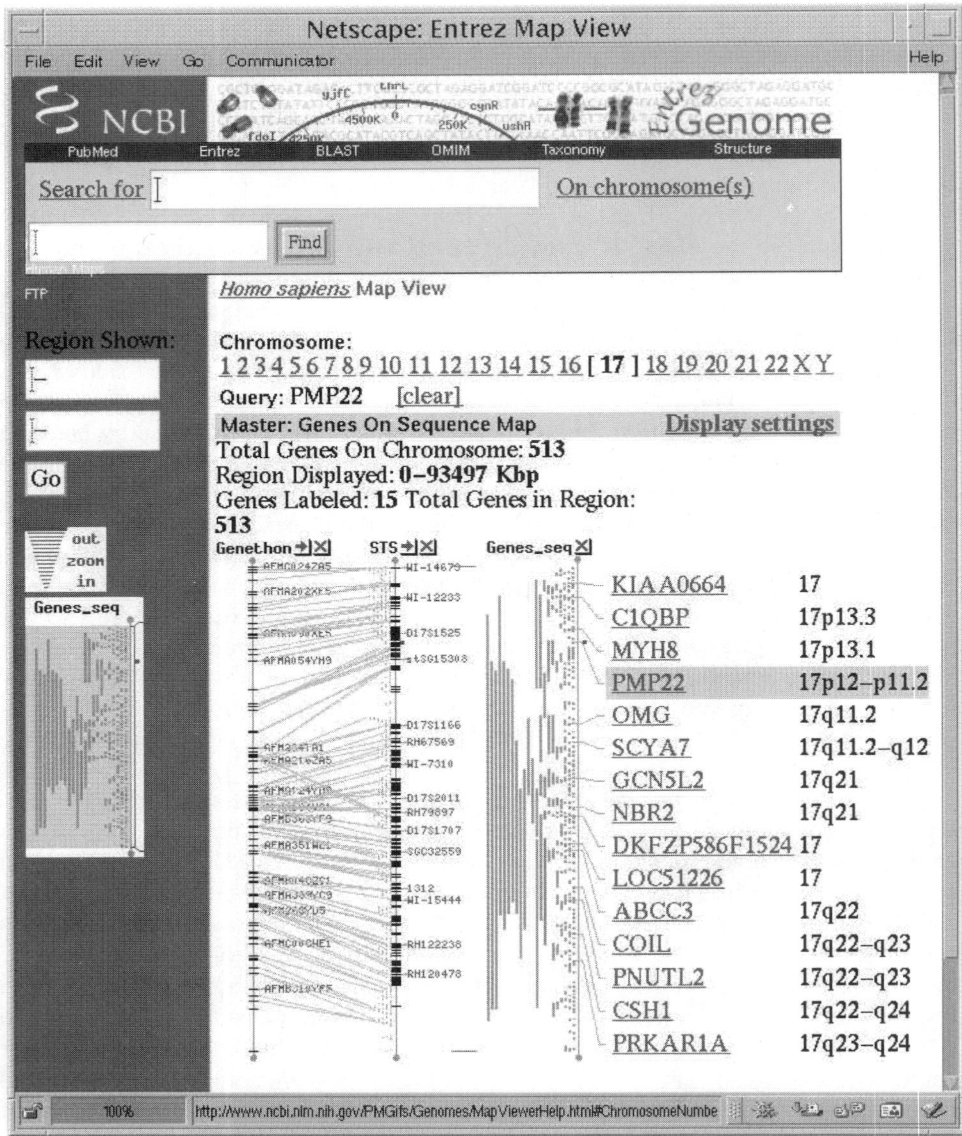

Figure 6.2. NCBI's Map View of the region surrounding the PMP22 gene. The Généthon, STS, and Genes_seq maps are displayed with lines connecting markers in common.

positions for EST markers, although additional fine-mapping should be performed to confirm order in critical regions. LocusLink, OMIM, and UniGene are good starting points for genome catalog information about genes and gene-based markers. LocusLink (Pruitt et al., 2000) presents information on official nomenclature, aliases, sequence accessions, phenotypes, EC numbers, MIM numbers, UniGene clusters, homology, map locations, and related Web sites. The dbSTS and dbEST databases themselves play a lesser role in human and mouse gene mapping endeavors as their relevant information has already been captured by other more detailed resources (such as LocusLink, GeneMap '99, UniGene, MGD, and eGenome) but are currently the primary source of genomic information for other organisms. The dbSNP database stores population-specific information on variation in humans, primarily for single nucleotide repeats but also for other types of polymorphisms. In addition, the NCBI's Genomic Biology page provides genomic resource home pages for many other organisms, including mouse, rat, *Drosophila*, and zebrafish.

MGI/MGD

The Mouse Genome Initiative Database (MGI) is the primary public mouse genomic catalogue resource. Located at The Jackson Laboratory, the MGI currently encompasses three cross-linked topic-specific databases: the Mouse Genome Database (MGD), the mouse Gene Expression Database (GXD), and the Mouse Genome Sequence project (MGS). The MGD has evolved from a mapping and genetics resource to include sequence and genome information and details on the functions and roles of genes and alleles (Blake et al., 2000). MGD includes information on mouse genetic markers and nomenclature, molecular segments (probes, primers, YACs and MIT primers), phenotypes, comparative mapping data, graphical displays of linkage, cytogenetic, and physical maps; experimental mapping data, and strain distribution patterns for recombinant inbred strains (RIs) and cross haplotypes. As of November 2000, there were over 29,500 genetic markers and 11,600 genes in MGD, with 85% and 70% of these placed onto the mouse genetic map, respectively. Over 4,800 genes have been matched with their human ortholog and over 1,800 matched with their rat ortholog.

Genes are easily searched through the Quick Gene Search box on the MGD home page. Markers and other map elements may also be accessed through several other search forms. The resulting pages contain summary information such as element type, official symbol, name, chromosome, map positions, MGI accession ID, references, and history. Additional element-specific information may also be displayed, including links to outside resources (Fig. 6.3). A thumbnail linkage map of the region is shown to the right, which can be clicked on for an expanded view.

The MGD contains many different types of maps and mapping data, including linkage data from 13 different experimental cross panels and the WICGR mouse physical maps, and cytogenetic band positions are available for some markers. The MGD also computes a linkage map that integrates markers mapped on the various panels. A very useful feature is the ability to build customized maps of specific regions using subsets of available data, incorporating private data, and showing homology information where available (see *Comparative Resources* section below). The MGD is storing radiation hybrid scores for mouse markers, but to date, no RH maps have been deposited at MGD.

MAPPING PROJECTS AND ASSOCIATED RESOURCES

In addition to the large-scale mapping data repositories outlined in the previous section, many invaluable and more focused resources also exist. Some of these are either not appropriate for storage at one of the larger-scale repositories or have never been deposited in them. These are often linked to specific mapping projects that primarily use only one or a few different types of markers or mapping approaches.

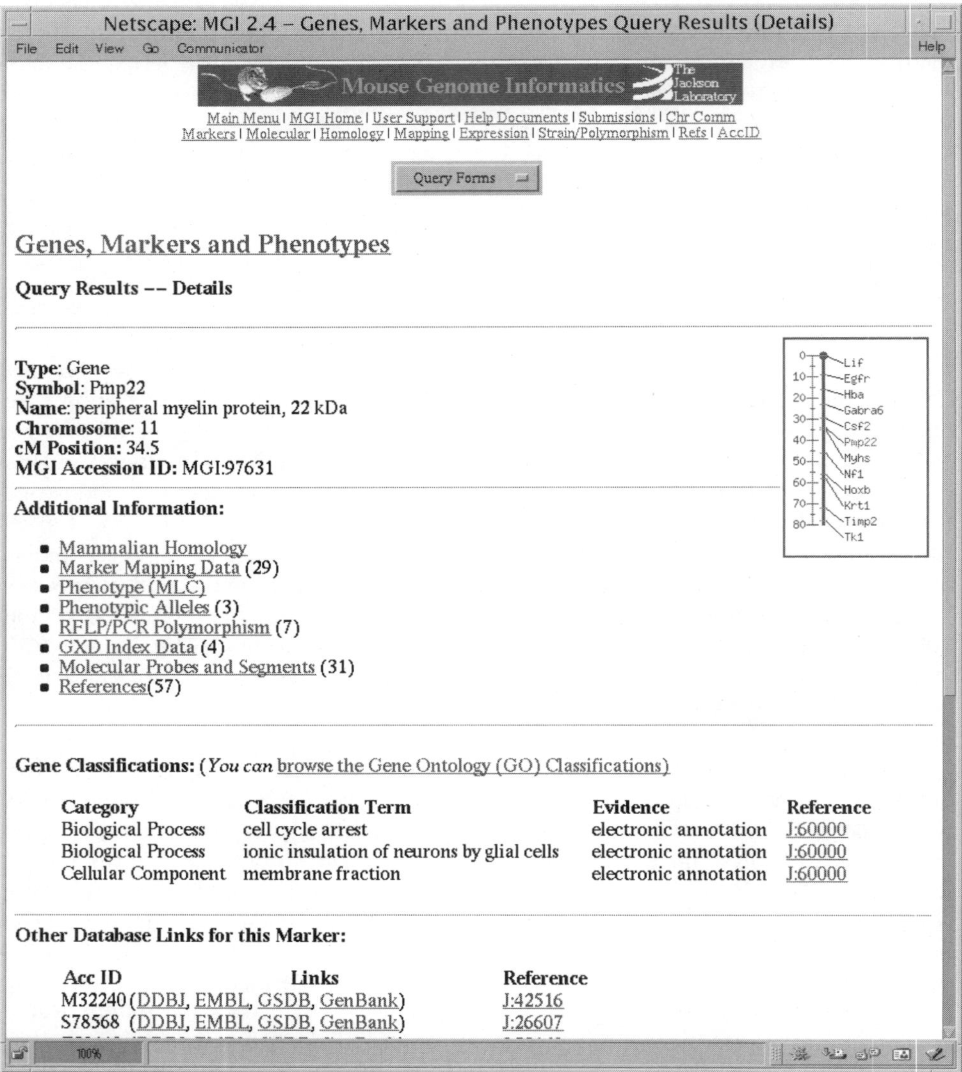

Figure 6.3. Results of an MGD Quick Gene Search for *pmp22*.

For most studies requiring the use of genome maps, it remains necessary to obtain maps or raw data from one or more of these additional resources. By visiting the resource-specific sites outlined in this section, it is usually possible to view maps in the form preferred by the originating laboratory, download the raw data, and review the laboratory protocols used for map construction.

Cytogenetic Resources

Cytogenetic-based methodologies are instrumental in defining inherited and acquired chromosome abnormalities, and (especially gene-based) chromosomal mapping data is often expressed in cytogenetic terms. However, because cytogenetic markers are not sequence based and the technique is less straightforward and usually more subjective than GL, RH, or physical mapping, there is only a modicum of integration between chromosomal band assignments and map coordinates derived from other techniques in humans and very little or none in other species. Thus, it is often difficult to determine the precise cytogenetic location of a gene or region. Useful human resources can be divided into displays of primary cytogenetic mapping data, efficient methods of integrating cytogenetic and other mapping data, and resources pertaining to specific chromosomal aberrations.

The central repository for human cytogenetic information is GDB, which offers several ways to query for marker and map information using cytogenetic coordinates (see above). GDB is a useful resource for cross-referencing cytogenetic positions with genes or regions of interest. NCBI's LocusLink and UniGene catalogues, as well as their other integrated mapping resources, are also valuable repositories of cytogenetic positions. LocusLink and NCBI's Online Mendelian Inheritance in Man (OMIM) list cytogenetic positions for all characterized genes and genetic abnormalities, respectively (McKusick, 1998; Pruitt et al., 2000). The National Cancer Institute (NCI)-sponsored project to identify GL-tagged BAC clones at 1 Mb density throughout the genome is nearing completion. This important resource, which is commercially available both as clone sets and as individual clones, provides the first complete integration of cytogenetic band information with other genome maps. At this site, BACs can be searched for individually by clone name, band position, or contained STS name, and chromosome sets are also listed. Each clone contains one or more microsatellite markers and has GL and/or RH mapping coordinates along with a FISH-determined cytogenetic band assignment. This information can be used to quickly determine the cytogenetic position of a gene or localized region and to map a cytogenetic observation such as a tumor-specific chromosomal rearrangement using the referenced GL and physical mapping reagents.

Three earlier genome-wide efforts to cytogenetically map large numbers of probes are complementary to the NCI site. The Lawrence Berkeley National Laboratory-University of California, San Francisco, Resource for Molecular Cytogenetics has mapped large-insert clones containing polymorphic and expressed markers using FISH to specific bands and also with fractional length (flpter) coordinates, in which the position of a marker is measured as a percentage of the length of the chromosome's karyotype. Similarly, the Genetics Institute at the University of Bari, Italy, and the Max Planck Institute for Molecular Genetics have independently localized large numbers of clones, mostly YACs containing GL-mapped microsatellite markers, onto chromosome bands by FISH. All three resources have also integrated the mapped probes relative to existing GL and/or RH maps.

Many data repositories and groups creating integrated genome maps list cytogenetic localizations for mapped genomic elements. These include GDB, NCBI, the Unified Database (UDB), the Genetic Location Database (LDB), and eGenome, all of which infer approximate band assignments to many or all markers in their databases. These assignments rely on determination of the approximate boundaries of each band using subsets of their marker sets for which accurate cytogenetic mapping data are available.

The NCI's Cancer Chromosome Aberration Project (CCAP; Wheeler et al., 2000), Infobiogen (Wheeler et al., 2000), the Southeastern Regional Genetics Group (SERGG), and the Coriell Cell Repositories all have Web sites that display cytogenetic maps or descriptions of characterized chromosomal rearrangements. These sites are useful resources for determining whether a specific genomic region is frequently disrupted in a particular disease or malignancy and for finding chromosomal cell lines and reagents for regional mapping. However, most of these rearrangements have only been mapped at the cytogenetic level.

Nonhuman resources are primarily limited to displays or simple integrations of chromosome idiograms. ArkDB is an advanced resource for displaying chromosomes of many amniotes; MGD incorporates mouse chromosome band assignments into queries of its database; and the Animal Genome Database has clickable chromosome idiograms for several mammalian genomes (Wada and Yasue, 1996). A recent work linking the mouse genetic and cytogenetic maps consists of 157 BAC clones distributed genome-wide (Korenberg et al., 1999) and an associated Web site is available for this resource at the Cedars-Sinai Medical Center.

Genetic Linkage Map Resources

Even with the "sequence era" approaching rapidly, linkage maps remain one of the most valuable and widely used genome mapping resources. Linkage maps are the starting point for many disease-gene mapping projects and have served as the backbone of many physical mapping efforts. Nearly all human linkage maps are based on genotypes from the standard CEPH reference pedigrees. There are three recent sets of genome-wide GL maps currently in use, all of which provide high-resolution, largely accurate, and convenient mapping information. These maps contain primarily the conveniently genotyped PCR-based microsatellite markers, use genotypes for only 8–15 of the 65 available CEPH pedigrees, and contain few, if any, gene-based or cytogenetically mapped markers. Many chromosome-specific linkage maps have also been constructed, many of which use a larger set of CEPH pedigrees and include hybridization- and gene-based markers. Over 11,000 markers have been genotyped in the CEPH pedigrees, and these genotypes have been deposited into the CEPH genotype database and are publicly available.

The first of the three genome-wide maps was produced by the Cooperative Human Linkage Center (CHLC; Murray et al., 1994). Last updated in 1997, the CHLC has identified, genotyped, and/or mapped over 3,300 microsatellite repeat markers. The CHLC Web site currently holds many linkage maps, including maps comprised solely of CHLC-derived markers and maps combining CHLC markers with those from other sources, including most markers in CEPHdb. CHLC markers can be recognized by unique identifiers that contain the nucleotide code for the tri- or tetranucleotide repeat units. For example, CHLC.GATA49A06 (D1S1608) contains a repeat unit of GATA, whereas CHLC.ATA28C07 (D1S1630) contains an ATA repeat. There are over 10,000 markers on the various linkage maps at CHLC, and most CHLC markers were genotyped in 15 CEPH pedigrees. The highest resolution CHLC maps have an average map distance of 1–2 cM between markers. Some of the maps contain markers in well-supported unique positions along with other markers placed into intervals.

Another set of genome-wide linkage maps was produced in 1996 by the group at Généthon (Dib et al., 1996). This group has identified and genotyped over 7,800 dinucleotide repeat markers and has produced maps containing only Généthon markers. These markers also have unique identifiers; each marker name has the symbols "AFM" at the beginning of the name. The Généthon map contains 5,264 genotyped in 8–20 CEPH pedigrees. These markers have been placed into 2,032 well-supported map positions, with an average map resolution of 2.2 cM. Because of homogeneity of their marker and linkage data and the RH and YAC-based mapping efforts at Généthon that incorporate many of their polymorphic markers, the Généthon map has become the most widely utilized human linkage map.

The third and most recent set of human maps was produced at the Center for Medical Genetics at the Marshfield Medical Research Foundation (Broman et al., 1998). This group has identified over 300 dinucleotide repeats and has constructed high-density maps using over 8,000 markers. Like the CHLC maps, the Marshfield maps include their own markers as well as others, such as markers from CHLC and Généthon. These maps have an average resolution of 2.3 cM per map interval. Markers developed at the Marshfield Foundation have an MFD identifier at the beginning of their names. The authors caution on their Web site that because only eight of the CEPH families were used for the map construction, the orders of some of the markers are not well determined. The Marshfield Web site provides a useful utility for displaying custom maps that contain user-specified subsets of markers.

Two additional linkage maps have been developed exclusively for use in performing efficient large-scale and/or genome-wide genotyping. The ABI PRISM linkage mapping sets are composed of dinucleotide repeat markers derived from the Généthon linkage map. The ABI marker sets are available at three different map resolutions (20, 10, and 5 cM), containing 811, 400, and 218 markers, respectively. The Center for Inherited Disease Research (CIDR), a joint program sponsored by The Johns Hopkins University and the National Institutes of Health, provides a genotyping service that uses 392 highly polymorphic tri- and tetranucleotide repeat markers spaced at an average resolution of 9 cM. The CIDR map is derived from the Weber v.9 marker set, with improved reverse primers and some additional markers added to fill gaps.

Although each of these maps is extremely valuable, it can be very difficult to determine marker order and intermarker distance between markers that are not all represented on the same linkage map. The MAP-O-MAT Web site at Rutgers University is a marker-based linkage map server that provides several map-specific queries. The server uses genotypes for over 12,000 markers (obtained from the CEPH database and from the Marshfield Foundation) and the CRI-MAP computer program to estimate map distances, perform two-point analyses, and assess statistical support for order for user-specified maps (Matise and Gitlin, 1999). Thus, rather than attempting to integrate markers from multiple maps by rough interpolation, likelihood analyses can be easily performed on any subset of markers from the CEPH database.

High-resolution linkage maps have also been constructed for many other species. These maps are often the most well-developed resource for animal species' whose genome projects are in early stages. The mouse and rat both have multiple genome-wide linkage maps (see MGD and the Rat Genome Database); other species with well-developed linkage maps include zebrafish, cat, dog, cow, pig, horse, sheep, goat, and chicken (O'Brien et al., 1999).

Radiation Hybrid Map Resources

Radiation hybrid maps provide an intermediate level of resolution between linkage and physical maps. Therefore, they are helpful for sequence alignment and will aid in completion of the human genome sequencing project. Three human whole-genome panels have been prepared with different levels of X-irradiation and are available for purchase from Research Genetics. Three high-resolution genome-wide maps have been constructed using these panels, each primarily utilizing EST markers. Mapping servers for each of the three human RH panels are available on-line to allow users to place their own markers on these maps. RH score data are deposited to, and publicly available from, The Radiation Hybrid Database (RHdb). Although this section covers RH mapping in humans, many RH mapping efforts are also underway in other species. More information regarding RH resources in all species are available at The Radiation Hybrid Mapping Information Web site.

In general, lower-resolution panels are most useful for more widely spaced markers over longer chromosomal regions, whereas higher-resolution panels are best for localizing very densely spaced markers over small regions. The lowest-resolution human RH panel is the Genebridge4 (GB4) panel (Gyapay et al., 1996). This panel contains 93 hybrids that were exposed to 3000 rads of irradiation. The maximum map resolution attainable by GB4 is 800–1,200 kb. An intermediate level panel was produced at the Stanford Human Genome Center (Stewart et al., 1997). The Stanford Generation 3 (G3) panel contains 83 hybrids exposed to 10,000 rads of irradiation. This panel can localize markers as close as 300–600 kb apart. The highest resolution panel ("The Next Generation," or TNG) was also developed at Stanford (Beasley et al., 1997). The TNG panel has 90 hybrids exposed to 50,000 rads of irradiation and can localize markers as close as 50–100 kb.

The Whitehead Institute/MIT Center for Genome Research constructed a map with approximately 6,000 markers using the GB4 panel (Hudson et al., 1995). Framework markers on this map were localized with odds ≥300:1, yielding a resolution of approximately 2.3 Mb between framework markers. Additional markers are localized to broader map intervals. A mapping server is provided for placing markers (scored in the GB4 panel) relative to the MIT maps.

The Stanford group has constructed a genome-wide map using the G3 RH panel (Stewart et al., 1997). This map contains 10,478 markers with an average resolution of 500 kb. Markers localized with odds = 1,000:1 are used to define "high-confidence bins," and additional markers are placed into these bins with lower odds. A mapping server is provided for placing markers scored in the G3 panel onto the SHGC G3 maps.

A fourth RH map has been constructed using both the G3 and GB4 panels. This combined map, the Transcript Map of the Human Genome (GeneMap '99; Fig. 6.4), was produced by the RH Consortium, an international collaboration between several groups (Deloukas et al., 1998). This map contains over 30,000 ESTs localized against a common framework of approximately 1,100 polymorphic Généthon markers. The markers were localized to the framework using the GB4 RH panel, the G3 panel, or both. The map includes the majority of human genes with known function. Most markers on the map represent transcribed sequences with unknown function. The order of the framework markers is well supported, but most ESTs are mapped relative to the framework with odds <1,000:1. The majority of markers on the GeneMap have a lod score <2.0, and many are <1.0. Such markers are localized with relatively low support for local order, and their map positions should be confirmed by other means if critical. A mapping server for placing markers on GeneMap '99 is available at the Sanger Centre.

The Radiation Hybrid Database (RHdb) is the central repository for all RH data. It is maintained at the European Bioinformatics Institute (EBI) in Cambridge, UK (Rodriguez-Tome and Lijnzaad, 2000). RHdb is a sophisticated Web- and FTP-based

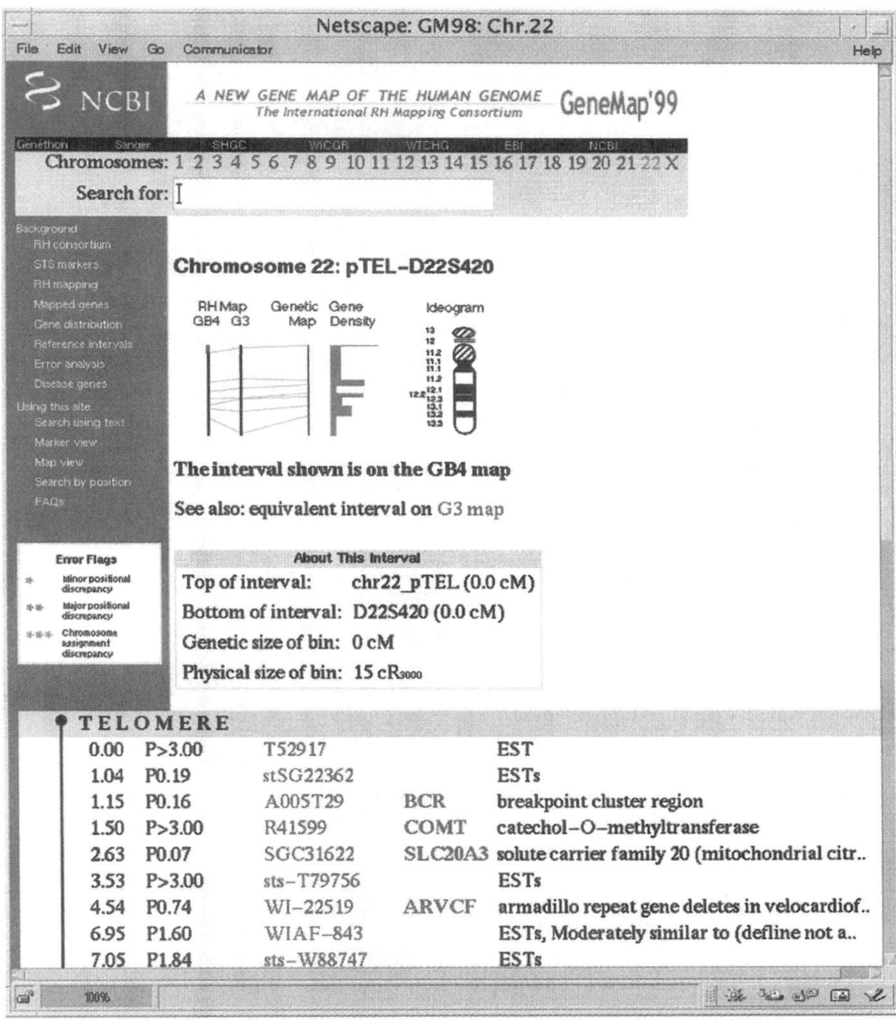

Figure 6.4. GeneMap '99. Example segment of The Human Gene Map, showing the first map interval on human chromosome 22q. Although the figure indicates that the map begins at a telomere, on this acrocentric chromosome, it actually begins near the centromere. The lower section of the figure contains 6 columns describing the elements mapped to this interval: column 1 gives cM linkage map positions for the polymorphic markers (none shown here); column 2 shows the computed cR position on either the GB4 or G3 portion of the GeneMap; column 3 contains either an `F` (for framework markers), or `P` followed by a number. This value represents the difference in statistical likelihood (lod score) for the given map position versus the next most likely position. A lod score of 3 is equivalent to odds of 1000:1 in favor of the reported marker position, 2 is equivalent to odds of 100:1, and a lod score of 1 represents odds of 10:1. Columns 4, 5, and 6 provide marker and gene names (if known).

searchable relational database that stores RH score data and RH maps. Data submission and retrieval are completely open to the public. Data are available in multiple formats or as flatfiles. Release 18.0 (September 2000) contained over 126,000 RH entries for 100,000 different STSs scored on 15 RH panels in 5 different species, as well as 91 RH maps.

STS Content Maps and Resources

Many physical mapping techniques have been used to order genomic segments for regional mammalian genome mapping projects. However, only RH and STS content/large-insert clone mapping methods have yielded the high throughput and automation necessary for whole-genome analysis to date, although advances in sequencing technology and capacity have recently made sequence-based mapping feasible. Two landmark achievements by the CEPH/Généthon and WICGR groups have mapped the entire human genome in YACs. The most comprehensive human physical mapping project is the collection of overlapping BAC and PAC clones being identified for the human DNA sequencing project, along with the now complete draft sequence of the human genome. This information is being generated by many different labs, and informatics tools to utilize the data are rapidly evolving.

The WICGR physical map is STS content based and contains more than 10,000 markers for which YAC clones have been identified, thus providing an average resolution of approximately 200 kb (Hudson et al., 1995). This map has been integrated with the Généthon GL and the WICGR RH maps. Together, the integration provides STS coverage of 150 kb, and approximately half the markers are expressed sequences also placed on GM99. The map was generated primarily by screening the CEPH MegaYAC library with primers specific for each marker and then by assembling the results by STS content analysis into sets of YAC contigs. Contigs are separately divided into "single-linked" and "double-linked," depending on the minimum number of YACs (one or two) required to simultaneously link markers within a contig. Predictably, the double-linked contigs are shorter and much more reliable than the single-linked ones, largely because of the high chimeric rate of the MegaYAC library. Thus, some skill is required for proper interpretation of the YAC-based data.

The WICGR Human Physical Mapping Project Home Page provides links to downloadable (but large) GIFs of the maps, a number of ways to search the maps, and access to the raw data. Maps can be searched by entering or selecting a marker name, keyword, YAC, or YAC contig. Text-based displays of markers list marker-specific information, YACs containing the marker, and details of the associated contig. Contig displays summarize the markers contained within them, along with their coordinates on the GL and RH maps, which is a very useful feature for assessing contig integrity. Details of which YACs contain which markers and the nature and source of each STS/YAC hit are also shown. Clickable STS content maps are also provided from the homepage, and users have the option of viewing the content map alone or integrated with the GL and RH maps. Although there are numerous conflicts between the GL, RH, and STS content maps that often require clarification with other techniques, this resource is very informative once its complexities and limitations are understood, especially where BAC/PAC/sequence coverage is not complete and in linking together BAC/PAC contigs.

The CEPH/Généthon YAC project is a similar resource to the WICGR project, also centered around screening of the CEPH MegaYAC library with a large set of STSs (Chumakov et al., 1995). Much of the CEPH YAC screening results have been incorporated into the WICGR data (those YAC/STS hits marked as C). However, the CEPH data includes YAC fingerprinting, hybridization of YACs to inter-Alu PCR products, and FISH localizations as complementary methods to confirm contig holdings. As with WICGR, these data suffer from the high YAC chimerism rate; long-range contig builds should be interpreted with caution, and the data are best used only as a supplement to other genomic data. The CEPH YAC Web site includes a rudimentary text search engine for STSs and YACs that is integrated with the Généthon GL map, and the entire data set can be downloaded and viewed using the associated QUICKMAP application (Sun OS only; Chumakov et al., 1995).

Much of the human draft sequence was determined from BAC libraries that have been whole-scale DNA fingerprinted and end sequenced. To date, over 346,000 clones have been fingerprinted by Washington University Genome Sequencing Center (WUGSC), and the clone coverage is sufficient to assemble large contigs spanning almost the entire human euchromatin. The fingerprinting data can be searched by clone name at the WUGSC Web site and provides a list of clones overlapping the input clone, along with a probability score for the likelihood of each overlap. Alternatively, users can download the clone database and analyze the raw data using the Unix platform software tools IMAGE (for fingerprint data) and FPC (for contig assembly), which are available from the Sanger Centre.

In parallel with the BAC fingerprinting, a joint project by The Institute for Genome Research (TIGR) and the University of Washington High-Throughput Sequencing Center (UWHTSC) has determined the insert-end sequences (STCs) of the WUGSC-fingerprinted clones (743,000 sequences). These data can be searched by entering a DNA sequence at the UWHTSC site or by entering a clone name at the TIGR site. Together with the fingerprinting data, this is a convenient way to build and analyze maps in silico. The fingerprinting and STC data have been widely used for draft sequence ordering by the human sequencing centers, and the BAC/PAC contigs displayed by the NCBI Map Viewer are largely assembled from these data.

Many human single-chromosome or regional physical maps are also available. Because other complex genome mapping projects are less well developed, the WICGR mouse YAC mapping project is the only whole-genome nonhuman physical map available. This map is arranged almost identically to its human counterpart and consists of 10,000 STSs screened against a mouse YAC library (Nusbaum et al., 1999). However, whole-genome mouse fingerprinting and STC generation projects similar to their human counterparts are currently in production by TIGR/UWHTSC and the British Columbia Genome Sequence Centre (BCGSC), respectively.

DNA Sequence

As mentioned above, the existing human and forthcoming mouse draft genomic sequences are excellent sources for confirming mapping information, positioning and orienting localized markers, and bottom-up mapping of interesting genomic regions. NCBI tools like BLAST (Chapter 8) can be very powerful in finding marker/sequence links. NCBI's LocusLink lists all homologous sequences, including genomic sequences, for each known human gene (genomic sequences are type "g" on the LocusLink Web site; Maglott et al., 2000). e-PCR results showing all sequences containing a specific marker are available at the GM99, dbSTS, GDB, and eGenome Web sites, where each sequence and the exact base pair position of the marker in the sequence are listed. Large sequence contigs can also be viewed schematically by NCBI's Entrez contig viewer and the Oakridge National Laboratory's Genome Channel web tool (Wheeler et al., 2000).

As the mammalian sequencing projects progress, a "sequence first" approach to mapping becomes more feasible. As an example, a researcher can go to the NCBI's human genome sequencing page and click on the idiogram of the chromosome of interest or on the chromosome number at the top of the page. Clicking on the idiogram shows an expanded idiogram graphically depicting all sequence contigs relative to the chromosome. Clicking on the chromosome number instead displays a list of all sequence contigs listed in order by cytogenetic and RH-extrapolated positions. These contigs can then be further viewed for clone, sequence, and marker content, and links to the relevant GenBank and dbSTS records are provided.

Integrated Maps and Genomic Cataloguing

GDB's Comprehensive Maps provide an estimated position of all genes, markers, and clones in GDB on a megabase scale. This estimate is generated by sequential pairwise comparison of shared marker positions between all publicly available genome-wide maps. This results in a consensus linear order of markers. At the GDB Web site, the Web page for each genomic element lists one or more maps on which the element has been placed, with the estimated Mb position of the marker on each map:

Element	Chromosome	Map	Coordinate	Units	EST MB	+/−
D1S228	1	GeneMap '99	782.0000	cR	32.2	0.0

This example shows that marker D1S228 has been placed 782 cR from the 1p telomere on GM99, and this calculates to 32.2 Mb from the telomere with the GDB mapping algorithm. Well-mapped markers such as the Généthon microsatellites generally have more reliable calculated positions than those that are mapped only once and/or by low-resolution techniques such as standard karyotype-based FISH. For chromosomes with complete DNA sequence available, the Mb estimates are very precise.

LDB and UDB are two additional sites that infer physical positions of a large, heterogeneous set of markers from existing maps using algorithms analogous to GDB's. Both Web sites have query pages where a map region can be selected by Mb coordinates, cytogenetic band, or specific marker names. The query results show a text-based list of all markers in the region ordered by their most likely positions, along with an estimated physical distance in Mb from the p telomere. LDB also displays the type of mapping technique(s) used to determine the comprehensive position, the position of the marker in each underlying single-dimension map, and appropriate references. An added feature of the UDB site is its provision of marker-specific links to other genomic databases. At present, there are no graphical depictions for either map.

Physical map positions derived from the computationally based algorithms used by GDB, LDB, and UDB are reliant on the accuracy and integrity of the underlying maps used to determine the positions. Therefore, these estimates serve better as initial localization guides and as supportive ordering information rather than as a primary ordering mechanism. For instance, a researcher defining a disease locus to a chromosome band or between two flanking markers can utilize these databases to quickly collect virtually all mapped elements in the defined region, and the inferred physical positions serve as an approximate order of the markers. This information would then be supplanted by more precise ordering information present in single-dimension maps and/or from the researcher's own experimental data.

The eGenome project uses a slightly different approach for creating integrated maps of the human genome (White et al., 1999). All data from RHdb are used to generate an RH framework map of each chromosome by a process that maximizes the number of markers ordered with high confidence (1,000:1 odds). This extended, high-resolution RH framework is then used as the central map scale from which the high-confidence intervals for additional RH and GL markers are positioned. As with GDB, the absolute base pair positions of all markers are calculated for chromosomes that have been fully sequenced. eGenome also integrates UniGene EST clusters, large-insert clones, and DNA sequences associated with mapped markers, and it also infers cytogenetic positions for all markers. The eGenome search page allows querying by marker name or GenBank accession ID or by defining a region with cytogenetic band or flanking marker coordinates. The marker displays include the RH and GL (if applicable) positions, large-insert clones containing the marker, cytogenetic position, and representative DNA sequences and UniGene clusters. Advantages of eGenome include the ability to view regions graphically using GDB's Mapview, exhaustive cataloguing of marker names, and an extensive collection of marker-specific hypertext links to related database sites. eGenome's maps are more conservative than GDB, LDB, and UDB as they show only the high-confidence locations of markers (often quite large intervals). Researchers determining a regional order *de novo* would be best advised to use a combination of these integrated resources for initial data collection and ordering.

Because of the large number of primary data sources available for human genome mapping, ensuring that the data collected for a specific region of interest are both current and all-inclusive is a significant task. Genomic catalogues help in this regard, both to provide a single initial source containing most of the publicly available genomic information for a region and to make the task of monitoring new information easier. Human genomic catalogues include the NCBI, GDB, and eGenome Web sites. NCBI's wide array of genomic data sets and analysis tools are extremely well integrated, allowing a researcher to easily transition between marker, sequence, gene, and functional information. GDB's concentration on mapped genomic elements makes it the most extensive source of positional information, and its inclusion of most genomic maps provides a useful mechanism to collect information about a defined region. eGenome also has powerful "query-by-position" tools to allow rapid collection of regional information. No existing database is capable of effectively organizing and disseminating all available human genomic information. However, the eGenome, GDB, and NCBI Web sites faithfully serve as genomic Web portals by providing hyperlinks to the majority of data available for a given genomic locus.

WICGR's mouse mapping project and the University of Wisconsin's Rat Genome Database (RGD; Steen et al., 1999) have aligned the GL and RH maps for the respective species in a comparative manner. MGD's function as a central repository for mouse genomic information makes it useful as a mouse genomic catalogue, and, increasingly, RGD can be utilized as a rat catalogue. Unfortunately, other complex species' genome projects have not yet progressed to the point of offering true integrated maps or catalogues.

Comparative Resources

Comparative maps provide extremely valuable tools for studying the evolution and relatedness of genes between species and finding disease genes through position-based orthology. There are several multispecies comparative mapping resources available that include various combinations of most animal species for which linkage maps are available. In addition, there are also many sequence-based comparative analysis resources (Chapter 15). Each resource has different coverage and features. Presently, it is necessary to search multiple resources, as no single site contains all of the currently available homology information. Only the most notable resources will be described here.

A good starting point for homology information is NCBI's LocusLink database. The LocusLink reports include links to HomoloGene, a resource of curated and computed cross-species gene homologies (Zhang et al., 2000). Currently, HomoloGene contains human, mouse, rat, and zebrafish homology data. For example, a LocusLink search of all organisms for the gene PMP22 (peripheral myelin protein) returns three entries, one each for human, mouse, and rat. At the top of the human PMP22 page is a link to HOMOL (HomoloGene). HomoloGene lists six homologous elements, including the rat and mouse Pmp22 genes, as well as additional mouse UniGene cluster and a weakly similar zebrafish UniGene cluster. The availability of both curated and computed homology makes this a unique resource. However, the lack of integrated corresponding homology maps is a disadvantage.

The MGD does provide homology maps that simplify the task of studying conserved chromosome segments. Homologies are taken from the reported literature for mouse, human, rat, and 17 other species. Homology information can be obtained in one of three manners: searching for genes with homology information, building a comparative linkage map, or viewing an Oxford Grid. The simple search returns detailed information about homologous genes in other species, including map positions and codes for how the homology was identified, links to the relevant references, and links for viewing comparative maps of the surrounding regions in any two species. For example, a homology search for the Pmp22 gene returns a table listing homologous genes in cattle, dog, human, mouse, and rat. Figure 6.5 shows the mouse-human comparative map for the region surrounding Pmp22 in the mouse. A comparative map can also be obtained by using the linkage map-building tool to specify a region of the mouse as the index map and to select a second, comparison, species. The resulting display is similar to that shown in Figure 6.5. An Oxford Grid can also be used to view a genome-wide matrix in which the number of gene homologies between each pair of chromosomes between two species is shown. This view is currently available for seven species. Further details on the gene homologies can be obtained via the links for each chromosome pair shown on the grid. The map-viewing feature of MGD is quite useful; however, the positions of homologous nonmouse genes are only cytogenetic, so confirmation of relative marker order within

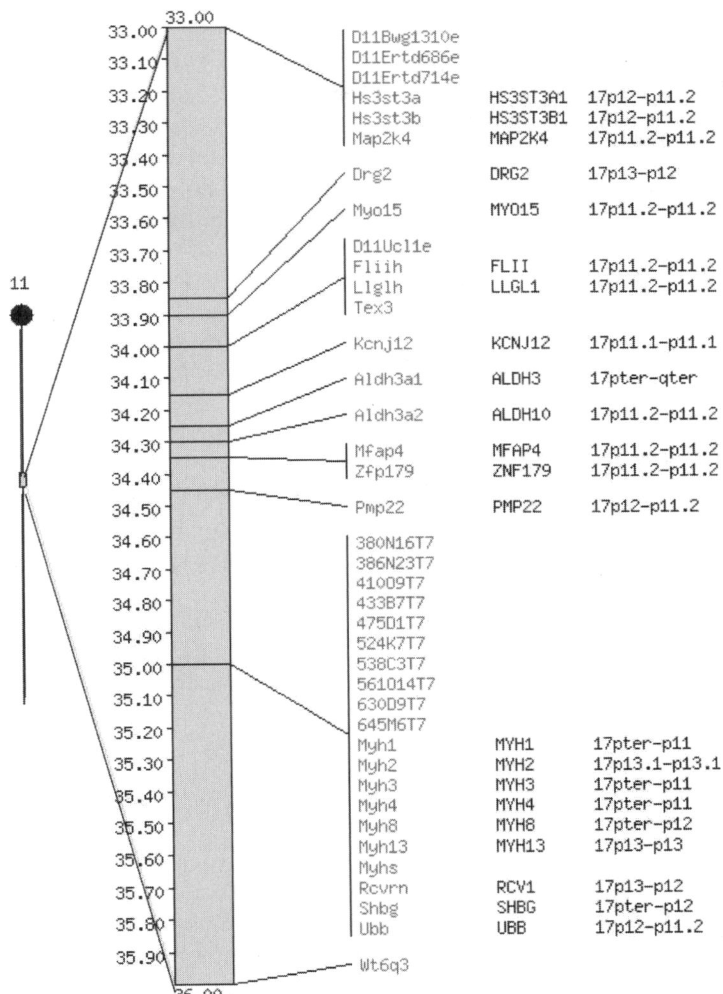

Figure 6.5. MGD mouse-human comparative map of the region surrounding the mouse Pmp22 gene. Pmp22 is on mouse chromosome 11 at the position 34.5 cM on the mouse linkage map. As shown by the human genes displayed on the right, a segment of human chromosome 17 is homologous to this mouse region.

small regions is not possible. It is also possible to view MGD homology information using GDB (Gatewood and Cottingham, 2000).

In silico mapping is proving to be a very valuable tool for comparative mapping. The Comparative Mapping by Annotation and Sequence Similarity (COMPASS) approach (Ma et al., 1998) has been used by researchers studying the cattle genome to construct cattle-human comparative maps with 638 identified human orthologs (Band et al., 2000). Automated comparison of cattle and human DNA sequences, along with the available human mapping information, facilitated localization predictions for tens of thousands of unmapped cattle ESTs. The COMPASS approach has been shown to have 95% accuracy. The Bovine Genome Database displays the gene-based comparative maps, which also integrate mouse homologies. A similar approach is being used at the Bioinformatics Research Center at the Medical College of Wisconsin. Here, human, rat, and mouse radiation hybrid maps are coupled with theoretical gene assemblies based on EST and cDNA data (such as the UniGene set at NCBI) for all three species and provide the fundamental resources allowing for the creation of iteratively built comparative maps (Tonellato et al., 1999). Homologies with uniformly mapped ESTs form the anchor points for the comparative maps. This work has, so far, identified 8,036 rat-human, 13,720 rat-mouse, and 9,745 mouse-human UniGene homologies, most mapped on one or all of the organisms. The creation of these comparative maps is an iterative exercise that is repeated as the radiation hybrid maps, ESTs, and UniGene rebuilds are developed. In addition, the algorithm predicts the placement of unmapped assemblies relative to the anchor information, providing a powerful environment for "virtual mapping" before radiation hybrid or other wet-lab methods are used to confirm the predictions.

Another project utilizing electronic mapping has developed a high-resolution human/mouse comparative map for human chromosome 7. Recent efforts have greatly increased the number of identified gene homologies and have facilitated the construction of sequence-ready BAC-based physical maps of the corresponding mouse regions (Thomas et al., 2000).

An additional notable resource details homology relationships between human, mouse, and rat. Derived from a high-resolution RH maps, homologies for over 500 genes have been identified and are available in tabular format at a user-friendly Web site (Watanabe et al., 1999).

Single-Chromosome and Regional Map Resources

Although whole-genome mapping resources are convenient for initial collection and characterization of a region of interest, data generated for only a single chromosome or a subchromosomal region are often important for fine mapping. In many cases, these regional maps contain more detailed, better integrated, and higher resolution data than the whole-genome maps can provide. There are numerous such data sets, databases, and maps available for each human chromosome, although little regional information is yet available on-line for other complex genomes. Most published human chromosome maps are listed and can be viewed at GDB's Web site (see above).

Another excellent resource is the set of human chromosome-specific Web sites that have been created by various groups. Recently, the Human Genome Organization (HUGO) has developed individual human chromosome Web pages, each of which is maintained by the corresponding HUGO chromosome committees. Each page has links to chromosome-specific information from a variety of mapping sources, most of them being chromosome-specific subsets of data derived from whole-genome resources (such as the chromosome 7 GL map from Généthon). At the top of most HUGO chromosome pages are links to other chromosome pages designed by groups mapping the specific chromosome. These sites vary widely in their utility and content; some of the most useful are briefly mentioned below. The sites offer a range of resources, including chromosome- and/or region-specific GL, RH, cytogenetic, and physical maps; DNA sequence data and sequencing progress, single chromosome databases and catalogues of markers, clones, and genomic elements; and links to related data and resources at other sites, single chromosome workshop reports, and chromosome E-mail lists and discussion forums.

The major genome centers often include detailed mapping and sequence annotation for particular chromosomes at their sites. The Sanger Centre and the WUGSC have two of the most advanced collections of chromosome-specific genomic data, informatics tools, and resources. Sanger has collected and generated most available mapping data and reagents for human chromosomes 1, 6, 9, 10, 13, 20, 22, and X. These data are stored and displayed using ACeDB, which can be utilized through a Web interface (WEBACE) at the Sanger Web site or, alternatively, downloaded onto a local machine (Unix OS). ACeDB is an object-oriented database that provides a convenient, relational organizational scheme for storing and managing genomic data, as well as for viewing the information in both text-based and graphical formats. ACeDB is the database of choice for most researchers tackling large genomic mapping projects. WUGSC has recently implemented single-chromosome ACeDB sequence and mapping databases for most human chromosomes, each of which has a Web interface.

The Human Chromosome 1 Web site is an example of a community-based approach to genomic research. This site includes a repository for chromosome data generated by several labs, an extensive list of hyperlinks to chromosome 1 data, an E-mail list and discussion forum, a listing of chromosome 1 researchers and their interests, and several workshop reports. The University of Texas at San Antonio's chromosome 3 site contains a database of large-insert clones and markers along with GL, RH, cytogenetic, and comparative maps. The University of California-Irvine has an on-line chromosome 5 ACeDB database, whereas the Joint Genome Institute (JGI) maintains chromosome 5 large-insert clone maps and some external Web links at their site. The University of Toronto chromosome 7 Web site includes a searchable comprehensive chromosome 7 database containing markers, clones, and cytogenetic information; this site also has a long list of chromosome links. Also, the National Human Genome Research Institute's chromosome 7 Web site contains a YAC/STS map, a list of ESTs, and integration with chromosome 7 sequence files. The University College London maintains a good comprehensive resource of chromosome 9 genomic links, an E-mail group, workshop reports, and a searchable chromosome 9 database. Genome Therapeutics Corporation has developed an inclusive Web site for chromosome 10. This site has both GL/physical and integrated sequence-based maps, links to related data, and workshop reports.

Imperial College maintains a searchable chromosome 11 database at their chromosome 11 Web site, whereas the chromosome 16 Web site at JGI contains restriction-mapped BAC and cosmid contigs and determined sequence, along with a list of chromosome 16 hyperlinks. A similar JGI resource for chromosome 19 includes a completely integrated physical map with sequence links and a list of external resources. The University of Colorado, the RIKEN Genomic Sciences Center, and the Max Planck Institute for Molecular Genetics (MPIMG) have an interconnected set of resources that together integrate genomic clones, markers, and sequence for the completely sequenced chromosome 21. The Sanger Centre and the LDB have comprehensive resources for the viewing and analysis of chromosome 22. It is expected that additional resources for all completely sequenced chromosomes will be available soon. The resources for the X chromosome are most impressive. The MPIMG has established a complete genomic catalogue of this chromosome that features integration of genomic mapping and sequence data derived from many sources and experimental techniques. These data can be viewed graphically with the powerful online Java application derBrowser. Finally, the sequenced and well-characterized mitochondrial genome is well displayed at Emory University, where a highly advanced catalogue encompassing both genomic and functional information has been established.

PRACTICAL USES OF MAPPING RESOURCES

Potential applications of genomic data are numerous and, to a certain extent, depend on the creativity and imagination of the researcher. However, most researchers utilize genomic information in one of three ways: to find out what genomic elements—usually transcribed elements—are contained within a genomic region, to determine the order of defined elements within a region, or to determine the chromosomal position of a particular element. Each of these goals can be accomplished by various means, and the probability of efficient success is often enhanced by familiarity with many of the resources discussed in this chapter. It is prudent to follow a logical course when using genomic data. During the initial data acquisition step, in which genomic data are either generated experimentally or retrieved from publicly available data sources, simultaneous evaluation of multiple data sets will ensure both higher resolution and greater confidence while increasing the likelihood that the genomic elements of interest are represented. Second, the interrelationships and limitations of the data sets must be sufficiently understood, as it is easy to overinterpret or underrepresent the data. Finally, it is important to verify critical assignments independently, especially when using mapping data that are not ordered with high confidence. Below, we give some brief suggestions on how to approach specific map-related tasks, but many modifications or alternative approaches are also viable. The section is organized in a manner similar to a positional cloning project, starting with definition of the region's boundaries, determining the content and order of elements in the region, and defining a precise map position of the targeted element.

Defining a Genomic Region

A genomic region of interest is best defined by two flanking markers that are commonly used for mapping purposes, such as polymorphic Généthon markers in humans or MIT microsatellites in mice. Starting with a cytogenetically defined region is more difficult due to the subjective nature of defining chromosomal band boundaries. Conversion of cytogenetic boundaries to representative markers can be approximated by viewing the inferred cytogenetic positions of markers in comprehensive maps such as GDB's universal map, UDB, LDB, or eGenome. Because these cytogenetic positions are inferred and approximate, a conservative approach is recommended when using cytogenetic positions for region definition. The choice of flanking markers will impact how precisely a region's size and exact boundary locations can be defined. Commonly used markers are often present on multiple, independently derived maps, so their "position" on the chromosome provides greater confidence for anchoring a regional endpoint. In contrast, the exact location of less commonly used markers is often locally ambiguous. These markers can sometimes be physically tethered to other markers if a large sequence tract that contains multiple markers can be found.

This can be performed by BLASTing marker sequences against GenBank or by scanning e-PCR results in UniGene or eGenome for a particular marker.

Determining and Ordering the Contents of a Defined Region

Once a region has been defined, there are a number of resources available for determining what lies within the region. A good way to start is to identify a map that contains both flanking markers, either from a chromosome-wide or genome-wide map from the sources listed above, from a genomic catalogue, or from a local map that has been generated by a laboratory interested in this particular region. For humans, GDB is the most inclusive map repository, although many regional maps have not been deposited in GDB but can be found with a literature search of the corresponding cytogenetic band or a gene known to map to the region. Many localized maps are physically based and are more accurate than their computationally derived, whole-chromosome counterparts. For other species, the number of maps to choose from is usually limited, so it is useful to first define flanking markers known to be contained in the available maps.

The map or maps containing the flanking markers can then be used to create a consensus integrated map of the region. This is often an inexact and tedious process. To begin, it is useful to identify from the available maps an index map that contains many markers, high map resolution, and good reliability. Integration of markers from additional maps relative to the index map proceeds by comparing the positions of markers placed on each map. For example, if an index map contains markers in the order A-B-C-D and a second map has markers in the order B-E-D, then marker E can be localized to the interval between markers B and D on the index map. Importantly, however, the relative position of marker E with respect to marker C usually cannot be accurately determined by this method. Repeated iterations of this process should allow localization of all markers from multiple maps relative to the index map. This process is of course significantly reinforced by experimental verification, such as with STS content mapping of large-insert clones identified for the region-specific markers or, ideally, by sequence-determined order.

Each marker represents some type of genomic element: a gene, an EST, a polymorphism, a large-insert clone end, or a random genomic stretch. In humans, identifying what a marker represents is relatively straightforward. Simply search for the marker name in GDB or eGenome, and, in most cases, the resulting Web display will provide a summary of what the marker represents, usually along with hyperlinks to relevant functional information. For mice, MGD provides a similar function to GDB. For other organisms, the best source is usually either dbSTS or, if present, Web sites or publications associated with the underlying maps. GenBank and dbSTS are alternatives for finding markers, but, because these repositories are passive (requiring researchers to submit their markers rather than actively collecting markers), many marker sets are not represented. If a marker is known to be expressed, UniGene, LocusLink, and dbEST are excellent sources of additional information. Many genes and some polymorphisms have been independently discovered and developed as markers multiple times, and creating a nonredundant set from a collection of markers is often challenging. GDB, eGenome, MGD, and (for genes) UniGene are good sources to use for finding whether two markers are considered equivalent but even more reliable is a DNA sequence or sequence contig containing both marker's primers. BLAST and the related BLAST2 are efficient for quickly determining sequence relatedness (Chapter 8).

Obviously, the most reliable tool for marker ordering is a DNA sequence or sequence contig. For expressed human markers, searching with the marker name in UniGene or Entrez Genomes returns a page stating where (or if) the marker has been mapped in GeneMap '99 and other maps, a list of mRNA, genomic, and EST sequences, and with Entrez Genomes, a Mapviewer-based graphical depiction of the maps, sequence-ready contigs, and available sequence of the region. Similarly, GDB and eGenome show which DNA sequences contain each displayed marker. For other markers, the sequence from which the marker is derived, or alternatively one of the primer sequences, may be used to perform a BLAST search that can identify completely or nearly homologous sequences. The nonredundant, EST, GSS, and HTGS divisions of GenBank are all potentially relevant sources of matching sequence, depending on the aim of the project. Only long sequences are likely to have worthwhile marker-ordering capabilities. Finished genomic sequence tracts have at least some degree of annotation, and scanning the GenBank record for the large sequence will often yield an annotated list of what markers lie within the sequence and where they are. Keep in mind that such annotations vary considerably in their thoroughness and most are fixed in time; that is, they only recognize markers that were known at the time of the annotation. BLAST, BLAST2, or other sequence-alignment programs are helpful in identification or confirmation of what might lie in a large sequence. Also, the NCBI e-PCR Web interface can be used to identify all markers in dbSTS contained within a given sequence, and this program can be installed locally to query customized marker sets with DNA sequences (Schuler, 1997).

For genomes for which DNA sequencing is complete or is substantially underway, it may be possible to construct local clone or sequence contigs. Among higher organisms, this is currently possible only for the human and mouse genomes. Although individual clone sequences can be found in GenBank, larger sequence contigs —sequence tracts comprising more than one BAC or PAC—are more accessible using the Entrez Genomes Web site (see above). Here, by entering a marker or DNA accession number into the contigs search box, researchers can identify sequence contigs containing that marker or element. This site also provides a graphical view of all other markers contained in that sequence, the base pair position of the markers in the sequence, and, with the Mapviewer utility, graphical representations of clone contigs. This process can also be performed using BLAST or e-PCR, although it is somewhat more laborious.

Once a sequence has been identified for markers in a given region, YAC clone, DNA fingerprinting, and STC data can be used to bridge gaps. For humans and mice, the WICGR YAC data provide a mechanism for identifying YAC clones linking adjacent markers. However, caution should be exercised to rely mainly on double-linked contigs and/or to experimentally confirm YAC/marker links. Also for human genome regions, the UWHTSC and TIGR Web sites for identifying STCs from DNA sequence or BAC clones are very useful. For example, researchers with a sequence tract can go to the UWHTSC TSC search page, enter their sequence, and find STCs contained in the sequence. Any listed STC represents the end of a BAC clone whose insert contains a portion of the input sequence (Venter et al., 1996). The TIGR search tool is complementary to the UWHTSC search, as the TIGR site requires input of a large-insert clone name, which yields STC sequences. STCs represent large-insert clones that potentially extend a contig or link two adjacent, nonoverlapping contigs.

Similarly, ~375,000 human BAC clones have been fingerprinted for rapid identification of overlapping clones (Marra et al., 1997). The fingerprinting data are available for searching at the Washington University Human Genome Sequencing Center (WUGSC). Combined use of Entrez, BLAST, the STC resources, and the BAC fingerprinting data can often provide quick and reliable contig assembly by in silico sequence and clone walking.

Defining a Map Position From a Clone or DNA Sequence

Expressing the chromosomal position of a gene or genomic element in physical, RH, GL, or cytogenetic terms is not always straightforward. The first approach is to determine whether the element of interest has already been localized. The great majority of human transcripts are precisely mapped, and many genes have been well localized in other organisms as well. For species with advanced DNA sequencing projects, it is helpful to identify a large DNA sequence tract containing the genomic element of interest and then determine what markers it contains by looking at the sequence annotation record in GenBank or by e-PCR. Identified human and mouse genes are catalogued in GDB and LocusLink or MGD, respectively, and searching UniGene with a marker name, mRNA or EST sequence accession number, or gene name will often provide a localization if one is known. Here again, nomenclature difficulties impede such searches, making it necessary to search each database with one or more alternate names in some cases. Another alternative is to determine if the genomic element is contained in a genomic sequence by a simple BLAST search. Most large genomic sequences have been cytogenetically localized, and this information is contained in the sequence annotation record (usually in the title).

If gene-specific or closely linked markers have been used previously for mapping, a position can usually be described in terms specific to the mapping method that was employed. For example, if an unknown gene is found to map very close to a Généthon marker, then the gene position can be reported relative to the Généthon GL centiMorgan coordinates. Most human markers and many maps have been placed in GDB, so this is a good first step in determining whether a marker has been mapped. Simply search for the relevant marker and see where it has been placed on one or several maps listed under "cytogenetic localizations" and "other localizations." Inferred cytogenetic positions of human genes and markers are usually listed in GDB, UniGene, and eGenome if the elements have been previously mapped. If not, band or band range assignments can usually be approximated by finding the cytogenetic positions of flanking or closely linked markers and genes. Many sequenced large-insert clones have been assigned by FISH to a cytogenetic position; this information can usually be found in the sequence annotation or at the clone originator's Web site. The process of determining whether a transcript or genomic element from another organism has been mapped varies somewhat due to the lack of extensive genomic catalogs, making it usually necessary to cross-reference a marker with the GL and/or RH maps available for the species.

If no previous localization exists for a genomic element, some experimental work must be undertaken. For human and mouse markers, an efficient and precise way to map a sequence-based element is to develop and map an STS derived from the element by RH analysis. A set of primers should be designed that uniquely amplify a product in the species of interest, but not in the RH panel background genome. An STS is usually unique if at least one primer is intronic. Primers designed from an mRNA sequence will not amplify the correct-sized product in genomic DNA if they span an intron, but a good approximation is to use primers from the 3′ untranslated region, as these stretches only rarely contain introns and usually comprise sequences divergent enough from orthologous or paralogous sequences. However, beware of pseudogenes and repetitive sequences, and genomic sequence stretches are superior for primer design. Suitable primers can then be used to type an appropriate RH panel; currently, human (G3, GB4, or TNG), mouse, rat, baboon, zebrafish, dog, horse, cow and pig panels are available commercially. After the relevant panel is typed, the resulting data can be submitted to a panel-specific on-line RH server (see above) for the human, mouse, rat, and zebrafish panels. For other species, isolation and FISH of a large-insert clone or GL mapping with an identified or known flanking polymorphism may be necessary.

INTERNET RESOURCES FOR TOPICS PRESENTED IN CHAPTER 6

Data Repositories

The Genome DataBase (GDB)	*http://www.gdb.org/*
National Center for Biotechnology Information (NCBI)	
Home Page	*http://www.ncbi.nlm.nih.gov*
Entrez Genomes Division	*http://www.ncbi.nlm.nih.gov/Entrez/Genome/main_genomes.html*
LocusLink	*http://www.ncbi.nlm.nih.gov/LocusLink/*
GeneMap'99	*http://www.ncbi.nlm.nih.gov/genemap99/*
OMIM	*http://www.ncbi.nlm.nih.gov/Omim/*
HomoloGene	*http://www.ncbi.nlm.nih.gov/HomoloGene/*
BLAST	*http://www.ncbi.nlm.nih.gov/BLAST/*
ePCR	*http://www.ncbi.nlm.nih.gov/STS/*
Entrez sequence viewer	*http://www.ncbi.nlm.nih.gov/genome/seq/*
GenBank	*http://www.ncbi.nlm.nih.gov/Genbank*
Genomic Biology page	*http://www.ncbi.nlm.nih.gov/Genomes*
dbSTS	*http://www.ncbi.nlm.nih.gov/dbSTS*
Mouse Genome Informatics (MGD/MGI)	*http://www.informatics.jax.org/*

Resources and Projects

Cytogenetic	
BAC	*http://bacpac.med.buffalo.edu/human/overview.html*
LBNL/UCSF RMC	*http://ioerror.ucsf.edu:8080/~dfdavy/rmc/OUTSIDE.html*
U. of Bari	*http://bioserver.uniba.it/fish/rocchi*
Cytogenetic/YAC data	*http://www.mpimg-berlin-dahlem.mpg.de/~cytogen/probedat.htm*
NCI	*http://www.ncbi.nlm.nih.gov/CGAP/*
CCAP	*http://www.ncbi.nlm.nih.gov/CCAP/mitelsum.cgi*
Infobiogen	*http://www.infobiogen.fr/services/chromcancer/*
SERGG	*http://www.ir.miami.edu/genetics/sergg/chromosome.html*

Coriell	*http://locus.umdnj.edu/nigms/ideograms/1.html*
ARKdb	*http://www.ri.bbsrc.ac.uk/bioinformatics/ark_overview.html*
Animal Genome Database	*http://ws4.niai.affrc.go.jp/jgbase.html*
Cedars-Sinai	*http://www.csmc.edu/genetics/korenberg/korenberg.html#A*
Genetic Linkage	
CEPH Genotype Database	*http://www.cephb.fr/cephdb/*
CHLC	*http://lpg.nci.nih.gov/CHLC/*
Généthon	*http://www.genethon.fr/genethon_en.html*
Marshfield	*http://www.marshmed.org/genetics/*
MAP-O-MAT	*http://compgen.rutgers.edu/mapomat*
Rat Genome Database	*http://www.lgr.mcw.edu/projects/rgd.html*
Radiation Hybrid	
RHdb	*http://www.ebi.ac.uk/RHdb/*
RH Information Page	*http://compgen.rutgers.edu/rhmap/*
Research Genetics	*http://www.resgen.com*
WICGR RH Maps	*http://www-genome.wi.mit.edu/cgi-bin/contig/phys_map*
WICGR GB4 RH Map Server	*http://www-genome.wi.mit.edu/cgi-bin/contig/rhmapper.pl*
SHGC RH Maps	*http://shgc-www.stanford.edu/Mapping/rh/*
SHGC G3 Map Server	*http://shgc-www.stanford.edu/RH/*
Sanger Centre GB4/GM Map server	*http://www.sanger.ac.uk/Software/RHserver/*
STS content	
WICGR human physical map	*http://carbon.wi.mit.edu:8000/cgi-bin/contig/phys_map*
CEPH/Généthon YAC map	*http://www.cephb.fr/bio/ceph-genethon-map.html*
WUGSC home	*http://genome.wustl.edu/gsc/index.shtml*
TIGR STCs	*http://www.tigr.org/tdb/humgen/bac_end_search/bac_end_intro.html*
UWHTSC STCs	*http://www.htsc.washington.edu/human/info/index.cfm*
WUGSC BAC fingerprints	*http://genome.wustl.edu/gsc/human/human_database.shtml*
UBGSC mouse BAC fingerprints	*http://www.bcgsc.bc.ca/projects/mouse_mapping/*
Trask	*http://fishfarm.biotech.washington.edu/BACResource/Random/index.html*
WICGR mouse physical/genetic map	*http://carbon.wi.mit.edu:8000/cgi-bin/mouse/index*
DNA Sequence	
see NCBI links	
ORNL Genome Channel	*http://compbio.ornl.gov/tools/channel/*
Integrated and Catalogs	
UDB	*http://bioinformatics.weizmann.ac.il/udb/*
LDB	*http://cedar.genetics.soton.ac.uk/public_html/ldb.html*

LDB Sequence-based maps	*http://cedar.genetics.soton.ac.uk/public_html/LDB2000.html*
eGenome	*http://genome.chop.edu*
Comparative	
Mouse Homology	*http://www.informatics.jax.org/menus/homology_menu.shtml*
Otsuka/Oxford rat-mouse-human	*http://ratmap.ims.u-tokyo.ac.jp/*
Human Chromosome 7–mouse map	*http://genome.nhgri.nih.gov/chr7/comparative/*
Bovine Genome Database	*http://bos.cvm.tamu.edu/bovgbase.html*
MCW Rat-Mouse-Human	*http://rgd.mcw.edu*
Single-chromosome/regional	
1 Rutgers	*http://linkage.rockefeller.edu/chr1/*
3 UTSA	*http://apollo.uthscsa.edu/*
5 UCI	*http://chrom5.hsis.uci.edu*
5 JGI	*http://jgi.doe.gov/Data/JGI_mapping.html*
7 HSC	*http://www.genet.sickkids.on.ca/chromosome7/*
7 NHGRI	*http://www.nhgri.nih.gov/DIR/GTB/CHR7*
9 UCL	*http://www.gene.ucl.ac.uk/chr9/*
10 GTC	*http://www.cric.com/sequence_center/chromosome10/*
11 Imperial College	*http://chr11.bc.ic.ac.uk/*
16 JGI	*http://jgi.doe.gov/Data/JGI_mapping.html*
19 JGI	*http://jgi.doe.gov/Data/JGI_mapping.html*
21 Colorado	*http://www-eri.uchsc.edu/chromosome21/frames.html*
21 RIKEN	*http://hgp.gsc.riken.go.jp/chr21/index.html*
21 MPIMG	*http://chr21.rz-berlin.mpg.de/*
X	*http://www.mpimg-berlin-dahlem.mpg.de/~xteam/*
Mito Emory	*http://infinity.gen.emory.edu/mitomap.html*
HUGO Chromosome resources	*http://www.gdb.org/hugo/*
Sanger Centre	*http://www.sanger.ac.uk/HGP/*
ACEDB	*http://www.acedb.org/*

PROBLEM SET

You have performed a large-scale genome-wide search for the gene for the inherited disorder Bioinformatosis. Initial analyses have identified one region with significant results, flanked by the markers D21S260–D21S262. There are many genes mapping within this region, one of which is particularly interesting, superoxide dismutase 1 (SOD1).

1. What is the cytogenetic location of this gene (and hence, at least part of the region of interest)?

2. How large is this region in cM?
3. What polymorphic markers can be identified in this region (that you might use to try to narrow the region)? Choose six of these polymorphic markers. Based on the chosen markers, can a map based on these markers be identified or constructed?
4. What STS markers have been developed for SOD1? What are their map positions on the Human Transcript Map (GeneMap '99)? Are these positions statistically well-supported? Have any SNP markers been identified within SOD1?
5. What other genes are in this region?
6. Has the region including the SOD1 gene been sequenced? What contigs and/or clones contain SOD1?
7. Have orthologous regions been identified in any other species?
8. Have mutations in SOD1 been associated with any diseases other than Bioinformatosis?

REFERENCES

Adams, M. D., Kelley, J. M., Gocayne, J. D., Dubnick, M., Polymeropoulos, M. H., Xiao, H., Merril, C. R., Wu, A., Olde, B., Moreno, R. F., et al. (1991). Complementary DNA sequencing: expressed sequence tags and human genome project. *Science* 252, 1651–1656.

Aston, C., Mishra, B., and Schwartz, D. C. (1999). Optical mapping and its potential for large-scale sequencing projects. *Trends Biotechnol.* 17, 297–302.

Band, M. R., Larson, J. H., Rebeiz, M., Green, C. A., Heyen, D. W., Donovan, J., Windish, R., Steining, C., Mahyuddin, P., Womack, J. E., and Lewin, H. A. (2000). An ordered comparative map of the cattle and human genomes. *Genome Res.* 10, 1359–1368.

Beasley, E., Stewart, E., McKusick, K., Aggarwal, A., Brady-Hebert, S., Fang, N., Lewis, S., Lopez, F., Norton, J., Pabla, H., Perkins, S., Piercy, M., Qin, F., Reif, T., Sun, W., Vo, N., Myers, R., and Cox, D. (1997). The TNG4 radiation hybrids improve the resolution of the G3 panel. *Am. J. Hum. Genet.* 61(Suppl.), A231.

Blake, J. A., Eppig, J. T., Richardson, J. E., and Davisson, M. T. (2000). The mouse genome database (MGD): expanding genetic and genomic resources for the laboratory mouse. *Nucleic Acids Res.* 28, 108–111.

Boehnke, M., Lange, K., and Cox, D. R. (1991). Statistical methods for multipoint radiation hybrid mapping. *Am. J. Hum. Genet.* 49, 1174–1188.

Broman, K. W., Murray, J. C., Sheffield, V. C., White, R. L., and Weber, J. L. (1998). Comprehensive human genetic maps: individual and sex-specific variation in recombination. *Am. J. Hum. Genet.* 63, 861–869.

Burke, D. T., Carle, G. F., and Olson, M. V. (1987). Cloning of large segments of exogenous DNA into yeast by means of artificial chromosome vectors. *Science* 236, 806–812.

Chakravarti, A., and Lynn, A. (1999). Meiotic mapping in humans. In *Genome Analysis: A Laboratory Manual, Vol. 4, Mapping Genomes*, B. Birren, E. Green, P. Hieter, S. Klapholz, R. Myers, H. Riethman, and J. Roskams, eds. (Cold Spring Harbor: Cold Spring Harbor Laboratory Press).

Chumakov, I. M., Rigault, P., Le Gall, I., Bellanne-Chantelot, C., Billault, A., Guillou, S., Soularue, P., Guasconi, G., Poullier, E., Gros, I., Belova, M., Sambucy, J.-L., Susini, L., Gervy, P., Glibert, F., Beaufils, S., Bul, H., Massart, C., De Tand, M.-F., Dukasz, F., Lecoulant, S., Ougen, P., Perrot, V., Saumier, M., Soravito, C., Bahouayila, R., Cohen-

Akenine, A., Barillot, E., Bertrand, S., Codani, J.-J., Caterina, D., Georges, I., Lacroix, B., Lucotte, G., Sahbatou, M., Schmit, C., Sangouard, M., Tubacher, E., Dib, C., Faure, S., Fizames, C., Cyapay, G., Millasseau, P., NGuyen, S., Muselet, D., Vignal, A., Morissette, J., Menninger, J., Lieman, J., Desai, T., Banks, A., Bray-Ward, P., Ward, D., Hudson, T., Gerety, S., Foote, S., Stein, L., Page, D. C., Lander, E. S., Weissenbach, J., Le Paslier, D., and Cohen, D. (1995). A YAC contig map of the human genome. *Nature* 377, 175–297.

Collins, A., Frezal, J., Teague, J., and Morton, N. E. (1996). A metric map of humans: 23,500 loci in 850 bands. *Proc. Natl. Acad. Sci. USA* 93, 14771–14775.

Collins, A., Teague, J., Keats, B., and Morton, N. (1996). Linkage map integration. *Genomics* 35, 157–162.

Dausset, J., Cann, H., Cohen, D., Lathrop, M., Lalouel, J.-M., and White, R. (1990). Centre d'Etude du Polymorphisme Humain (CEPH): Collaborative genetic mapping of the human genome. *Genomics* 6, 575–577.

Deloukas, P., Schuler, G. D., Gyapay, G., Beasley, E. M., Soderlund, C., Rodriguez-Tomé, P., Hui, L., Matise, T. C., McKusick, K. B., Beckmann, J. S., Benolila, S., Bihoreau, M.-T., Birren, B. B., Browne, J., Butler, A., Castle, A. B., Chiannikulchai, N., Clee, C., Day, P. J. R., Dehejia, A., Dibling, T., Drouot, N., Duprat, S., Fizames, C., Fox, S., Gelling, S., Green, L., Harison, P., Hocking, R., Holloway, E., Hunt, S., Keil, S., Lijnzaad, P., Louis-Dit-Sully, C., Ma, J., Mendis, A., Miller, J., Morissette, J., Muselet, D., Nusbaum, H. C., Peck, A., Rozen, S., Simon, D., Slonim, D. K., Staples, R., Stein, L. D., Stewart, E. A., Suchard, M. A., Thangarajah, T., Vega-Czarny, N., Webber, C., Wu, X., Auffray, C., Nomura, N., Sikela, J. M., Polymeropoulos, M. H., James, M. R., Lander, E. S., Hudson, T. J., Myers, R. M., Cox, D. R., Weissenbach, J., Boguski, M. S., and Bentley, D. R. (1998). A physical map of 30,000 human genes. *Science* 282, 744–746.

Dib, C., Fauré, S., Fizames, C., Samson, D., Drouot, N., Vignal, A., Millasseau, P., Marc, S., Hazan, J., Seboun, E., Lathrop, M., Gyapay, G., Morissette, J., and Weissenbach, J. (1996). A comprehensive genetic map of the human genome based on 5,264 microsatellites. *Nature* 380, 152–154.

Dietrich, W., Weber, J., Nickerson, D., and Kwok, P.-Y. (1999). Identification and Analysis of DNA Polymorphisms. In *Genome Analysis: A Laboratory Manual, Vol. 4, Mapping Genomes*, B. Birren, E. Green, P. Hieter, S. Klapholz, R. Myers, H. Riethman and J. Roskams, eds. (Cold Spring Harbor: Cold Spring Harbor Laboratory Press).

Gatewood, B., and Cottingham, R. (2000). Mouse-human comparative map resources on the web. *Briefings in Bioinformatics* 1, 60–75.

Green, E. D., and Olson, M. V. (1990). Chromosomal region of the cystic fibrosis gene in yeast artificial chromosomes: a model for human genome mapping. *Science* 250, 94–98.

Gyapay, G., Schmitt, K., Fizames, C., Jones, H., Vega-Czarny, N., Spillet, D., Muselet, D., Prud'homme, J., Dib, C., Auffray, C., Morissette, J., Weissenbach, J., and Goodfellow, P. N. (1996). A radiation hybrid map of the human genome. *Hum. Mol. Genet.* 5, 339–358.

Hillier, L. D., Lennon, G., Becker, M., Bonaldo, M. F., Chiapelli, B., Chissoe, S., Dietrich, N., DuBuque, T., Favello, A., Gish, W., Hawkins, M., Hultman, M., Kucaba, T., Lacy, M., Le, M., Le, N., Mardis, E., Moore, B., Morris, M., Parsons, J., Prange, C., Rifkin, L., Rohlfing, T., Schellenberg, K., Marra, M., and et al. (1996). Generation and analysis of 280,000 human expressed sequence tags. *Genome Res.* 6, 807–828.

Houlgatte, R., Mariage-Samson, R., Duprat, S., Tessier, A., Bentolila, S., Lamy, B., and Auffray, C. (1995). The Genexpress Index: a resource for gene discovery and the genic map of the human genome. *Genome Res.* 5, 272–304.

Hudson, T. J., Stein, L. D., Gerety, S. S., Ma, J., Castle, A. B., Silva, J., Slonim, D. K., Baptista, R., Kruglyak, L., Xu, S.-H., Hu, X., Colbert, A. M. E., Rosenberg, C., Reeve-Daly, M. P., Rozen, S., Hui, L., Wu, X., Vestergaard, C., Wilson, K. M., Bae, J. S., Maitra, S., Ganiatsas, S., Evans, C. A., DeAngelis, M. M., Kngalls, K. A., Nahf, R. W., Horton

Jr., L. T., Anderson, M. O., Collymore, A. J., Ye, W., Kouyoumijan, V., Zemsteva, I. S., Tam, J., Devine, R., Courtney, D. F., Renaud, M. T., Nguyen, H., O'Connor, T. J., Fizames, C., Fauré, S., Gyapay, G., Dib, C., Morissette, J., Orlin, J. B., Birren, B. W., Goodman, N., Weissenbach, J., Hawkins, T. L., Foote, S., Page, D. C., and Lander, E. S. (1995). An STS-based map of the human genome. *Science* 270, 1945–1954.

Iaonnou, P. A., Amemiya, C. T., Garnes, J., Kroisel, P. M., Shizuya, H., Chen, C., Batzer, M. A., and de Jong, P. J. (1994). A new bacteriophage P1-derived vector for the propagation of large human DNA fragments. *Nat. Genet.* 6, 84–89.

Jensen, S. J., Sulman, E. P., Maris, J. M., Matise, T. C., Vojta, P. J., Barrett, J. C., Brodeur, G. M., and White, P. S. (1997). An integrated transcript map of human chromosome 1p35–36. *Genomics* 42, 126–136.

Jing, J., Lai, Z., Aston, C., Lin, J., Carucci, D. J., Gardner, M. J., Mishra, B., Anantharaman, T. S., Tettelin, H., Cummings, L. M., Hoffman, S. L., Venter, J. C., and Schwartz, D. C. (1999). Optical mapping of *Plasmodium falciparum* chromosome 2. *Genome Res.* 9, 175–181.

Korenberg, J. R., Chen, X.-N., Devon, K. L., Noya, D., Oster-Granite, M. L., and Birren, B. W. (1999). Mouse Molecular Cytogenetic Resource: 157 BACs link the chromosomal and genetic maps. *Genome Res.* 9, 514–523.

Lander, E. S., and Green, P. (1987). Construction of multi-locus genetic linkage maps in humans. *Proc. Natl. Acad. Sci. USA* 84, 2363–2367.

Lander, E. S., Green, P., Abrahamson, J., Barlow, A., Daly, M. J., Lincoln, S. E., and Newburg, L. (1987). MAPMAKER: An interactive computer package for constructing primary genetic linkage maps of experimental and natural populations. *Genomics* 1, 174–181.

Lathrop, G. M., Lalouel, J. M., Julier, C., and Ott, J. (1984). Strategies for multilocus linkage analysis in humans. *Proc. Natl. Acad. Sci. USA* 81, 3443–3446.

Lawrence, J. B., Singer, R. H., and NcNeil, J. A. (1990). Interphase and metaphase resolution of different distances within the human dystrophin gene. *Science* 249, 928–932.

Letovsky, S. I., Cottingham, R. W., Porter, C. J., and Li, P. W. D. (1998). GDB: the Human Genome Database. *Nucleic Acids Res.* 26, 94–99.

Lewis, T. B., Nelson, L., Ward, K., and Leach, R. J. (1995). A radiation hybrid map of 40 loci for the distal long arm of human chromosome 8. *Genome Res.* 5, 334–341.

Ma, R. Z., van Eijk, M. J., Beever, J. E., Guerin, G., Mummery, C. L., and Lewin, H. A. (1998). Comparative analysis of 82 expressed sequence tags from a cattle ovary cDNA library. *Mamm Genome* 9, 545–549.

Maglott, D. R., Katz, K. S., Sicotte, H., and Pruitt, K. D. (2000). NCBI's LocusLink and RefSeq. *Nucleic Acids Res.* 28, 126–128.

Marra, M. A., Kucaba, T. A., Dietrich, N. L., Green, E. D., Brownstein, B., Wilson, R. K., McDonald, K. M., Hillier, L. W., McPherson, J. D., and Waterston, R. H. (1997). High throughput fingerprint analysis of large-insert clones. *Genome Res.* 7, 1072–1084.

Matise, T., and Gitlin, J. (1999). MAP-O-MAT: marker-based linkage mapping on the World Wide Web. *Am. J. Hum. Genet.* 65, A2464.

Matise, T. C., Perlin, M., and Chakravarti, A. (1994). Automated construction of genetic linkage maps using an expert system (MultiMap): a human genome linkage map. *Nat. Genet.* 6, 384–390.

McKusick, V. A. (1998). *Mendelian Inheritance in Man. Catalogs of Human Genes and Genetic Disorders*, 12th Edition (Baltimore: Johns Hopkins University Press).

Murray, J. C., Buetow, K. H., Weber, J. L., Ludwigsen, S., Scherpbier-Heddema, T., Manion, F., Quillen, J., Sheffield, V. C., Sunden, S., Duyk, G. M., Weissenbach, J., Gyapay, G., Dib, C., Morrissette, J., Lathrop, G. M., Vignal, A., White, R., Matsunami, N., Gerken, S., Melis, R., Albertsen, H., Plaetke, R., Odelberg, O., Ward, D., Dausset, J., Cohen, D., and

Cann, H. (1994). A comprehensive human linkage map with centimorgan density. *Science* 265, 2049–2054.

Nusbaum, C., Slonim, D., Harris, K., Birren, B., Steen, R., Stein, L., Miller, J., Dietrich, W., Nahf, R., Wang, V., Merport, O., Castle, A., Husain, Z., Farino, G., Gray, D., Anderson, M., Devine, R., Horton, L., Ye, W., Kouyoumjian, V., Zemsteva, I., Wu, Y., Collymore, A., Courtney, D., Tam, J., Cadman, M., Haynes, A., Heuston, C., Marsland, T., Southwell, A., Trickett, P., Strivens, M., Ross, M., Makalowski, W., Wu, Y., Boguski, M., Carter, N., Denny, P., Brown, S., Hudson, T., and Lander, E. (1999). A YAC-based physical map of the mouse genome. *Nat. Genet.* 22, 388–393.

O'Brien, S. J., Menotti-Raymond, M., Murphy, W. J., Nash, W. G., Wienberg, J., Stanyon, R., Copeland, N. G., Jenkins, N. A., Womack, J. E., and Marshall Graves, J. A. (1999). The promise of comparative genomics in mammals. *Science* 286, 458–462.

Parra, I., and Windle, B. (1993). High resolution visual mapping of stretched DNA by fluorescent hybridization. *Nat. Genet.* 5, 17–21.

Pearson, P. L. (1991). The genome data base (GDB)—a human gene mapping repository. *Nucleic Acids Res.* 19 Suppl, 2237–9.

Pinkel, D., Straume, T., and Gray, J. W. (1986). Cytogenetic analysis using quantitative, high-sensitivity, fluorescence hybridization. *Proc. Natl. Acad. Sci. USA* 83, 2934–2938.

Pruitt, K., Katz, K., Sicotte, H., and Maglott, D. (2000). Introducing RefSeq and LocusLink: curated human genome resources at the NCBI. *Trends Genet.* 16, 44–47.

Roberts, T., Auffray, C., and Cowell, J. K. (1996). Regional localization of 192 genic markers on human chromosome 1. *Genomics* 36, 337–340.

Rodriguez-Tome, P., and Lijnzaad, P. (2000). RHdb: the radiation hybrid database. *Nucleic Acids Res.* 28, 146–147.

Schuler, G. D. (1997). Sequence mapping by electronic PCR. *Genome Res.* 7, 541–550.

Schwartz, D. C., Li, X., Hernandez, L. I., Ramnarain, S. P., Huff, E. J., and Wang, Y. K. (1993). Ordered restriction maps of Saccharomyces cerevisiae chromosomes constructed by optical mapping. *Science* 262, 110–114.

Shizuya, H., Birren, B., Kim, U. J., Mancino, V., Slepak, T., Tachiiri, Y., and Simon, M. (1992). Cloning and stable maintenance of 300-kilobase-pair fragments of human DNA in Escherichia coli using an F-factor-based vector. *Proc. Natl. Acad. Sci. USA* 89, 8794–8797.

Slonim, D., Kruglyak, L., Stein, L., and Lander, E. (1997). Building human genome maps with radiation hybrids. *J. Comput. Biol.* 4, 487–504.

Steen, R., Kwitek-Black, A., Glenn, C., Gullings-Handley, J., Etten, W., Atkinson, S., Appel, D., Twigger, S., Muir, M., Mull, T., Granados, M., Kissebah, M., Russo, K., Crane, R., Popp, M., Peden, M., Matise, T., Brown, D., Lu, J., Kingsmore, S., Tonellato, P., Rozen, S., Slonim, D., Young, P., Knoblauch, M., Provoost, A., Ganten, D., Colman, S., Rothberg, J., Lander, E., and Jacob, H. (1999). A high-density integrated genetic linkage and radiation hybrid map of the laboratory rat. *Genome Res.* 9, AP1–AP8.

Stewart, E. A., McKusick, K. B., Aggarwal, A., Bajorek, E., Brady, S., Chu, A., Fang, N., Hadley, D., Harris, M., Hussain, S., Lee, R., Maratukulam, A., O'Connor, K., Perkins, S., Piercy, M., Qin, F., Reif, T., Sanders, C., She, X., Sun, W., Tabar, P., Voyticky, S., Cowles, S., Fan, J., Mader, C., Quackenbush, J., Myers, R. M., and Cox, D. R. (1997). An STS-based radiation hybrid map of the human genome. *Genome Res.* 7, 422–433.

Talbot, C. A., and Cuticchia, A. J. (1994). Human Mapping Databases. In *Current Protocols in Human Genetics*, N. Dracopoli, J. Haines, B. Korf, D. Moir, C. Morton, C. Seidman, J. Seidman and D. Smith, eds. (New York: J. Wiley), p. 1.13.1–1.13.21.

Thomas, J. W., Summers, T. J., Lee-Lin, S. Q., Maduro, V. V., Idol, J. R., Mastrian, S. D., Ryan, J. F., Jamison, D. C., and Green, E. D. (2000). Comparative genome mapping in the sequence-based era: early experience with human chromosome 7. *Genome Res.* 10, 624–633.

Tonellato, P. J., Zho, H., Chen, D., Wang, Z., Stoll, M., Kwitek-Black, A., and Jacob, H. *Comparative Mapping of the Human and Rat Genomes with Radiation Hybrid Maps*, RECOMB '99, Lyon, France, April 1999.

van den Engh, G., Sachs, R., and Trask, B. J. (1992). Estimating genomic distance from DNA sequence location in cell nuclei by a random walk model. *Science* 257, 1410–1412.

Venter, J. C., Smith, H. O., and Hood, L. (1996). A new strategy for genome sequencing. *Nature* 381, 364–366.

Vollrath, D., Foote, S., Hilton, A., Brown, L. G., Beer-Romero, P., Bogan, J. S., and Page, D. C. (1992). The human Y chromosome: a 43-interval map based on naturally occurring deletions. *Science* 258, 52–59.

Wada, Y., and Yasue, H. (1996). Development of an animal genome database and its search system. Comput. Appl. Biosci. 12, 231–235.

Wang, D. G., Fan, J. B., Siao, C. J., Berno, A., Young, P., Sapolsky, R., Ghandour, G., Perkins, N., Winchester, E., Spencer, J., Kruglyak, L., Stein, L., Hsie, L., Topaloglou, T., Hubbell, E., Robinson, E., Mittmann, M., Morris, M. S., Shen, N., Kilburn, D., Rioux, J., Nusbaum, C., Rozen, S., Hudson, T. J., Lander, E. S., and et al. (1998). Large-scale identification, mapping, and genotyping of single- nucleotide polymorphisms in the human genome. *Science* 280, 1077–1082.

Watanabe, T. K., Bihoreau, M. T., McCarthy, L. C., Kiguwa, S. L., Hishigaki, H., Tsuji, A., Browne, J., Yamasaki, Y., Mizoguchi-Miyakita, A., Oga, K., Ono, T., Okuno, S., Kanemoto, N., Takahashi, E., Tomita, K., Hayashi, H., Adachi, M., Webber, C., Davis, M., Kiel, S., Knights, C., Smith, A., Critcher, R., Miller, J., James, M. R., and et al. (1999). A radiation hybrid map of the rat genome containing 5,255 markers. *Nat. Genet.* 22, 27–36.

Wheeler, D. L., Chappey, C., Lash, A. E., Leipe, D. D., Madden, T. L., Schuler, G. D., Tatusova, T. A., and Rapp, B. A. (2000). Database resources of the national center for biotechnology information. *Nucleic Acids Res.* 28, 10–14.

White, P. S., Sulman, E. P., Porter, C. J., and Matise, T. C. (1999). A comprehensive view of human chromosome 1. *Genome Res.* 9, 978–988.

Zhang, Z., Schwartz, S., Wagner, L., and Miller, W. (2000). A greedy algorithm for aligning DNA sequences. *J. Comput. Biol.* 7, 203–214.

7

INFORMATION RETRIEVAL FROM BIOLOGICAL DATABASES

Andreas D. Baxevanis

Genome Technology Branch
National Human Genome Research Institute
National Institutes of Health
Bethesda, Maryland

As discussed earlier in this book, GenBank was created in response to the explosion in sequence information resulting from a panoply of scientific efforts such as the Human Genome Project. To review, GenBank is an annotated collection of all publicly available DNA and protein sequences and is maintained by the National Center for Biotechnology Information (NCBI). As of this writing, GenBank contains 7 million sequence records covering almost 9 billion nucleotide bases. Sequences find their way into GenBank in several ways, most often by direct submission by individual investigators through tools such as Sequin or through ‘‘direct deposit’’ by large genome sequencing centers.

GenBank, or any other biological database for that matter, serves little purpose unless the database can be easily searched and entries retrieved in a usable, meaningful format. Otherwise, sequencing efforts have no useful end, since the biological community as a whole cannot make use of the information hidden within these millions of bases and amino acids. Much effort has gone into making such data accessible to the average user, and the programs and interfaces resulting from these efforts are the focus of this chapter. The discussion centers on querying the NCBI databases because these more ‘‘general’’ repositories are far and away the ones most often accessed by biologists, but attention is also given to a number of smaller, specialized databases that provide information not necessarily found in GenBank.

Bioinformatics: A Practical Guide to the Analysis of Genes and Proteins
Edited by A. D. Baxevanis and B. F. F. Ouellette
ISBN 0-471-38390-2 (cloth), ISBN 0-471-383910 (paper)

INTEGRATED INFORMATION RETRIEVAL: THE ENTREZ SYSTEM

The most widely used interface for the retrieval of information from biological databases is the NCBI Entrez system. Entrez capitalizes on the fact that there are preexisting, logical relationships between the individual entries found in numerous public databases. For example, a paper in MEDLINE (or, more properly, PubMed) may describe the sequencing of a gene whose sequence appears in GenBank. The nucleotide sequence, in turn, may code for a protein product whose sequence is stored in the protein databases. The three-dimensional structure of that protein may be known, and the coordinates for that structure may appear in the structure database. Finally, the gene may have been mapped to a specific region of a given chromosome, with that information being stored in a mapping database. The existence of such natural connections, mostly biological in nature, argued for the development of a method through which all information about a particular biological entity could be found without having to sequentially visit and query disparate databases.

Entrez, to be clear, is not a database itself—it is the interface through which all of its component databases can be accessed and traversed. The Entrez information space includes PubMed records, nucleotide and protein sequence data, three-dimensional structure information, and mapping information. The strength of Entrez lies in the fact that *all* of this information can be accessed by issuing one and only one query. Entrez is able to offer integrated information retrieval through the use of two types of connection between database entries: *neighboring* and *hard links*.

Neighboring

The concept of neighboring allows for entries *within* a given database to be connected to one another. If a user is looking at a particular PubMed entry, the user can ask Entrez to find all other papers in PubMed that are similar in subject matter to the original paper. Similary, if a user is looking at a sequence entry, Entrez can return a list of all other sequences that bear similarity to the original sequence. The establishment of neighboring relationships within a database is based on statistical measures of similarity, as follows.

BLAST. Sequence data are compared with one another using the Basic Local Alignment Search Tool or BLAST (Altschul et al., 1990). This algorithm attempts to find "high-scoring segment pairs" (HSPs), which are pairs of sequences that can be aligned with one another and, when aligned, meet certain scoring and statistical criteria. Chapter 8 discusses the family of BLAST algorithms and their application at length.

VAST. Sets of coordinate data are compared using a vector-based method known as VAST, for Vector Alignment Search Tool (Madej et al., 1995; Gibrat et al., 1996). There are three major steps that take place in the course of a VAST comparison:

- First, based on known three-dimensional coordinate data, all of the α-helices and β-sheets that comprise the core of the protein are identified. Straight-line vectors are then calculated based on the position of these secondary structure elements. VAST keeps track of how one vector is connected to the next (that is, how the C-terminal end of one vector connects to the N-terminal end of the next vector), as well as whether a particular vector represents an α-helix or a β-sheet. Subsequent steps use *only* these vectors in making comparisons to other proteins. In effect, most of the coordinate data is discarded at this step. The reason for this apparent oversimplification is simply due to the scale of the problem at hand; with over 11,000 structures in PDB, the time that it would take to do an in-depth comparison of each and every structure with all of the other structures in the database would make the calculations both impractical and intractable. The user should keep this simplification in mind when making biological inferences based on the results presented in a VAST table.

- Next, the algorithm attempts to optimally align these sets of vectors, looking for pairs of structural elements that are of the same type and relative orientation, with consistent connectivity between the individual elements. The object is to identify highly similar "core substructures," pairs that represent a statistically significant match above that which would be obtained by comparing randomly chosen proteins with one another.
- Finally, a refinement is done using Monte Carlo methods at each residue position in an attempt to optimize the structural alignment.

Through this method, it is possible to find structural (and, presumably, functional) relationships between proteins in cases that may lack overt sequence similarity. The resultant alignment need not be global; matches may be between individual domains of different proteins.

It is important to note here that VAST is not the best method for determining structural similarities. More robust methods, such as homology model building, provide much greater resolving power in determining such relationships, since the raw information within the three-dimensional coordinate file is used to perform more advanced calculations regarding the positions of side chains and the thermodynamic nature of the interactions between side chains. Reducing a structure to a series of vectors *necessarily* results in a loss of information. However, considering the magnitude of the problem here—again, the number of pairwise comparisons that need to be made—and both the computing power and time needed to employ any of the more advanced methods, VAST provides a simple and fast first answer to the question of structural similarity. More information on other structure prediction methods based on X-ray or NMR coordinate data can be found in Chapter 11.

Weighted Key Terms. The problem of comparing sequence data somewhat pales next to that of comparing PubMed entries, free text whose rules of syntax are not necessarily fixed. Given that no two people's writing styles are exactly the same, finding a way to compare seemingly disparate blocks of text poses a substantial problem. Entrez employs a method known as the relevance pairs model of retrieval to make such comparisons, relying on what are known as weighted key terms (Wilbur and Coffee, 1994; Wilbur and Yang, 1996). This concept is best described by example. Consider two manuscripts with the following titles:

```
BRCA1 as a Genetic Marker for Breast Cancer
Genetic Factors in the Familial Transmission of the
Breast Cancer BRCA1 Gene
```

Both titles contain the terms `BRCA1`, `Breast`, and `Cancer`, and the presence of these common terms may indicate that the manuscripts are similar in their subject matter. The proximity between the words is also taken into account, so that words common to two records that are closer together are scored higher than common words that are further apart. In the current example, the terms `Breast` and `Cancer` would score higher based on proximity than either of those words would against `BRCA1`, since the words are next to each other. Common words found in a title are scored higher than those found in an abstract, since title words are presumed to be "more important" than those found in the body of an abstract. Overall weighting depends on the frequency of a given word among all the entries in PubMed, with words that occur infrequently in the database as a whole carrying a higher weight.

Regardless of the method by which the neighboring relationships are established, the ability to actually code and maintain these relationships is rooted in the format underlying all of the constituent databases. This format, called Abstract Syntax Notation (ASN.1), provides a format in which all similar fields (e.g., those for a bibliographic citation) are all structured identically, regardless of whether the entry is in a protein database, nucleotide database, and so forth. This NCBI data model is discussed in depth in Chapter 2.

Hard Links

The hard link concept is much easier conceptually than neighboring. Hard links are applied between entries in different databases and exist everywhere there is a logical connection between entries. For instance, if a PubMed entry talks about the sequencing of a cosmid, a hard link is established between the PubMed entry and the corresponding nucleotide entry. If an open reading frame in that cosmid codes for a known protein, a hard link is established between the nucleotide entry and the protein entry. If, by sheer luck, the protein entry has an experimentally deduced structure, a hard link would be placed between the protein entry and the structural entry. The hard link relationships between databases is illustrated in Figure 7.1.

As suggested by the figure, searches can, in essence, begin anywhere within Entrez—the user has no constraints with respect to where the foray into this information space must begin. However, depending on which database is used as the jumping-off point, different fields are available for searching. This stands to reason, inasmuch as the entries in databases of different types are necessarily organized differently, reflecting the biological nature of the entity they are trying to catalog.

Implementations

Regardless of platform, Entrez searches can be performed using one of two interfaces. The first is a client-server implementation known as Network Entrez. This is the fastest of the Entrez programs in that it makes a direct connection to an NCBI "dispatcher." The graphical user interface features a series of windows, and each time a new piece of information is requested, a new window appears on the user's screen. Because the client software resides on the user's machine, it is up to the user to obtain, install, and maintain the software, downloading periodic updates as new features are introduced. The installation process itself is fairly trivial. Network Entrez also comes bundled with interactive, graphical viewers for both genome sequences and three-dimensional structures (Cn3D, cf. Chapter 5).

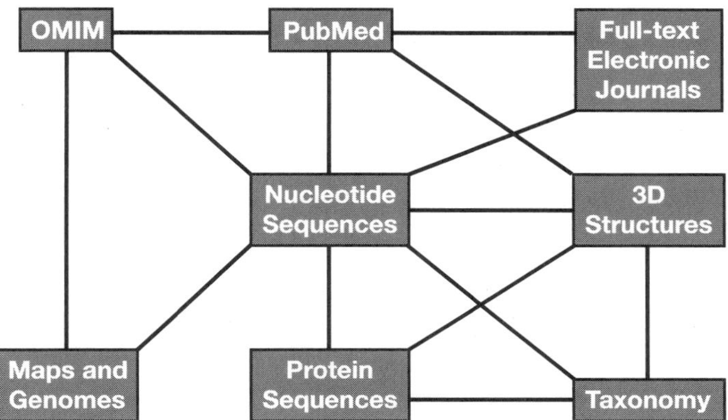

Figure 7.1. Overview of the relationships in the Entrez integrated information retrieval system. Each square represents one of the elements that can be accessed through Entrez, and the lines represent how each element connects to the other elements. Entrez is under continuous evolution, with both new components being added and the interrelationships between the elements changing dynamically.

The second and more widely used implementation is through the World Wide Web. This option makes use of available Web browsers, such as Internet Explorer or Netscape, to deliver search results to the desktop. The use of a Web browser relieves the user of having to make sure that the most current version of Entrez is installed—as long as the browser is of relatively recent vintage, results will always be displayed via the latest Entrez release. The Web naturally lends itself to an application such as this, since all the neighboring and hard link relationships described above can easily be expressed as hypertext, allowing the user to navigate by clicking on selected words in an entry.

The advantage of the Web implementation over the Network version is that the Web allows for the ability to link to external data sources, such as full-text versions of papers maintained by a given journal or press or specialized databases that are not part of Entrez proper. The speed advantage that is gained by the network version causes its limitation in this respect; the direct connection to the NCBI dispatcher means that the user, once connected to NCBI, cannot travel anywhere else. The other main difference between the two methods lies simply in the presentation: the Network version uses a series of windows in presenting search results, whereas the Web version is formatted as sequential pages, following the standard Web paradigm. The final decision is one of personal preference, for both methods will produce the same results *within* the Entrez search space. However, given that Web Entrez can link to external data sources, the remainder of this discussion will focus on the Web implementation.

The Entrez Discovery Pathway: Examples

The best way to illustrate the integrated nature of the Entrez system and to drive home the power of neighboring is by considering two biological examples, using the Web version of Entrez as the interface.

The simplest way to query Entrez is through the use of individual search terms, coupled together by Boolean operators such as AND, OR, or NOT. Consider the case in which one wants to retrieve all papers that discuss aspirin in the context of treating or preventing atherosclerosis. Beginning at the Entrez home page, one would select PubMed from the Search pull-down menu to indicate that the search is to take place in the bibliographic portion of the Entrez search space. Within the text box to the right, one would simply type `atherosclerosis [MH] AND aspirin [NM]`. The `[MH]` qualifying the first term indicates to Entrez that this is a *MeSH term*; MeSH stands for *medical subject heading* and is the qualifier that should be used when searching by subject. The `[NM]` qualifying the second term indicates that this is the name of a substance or chemical. In this case, the query returned 197 papers (Fig. 7.2; the query is echoed at the top of the new Web page). The user can further narrow down the query by adding additional terms, if the user is interested in a more specific aspect of the pharmacology or if there are quite simply too many papers to read. A list of all available qualifiers is given in Table 7.1.

At this point, to actually look at one of the papers resulting from the search, the user can click on a hyperlinked author's name. By doing so, the user is taken to the Abstract view for the selected paper. Figure 7.3 shows the Abstract view for the first paper in the hit list, by Cayatte et al. The Abstract view presents the name of the paper, the list of authors, their institutional affiliation, and the abstract itself, in standard format. A number of alternative formats are available for displaying this information, and these various formats can be selected using the pull-down menu next to the Display button. Switching to Citation format would produce a very similar-looking entry, the difference being that cataloguing information such as MeSH terms and indexed substances relating to the entry are shown below the abstract. MEDLINE format produces the MEDLINE/MEDLARS layout, with two-letter codes corresponding to the contents of each field going down the left-hand side of the entry (e.g., the author field is denoted by the code AU). Entries in this format can be saved and easily imported into third-party bibliography management programs, such as EndNote and Reference Manager.

At the top of the entry are a number of links that are worth mentioning. First, on the right-hand side is a hyperlink called Related Articles. This is one of the entry points from which the user can take advantage of the neighboring and hard link relationships described earlier. If the user clicks on Related Articles, Entrez will indicate that there are 101 neighbors associated with the original Cayatte reference —that is, 101 references of similar subject matter—and the first six of these papers are shown in Figure 7.4. The first reference in the list is the same Cayatte paper because, by definition, it is most related to itself (the "parent"). The order in which the neighbored entries follows is from most statistically similar downward. Thus, the entry closest to the parent is deemed to be the closest in subject matter to the parent. By scanning the titles, the user can easily find related information on other studies that look at the pharmacology of aspirin in atherosclerosis as well as quickly amass a bibliography of relevant references. This is a particularly useful and time-saving functionality when one is writing grants or papers because abstracts can be scanned and papers of real interest identified before one heads off for the library stacks.

The next link in the series is labeled Books, and clicking on that link will take the user to a heavily hyperlinked version of the original citation. The highlighted words in this view correspond to keywords that can take the user to full-text books that are available through NCBI. The first of these books to be made available is

Figure 7.2. A text-based Entrez query using Boolean operators against PubMed. The initial query is shown in the search box near the top of the window. Each entry gives the names of the authors, the title of the paper, and the citation information. The actual record can be retrieved by clicking on the author list.

TABLE 7.1. Entrez Boolean Search Statements

General syntax:

`search term [tag]` ***`boolean operator`*** **`search term [tag] . . .`**

where **`[tag]`** =

`[AD]`	Affiliation
`[ALL]`	All fields
`[AU]`	Author name `O'Brien J [AU]` *yields all of* O'Brien JA, O'Brien JB, etc. `''O'Brien J'' [AU]` *yields only* O'Brien J
`[RN]`	Enzyme Commission or Chemical Abstract Service numbers
`[EDAT]`	Entrez date `YYYY/MM/DD, YYYY/MM,` or `YYYY`
`[IP]`	Issue of journal

`[TA]`	Journal title, official abbreviation, or ISSN number
	`Journal of Biological Chemistry`
	`J Biol Chem`
	`0021-9258`

`[LA]`	Language
`[MAJR]`	MeSH major ropic *One of the* ***major*** *topics discussed in the article*
`[MH]`	MeSH terms *Controlled vocabulary of biomedical terms (**subject**)*
`[PS]`	Personal name as subject *Use when name is subject of article*, e.g., `Varmus H [PS]`
`[DP]`	Publication date `YYYY/MM/DD, YYYY/MM,` or `YYYY`

`[PT]`	Publication type
	`Review`
	`Clinical Trial`
	`Lectures`
	`Letter`
	`Technical Publication`

`[SH]`	Subheading *Used to modify MeSH Terms* `hypertension [MH] AND toxicity [SH]`
`[NM]`	Substance name *Name of chemical discussed in article*
`[TW]`	Text words *All words and numbers in the title and abstract, MeSH terms, subheadings, chemical substance names, personal name as subject, and MEDLINE secondary sources*
`[UID]`	Unique identifiers (PMID/MEDLINE numbers)
`[VI]`	Volume of journal

and ***`boolean operator`*** = `AND, OR,` or `NOT`

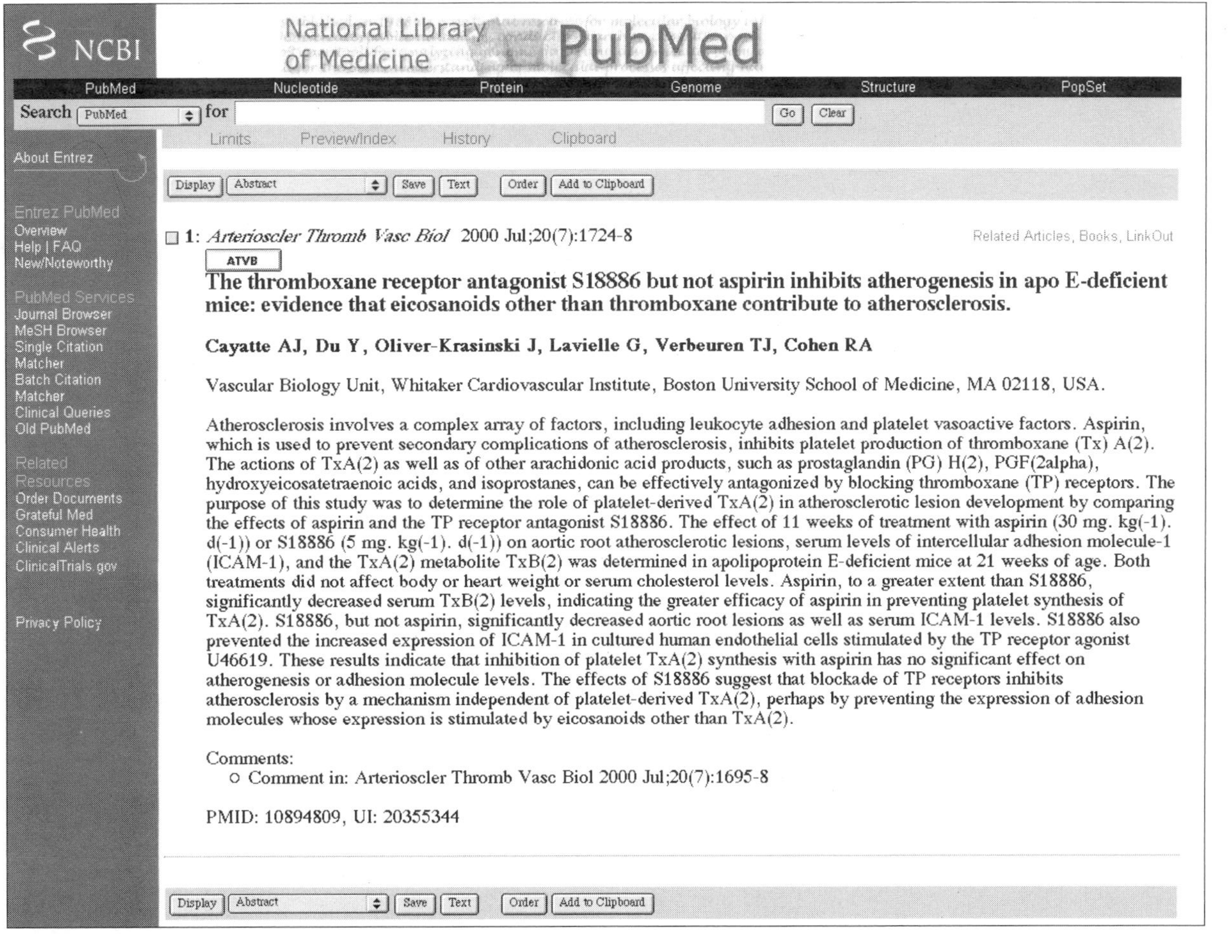

NCBI | National Library of Medicine | PubMed

PubMed | Nucleotide | Protein | Genome | Structure | PopSet

Search PubMed for [Go] [Clear]

Limits | Preview/Index | History | Clipboard

About Entrez

Entrez PubMed
Overview
Help | FAQ
New/Noteworthy

PubMed Services
Journal Browser
MeSH Browser
Single Citation Matcher
Batch Citation Matcher
Clinical Queries
Old PubMed

Related Resources
Order Documents
Grateful Med
Consumer Health
Clinical Alerts
ClinicalTrials.gov

Privacy Policy

[Display] [Abstract] [Save] [Text] [Order] [Add to Clipboard]

1: *Arterioscler Thromb Vasc Biol* 2000 Jul;20(7):1724-8 — Related Articles, Books, LinkOut

ATVB

The thromboxane receptor antagonist S18886 but not aspirin inhibits atherogenesis in apo E-deficient mice: evidence that eicosanoids other than thromboxane contribute to atherosclerosis.

Cayatte AJ, Du Y, Oliver-Krasinski J, Lavielle G, Verbeuren TJ, Cohen RA

Vascular Biology Unit, Whitaker Cardiovascular Institute, Boston University School of Medicine, MA 02118, USA.

Atherosclerosis involves a complex array of factors, including leukocyte adhesion and platelet vasoactive factors. Aspirin, which is used to prevent secondary complications of atherosclerosis, inhibits platelet production of thromboxane (Tx) A(2). The actions of TxA(2) as well as of other arachidonic acid products, such as prostaglandin (PG) H(2), PGF(2alpha), hydroxyeicosatetraenoic acids, and isoprostanes, can be effectively antagonized by blocking thromboxane (TP) receptors. The purpose of this study was to determine the role of platelet-derived TxA(2) in atherosclerotic lesion development by comparing the effects of aspirin and the TP receptor antagonist S18886. The effect of 11 weeks of treatment with aspirin (30 mg. kg(-1). d(-1)) or S18886 (5 mg. kg(-1). d(-1)) on aortic root atherosclerotic lesions, serum levels of intercellular adhesion molecule-1 (ICAM-1), and the TxA(2) metabolite TxB(2) was determined in apolipoprotein E-deficient mice at 21 weeks of age. Both treatments did not affect body or heart weight or serum cholesterol levels. Aspirin, to a greater extent than S18886, significantly decreased serum TxB(2) levels, indicating the greater efficacy of aspirin in preventing platelet synthesis of TxA(2). S18886, but not aspirin, significantly decreased aortic root lesions as well as serum ICAM-1 levels. S18886 also prevented the increased expression of ICAM-1 in cultured human endothelial cells stimulated by the TP receptor agonist U46619. These results indicate that inhibition of platelet TxA(2) synthesis with aspirin has no significant effect on atherogenesis or adhesion molecule levels. The effects of S18886 suggest that blockade of TP receptors inhibits atherosclerosis by a mechanism independent of platelet-derived TxA(2), perhaps by preventing the expression of adhesion molecules whose expression is stimulated by eicosanoids other than TxA(2).

Comments:

- Comment in: Arterioscler Thromb Vasc Biol 2000 Jul;20(7):1695-8

PMID: 10894809, UI: 20355344

[Display] [Abstract] [Save] [Text] [Order] [Add to Clipboard]

Figure 7.3. An example of a PubMed record in Abstract format as returned through Entrez. This Abstract view for the first reference shown in Figure 7.2. This view provides links to Related Articles, Books, LinkOut, and the actual, full-text journal paper. See text for details.

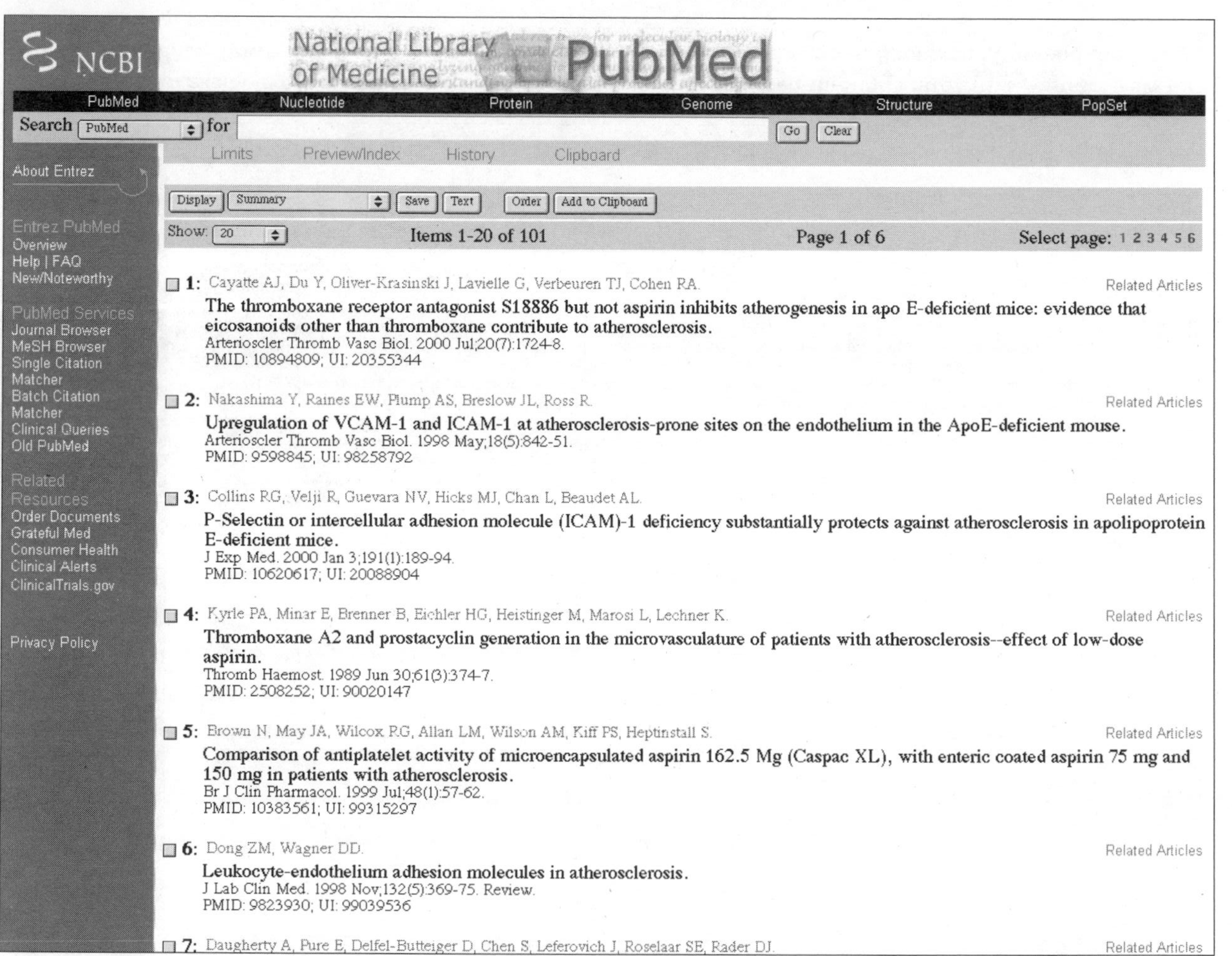

Figure 7.4. Neighbors to an entry found in PubMed. The original entry (Cayette et al., 2000) is at the top of the list, indicating that this is the parent entry. See text for details.

Molecular Biology of the Cell (Alberts et al., 1994). Following the Cayette example, if the user clicks on `atherosclerosis` at this point, it will take them to the relevant part of the textbook, a section devoted to how cells import cholesterol by receptor-mediated endocytosis (Fig. 7.5). From this page, the user can navigate through this particular unit, gathering more general information on transport from the plasma membrane via endosomes and vesicular traffic in the secretory and endocytic pathways.

The final link in the series in the upper right is LinkOut. This feature provides a list of third-party Web sites and resources relating to the Entrez entry being viewed, such as full-text of articles that can be displayed directly through the Web browser, or the capability of ordering the document through services such as Loansome Doc. A "cubby" service for LinkOut enables users to customize which links are displayed in a LinkOut view. Another way of getting to the full text of an article is by following a direct link to the publisher's Web site. In the Abstract view for the Cayette example (Fig. 7.3), a button directly under the citation is marked ATVB, for *Arteriosclerosis, Thrombosis, and Vascular Biology*, the journal in which the paper is published. With the proper individual or institutional subscription privileges, one would be able to immediately see the full text of the paper, including all figures and tables.

There is another way to perform an Entrez query, involving some built-in features of the system. Consider an example in which one is attempting to find all genes coding for DNA-binding proteins in methanobacteria. In this case, the search would begin by setting the Search pull-down menu to `Nucleotide` and then typing the term `DNA-binding` into the text box. This search returns 23,797 entries in which the search term appears (Fig. 7.6). At this point, to narrow down the search, the user would click on the `Limits` hyperlink, directly below the text box. This brings the user to a new page that allows the search to be limited, as implied by the name of the hyperlink. Here, the search will be limited by organism, so the Limited To pull-down is changed to `Organism`, and the word `methanobacterium` is typed into the search box (Fig. 7.7). Clicking `Go` will now return all of the entries in which *Methanobacterium* is the organism (303). The results from the first search can also be combined with those from the second by clinking on the `History` hyperlink below the text box, resulting in a list of recent queries (Fig. 7.8). The list shows the individual queries, whether those queries were field-limited, the time at which the query was performed, and how many entries that individual query returned. To combine two separate queries into one, the user simply combines the queries by number; in this case, because the queries are numbered #8 and #9, the syntax would be `#8 AND #9`. Clicking `Preview` regenerates a table, showing the new, combined query as #10, containing three entries. Alternatively, clicking `Go` shows the user the three entries, in the now-familiar nucleotide summary format (Fig. 7.9).

As before, there are a series of hyperlinks to the upper right of each entry; three are shown for the first entry, which is for the *M. thermoautotrophicum* tfx gene. The PubMed link takes the user back to the bibliographic entry or entries corresponding to this GenBank entry. Clicking on `Protein` brings into play one of the hard link relationships, showing the GenPept entry that corresponds to the tfx gene's conceptual translation (Fig. 7.10). Notice that, within the entry itself, the scientific name of the organism is represented as hypertext; clicking on that link takes the user to the NCBI Taxonomy database, where information on that particular organism's lineage is available. One of the useful views at this level is the Graphics view; this view

Vesicular Traffic in the Secretory and Endocytic Pathways

Transport from the Plasma Membrane via Endosomes: Endocytosis[19]

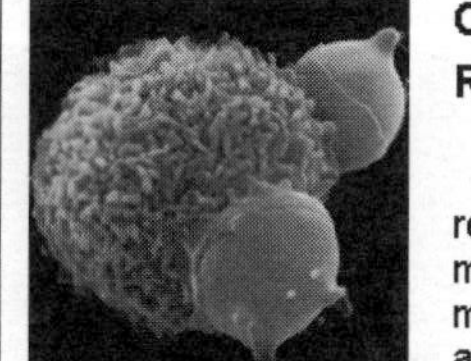

Outline

Introduction

Specialized Phagocytic Cells Can Ingest Large Particles

Pinocytic Vesicles Form from Coated Pits in the Plasma Membrane

Clathrin-coated Pits Can Serve as a Concentrating Device for Internalizing Specific Extracellular Macromolecules

Cells Import Cholesterol by Receptor-mediated Endocytosis

Endocytosed Materials Often End Up in Lysosomes

Cells Import Cholesterol by Receptor-mediated Endocytosis[23]

Many animal cells take up cholesterol through receptor-mediated endocytosis and in this way acquire most of the cholesterol they require to make new membrane. If the uptake is blocked, cholesterol accumulates in the blood and can contribute to the formation in blood vessel walls of atherosclerotic plaques - the deposits of lipid and fibrous tissue that cause strokes and heart attacks by blocking blood flow. In fact, it was through a study of humans with a strong genetic predisposition for atherosclerosis that the mechanism of receptor-mediated endocytosis was first clearly revealed.

Most cholesterol is transported in the blood bound to protein in the form of particles known as low-density lipoproteins, or LDL (Figure 13-29). When a cell needs cholesterol for membrane synthesis, it makes transmembrane receptor proteins for LDL and inserts them into its plasma membrane. Once in the plasma membrane, the LDL receptors diffuse until they associate with clathrin-coated pits that are in the process of forming (Figure 13-30A). Since coated pits constantly pinch off to form coated vesicles, any LDL particles bound to LDL receptors in the coated pits are rapidly internalized in coated vesicles. After shedding their clathrin coats, these vesicles deliver their contents to early endosomes, which are located near the cell periphery. Once in the endosomal compartment, the

Figure 7.5. Text related to the original Cayette et al. (2000) entry from *Molecular Biology of the Cell* (Alberts et al., 1994). See text for details.

Figure 7.6. Formulating a search against the nucleotide portion of Entrez. The initial query is shown in the text box towards the top of the window, and the nucleotide entries matching the query are shown below. See text for details.

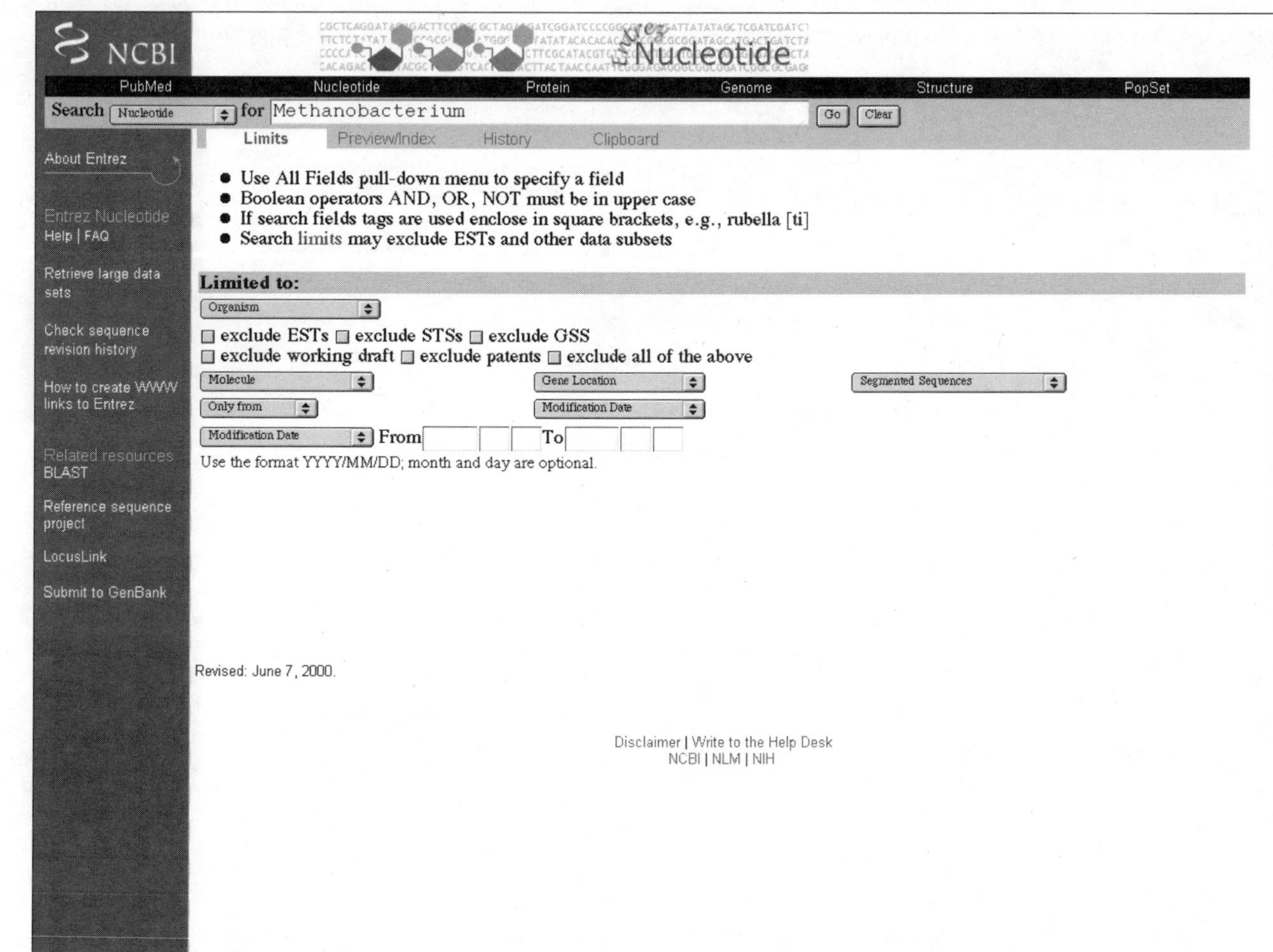

Figure 7.7. Using the Limits feature of Entrez to limit a search to a particular organism. See text for details.

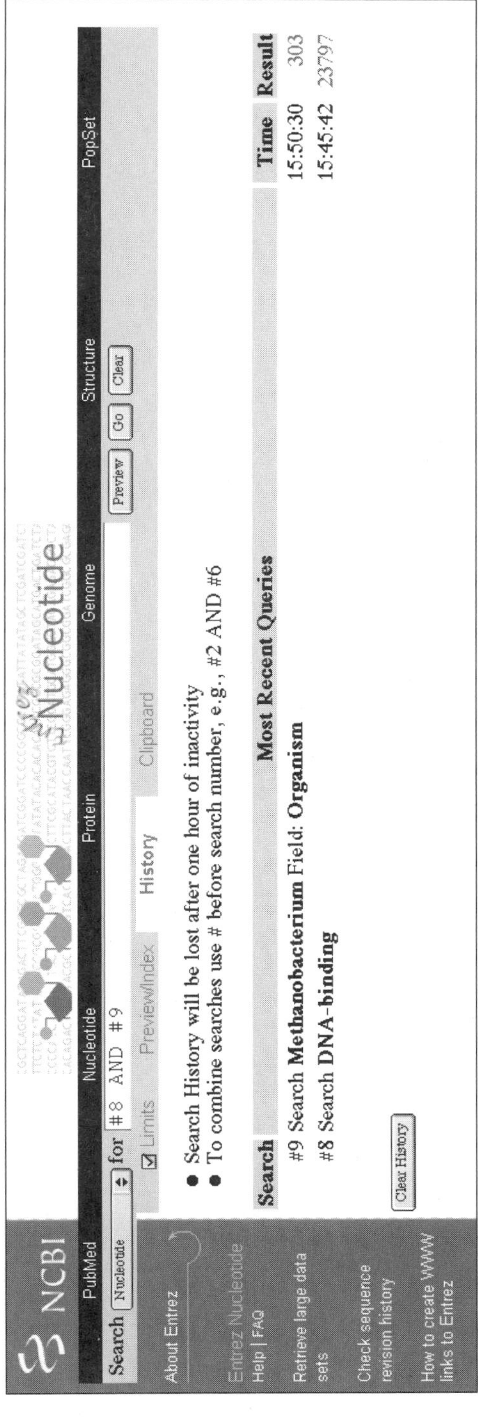

Figure 7.8. Combining individual queries using the History feature of Entrez. See text for details.

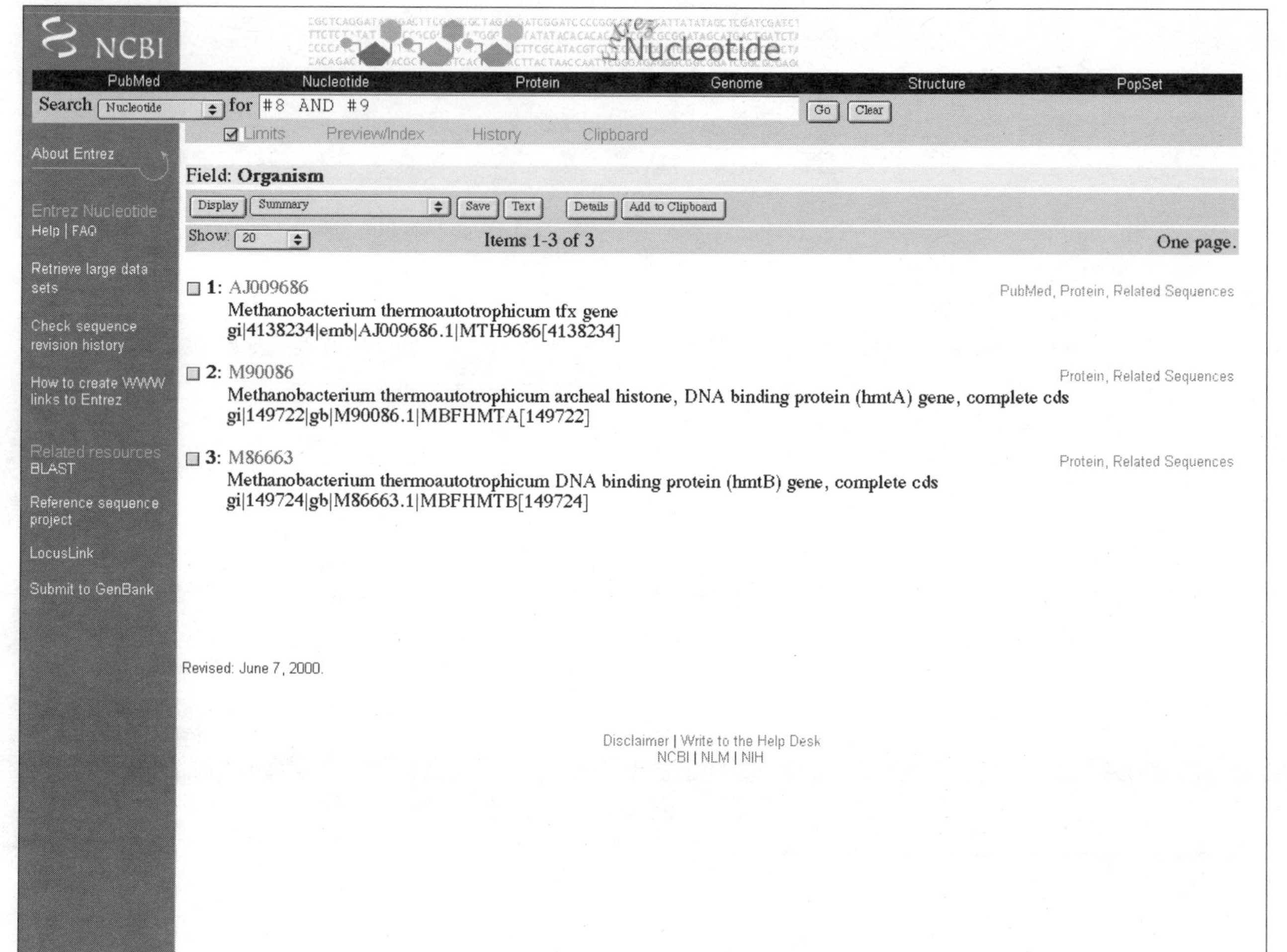

Figure 7.9. Entries resulting from the combination of two individual Entrez queries. The command producing this Entrez is shown in the text box at the top of the figure, and information on the individual queries that were combined is given in Figure 7.8.

1 : GI = "4138235" [GenPept] Tfx protein [Methanobacteri... Related Articles, Protein, Nucleotide

```
LOCUS       CAA08778      138 aa                    BCT       22-JAN-1999
DEFINITION  Tfx protein [Methanobacterium thermoautotrophicum].
ACCESSION   CAA08778
PID         g4138235
VERSION     CAA08778.1  GI:4138235
DBSOURCE    embl locus MTH9686, accession AJ009686.1
KEYWORDS    .
SOURCE      Methanobacterium thermoautotrophicum.
  ORGANISM  Methanobacterium thermoautotrophicum
            Archaea; Euryarchaeota; Methanobacteriales; Methanobacteriaceae;
            Methanobacterium.
REFERENCE   1  (residues 1 to 138)
  AUTHORS   Hochheimer,A.
  TITLE     Direct Submission
  JOURNAL   Submitted (21-JUL-1998) Hochheimer A., Abt. Biochemie,
            Max-Planck-Institut fuer terrestrische Mikrobiologie, Karl von
            Frisch Strasse, 35043 Marburg, GERMANY
REFERENCE   2  (residues 1 to 138)
  AUTHORS   Hochheimer,A., Hedderich,R. and Thauer,R.K.
  TITLE     The DNA binding protein Tfx from Methanobacterium
            thermoautotrophicum: structure, DNA binding properties and
            transcriptional regulation
  JOURNAL   Mol. Microbiol. 31 (2), 641-650 (1999)
  MEDLINE   99157570
FEATURES             Location/Qualifiers
     source          1..138
                     /organism="Methanobacterium thermoautotrophicum"
                     /strain="Marburg"
                     /db_xref="taxon:2166"
     Protein         1..138
                     /function="DNA-binding protein"
                     /product="Tfx protein"
     mat_peptide     2..138
                     /product="Tfx protein"
     CDS             1..138
                     /gene="tfx"
                     /coded_by="AJ009686.1:1..417"
                     /transl_table=11
ORIGIN
        1 mskktflter qktvlemrer gwsqkkiare lkttrqnvsa ierkamenie ksrntldfvk
       61 flkspvrilc rrgdtldeii krlleesnke gihvihdsit laflirekas hrivhrvvks
      121 dfeigvtrdg eiivdlns
//
```

Figure 7.10. The protein neighbor for the *M. thermoautotrophicum* tfx gene. Clicking on the Protein hyperlink next to the first entry in Figure 7.9 leads the user to this GenPept entry. See text for details.

attempts to show graphically all of the features described within the entry's feature table, providing a very useful overview, particularly when the feature table is very long. The Related Sequences link shows all sequences similar to that of the tfx gene at the nucleotide level, in essence showing the results of a precomputed BLAST search.

The last part of Entrez to be discussed deals with structures. Structure queries can be done directly by specifying `Structure` in the Search pull-down menu. For example, suppose that one wishes to find out information about the structure of HMG-box B from rat, whose accession number is 1HMF. Typing `1HMF` into the query box leads the user to the structure summary page for 1HMF, which has a decidedly different format than any of the pages seen so far (Fig. 7.11). This page shows details from the header of the source MMDB document (which is derived from PDB), links to PubMed and to the taxonomy of the source organism, and links to both sequence and structure neighbors. The Sequence Neighbors links show neighbors to 1HMF on the basis of sequence—that is, by BLAST search—thus, although this is a *structure* entry, it is important to realize that sequence neighbors have nothing to do with the structural information, at least not directly. To get information about related structures, one of the Structure Neighbor links can be followed, producing a table of neighbors as assessed by VAST. For a user interested in gleaning initial impressions about the shape of a protein, the Cn3D plug-in, invoked by clicking on `View/Save Structure`, provides a powerful interface, giving far more information than anyone could deduce from simply examining a string of letters (the sequence of the protein). The protein may be rotated along its axes by means of the scroll bars on the bottom, top, and right-hand side of the window or may be freely rotated by clicking and holding down the mouse key while the cursor is within the structure window and then dragging. Users are able to zoom in on particular parts of the structure or change the coloration of the figure, to determine specific structural features about the protein. In Figure 7.12, for instance, `Spacefilling` and `Hydrophobicity` were chosen as the Render and Color options, respectively. More information on Cn3D is presented in Chapter 5 as well as in the online Cn3D documentation. In addition, users can save coordinate information to a file and view the data using third-party applications such as Kinemage (Richardson and Richardson, 1992) and RasMol (Sayle and Milner-White, 1995).

Finally, at any point along the way in using Entrez, if there are partial or complete search results that the user wishes to retain while moving onto a new query, the Add to Clipboard button can be pushed. This stores the results of the current query, which the user can return to by clicking the Clipboard hyperlink directly under the text box. The clipboard holds a maximum of 500 items, and the items are held in memory for 1 h.

LOCUSLINK

The Entrez system revolves necessarily around the individual entries making up the various component databases that are part of the Entrez search space. Another way to think about this search space is to organize it around discrete genetic loci. NCBI LocusLink does just this, providing a single query interface to various types of information regarding a given genetic locus, such as phenotypes, map locations, and

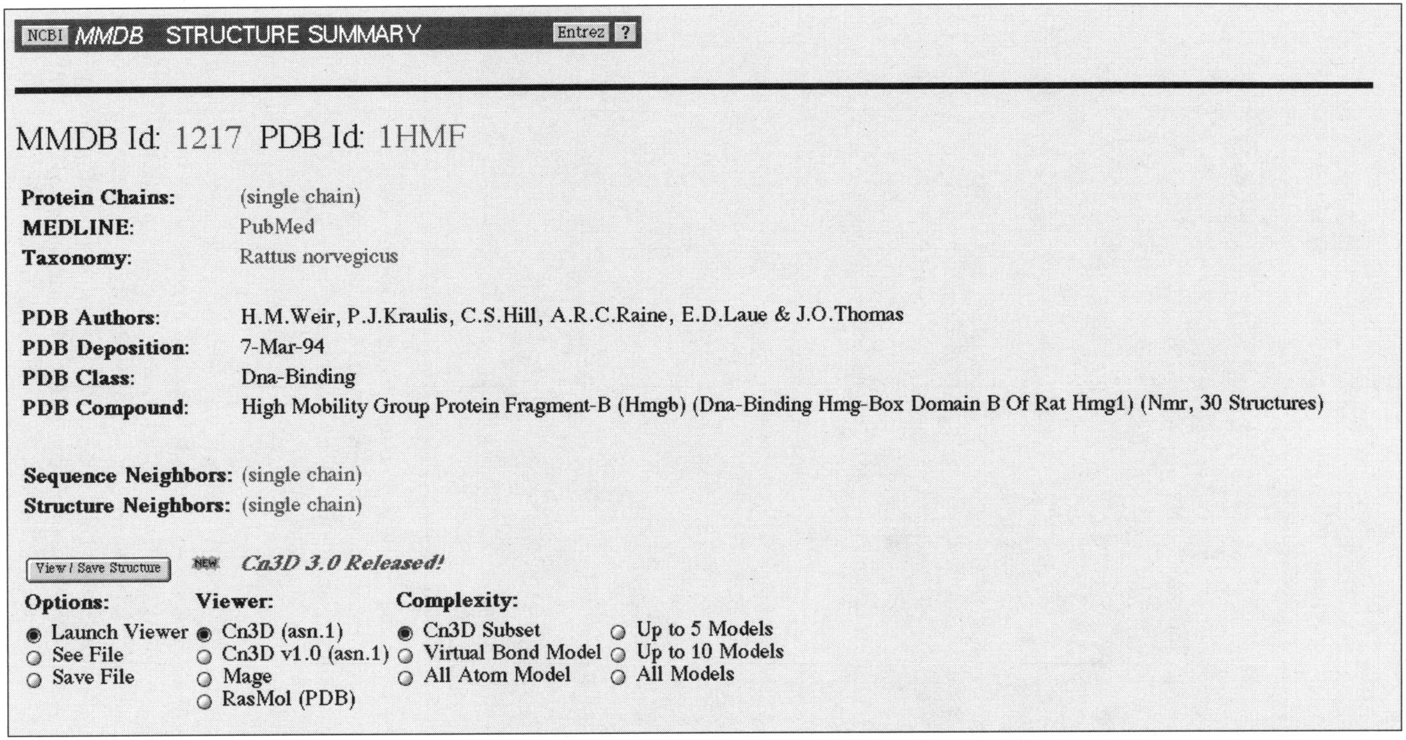

NCBI *MMDB* STRUCTURE SUMMARY Entrez ?

MMDB Id: 1217 PDB Id: 1HMF

Protein Chains: (single chain)
MEDLINE: PubMed
Taxonomy: Rattus norvegicus

PDB Authors: H.M.Weir, P.J.Kraulis, C.S.Hill, A.R.C.Raine, E.D.Laue & J.O.Thomas
PDB Deposition: 7-Mar-94
PDB Class: Dna-Binding
PDB Compound: High Mobility Group Protein Fragment-B (Hmgb) (Dna-Binding Hmg-Box Domain B Of Rat Hmg1) (Nmr, 30 Structures)

Sequence Neighbors: (single chain)
Structure Neighbors: (single chain)

View / Save Structure NEW ***Cn3D 3.0 Released!***

Options:	**Viewer:**	**Complexity:**	
◉ Launch Viewer	◉ Cn3D (asn.1)	◉ Cn3D Subset	○ Up to 5 Models
○ See File	○ Cn3D v1.0 (asn.1)	○ Virtual Bond Model	○ Up to 10 Models
○ Save File	○ Mage	○ All Atom Model	○ All Models
	○ RasMol (PDB)		

Figure 7.11. The structure summary for 1HMF, resulting from a direct query of the structures accessible through the Entrez system. The entry shows header information from the corresponding MMDB entry, links to PubMed and to the taxonomy of the source organism, and links to sequence and structure neighbors.

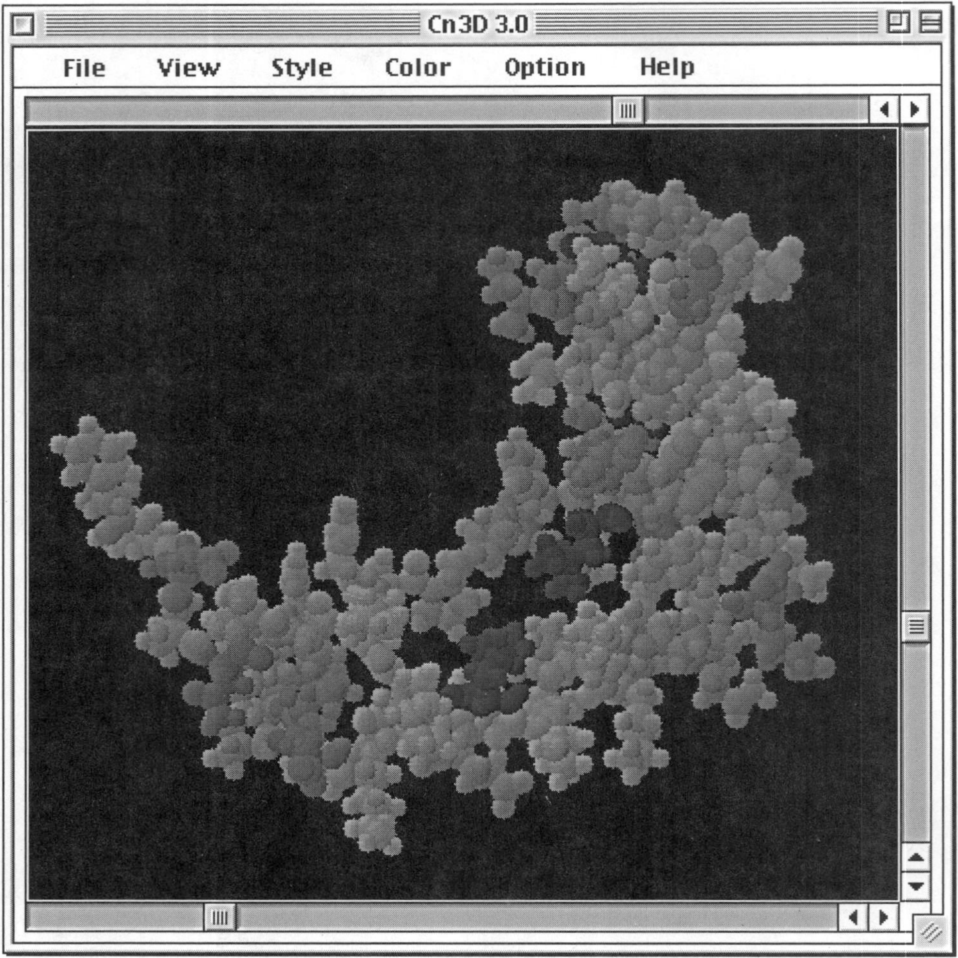

Figure 7.12. The structure of 1HMF rendered using Cn3D version 3.0, an interactive molecular viewer that acts as a plug-in to Web Entrez. Cn3D is also bundled with and can be used with Network Entrez. Details are given in the text.

homologies to other genes. The LocusLink search space currently includes information from humans, mice, rats, fruit flies, and zebrafish.

With the use of the gene for the high-mobility group protein HMG1 as an example, the LocusLink query begins by the user simply typing the name of the gene into the query box appearing at the top of the LocusLink home page. Alternatively, the user could select the gene of interest from an alphabetical list. The query on `HMG1` returns three LocusLink entries, from human, mouse, and rat (Fig. 7.13). In this view, the user is given the Locus ID in the first column; the Locus ID is intended to be a unique, stable identifier that is associated with this gene locus. Clicking on the Locus ID for the human (3146) produces the LocusLink Report view, as shown in Figure 7.14. The Report view begins with links to external sources of information, shown as colored buttons at the top of the page. In this particular report, the links

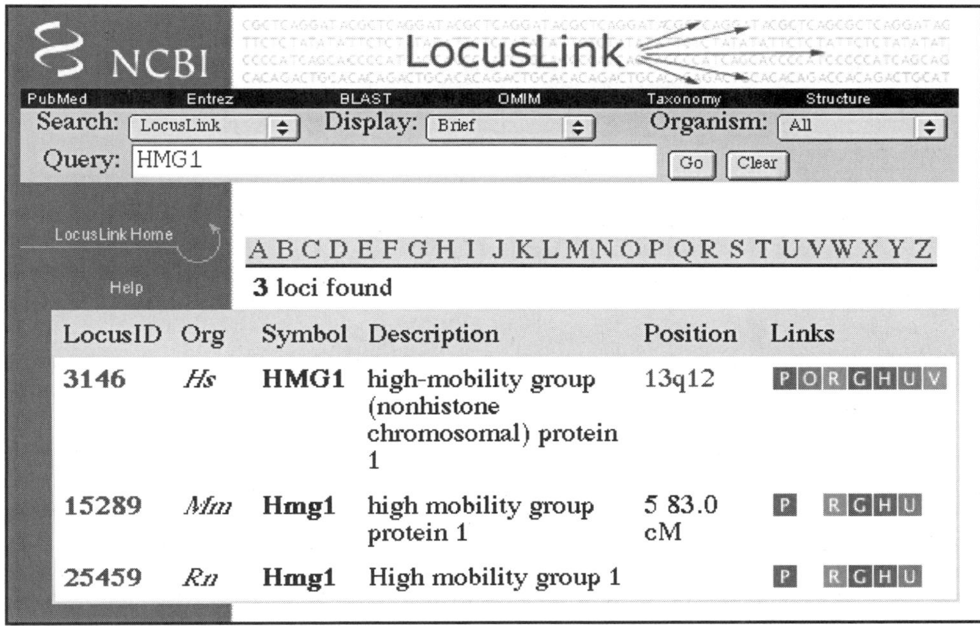

Figure 7.13. Results of a LocusLink query, using HMG1 as the search term. The report shows three entries corresponding to HMG1 in human (*Hs*), mouse (*Mm*), and rat (*Rn*). A brief description is given for each found locus, as well as its chromosomal location. A series of blocks is found to the right of each entry, providing a jumping-off point to numerous other sources of data; these links are described in the text.

would lead the user to PubMed (Pub), UniGene (UG, cf. Chapter 12), the dbSNP variation database (VAR, cf. Chapter 12), HomoloGene (HOMOL, see below), and the Genome Database (GDB). These offsite links will change from entry to entry, depending on what information is available for the gene of interest. A complete list of offsite data sources is given in the LocusLink online documentation.

Continuing down the Report view, in the section marked Locus Information, the user is presented with the official gene symbol, along with any alternate symbols that may have traditionally been used in the literature or in sequence records. This section would also include information on the name of the gene product, any aliases for the name of the gene product, the Enzyme Commission number, the name of any diseases that result from variants at this gene locus, and links to OMIM and UniGene. Only those pieces of information that are known or are applicable to this particular gene locus are shown.

In the section labeled Map Information, the report shows what chromosome this locus is on, the cytogenetic and genetic map positions, when known, and any STS markers that are found within the mRNA corresponding to this locus. There is a hyperlink that can take the user to the Entrez Map Viewer, showing the position of this locus and the relationship of this locus to surrounding loci (Fig. 7.15). The Map Viewer shows the chromosomal ideogram to the left, with the highlighted region marked by a thick bar to the right of the ideogram. The user can zoom in or out by clicking on the icon above the ideogram. In the main window, the user is presented

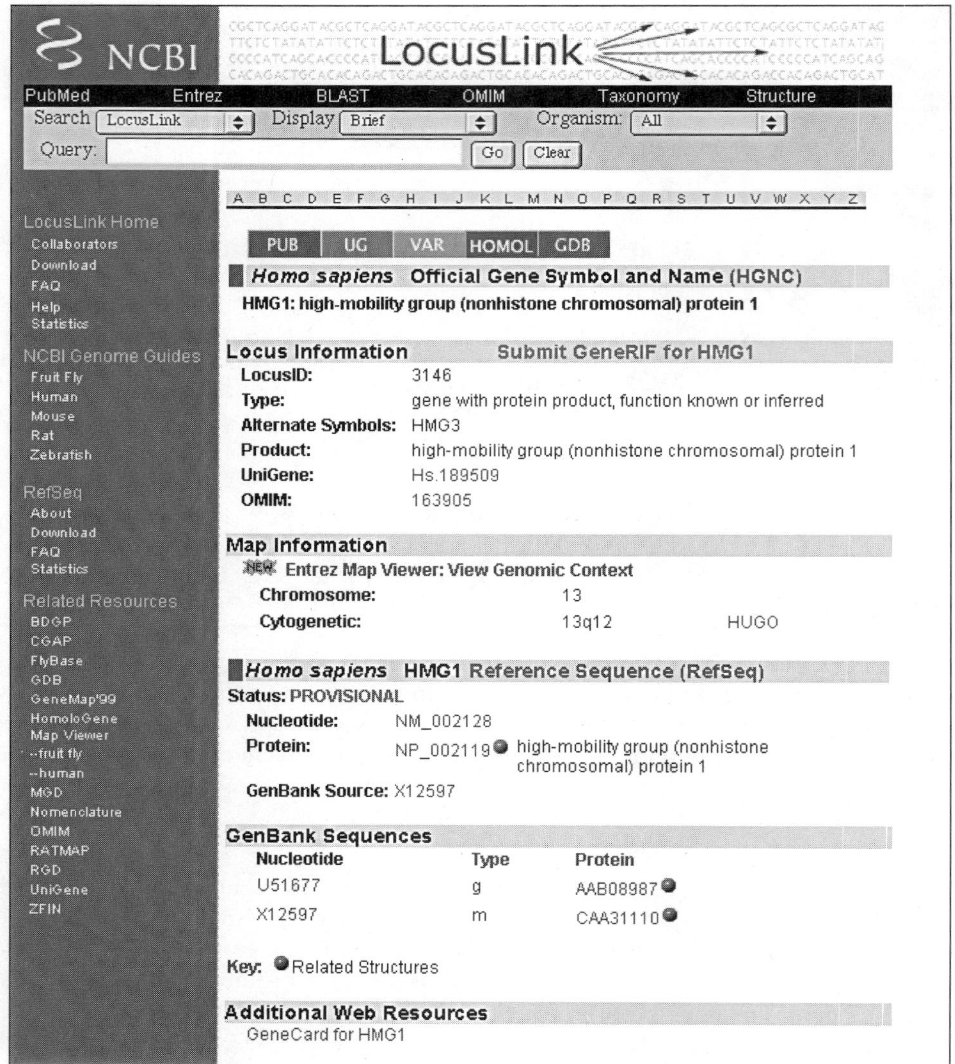

Figure 7.14. The LocusLink report view for human HMG1. The report is divided into six sections, providing gene symbol, locus, map, RefSeq, and GenBank information, as well as links to external data sources. See text for details.

with both the cytogenetic and sequence map. In this particular view, 20 genes are shown, with the original locus of interest highlighted. As with most graphical views of this type, the majority of the elements in this view are clickable, taking the user to more information about that particular part of either the cytogenetic or sequence map.

The next section deals with RefSeq information. RefSeq is short for the NCBI Reference Sequence Project, which is an effort aimed at providing a stable, reference sequence for all of the molecules corresponding to the central dogma of biology. The intention is that these reference sequences will provide a framework for making

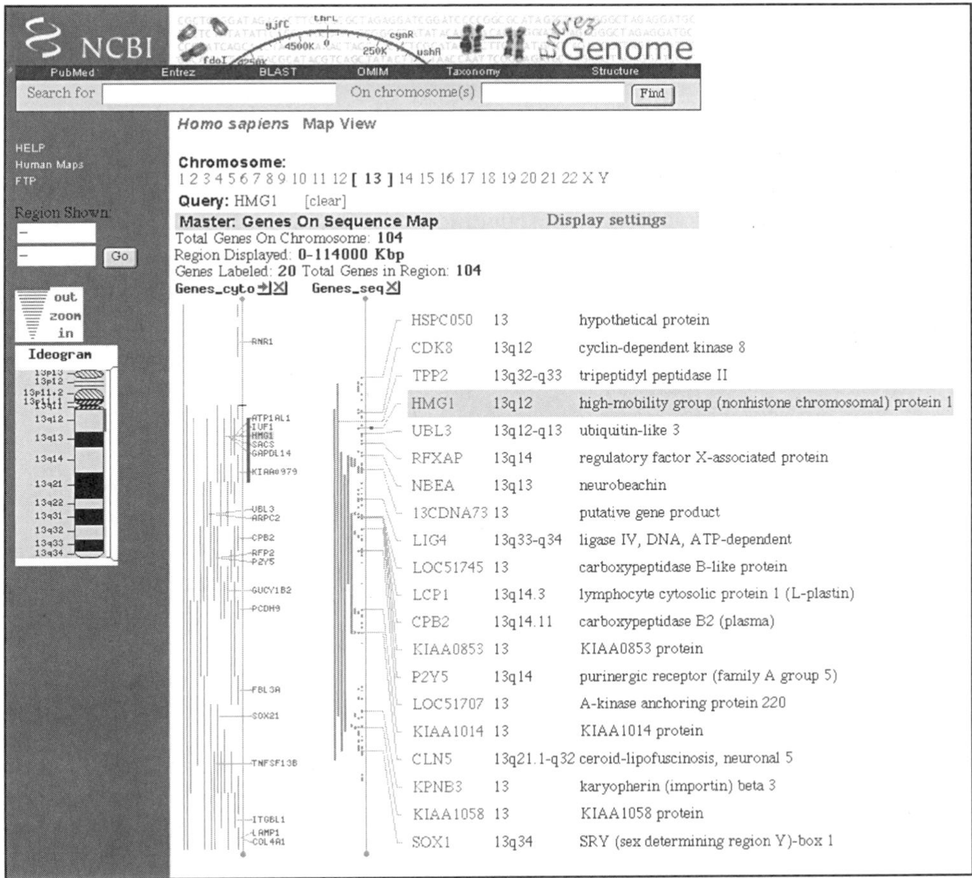

Figure 7.15. The Entrez map view for human HMG1. The chromosomal position is indicated by the ideogram at the left of the window. The main window contains a depiction of both the cytogenetic and sequence map, with the HMG1 gene highlighted. Interestingly, the gene shown at the very bottom of this view (SOX1), like HMG1, is also a member of the high mobility group family of proteins (Baxevanis and Landsman, 1995).

functional annotations, as well as for information regarding mutations, gene expression, and polymorphisms. The sequences listed in this section represent the sequences that were used to build the corresponding RefSeq record. Notice that RefSeq nucleotide records are preceded by NM and that protein records are preceded by NP. A blue button appears next to protein information if there is also structural information available about the protein. The final portion of the LocusLink report shows the GenBank accession numbers that correspond to this locus. The middle column indicates the molecule type for these GenBank entries, with m standing for mRNA, g for genomic DNA, e for an EST, and u for undetermined. In this particular case, there is also a link to the GeneCard for HMG1; clicking on that hyperlink takes the user to the GeneCards database at the Weizmann Institute, providing a concise summary of the function of this gene.

Another way to proceed through the information is to return to the query result screen shown in Figure 7.13 and use the linked alphabet blocks shown to the right of each of the entries. In turn, these links are

- P, for PubMed bibliographic entries corresponding to the locus;
- O, for the Online Mendelian Inheritance in Man summary on this locus;
- R, for the RefSeq entries corresponding to the locus;
- G, for the individual GenBank entries related to the locus; these entries will correspond to the RefSeq entry, as shown in the LocusLink Report view;
- H, for HomoloGene. HomoloGene, which is discussed in greater detail in Chapter 12, allows the user to find both orthologs and homologs for genes represented in LocusLink and UniGene, the assignments being made based on sequence similarity;
- U, whether this locus is part of a UniGene cluster; and
- V, for variation data on this locus contained within dbSNP.

When following either the PubMed or GenBank links, the user is, in essence, returned to the Entrez search space, enabling the user to take advantage of Entrez's navigational features once again.

SEQUENCE DATABASES BEYOND NCBI

Although it may appear from this discussion that NCBI is the center of the sequence universe, many specialized sequence databases throughout the world serve specific groups in the scientific community. Often, these databases provide additional information such as phenotypes, experimental conditions, strain crosses, and map features. The data are of great importance to these subsets of the scientific community, inasmuch as they can influence rational experimental design, but such types of data do not always fit neatly within the confines of the NCBI data model. Development of specialized databases necessarily ensued, but they are intended to be used as an adjunct to GenBank, not in place of it. It is impossible to discuss all of these kinds of databases here, but, to emphasize the sheer number of such databases that exist, *Nucleic Acids Research* devotes its first issue every year to papers describing these databases (cf. Baxevanis, 2001).

An example of a specialized organismal database is the *Saccharomyces* Genome Database (SGD), housed at the Stanford Human Genome Center. The database provides a very simple search interface that allows text-based searches by gene name, gene information, clone, protein information, sequence name, author name, or full text. For example, using Gene Name as the search topic and `hho1` as the name of the gene to be searched for produces a SacchDB information window showing all known information on locus HHO1 (Fig. 7.16). This window provides jumping-off points to other databases, such as GenBank/GenPept, MIPS, and the Yeast Protein Database (YPD). Following the link to Sacch3D for this entry provides information on structural homologs of the HHO1 protein product found in PDB, links to secondary and tertiary structure prediction sites, and precomputed BLAST reports against a number of query databases. Returning to the Locus window and clicking on the map in the upper right-hand corner, the user finds a graphical view of the

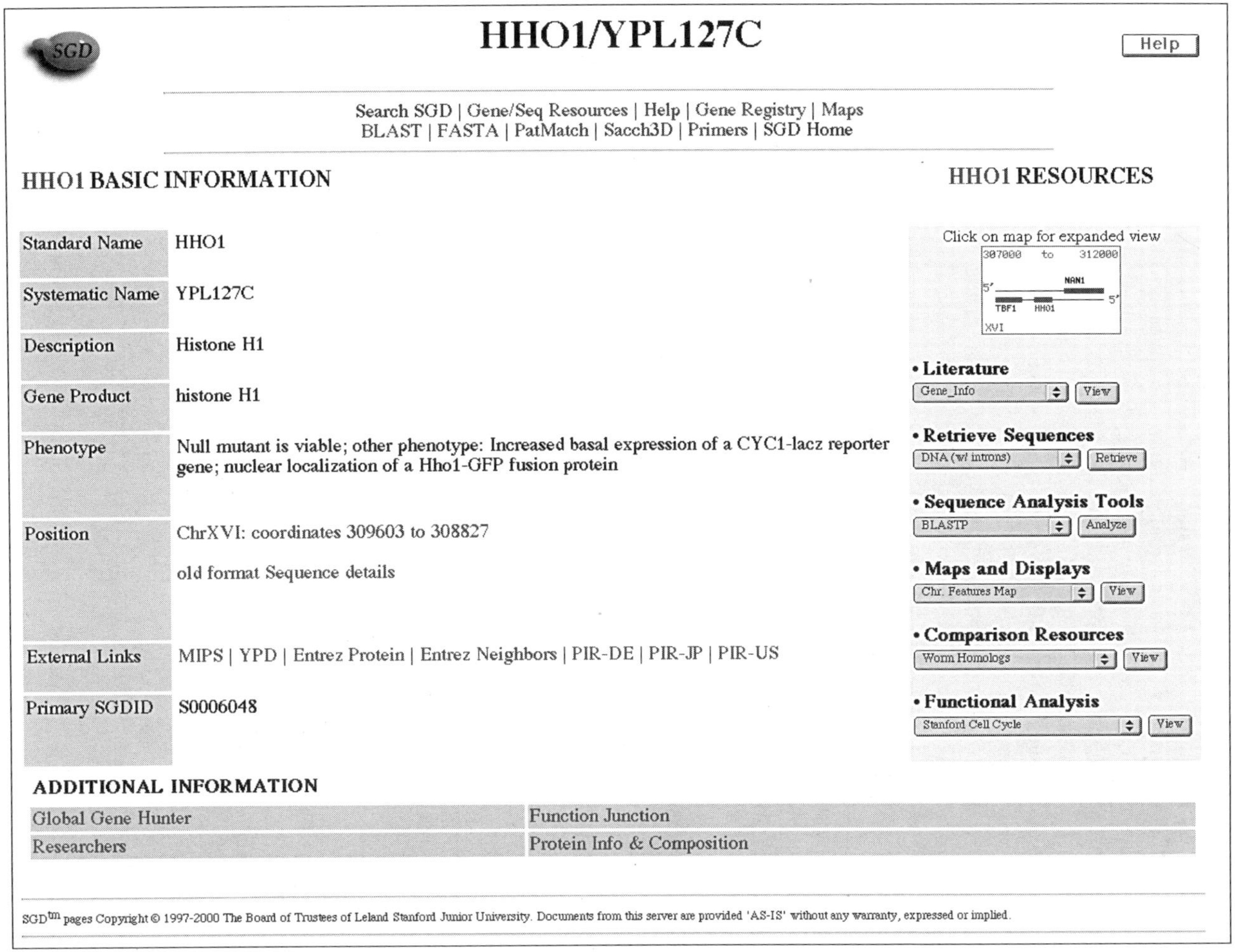
SGD
HHO1/YPL127C
Help
Search SGD | Gene/Seq Resources | Help | Gene Registry | Maps
BLAST | FASTA | PatMatch | Sacch3D | Primers | SGD Home
HHO1 BASIC INFORMATION
Standard Name HHO1
Systematic Name YPL127C
Description Histone H1
Gene Product histone H1
Phenotype Null mutant is viable; other phenotype: Increased basal expression of a CYC1-lacz reporter gene; nuclear localization of a Hho1-GFP fusion protein
Position ChrXVI: coordinates 309603 to 308827
old format Sequence details
External Links MIPS | YPD | Entrez Protein | Entrez Neighbors | PIR-DE | PIR-JP | PIR-US
Primary SGDID S0006048
HHO1 RESOURCES
Click on map for expanded view
307000 to 312000
5' 5'
NAN1
TBF1 HHO1
XVI
• Literature
Gene_Info View
• Retrieve Sequences
DNA (w/ introns) Retrieve
• Sequence Analysis Tools
BLASTP Analyze
• Maps and Displays
Chr. Features Map View
• Comparison Resources
Worm Homologs View
• Functional Analysis
Stanford Cell Cycle View
ADDITIONAL INFORMATION
Global Gene Hunter
Function Junction
Researchers
Protein Info & Composition
SGDtm pages Copyright © 1997-2000 The Board of Trustees of Leland Stanford Junior University. Documents from this server are provided 'AS-IS' without any warranty, expressed or implied.

Figure 7.16. A SacchDB Locus view resulting from an SGD query using hho1 as the gene name. The information returned is extensive; it includes the name of the gene product, phenotypic data, and the position of HHO1 within the yeast genome. There are numerous hyperlinks providing access to graphical views and to external database entries related to the current query.

area surrounding the locus in question. Available views include physical maps, genetic maps, and chromosomal physical maps, among others. The chromosomal features map view for HHO1 is shown in Figure 7.17. Note the thick bar at the top of the figure, which gives the position of the current view with respect to the centromere. Clicking on that bar allows the user to move along the chromosome, and clicking on individual gene or ORF name (or, as the authors cite in the figure legend, "any little colorful bar") gives more detailed information about that particular region.

Another example of an organism-specific database is FlyBase, whose goal is to maintain comprehensive information on the genetics and molecular biology of *Drosophila*. The information found in FlyBase includes an extensive *Drosophila* bibliography, addresses of over 6,000 researchers involved in *Drosophila* projects, a compilation of information on over 51,500 alleles of more than 13,200 genes, information about the expression and properties of over 4,800 transcripts and 2,500 proteins, and descriptions of over 16,700 chromosomal aberrations. Also included is relevant mapping information, functional information on gene products, lists of stock centers and genomic clones, and information from allied databases. Searches on any of these "fields" can be done through a simple search mechanism. For example, searching by gene symbol using `capu` as the search term brings up a record for a gene named cappuccino, which is required for the proper polarity of the developing *Drosophila* oocyte (Emmons et al., 1995). Calling up the cytogenetic map view generates a map showing the gene and cytologic location of cappuccino and other genes in that immediate area, and users can click on any of the gene bars to bring up detailed information on that particular gene (Fig. 7.18). The view can be changed by selecting

Features around YPL127C on chromosome XVI

Spanning a region 10 kb left and 10 kb right (coordinates 298827 to 319603)

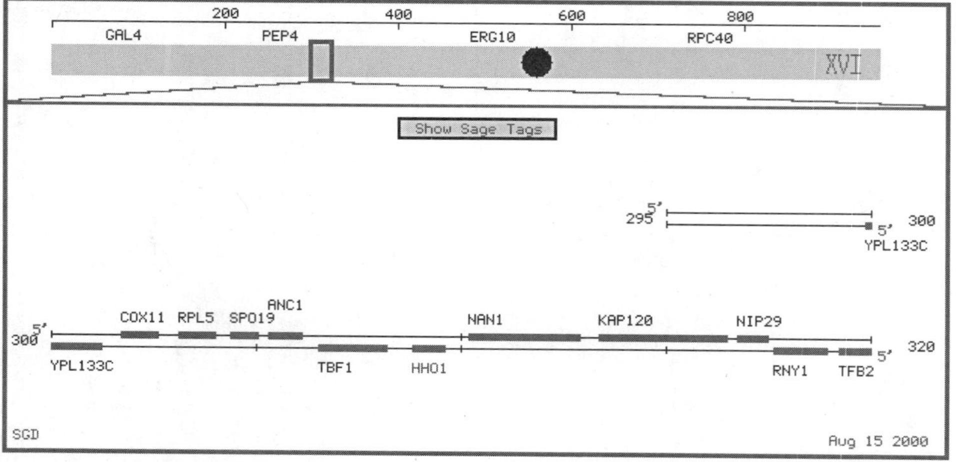

Figure 7.17. A chromosomal features map resulting from the query used to generate the Locus view shown in Figure 7.16. Chromosome XVI is shown at the top of the figure, with the exploded region highlighted by a box. Most items are clickable, returning detailed information about that particular gene, ORF, or construct.

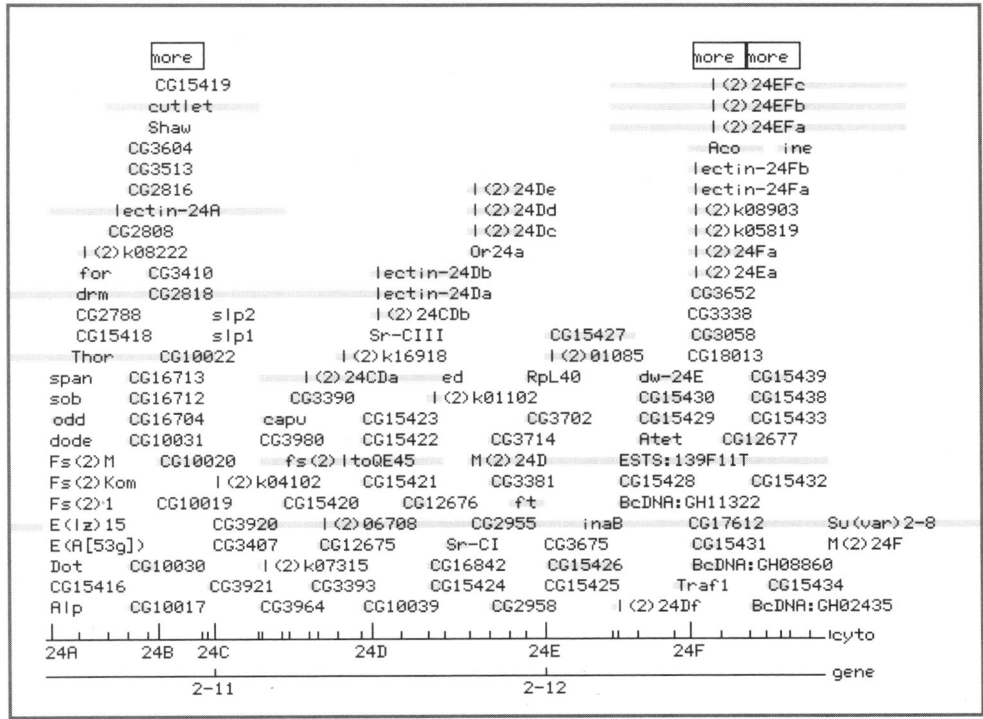

Figure 7.18. Genes view resulting from querying FlyBase for the *cappucino* gene (*capu* in the figure, between positions 24C and 24D on the cytologic map). The graphical view can be changed by clicking on any of the Class buttons that appear below the figure, as described in the text. Information on any of the genes shown can be obtained by clicking on the appropriate bar.

one of the Class buttons at the bottom of the window, so that a graphical view of cosmids, deficiencies, duplications, inversions, transpositions, translocations, or other aberrations can be examined instead.

MEDICAL DATABASES

Although the focus of this chapter (and the book in general) is on sequences, databases cataloguing and organizing sequence information are not the only kinds of databases useful to the biologist. An example of such a non-sequence-based information resource that is tremendously useful in genomics is Online Mendelian Inheritance in Man (OMIM), the electronic version of the catalog of human genes and genetic disorders founded by Victor McKusick at The Johns Hopkins University (McKusick, 1998; Hamosh et al., 2000). OMIM provides concise textual information from the published literature on most human conditions having a genetic basis, as well as pictures illustrating the condition or disorder (where appropriate) and full citation information. Because the online version of OMIM is housed at NCBI, links to Entrez are provided from all references cited within each OMIM entry.

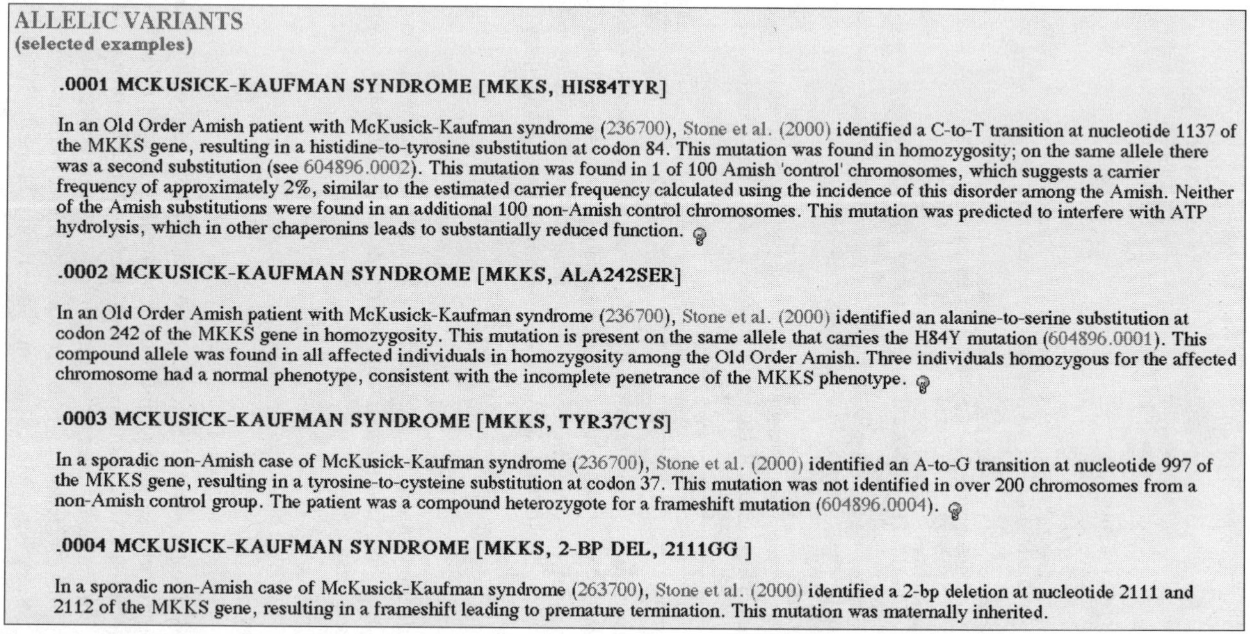

ALLELIC VARIANTS
(selected examples)

.0001 MCKUSICK-KAUFMAN SYNDROME [MKKS, HIS84TYR]

In an Old Order Amish patient with McKusick-Kaufman syndrome (236700), Stone et al. (2000) identified a C-to-T transition at nucleotide 1137 of the MKKS gene, resulting in a histidine-to-tyrosine substitution at codon 84. This mutation was found in homozygosity; on the same allele there was a second substitution (see 604896.0002). This mutation was found in 1 of 100 Amish 'control' chromosomes, which suggests a carrier frequency of approximately 2%, similar to the estimated carrier frequency calculated using the incidence of this disorder among the Amish. Neither of the Amish substitutions were found in an additional 100 non-Amish control chromosomes. This mutation was predicted to interfere with ATP hydrolysis, which in other chaperonins leads to substantially reduced function.

.0002 MCKUSICK-KAUFMAN SYNDROME [MKKS, ALA242SER]

In an Old Order Amish patient with McKusick-Kaufman syndrome (236700), Stone et al. (2000) identified an alanine-to-serine substitution at codon 242 of the MKKS gene in homozygosity. This mutation is present on the same allele that carries the H84Y mutation (604896.0001). This compound allele was found in all affected individuals in homozygosity among the Old Order Amish. Three individuals homozygous for the affected chromosome had a normal phenotype, consistent with the incomplete penetrance of the MKKS phenotype.

.0003 MCKUSICK-KAUFMAN SYNDROME [MKKS, TYR37CYS]

In a sporadic non-Amish case of McKusick-Kaufman syndrome (236700), Stone et al. (2000) identified an A-to-G transition at nucleotide 997 of the MKKS gene, resulting in a tyrosine-to-cysteine substitution at codon 37. This mutation was not identified in over 200 chromosomes from a non-Amish control group. The patient was a compound heterozygote for a frameshift mutation (604896.0004).

.0004 MCKUSICK-KAUFMAN SYNDROME [MKKS, 2-BP DEL, 2111GG]

In a sporadic non-Amish case of McKusick-Kaufman syndrome (263700), Stone et al. (2000) identified a 2-bp deletion at nucleotide 2111 and 2112 of the MKKS gene, resulting in a frameshift leading to premature termination. This mutation was maternally inherited.

Figure 7.19. An example of a list of allelic variants that can be obtained through OMIM. The figure shows the list of allelic variants for McKusick-Kaufman syndrome.

OMIM has a defined numbering system in which each entry is assigned a unique number, similar to an accession number, but certain positions within that number indicate information about the genetic disorder itself. For example, the first digit represents the mode of inheritance of the disorder: 1 stands for autosomal dominant, 2 for autosomal recessive, 3 for X-linked locus or phenotype, 4 for Y-linked locus or phenotype, 5 for mitochondrial, and 6 for autosomal locus or phenotype. (The distinction between 1 or 2 and 6 is that entries catalogued before May 1994 were assigned either a 1 or 2, whereas entries after that date were assigned a 6 regardless of whether the mode of inheritance was dominant or recessive.) An asterisk preceding a number indicates that the phenotype caused by the gene at this locus is *not* influenced by genes at other loci; however, the disorder itself may be caused by mutations at multiple loci. Disorders for which no mode of inheritance has been determined do not carry asterisks. Finally, a pound sign (#) indicates that the phenotype is caused by two or more genetic mutations.

OMIM searches are very easy to perform. The search engine performs a simple query based on one or more words typed into a search window. A list of documents containing the query words is returned, and users can select one or more disorders from this list to look at the full text of the OMIM entry. The entries include information such as the gene symbol, alternate names for the disease, a description of the disease (including clinical, biochemical, and cytogenetic features), details on the mode of inheritance (including mapping information), a clinical synopsis, and references. A particularly useful feature is lists of allelic variants; a short description is given after each allelic variant of the clinical or biochemical outcome of that particular mutation. There are currently over 1,000 gene entries containing at least one allelic variant that either causes or is associated with a discrete phenotype in humans. Figure 7.19 shows an example of an allelic variant list, in this case for mutations observed in patients with McKusick-Kaufman syndrome (MKKS).

INTERNET RESOURCES FOR TOPICS PRESENTED IN CHAPTER 7

BLAST	*http://www.ncbi.nlm.nih.gov/BLAST/*
Cn3D	*http://www.ncbi.nlm.nih.gov/Structure/CN3D/cn3d.shtml*
EndNote	*http://www.niles.com/*
Entrez	*http://www.ncbi.nlm.nih.gov/Entrez/*
FlyBase	*http://flybase.bio.indiana.edu*
GDB	*http://www.gdb.org/*
GeneCards	*http://bioinfo.weizmann.ac.il/cards/*
HomoloGene	*http://www.ncbi.nlm.nih.gov/HomoloGene/*
Kinemage	*http://www.umass.edu/microbio/rasmol/mage.htm*
LocusLink	*http://www.ncbi.nlm.nih.gov/LocusLink/*
MIPS	*http://www.mips.biochem.mpg.de/*
MMDB	*http://www.ncbi.nlm.nih.gov/Structure/MMDB/mmdb.shtml*
OMIM	*http://www.ncbi.nlm.nih.gov/Omim*
PDB	*http://www.rcsb.org/pdb/*
RasMol	*http://www.umass.edu/microbio/rasmol/*
Reference Manager	*http://www.risinc.com/*
Sacch3D	*http://www-genome.stanford.edu/Sacch3D/*

SGD	*http://genome-www.stanford.edu/Saccharomyces/*
VAST	*http://www.ncbi.nlm.nih.gov/Structure/VAST/vast.shtml*
YPD	*http://www.proteome.com/databases/index.html*

PROBLEM SET

1. You have been watching the evening news and have just heard an interesting story regarding recent developments on the genetics of colorectal cancer. You would like to get some more information on this research, but the news story was short on details. The only hard information you have is that the principal investigator was Bert Vogelstein at the Johns Hopkins School of Medicine.

a. How many of the papers that Dr. Vogelstein has written on the subject of colorectal neoplasms are available through PubMed?

b. A paper by Hedrick and colleagues describes the role of the DCC gene product in cellular differentiation and colorectal tumorigenesis. Based on this study, what is the chromosomal location of the DCC gene?

c. DCC codes for a cell-surface-localized protein involved in tumor suppression. From what cell line and tissue type was the human tumor suppressor protein (*not* the precursor) isolated?

d. In the DCC human tumor suppressor protein *precursor*, what range of amino acids comprise the signal sequence?

2. Online Mendelian Inheritance in Man (OMIM) indicates that the development of colorectal carcinomas involves a dominantly acting oncogene coupled with the loss of several genes (such as DCC) that normally suppress tumorigenesis.

a. An allelic variant of DCC also involved in esophageal carcinoma has been cataloged in OMIM. What was the mutation at the amino acid level, and what biological effect did it have in patients?

b. Based on the MIM gene map, how many other genes have been mapped to the exact cytogenetic map location as DCC *by PCR of somatic cell hybrid DNA*?

c. The OMIM entry for DCC is coupled to the Mouse Genome Database at The Jackson Laboratory, showing that the corresponding mouse gene is located on mouse chromosome 18. What is the resultant phenotype of a *null* mutation of *Dcc* in the mouse?

3. A very active area of commercial research involves the identification and development of new sweeteners for use by the food industry. Whereas traditional sweeteners such as table sugar (sucrose) are carbohydrates, most current research is instead focusing on proteins which have an intrinsically sweet taste. Because these "sweet-tasting proteins" are much sweeter than their carbohydrate counterparts, they are, in essence, calorie free, since so little is used to achieve a sweet taste in food. The most successful example of such a protein is aspartame; however, aspartame is synthetic and does not occur in nature. Alternate, natural protein sources are being investigated, including a sweet tasting protein called monellin.

a. According to Ogata and colleagues, how much sweeter than ordinary sugar is monellin on both a molar and weight bases?

b. Based on the SWISS-PROT entry for monellin chain B from serendipity berry, how many α-helices and β-strands does this protein possess?
c. What residue (amino acid *and* position), when blocked, abolishes monellin's sweet taste?
d. Three-dimensional structures are available for monellin. What *other* structure is most closely related to monellin structure 1MOL, as assessed by VAST P-value? Does this structure have the highest *sequence* similarity to 1MOL as well?
e. The monellin structure is based on a single-chain fusion product. How do the stability and renaturation properties of the fusion product differ from that of the native protein?

REFERENCES

Alberts, B., Bray, D., Lewis, J., Raff, M., Roberts, K., and Watson, J. D. (1994). *Molecular Biology of the Cell*, Garland Publishing, New York.

Altschul, S., Gish, W., Miller, W., Myers, E., and Lipman, D. (1990). Basic local alignment search tool. *J. Mol. Biol.* 215, 403–410.

Baxevanis, A. D., and Landsman, D. (1995). The HMG-1 box protein family: classification and functional relationships. *Nucleic Acids Res.* 23, 1604–1613.

Baxevanis, A. D. (2001). The Molecular Biology Database Collection: An Updated Compilation of Biological Database Resources. *Nucleic Acids Res.* 29, 1–7.

Emmons, S., Phan, H., Calley, J., Chen, W., James, B., and Manseau, L. (1995). Cappuccino, a *Drosophila* maternal effect gene required for polarity of the egg and embryo, is related to the vertebrate limb deformity locus. *Genes Dev.* 9, 2484–2494.

Gibrat, J.-F., Madej, T., and Bryant, S. (1996). Surprising similarities instructure comparison. *Curr. Opin. Struct. Biol.* 6, 377–385.

Hamosh, A., Scott, A. F., Amberger, J., Valle, D., and McKusick, V. A. (2000). Online Mendelian Inheritance in Man (OMIM). *Human Mutation* 15, 57–61.

Madej, T., Gibrat, J.-F., and Bryant, S. (1995). Threading a database ofprotein cores. *Proteins.* 23, 356–369.

McKusick, V. A. (1998). *Mendelian Inheritance in Man. Catalogs of Human Genes and Genetic Disorders* (The Johns Hopkins University Press, Baltimore).

Richardson, D., and Richardson, J. (1992). The kinemage: A tool for scientific communication. *Protein Sci.* 1, 3–9.

Sayle, R., and Milner-White, E. (1995). RasMol: biomolecular graphics for all. *Trends Biochem. Sci.* 20, 374.

Wilbur, W., and Coffee, L. (1994). The effectiveness of document neighboring in search enhancement. *Process Manage.* 30, 253–266.

Wilbur, W., and Yang, Y. (1996). An analysis of statistical term strength and its use in the indexing and retrieval of molecular biology texts. *Comput. Biol. Med.* 26, 209–222.

8

SEQUENCE ALIGNMENT AND DATABASE SEARCHING

Gregory D. Schuler

National Center for Biotechnology Information
National Library of Medicine
National Institutes of Health
Bethesda, Maryland

INTRODUCTION

There is a long tradition in biology of comparative analysis leading to discovery. For instance, Darwin's comparison of morphological features of the Galapagos finches and other species ultimately led him to postulate the theory of natural selection. In essence, we are performing the same type of analysis today when we compare the sequences of genes and proteins but in much greater detail. In this activity, the similarities and differences—at the level of individual bases or amino acids—are analyzed, with the aim of inferring structural, functional, and evolutionary relationships among the sequences under study. The most common comparative method is *sequence alignment*, which provides an explicit mapping between the residues of two or more sequences. In this chapter, only *pairwise alignments*, in which only two sequences are compared, will be discussed; the process of constructing *multiple alignments*, which involves more than two sequences, is discussed in Chapter 9. The number of sequences available for comparison has grown explosively since the 1970s, when development of rapid DNA sequencing methodology sparked the "big bang" of sequence information expansion. Comparison of one sequence to the entire database of known sequences is an important discovery technique that should be at the disposal of all molecular biologists. Over the past 30 years, improvements in the speed and sophistication of sequence alignment algorithms, not to mention perfor-

Bioinformatics: A Practical Guide to the Analysis of Genes and Proteins
Edited by A. D. Baxevanis and B. F. F. Ouellette
ISBN 0-471-38390-2 (cloth), ISBN 0-471-383910 (paper)

mance of computers, have more than kept pace with the growth in the size of the sequence databases. Today, with the complete genomes and large cDNA sequence collections available for many organisms, we are in the era of "comparative genomics," in which the full gene complement of two organisms can be compared with one another.

THE EVOLUTIONARY BASIS OF SEQUENCE ALIGNMENT

One goal of sequence alignment is to enable the researcher to determine whether two sequences display sufficient similarity such that an inference of homology is justified. Although these two terms are often interchanged in popular usage, let us distinguish them to avoid confusion in the current discussion. *Similarity* is an observable quantity that might be expressed as, say, percent identity or some other suitable measure. *Homology*, on the other hand, refers to a conclusion drawn from these data that two genes share a common evolutionary history. Genes either are or are not homologous—there are no degrees for homology as there are for similarity. For example, Figure 8.1 shows an alignment between the homologous trypsin proteins from *Mus musculus* (house mouse) and *Astracus astracus* (broad-fingered crayfish), from which it can be calculated that these two sequences have 41% identity.

Bearing in mind the goal of inferring evolutionary relationships, it is fitting that most alignment methods try, at least to some extent, to model the molecular mechanisms by which sequences evolve. Although it is presumed that homologous sequences have diverged from a common ancestral sequence through iterative molecular changes, it is actually known what the ancestral sequence was (barring the possibility that DNA could be recovered from a fossil); all that can be observed are

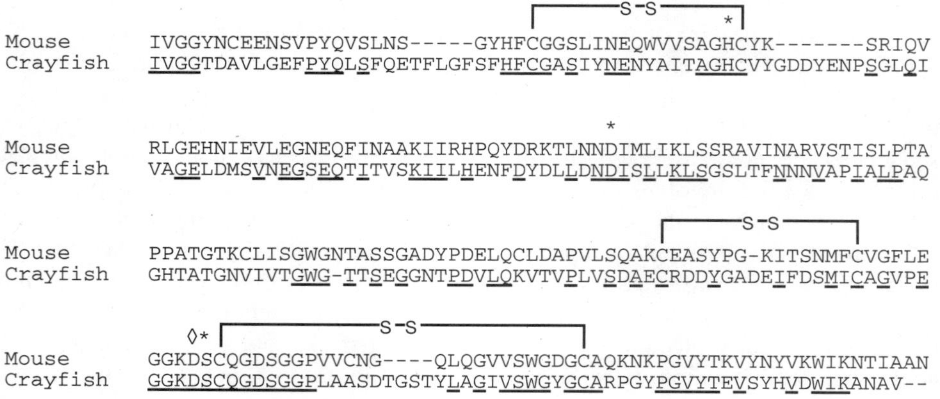

Figure 8.1. Conserved positions are often of functional importance. Alignment of trypsin proteins of mouse (SWISS-PROT P07146) and crayfish (SWISS-PROT P00765). Identical residues are underlined. Indicated above the alignments are three disulfide bonds (–S–S–), with participating cysteine residues conserved, amino acids side chains involved in the charge relay system (asterisk), and active side residue governing substrate specificity (diamond).

the raw sequences from extant organisms. The changes that occur during divergence from the common ancestor can be categorized as substitutions, insertions, and deletions. In the ideal case, in which a sequence alignment genuinely reflects the evolutionary history of two genes or proteins, residues that have been aligned but are not identical would represent substitutions. Regions where the residues of one sequence correspond to nothing in the other would be interpreted as either an insertion into one sequence or a deletion from the other. These *gaps* are usually represented in the alignment as consecutive dashes (or other punctuation character) aligned with letters. For example, the alignment in Figure 8.1 contains five gaps.

In a residue-by-residue alignment, it is often apparent that certain regions of a protein, or perhaps specific amino acids, are more highly conserved than others. This information may be suggestive as to which residues are most crucial for maintaining a protein's structure or function. In the trypsin alignment of Figure 8.1, the active site residues that determine substrate specificity and provide the "charge relay system" of serine proteases correspond to conserved positions, as do the cysteines residues that form several disulfide bonds important for maintaining the enzyme's structure. On the other hand, there may be other positions that do not play a significant functional role yet happen to be identical for historical reasons. Particular caution should be taken when the sequences are taken from very closely related species because similarities may be more reflective of history than of function. For example, regions of high sequence similarity between mouse and rat homologs may simply be those that have not had sufficient time to diverge. Nevertheless, sequence alignments provide a useful way to gain new insights by leveraging existing knowledge, such as deducing structural and functional properties of a novel protein from comparisons to those that have been well studied. It must be emphasized, however, that these inferences should always be tested experimentally and *not* assumed to be correct based on computational analysis alone.

By observing a surprisingly high degree of sequence similarity between two genes or proteins, we might infer that they share a common evolutionary history, and from this it might be anticipated that they should also have similar biological functions. But again, this should be treated as hypothetical until tested experimentally. Zeta-crystallin, for instance, is a component of the transparent lens matrix of the vertebrate eye. However, on the basis of extended sequence similarity, it can be inferred that its homolog in *E. coli* is the metabolic enzyme quinone oxidoreductase (Fig. 8.2). Despite the common ancestry, the function has changed during evolution (Gonzalez et al., 1994). This is analogous to a railroad car that has been converted into a diner: inspection of the exterior structure reveals the structure's history, but relying exclusively on this information may lead to an erroneous conclusion about its current function. When a gene adapts to a new niche, it might also be anticipated that the pattern of conserved positions would change. For example, active site residues should be conserved so long as the protein plays a role in catalysis but could drift once the protein takes on a different function.

The earliest sequence alignment methods were applicable to a simple type of relationship in which the sequences show easily detectable similarity along their entire lengths. An alignment that essentially spans the full extents of the input sequences is called a *global alignment*. The trypsin and quinone oxidoreductase/zeta-crystallin alignments discussed above are both examples of global alignments. Proteins consisting of a single globular domain can often be aligned using a global strategy as can any homologous sequences that have not diverged substantially.

```
Human-ZCr    MATGQKLMRAVRVFEFGGPEVLKLRSDIAVPIPKDHQVLIKVHACGVNPVETYIRSGTYS
Ecoli-QOR    ------MATRIEFHKHGGPEVLQA-VEFTPADPAENEIQVENKAIGINFIDTYIRSGLYP
                   .   .     ******.   . .   * .... .  .* *.* ..****** *

Human-ZCr    RKPLLPYTPGSDVAGVIEAVGDNASAFKKGDRVFTSSTISGGYAEYALAADHTVYKLPEK
Ecoli-QOR    -PPSLPSGLGTEAAGIVSKVGSGVKHIKAGDRVVYAQSALGAYSSVHNIIADKAAILPAA
               * **   *.. **..  ** .    * ****  . .  * *.            **

Human-ZCr    LDFKQGAAIGIPYFTAYRALIHSACVKAGESVLVHGASGGVGLAACQIARAYGLKILGTA
Ecoli-QOR    ISFEQAAASFLKGLTVYYLLRKTYEIKPDEQFLFHAAAGGVGLIACQWAKALGAKLIGTV
             . * * **  .   * *  * ..  .*  *  * * *.***** *** *.* * *..**

Human-ZCr    GTEEGQKIVLQNGAHEVFNHREVNYIDKIKKYVGEKGIDIIIEMLANVNLSKDLSLLSHG
Ecoli-QOR    GTAQKAQSALKAGAWQVINYREEDLVERLKEITGGKKVRVVYDSVGRDTWERSLDCLQRR
             ** .  .  *. ** .* * **   ....*   * * . .. . .   .  . *  * .

Human-ZCr    GRVIVVG-SRGTIEINPRDTMAKES----SIIGVTLFSSTKEEFQQYAAALQAGMEIGWL
Ecoli-QOR    GLMVSFGNSSGAVTGVNLGILNQKGSLYVTRPSLQGYITTREELTEASNELFSLIASGVI
             * ..  * * *..       . .      .   .   ..*.**  . .  * . .  * .

Human-ZCr    KPVIGSQ--YPLEKVAEAHENIIHGSGATGKMILLL
Ecoli-QOR    KVDVAEQQKYPLKDAQRAHE-ILESRATQGSSLLIP
             *  .  *  ***     *** *.    . *  .*.
```

Figure 8.2. Optimal global sequence alignment. Alignment of the amino acid sequences of human zeta-crystallin (SWISS-PROT Q08257) and *E. coli* quinone oxidoreductase (SWISS-PROT P28304). It is an optimal global alignment produced by the CLUSTAL W program (Higgins et al., 1996). Identical residues are marked by asterisks below the alignment, and dots indicate conserved residues.

THE MODULAR NATURE OF PROTEINS

Many proteins do not display global patterns of similarity but instead appear to be mosaics of modular domains (Baron et al., 1991; Doolittle and Bork, 1993; Patthy, 1991). One example of this is illustrated in Figure 8.3, which shows the modular structure of two proteins involved in blood clotting: coagulation factor XII (F12) and tissue-type plasminogen activator (PLAT). Besides the catalytic domain, which provides the serine protease activity, these proteins have different numbers of other structural modules: two types of fibronectin repeats, a domain with similarity to epidermal growth factor, and a module that is called a "kringle" domain. These modules can be repeated or appear in different orders. Patterns of modularity often arises by in-frame exchange of whole exons (Patthy, 1991). Global alignment methods do not take this phenomenon into account, which is understandable considering that they were developed before the exon/intron structure of genes had been discov-

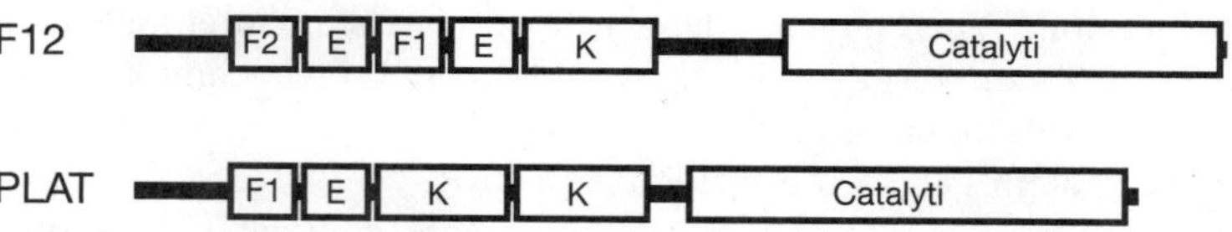

Figure 8.3. Modular structure of two proteins involved in blood clotting. Schematic representation of the modular structure of human tissue plasminogen activator and coagulation factor XII. A module labeled C is shared by several proteins involved in blood clotting. F1 and F2 are frequently repeated units that were first seen in fibronectin. E is a module resembling epidermal growth factor. A module known as a "kringle domain" is denoted K.

ered. In most cases, it is advisable to instead use a sequence comparison method that can produce a *local alignment*. Such an alignment consists of paired subsequences that may be surrounded by residues that are completely unrelated. Consequently, users should bear in mind that some local similarities could be missed if a global alignment strategy is applied inappropriately. Another obvious case in which local alignments are desired is the alignment of the nucleotide sequence of a spliced mRNA to its genomic sequence, where each exon would be a distinct local alignment.

Dot-matrix representations have enjoyed a widespread popularity, in part because of their ability to reveal complex relationships involving multiple regions of local similarity (Fitch, 1969; Gibbs and McIntyre, 1970). An example of this approach is shown in Figure 8.4, in which the F12 and PLAT protein sequences have been compared using dotter (Sonnhammer and Durban, 1996). The basic idea is to use the sequences as the coordinates of a two-dimensional graph and then plot points of correspondence within its interior. Each dot usually indicates that, within some small window, the sequence similarity is above some cutoff (or a range of cutoffs with the use of dotter, each plotted using a different shade of gray). When two sequences are

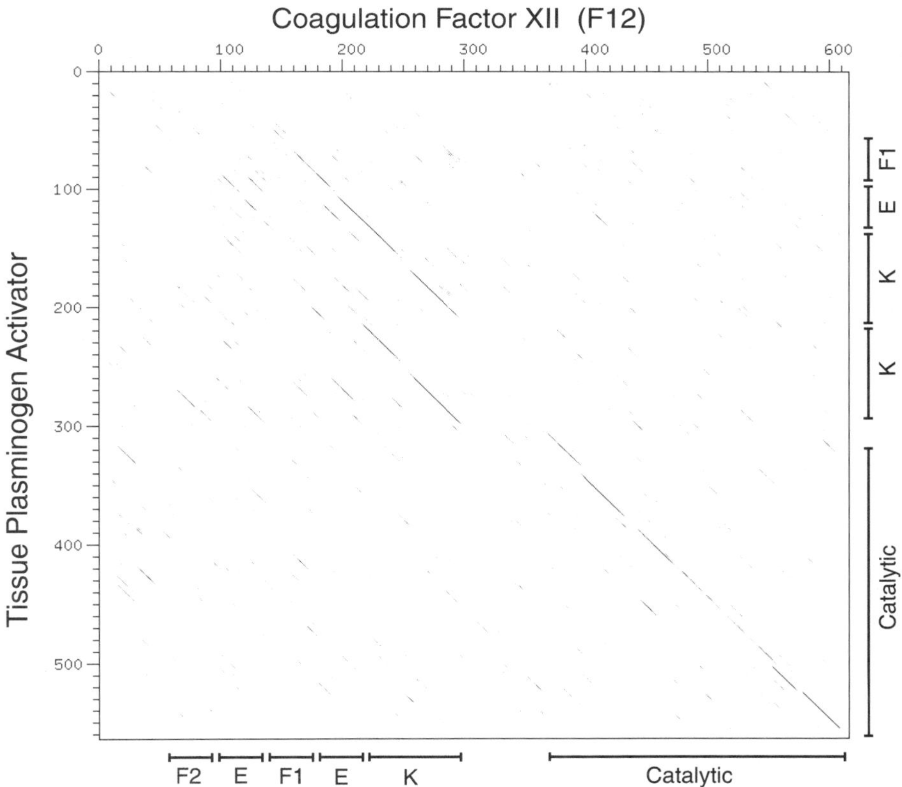

Figure 8.4. Dot matrix sequence comparison. Dot matrix comparison of the amino acid sequences of human coagulation factor XII (F12; SWISS-PROT P00748) and tissue plasminogen activator (PLAT; SWISS-PROT P00750). The figure was generated using the dotter program (Sonnhammer and Durban, 1996).

consistently matching over an extended region, the dots will merge to form a diagonal line segment. It is instructive to compare the positions of the diagonals in dot-matrix of Figure 8.4 with the known modular structure of the two proteins. In particular, note the way in which repeated domains appear: starting with the kringle domain in the PLAT and scanning horizontally, two diagonal segments may be seen, corresponding to the two kringle domains present in the F12 sequence. Although more sophisticated methods for finding local similarities are now available (discussed below), dot-matrix representations have remained popular as illustrative tools.

In a dot-matrix representation, certain patterns of dots may appear to sketch out a "path," but it is up to the viewer to deduce the alignment from this information. Another graphical representation known as a path graph provides an explicit representation of an alignment. Figure 8.5 illustrates the relationship between the dot-matrix, path graph, and alignment representations for the EGF similarity domain present in both the tissue-type plasminogen activator (PLAT) and the urokinase-type plasminogen activator (PLAU) proteins. To understand a path graph, imagine a two-dimensional lattice in which the vertices represent points between the sequence residues (as opposed to the residues themselves, as in the case of the dot-matrix). An edge that connects two vertices along a diagonal corresponds to the pairing of one

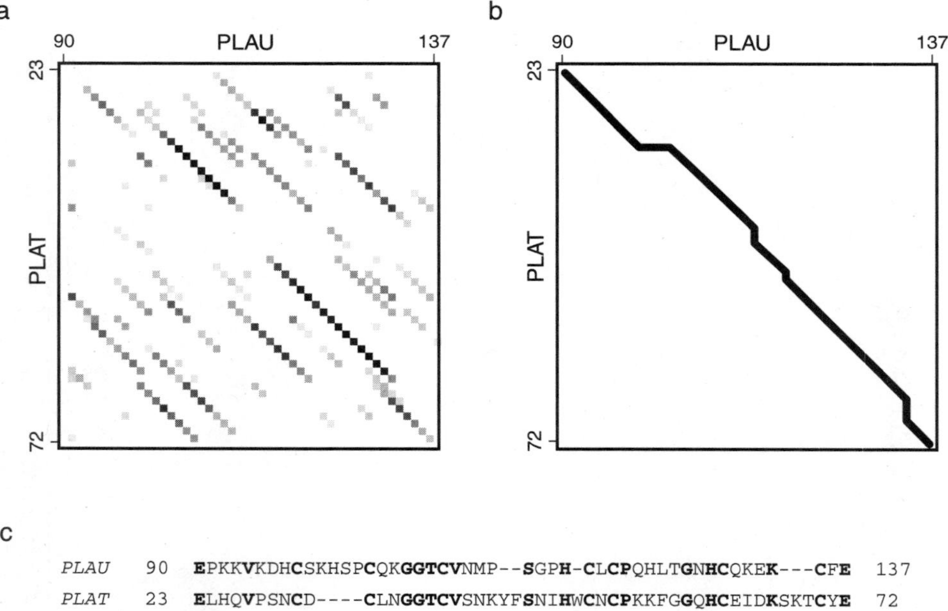

Figure 8.5. Dot-matrix, path graph, and alignment. All three views represent the alignment of the EGF similarity domains in the human urokinase plasminogen activator (PLAU; SWISS-PROT P00749) and tissue plasminogen activator (PLAT; SWISS-PROT P00750) proteins. (a) The entire proteins were compared with dotter and an enlargement of the small region corresponding to the EGF domain is shown here. (b) The path graph representation of the alignment found by BLASTP. (c) The BLASTpgp alignment represented in the familiar text form.

residue from each sequence. Horizontal and vertical edges pair a residue from one sequence with nothing in the other; in other words, these edges constitute a gap in the alignment. The entire graph corresponds to the *search space*, which must be examined for potential alignments. Each possible path through this space corresponds to exactly one alignment.

OPTIMAL ALIGNMENT METHODS

For any but the most trivial problems, the total number of distinct alignments is extraordinarily large, so it is usually of interest to identify the "best" one among them (or the several best ones). This is where the concept of representing an alignment as a path pays off. Many problems in computer science can be reduced to the task of finding the optimal path through a graph (for instance, the problem of finding the most efficient way to route a telephone call from New York to San Francisco), and efficient algorithms have been developed for this purpose. One requirement is a means of assigning a quality score to each possible path (alignment). Normally, this is accomplished by summing the incremental contributions of each step along its route. More sophisticated scoring schemes are discussed below, but for now let us assume that some positive incremental scores will be used for aligning identical residues, with negative scores used for substitutions and gaps. According to this definition of alignment quality, finding the path whose total score is maximal will give us the best sequence alignment.

What is today known as the Needleman-Wunsch algorithm is an application of a best-path strategy called *dynamic programming* to the problem of finding optimal sequence alignments (Needleman and Wunsch, 1970). The basic idea behind dynamic programming comes from the observation that any partial subpath that ends at a point along the true optimal path must itself be the optimal path leading up to that point. Thus, the optimal path can be found by incremental extension of optimal subpaths. In the basic Needleman-Wunsch formulation, the optimal alignment must extend from beginning to end in both sequences, that is, from the top-left corner in the search space to bottom-right (as it is typically drawn). In other words, it seeks global alignments. A simple modification to the basic strategy allows the optimal local alignment to be found (Smith and Waterman, 1981). The path for this alignment does not need to reach the edges of the search graph but may begin and end internally. Such an alignment would be locally optimal if its score cannot be improved either by increasing or decreasing the extent of the alignment. The Smith-Waterman algorithm relies on a property of the scoring system in which the cumulative score for a path will decrease in regions of poorly matching sequences (the scoring systems described below satisfy this criterion). When the score drops to zero, extension of path is terminated and a new one can begin. There can be many individual paths bounded by regions of poorly matching sequence; the one with the highest score is reported as the optimal local alignment.

It is important to bear in mind that optimal methods always report the best alignment that can be achieved, even if it has no biological meaning. On the other hand, when searching for local alignments there may be several significant alignments, so it is a mistake to look only at the optimal one. Refinements to the Smith-Waterman algorithm were proposed for detecting the k best nonintersecting local alignments (Altschul and Erickson, 1986; Sellers, 1984; Waterman and Eggert, 1987).

These ideas were later extended in the development of the SIM algorithm (Huang et al., 1990). A program called lalign (distributed with the FASTA package) provides a useful implementation of SIM (Pearson, 1996). Looking for suboptimal alignments is especially important when comparing multimodule proteins. This is illustrated in Figure 8.6, in which the lalign program was used to find the three best local alignments of the human coagulation factor IX and factor XII proteins. The second and

```
 Comparison of:
 (A) f9-human.aa >F9  gi|119772|sp|P00740|FA9_HUMAN COAGULATION FA  - 461 aa
 (B) f12-hum.aa >F12   gi|119763|sp|P00748|FA12_HUMAN COAGULATION  - 615 aa
 using protein matrix

①   35.4% identity in 254 aa overlap; score:  358

          220       230       240       250       260       270
F9     QSFNDFTRVVGGEDAKPGQFPWQVVLNGKVDAFCGGSIVNEKWIVTAAHCVE---TGVKI
       .:....:::::  : .:. :. ..:      ..::.::...  :..::::::..    .  ..
F12    KSLSSMTRVVGGLVALRGAHPYIAALY-WGHSFCAGSLIAPCWVLTAAHCLQDRPAPEDL
          370       380       390       400       410       420

           280       290       300       310       320       330
F9     TVVAGEHNIEETEHTEQKRNVIRIIPHHNYNAAINKYNHDIALLELDEPL-----VLNSY
       ::: :...  ... .. :. .: .   :......  .:.::.::: :.:        .:..:
F12    TVVLGQERRNHSCEPCQTLAVRSYRLHEAFSPV--SYQHDLALLRLQEDADGSCALLSPY
           430       440       450         460       470       480

             340       350       360       370       380
F9     VTPICIADKEYTNIFLKFGSGYVSGWGRVFHKGRS-ALVLQYLRVPLVDRATCLRSTKF-
       : :.:... .          ..       :.:::. :. . . :  ::  .::... . :   ..
F12    VQPVCLPSGAARPSETTLCQ--VAGWGHQFEGAEEYASFLQEAQVPFLSLERCSAPDVHG
             490       500         510       520       530

         390       400       410       420          430       440
F9     -TIYNNMFCAGFHEGGRDSCQGDSGGPHVTEVEGTS---FLTGIISWGEECAMKGKYGIY
         .:  .:.:::: ::: :.:::::::: : :  ....     : ::::::..:.  ..: : :.:
F12    SSILPGMLCAGFLEGGTDACQGDSGGPLVCEDQAAERRLTLQGIISWGSGCGDRNKPGVY
      540       550       560       570       580       590

           450
F9     TKVSRYVNWIKEKT
       :.:.  :..::.:.:
F12    TDVAYYLAWIREHT
      600       610

----------

②   34.7% identity in 49 aa overlap; score:  120

              100       110       120       130       140
F9     VDGDQCESNPCLNGGSCKDDINSYECWCPFGFEGKNCELDVTCNIKNGR
       .....: .::::.::.:  .     .   :  :: :..:    :..:.. .   .::
F12    LASQACRTNPCLHGGRCLEVEGHRLCHCPVGYTGPFCDVDTKASCYDGR
            180       190       200       210       220

----------

③   33.3% identity in 36 aa overlap; score:   87

          100       110       120
F9     DQCESN-PCLNGGSCKDDINSYECWCPFGFEGKNCE
       :.:...  :: .::.:  .    .. .: ::     ..:..:.
F12    DHCSKHSPCQKGGTCVNMPSGPHCLCPQHLTGNHCQ
        100       110       120       130

----------
```

Figure 8.6. Optimal and suboptimal local alignments. The three best alignments found when using laligntо align the sequences of human coagulation factor IX (F9; SWISS-PROT 900740) and coagulation factor XII (F12; SWISS-PROT P00748).

third alignments represent functional modules that would have been missed by a standard Smith-Waterman search, which would have reported only the first (optimal) alignment.

SUBSTITUTION SCORES AND GAP PENALTIES

The scoring system described above made use of a simple match/mismatch scheme, but, when comparing proteins, we can increase sensitivity to weak alignments through the use of a *substitution matrix*. It is well known that certain amino acids can substitute easily for one another in related proteins, presumably because of their similar physicochemical properties. Examples of these "conservative substitutions" include isoleucine for valine (both small and hydrophobic) and serine for threonine (both polar). When calculating alignment scores, identical amino acids should be given greater value than substitutions, but conservative substitutions should also be greater than nonconservative changes. In other words, a range of values is desired. Furthermore, different sets of values may be desired for comparing very similar sequences (e.g., a mouse gene and its rat homolog) as opposed to highly divergent sequences (e.g., mouse and yeast genes). These considerations can be dealt with in a flexible manner through the use of a substitution matrix, in which the score for any pair of amino acids can be easily looked up.

The first substitution matrices to gain widespread usage were those based on the point accepted mutation (PAM) model of evolution (Dayhoff et al., 1978). One PAM is a unit of evolutionary divergence in which 1% of the amino acids have been changed. This does not imply that after 100 PAMs every amino acid will be different; some positions may change several times, perhaps even reverting to the original amino acid, whereas others may not change at all. If there were no selection for fitness, the frequencies of each possible substitution would be primarily influenced by the overall frequencies of the different amino acids (called the *background frequencies*). However, in related proteins, the observed substitution frequencies (called the *target frequencies*) are biased toward those that do not seriously disrupt the protein's function. In other words, these are point mutations that have been "accepted" during evolution. Dayhoff and coworkers were the first to explicitly use a *log-odds* approach, in which the substitution scores in the matrix are proportional to the natural log of the ratio of target frequencies to background frequencies. To estimate the target frequencies, pairs of very closely related sequences (which could be aligned unambiguously without the aid of a substitution matrix) were used to collect mutation frequencies corresponding to 1 PAM, and these data were then extrapolated to a distance of 250 PAMs. The resulting PAM250 matrix is shown in Figure 8.7. Although PAM250 was the only matrix published by Dayhoff et al. (1978), the underlying mutation data can be extrapolated to other PAM distances to produce a family of matrices. When aligning sequences that are highly divergent, best results are obtained at higher PAM values, such as PAM200 or PAM250. Matrices constructed from lower PAM values can be used if the sequences have a greater degree of similarity (Altschul, 1991).

The BLOSUM substitution matrices have been constructed in a similar fashion, but make use of a different strategy for estimating the target frequencies (Henikoff and Henikoff, 1992). The underlying data are derived from the BLOCKS database (Henikoff and Henikoff, 1991), which contains local multiple alignments ("blocks")

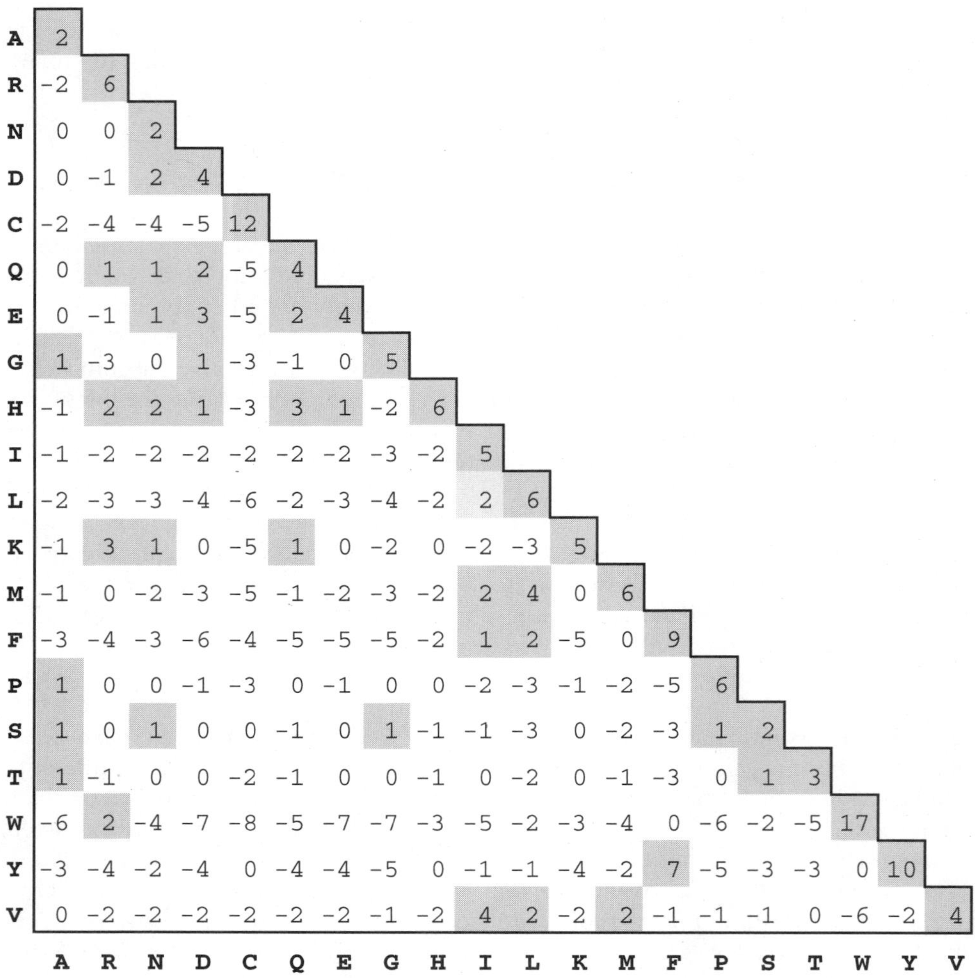

	A	R	N	D	C	Q	E	G	H	I	L	K	M	F	P	S	T	W	Y	V
A	2																			
R	-2	6																		
N	0	0	2																	
D	0	-1	2	4																
C	-2	-4	-4	-5	12															
Q	0	1	1	2	-5	4														
E	0	-1	1	3	-5	2	4													
G	1	-3	0	1	-3	-1	0	5												
H	-1	2	2	1	-3	3	1	-2	6											
I	-1	-2	-2	-2	-2	-2	-2	-3	-2	5										
L	-2	-3	-3	-4	-6	-2	-3	-4	-2	2	6									
K	-1	3	1	0	-5	1	0	-2	0	-2	-3	5								
M	-1	0	-2	-3	-5	-1	-2	-3	-2	2	4	0	6							
F	-3	-4	-3	-6	-4	-5	-5	-5	-2	1	2	-5	0	9						
P	1	0	0	-1	-3	0	-1	0	0	-2	-3	-1	-2	-5	6					
S	1	0	1	0	0	-1	0	1	-1	-1	-3	0	-2	-3	1	2				
T	1	-1	0	0	-2	-1	0	0	-1	0	-2	0	-1	-3	0	1	3			
W	-6	2	-4	-7	-8	-5	-7	-7	-3	-5	-2	-3	-4	0	-6	-2	-5	17		
Y	-3	-4	-2	-4	0	-4	-4	-5	0	-1	-1	-4	-2	7	-5	-3	-3	0	10	
V	0	-2	-2	-2	-2	-2	-2	-1	-2	4	2	-2	2	-1	-1	-1	0	-6	-2	4

Figure 8.7. The PAM250 scoring matrix.

involving distantly related sequences (as opposed to the closely related sequences used for PAM). Although there is no evolutionary model in this case, it is advantageous to have data generated by direct observation, rather than extrapolation. As with the PAM model, there is a numbered series of BLOSUM matrices, but the number in this case refers to the maximum level of identity that sequences may have and still contribute independently to the model. For example, with the BLOSUM62 matrix, sequences having at least 62% identity are merged into a single sequence, so that the substitution frequencies are more heavily influenced by sequences that are more divergent than this cutoff (see Fig. 8.8). Substitution matrices have been constructed using higher cutoffs (up to BLOSUM90) for comparing very similar sequences and lower cutoffs (down to BLOSUM30) for highly divergent sequences.

It is desirable to allow some gaps to be introduced into an alignment to compensate for insertions and deletions but not so many that the alignment asserts an

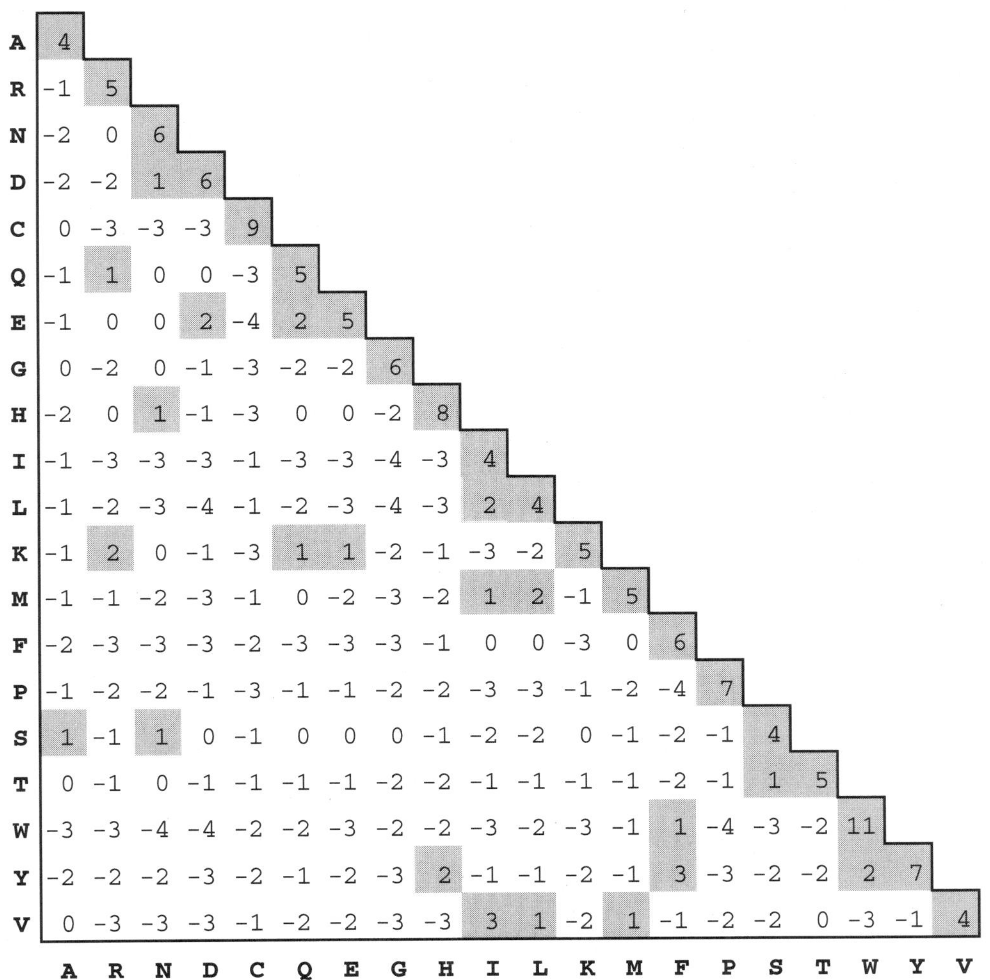

	A	R	N	D	C	Q	E	G	H	I	L	K	M	F	P	S	T	W	Y	V
A	4																			
R	-1	5																		
N	-2	0	6																	
D	-2	-2	1	6																
C	0	-3	-3	-3	9															
Q	-1	1	0	0	-3	5														
E	-1	0	0	2	-4	2	5													
G	0	-2	0	-1	-3	-2	-2	6												
H	-2	0	1	-1	-3	0	0	-2	8											
I	-1	-3	-3	-3	-1	-3	-3	-4	-3	4										
L	-1	-2	-3	-4	-1	-2	-3	-4	-3	2	4									
K	-1	2	0	-1	-3	1	1	-2	-1	-3	-2	5								
M	-1	-1	-2	-3	-1	0	-2	-3	-2	1	2	-1	5							
F	-2	-3	-3	-3	-2	-3	-3	-3	-1	0	0	-3	0	6						
P	-1	-2	-2	-1	-3	-1	-1	-2	-2	-3	-3	-1	-2	-4	7					
S	1	-1	1	0	-1	0	0	0	-1	-2	-2	0	-1	-2	-1	4				
T	0	-1	0	-1	-1	-1	-1	-2	-2	-1	-1	-1	-1	-2	-1	1	5			
W	-3	-3	-4	-4	-2	-2	-3	-2	-2	-3	-2	-3	-1	1	-4	-3	-2	11		
Y	-2	-2	-2	-3	-2	-1	-2	-3	2	-1	-1	-2	-1	3	-3	-2	-2	2	7	
V	0	-3	-3	-3	-1	-2	-2	-3	-3	3	1	-2	1	-1	-2	-2	0	-3	-1	4

Figure 8.8. The BLOSUM62 scoring matrix.

implausible series of molecular alterations. This is accomplished by deducting some amount from the alignment score for each gap introduced. Although a number of strategies have been proposed for penalizing gaps, the most common formulation, known as *affine gap penalties*, involves a fixed deduction for introducing a gap plus an additional deduction proportional to the length of the gap. This is governed by two parameters: G, sometimes called the gap-opening penalty, and L, the gap-extension penalty. For a gap of length n, the total deduction would be $G + Ln$. Unfortunately, the selection of gap parameters is highly empirical; there is little theory to support the choice of any particular set of values. However, it is common to use a high value for G (around 10–15, in the context of BLOSUM62) and a low value for L (around 1 or 2). The rationale for this is that insertion and mutation events are rare, but, when they do occur, several adjacent residues may be involved.

STATISTICAL SIGNIFICANCE OF ALIGNMENTS

For any given alignment, one can calculate a score representing the quality of the alignment, but an important question is whether or not this score is high enough to provide evidence of homology. In addressing this question, it is helpful to have some notion of how high of a score can be expected due purely to chance alone. Unfortunately, there is no mathematical theory to describe the expected distribution of scores for global alignments. One of the few methods available for assessing their significance is to compare the observed alignment score with those of many alignments made from random sequences of the same length and composition as those under study (Altschul and Erickson, 1985; Fitch, 1983).

However, for local alignments, the situation is much better. A statistical model advanced by Karlin and Altschul provides a mathematical theory to describe the expected distribution of random local alignment scores (Dembo et al., 1984; Karlin and Altschul, 1990). The form of the probability density function is known as the *extreme value distribution*. This is worth noting because application of the more familiar normal distribution can result in greatly exaggerated claims of significance. The extreme value distribution is characterized by two parameters, K and λ, which should be tailored for the particular set of alignment scoring rules and residue background frequencies at hand. Although analytical calculation of these parameters can currently be done only for alignments that lack gaps, methods have been developed to estimate appropriate values of K and λ for gapped alignments (Altschul and Gish, 1996; Waterman and Vingron, 1994). By relating an observed alignment score S to the expected distribution, it is possible to calculate statistical significance in the form of an *E value*. The simple interpretation of an E value is the number of alignments with scores at least equal to S that would be expected by chance alone. The significance of an alignment also depends on the size of the search space that was used; larger databases produce more chance alignments. The search space has typically been calculated as the product of the sequence lengths, but, for correct statistics, the lengths must be reduced by the expected length of a local alignment to avoid an "edge effect" (Altschul and Gish, 1996). This is due to the fact that an alignment that begins near the edge of the search space will run out of sequence before it can achieve a significant score.

DATABASE SIMILARITY SEARCHING

The discussion so far has focused on the alignment of specific pairs of sequences, but, for a newly determined sequence, one would generally have no way of knowing the appropriate sequence (or sequences) to use in such a comparison. Database similarity searching allows us to determine which of the hundreds of thousands of sequences present in the database are potentially related to a particular sequence of interest. This process sometimes leads to unexpected discoveries. The first "eureka moment" with this strategy came when the viral oncogene v-*sis* was found to be a modified form of the normal cellular gene that encodes platelet-derived growth factor (Doolittle et al., 1983; Waterfield et al., 1983). At the time of this discovery, sequence databases were small enough that such a finding might have been considered surprising. Today, however, it would be much more surprising to perform a database search and *not* get a hit. Large numbers of partial sequences representing novel human and mouse genes have been deposited in GenBank as a result of a number of expressed sequence tag projects (see Chapter 12). The genomes of *S. cerevisiae*, *C. elegans*, and *D. melanogaster* have been completely sequenced, as have many bacterial and viral genomes. More recently, an intermediate form of the human genome is now available in the form of a "working draft." Given this explosion of data, the new challenge has become how to focus searches in such a way as to reduce search times and limit the number of results that must be examined.

In database searching, the basic operation is to sequentially align a *query sequence* to each *subject sequence* in the database. The results are reported as a ranked *hit list* followed by a series of individual sequence alignments, plus various scores and statistics (e.g., Fig. 8.9). As will be discussed in more detail below, that choice of search program, sequence database, and various optional parameters can have an

a

```
  The best scores are:                                    initn init1 opt z-sc   E(59248)
  gi|1706794|sp|P49789|FHIT_HUMAN FRAGILE HISTIDINE  996 996   996 1350.4 0
  gi|1703339|sp|P49776|APH1_SCHPO BIS(5'-NUCLEOSYL)  431 395   395 536.2  2.8e-23
  gi|1723425|sp|P49775|YD15_YEAST HYPOTHETICAL 24.8  290 171   316 428.1  2.9e-17
  gi|1724021|sp|Q11066|YHIT_MYCTU HYPOTHETICAL 20.0  178 178   184 250.7  2.2e-07
  gi|417124|sp|Q04344|HIT_YEAST HIT1 PROTEIN (ORF U  159 104   157 216.2  1.8e-05
  gi|418447|sp|P32084|YHIT_SYNP7 HYPOTHETICAL 12.4   139 139   140 195.0  0.00028
  gi|1351828|sp|P47378|YHIT_MYCGE HYPOTHETICAL 15.6  132 132   133 183.9  0.0012
→ gi|1169826|sp|P43424|GAL7_RAT GALACTOSE-1-PHOSPHA   97  97   128 169.7  0.0072
  gi|418446|sp|P32083|YHIT_MYCHR HYPOTHETICAL 13.1   102 102   119 166.8  0.01
  gi|1708543|sp|P49773|IPK1_HUMAN PROTEIN KINASE C    87  87   118 164.5  0.014
  gi|1724020|sp|P49774|YHIT_MYCLE HYPOTHETICAL 17.0  131  82   117 161.6  0.02
  gi|1724019|sp|P53795|YHIT_CAEEL HYPOTHETICAL HIT-   98  98   116 161.5  0.02
  gi|1170581|sp|P16436|IPK1_BOVIN PROTEIN KINASE C    86  86   115 160.4  0.023
  gi|1730188|sp|Q03249|GAL7_MOUSE GALACTOSE-1-PHOSP   87  87   120 159.3  0.027
  gi|1177047|sp|P42856|ZB14_MAIZE 14 KD ZINC-BINDIN  132  79   112 156.3  0.04
  gi|120908|sp|P07902|GAL7_HUMAN GALACTOSE-1-PHOSPH   78  78   117 154.8  0.048
  gi|1177046|sp|P42855|ZB14_BRAJU 14 KD ZINC-BINDIN  115  76   110 154.5  0.05
  gi|140775|sp|P26724|YHIT_AZOBR HYPOTHETICAL 13.2   115  65   109 152.6  0.064
  gi|1169825|sp|P31764|GAL7_HAEIN GALACTOSE-1-PHOSP   62  62   104 137.9  0.42
  gi|113999|sp|P16550|APA1_YEAST 5',5'''-P-1,P-4-TE  108  66   103 137.1  0.47
```

b

```
>>gi|1169826|sp|P43424|GAL7_RAT GALACTOSE-1-PHOSPHATE UR (379 aa)
initn:   97  init1:   97  opt:  128 z-score: 169.7 E(): 0.0072
Smith-Waterman score: 128;    30.8% identity in 107 aa overlap

                                           10        20        30
HIT                                 MSFRFG-QHLIKPSVVFLKTELSFALVNRKPV
                                      ...: X.:..   . : .:  ..::    :
GalT   VWASNFLPDIAQREERSQQTYHNQHGKPLLLEYGHQELLRKERLVLTSEYWIVLVPFWAV
      190       200       210       220       230       240

              40        50        60        70        80
HIT    VPGHVLVCPLRPVERFHDLRPDEVADLFQTTQRVGTVVEKHFHGTSLTFSM--QDGP---
        : ..:. : : :.:. .: : :  :: .: ... :  .. X. ::. .::  . .:
GalT   WPFQTLLLPRRHVQRLPELTPAERDDLASTMKKLLTKYDNLFE-TSFPYSMGWHGAPMGL
      250       260       270       280       290        300

         90         100       110       120       130       140
HIT    EAGQTVKH--VHVHVLPRKAGDFHRNDSIYEELQKHDKEDFPASWRSEEEMAAEAAALRV
       ..: :  :  .:.:  :
GalT   KTGATCDHWQLHAHYYPPLLRSATVRKFMVGYEMLAQAQRDLTPEQAAERLRVLPEVHYC
       310       320       330       340       350       360
```

Figure 8.9. Output of a FASTA search. (a) Hit list from a FASTA search with human histidine triad (HIT) protein (SWISS-PROT P49789) as the query against the *swissprot* database. The search was performed using *ktup* = 1. (b) Optimal local alignment of the query to one of the database entries (marked by arrow in hit list) containing the sequence of rat galactose-1-phosphate uridylyltransferase (GalT). Although the sequence similarity is weak, these proteins have been shown to share structural similarity.

impact on the effectiveness of a search. Furthermore, there are various interfaces to these facilities such as console-style commands, Web-based forms, and E-mail. Figure 8.10 shows an example of performing a database search using the BLAST Web interface. One advantage of this approach is that, for any interesting alignment observed, complete annotation and literature citations can be obtained simply by following hypertext links to the original sequences entries and related on-line literature.

Current sequence databases are immense and have continued to increase at an exponential rate, making straightforward application of dynamic programming methods impractical for database searching. One solution is to use massively parallel computers and other specialized hardware, but, for the purposes of this discussion, we will consider only what can be done using general-purpose computers. With optimal methods being impractical, it is necessary to resort to heuristic methods, which make use of approximations to significantly speed up sequence comparisons, but with a small risk that true alignments can be missed. One heuristic method is based on the strategy of breaking a sequence up into short runs of consecutive letters called *words*. Word-based methods were introduced in the early 1980s and are used by virtually all popular search programs in use today (Wilbur and Lipman, 1983). The basic idea is that an alignment representing a true sequence relationship will contain at least one word that is common to both sequences. These *word hits* can be identified extremely rapidly by preindexing all words from the query and then consulting the index as the database is scanned.

FASTA

The first widely-used program for database similarity searching was FASTA (Lipman and Pearson, 1985; Pearson and Lipman, 1988; Pearson, 2000). To achieve a high degree of sensitivity, this program performs optimized searches for local alignments using a substitution matrix. However, as noted above, it would take a substantial amount of time to apply this strategy exhaustively. To improve speed, the program uses the observed pattern of word hits to identify potential matches before attempting the more time-consuming optimized search. The trade-off between speed and sensitivity is controlled by the *ktup* parameter, which specifies the size of a word. Increasing the value of *ktup* decreases the number of background word hits (i.e., those that do not mark the position of an optimal alignment). This, in turn, decreases the amount of optimized searching required and improves overall search speed. The default *ktup* value for comparing proteins is 2, but, for finding very distant relationships, it is recommended that it be reduced to 1.

The FASTA program does not investigate every word hit encountered, but instead looks initially for segments containing several nearby hits. By using a heuristic method, these segments are assigned scores and the score of the best segment found appears in the output as the `init1` score (Figure 8.9a). Several segments may then be combined and a new `initn` score is calculated from the ensemble. Most potential matches are then further evaluated by performing a search for a gapped local alignment that is constrained to a diagonal band centered around the best initial segment. The score of this optimized alignment is shown in the output as the `opt` score. For those alignments finally reported (a user-specified number from the top of the hit list), a full Smith-Waterman alignment search (without the constraining band) is performed. An example, of such an alignment is shown in Figure 8.9b. It should be noted that only the single optimal alignment is produced for each database sequence.

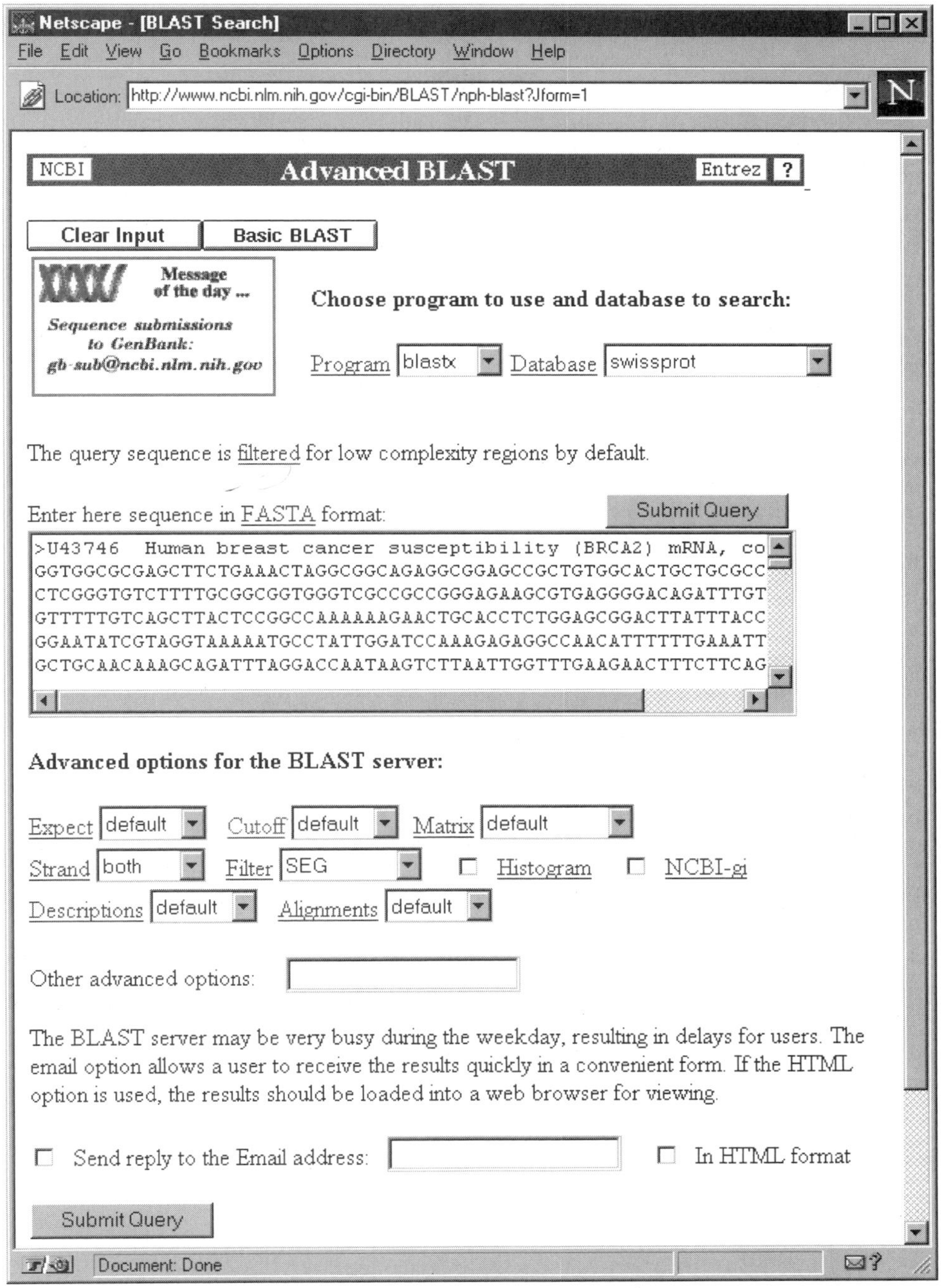

Figure 8.10. Database similarity search on the World Wide Web. The figure illustrates the use of the NCBI BLAST Web front end. The query sequence should be pasted from the clipboard into the large text field (where the sequence of U43746 is shown in this figure). Other essential elements of the search are the name of the search program and the database, both of which may be selected from drop-down lists. Additional optional parameters may be set if desired. In addition to this "Advanced BLAST" form, there is also a "Basic BLAST" form in which the advanced options are hidden. In either case, simply press the Submit Query button to begin the search.

As pointed out above, meaningful alignments can be missed by this approach if the proteins contain multiple modules. Consequently, it is recommended that matching sequences be further analyzed with the lalign program.

Beginning with version 2.0, FASTA provides an estimate of the statistical significance of each alignment found. The program assumes an extreme value distribution for random scores but with the use of a rewritten form of the probability density function in which the expected score is a linear function of the natural log of the length of the database sequence. Simple linear regression can then be used to calculate a normalized *Z*-score for each alignment. Finally, an expectation *E* is calculated, which gives the expected number of random alignments with *Z*-scores greater than or equal to the value observed.

BLAST

The BLAST programs introduced a number of refinements to database searching that improved overall search speed and put database searching on a firm statistical foundation (Altschul et al., 1990). One innovation introduced in BLAST is the idea of *neighborhood words*. Instead of requiring words to match exactly, a word hit is achieved if the word taken from the subject sequence has a score of at least *T* when a comparison is made using a substitution matrix to the word from the query. This strategy allows the word size (*W*) to be kept high (for speed) without sacrificing sensitivity. Thus, *T* becomes the critical parameter determining speed and sensitivity and *W* is rarely varied. If the value of *T* is increased, the number of background word hits will go down and the program will run faster. Reducing *T* allows more distant relationships to be found.

The occurrence of a word hit is followed by an attempt to find a locally optimal alignment whose score is at least equal to a score cutoff *S*. This is accomplished by iteratively extending the alignment both to the left and to the right, with accumulation of incremental scores for matches, mismatches, and the introduction of gaps. In practice, it is more convenient to specify an expectation cutoff *E*, which the program internally converts to an appropriate value of *S* (which would depend on the search context). In regions where matching residues are scarce, the cumulative score will begin to drop. As the mismatch and gap penalties mount, it becomes less likely that the score will rebound and ultimately reach *S*. This observation provides the basis for an additional heuristic whereby the extension of a hit is terminated when the reduction in score (relative to the maximum value encountered) exceeds the score dropoff threshold *X*. Using smaller values of *X* improves performance by reducing the time spent on unpromising hit extensions, at the expense of occasionally missing some true alignments.

There are several variants of BLAST, each distinguished by the type of sequence (DNA or protein) of the query and database sequences (see Table 8.1). The BLASTP program compares a protein query to a protein database. The corresponding program for nucleotide sequences is BLASTN. If the sequence types differ, the DNA sequence can be translated by the program (in all six reading frames) and compared to the protein sequence. BLASTX compares a DNA query sequence to the protein database, which is useful for analyzing new sequence data and ESTs. For a protein query against a nucleotide database, use the TBLASTN program. This is useful for finding unannotated coding regions in database sequences. A final variant is used only in

TABLE 8.1. BLAST Programs

Program	Query	Database	Comments
BLASTP	Protein	Protein	Uses substitution matrix for finding distant relationships; SEG filtering available
BLASTN	Nucleotide	Nucleotide	Tuned for very high-scoring matches, not distant relationships
BLASTX	Nucleotide (translated)	Protein	Useful for analysis of new DNA sequences and ESTs
TBLASTN	Protein	Nucleotide (translated)	Useful for finding unannotated coding regions in database sequences
TBLASTX	Nucleotide (translated)	Nucleotide (translated)	May be useful for EST analysis, but computationally intensive

specialized situations but is mentioned here for the sake of completeness: TBLASTX takes DNA query and database sequences, translates them both, and compares them as protein sequences. This program is mainly useful for comparisons of ESTs, where it is suspected that sequences may have coding potential even though the exact coding region has not been determined.

All of these programs make use of sequence databases located on server machines, which obviates the need for any local database maintenance. Some protein and nucleotide sequences databases currently available from the NCBI for BLAST searching are listed in Tables 8.2 and 8.3. For routine searches, the *nr* database provides comprehensive collections of both amino acid and nucleotide sequence data, with redundancy reduced by merging sequences that are completely identical. To examine all sequences submitted or updated within the last 30 days, a database called *month* is provided. Both *nr* and *month* are updated on a daily basis. Several other databases listed in Tables 8.2 and 8.3 are useful in more specialized situations, such as comparing against the complete genomes of model organisms (*ecoli* or *yeast*), searching specific classes of sequences (*est* or *sts*), or testing for the presence of contaminating or otherwise problematic sequences (*vector*, *alu*, or *mito*).

TABLE 8.2. Protein Sequence Databases for use with BLAST

Database	Description
nr	Non-redundant merge of SWISS-PROT, PIR, PRF, and proteins derived from GenBank coding sequences and PDB atomic coordinates
month	Subset of *nr* which is new or modified within the last 30 days
swissprot	The SWISS-PROT database
pdb	Amino acid sequences parsed from atomic coordinates of three-dimensional structures
ecoli	Complete set of proteins encoded by the *E. coli* genome
yeast	Complete set of proteins encoded by the *S. cerevisiae* genome
drosoph	Complete set of proteins encoded by the *D. melanogaster* genome

TABLE 8.3. Nucleotide Sequence Databases for use with BLAST

Database	Description
nr	Nonredundant GenBank, excluding the EST, STS, and GSS divisions
month	Subset of *nr*, which is new or modified within the last 30 days
est	GenBank EST division (expressed sequence tags)
sts	GenBank STS division (sequence tagged sites)
htgs	GenBank HTG division (high-throughput genomic sequences)
gss	GenBank GSS division (genome survey sequences)
ecoli	Complete genomic sequence of *E. coli*
yeast	Complete genomic sequence of *S. cerevisiae*
drosoph	Complete genomic sequence of *D. melanogaster*
mito	Complete genomic sequences of vertebrate mitochondria
alu	Collection of primate Alu repeat sequences
vector	Collection of popular cloning vectors

An example of a BLAST search will serve to introduce various elements of a search output. For the example in Figure 8.11, the amino acid sequence of one of the Alzheimer's disease susceptibility proteins (conceptual translation of GenBank L43964) was used as the query in a TBLASTN search of the *est* database. One goal of such a search would be to identify cDNA clones for potential homologs in model organisms, thereby opening the door for experimental studies that would not be practical in humans (the clones corresponding to EST sequences are readily available). Each of the EST sequences in the database is translated in all reading frames before they are compared against the Alzheimer's protein sequence. Figure 8.11a shows the hit list produced by this search. The first two columns give the identifiers and descriptions for each sequence having a significant match. Although the definitions are truncated in this overview, the figures shows that sequences from both mouse and *Drosophila* are represented. The next column gives the reading frame that produced the best alignment (although there may be hits to translations from other frames as well). The next three columns provide the score of the best alignment, the sum *P*-value, and the number of HSPs that were used in the *P*-value calculation. The alignment involving one of the *Drosophila* ESTs (marked by the arrow) is shown in Figure 8.11b. There are actually two alignments involved, and scores are provided for each. In each case, the conceptual translation of the EST is shown aligned with the query sequence. Identical amino acids are echoed to the text line in between the sequences, and plus (+) symbols are used to indicate nonidentical residues that have positive substitution scores (i.e., conservative substitutions). It is noteworthy that the two alignments arise from different reading frames and are adjacent to one another, as can be seen from the sequence coordinates. This pattern is indicative of a reading frame error in the EST sequence. When analyzing sequence single-pass data, it is extremely useful to have tools that are relatively error tolerant.

DATABASE SEARCHING ARTIFACTS

A query sequence that contains repetitive elements is likely to produce many false and confounding database matches. One clue that this may be a problem is the

a

```
                                                                         Smallest
                                                                            Sum
                                                          Reading  High  Probability
Sequences producing High-scoring Segment Pairs:              Frame Score  P(N)       N

gb|AA056325|AA056325 zf53a03.s1 Soares retina N2b4HR H...  +3    724  3.4e-102   2
gb|T03796|T03796     IB913 Infant brain, Bento Soares ...  +3    567  2.6e-78    2
gb|AA260597|AA260597 mx76g09.r1 Soares mouse NML Mus m...  +2    239  4.9e-53    4
gb|H86456|H86456     yt01b06.s1 Homo sapiens cDNA clon...  +2    323  4.3e-52    4
gb|N24576|N24576     yx72a04.s1 Homo sapiens cDNA clon...  +1    365  5.5e-47    2
gb|AA265273|AA265273 mx91c12.r1 Soares mouse NML Mus m...  +2    239  6.4e-41    2
gb|AA237206|AA237206 mx18e01.r1 Soares mouse NML Mus m...  +3    159  1.5e-40    3
gb|R14600|R14600     yf34b10.r1 Homo sapiens cDNA clon...  +1    278  1.5e-40    2
gb|AA200706|AA200706 mu03f12.r1 Soares mouse 3NbMS Mus...  +1    343  1.9e-40    1
gb|AA045064|AA045064 zk77f12.s1 Soares pregnant uterus...  -3    269  2.3e-37    2
gb|AA087434|AA087434 mm28a04.r1 Stratagene mouse skin ...  +3    322  3.6e-37    1
gb|R05907|R05907     ye93h02.r1 Homo sapiens cDNA clon...  +3    252  7.7e-37    2
gb|AA268820|AA268820 vb01c10.r1 Soares mouse NML Mus m...  +2    234  7.7e-35    2
gb|AA162310|AA162310 mn44a07.r1 Beddington mouse embry...  +1    134  8.3e-34    3
gb|N27820|N27820     yx54h10.r1 Homo sapiens cDNA clon...  +3    154  7.8e-29    2
gb|AA234907|AA234907 zs38f03.r1 Soares NhHMPu S1 Homo ...  +2    155  1.8e-28    2
gb|AA231081|AA231081 mw11d11.r1 Soares mouse 3NME12 5 ...  +3    134  8.8e-23    2
gb|H91652|H91652     ys80c04.s1 Homo sapiens cDNA clon...  -3    215  3.7e-22    1
gb|H50532|H50532     yo30h08.s1 Homo sapiens cDNA clon...  -2    211  1.2e-21    1
gb|AA150236|AA150236 zl03c01.r1 Soares pregnant uterus...  +1    159  5.0e-21    2
gb|AA144382|AA144382 mr15d12.r1 Soares mouse 3NbMS Mus...  +3    159  7.6e-21    2
→ gb|AA390557|AA390557 LD09473.5prime LD Drosophila Embr...  +3    130  1.6e-20    2
gb|AA210480|AA210480 mo86b03.r1 Beddington mouse embry...  +2    128  2.0e-20    3
gb|H19012|H19012     ym44b02.r1 Homo sapiens cDNA clon...  +2    134  5.9e-20    2
gb|AA283084|AA283084 zt14g09.s1 Soares NbHTGBC Homo sa...  -3    175  2.3e-19    2
gb|H25759|H25759     yl49d01.s1 Homo sapiens cDNA clon...  -2    185  5.0e-18    1
gb|H33787|H33787     EST110123 Rattus sp. cDNA 5' end ...  +1    137  6.7e-17    2
gb|AA201988|AA201988 LD05058.5prime LD Drosophila Embr...  +3    175  5.5e-15    1
gb|AA263526|AA263526 LD06652.5prime LD Drosophila Embr...  +1    167  7.0e-14    1
gb|R46340|R46340     yj52c04.s1 Homo sapiens cDNA clon...  -1    151  5.6e-13    1
gb|AA246675|AA246675 LD05588.5prime LD Drosophila Embr...  +2    117  2.8e-10    2
gb|AA282899|AA282899 zt14g09.r1 Soares NbHTGBC Homo sa...  +3    118  6.1e-07    1
gb|AA247705|AA247705 csh0941.seq.F Human fetal heart, ...  +3     56  0.0039     2
```

b

```
gb|AA390557|AA390557 LD09473.5prime LD Drosophila Embryo Drosophila
            melanogaster cDNA clone LD09473 5'
            Length = 659

 Score = 130 (60.4 bits), Expect = 1.6e-20, Sum P(2) = 1.6e-20
 Identities = 25/60 (41%), Positives = 40/60 (66%), Frame = +3

Query:   105 TIKSVRFYTEKNGQLIYTTFTEDTPSVGQRLLNSVLNTLIMISVIVVMTIFLVVLYKYRC 164
             +I S+ FY    +   L+YT F E +P     +   +++ ++LI++SV+VVMT   L+VLYK RC
Sbjct:   480 SINSISFYNSTDVYLLYTPFHEQSPEPSVKFWSALGSSLILMSVVVVMTFLLIVLYKKRC 659

 Score = 117 (54.3 bits), Expect = 1.6e-20, Sum P(2) = 1.6e-20
 Identities = 23/30 (76%), Positives = 27/30 (90%), Frame = +1

Query:    75 LEEELTLKYGAKHVIMLFVPVTLCMIVVVA 104
             +EEE  LKYGA+HVI LFVPV+LCM+VVVA
Sbjct:   391 MEEEQGLKYGAQHVIKLFVPVSLCMLVVVA 480
```

Figure 8.11. Output of a TBLASTN search. The protein product of the Alzheimer's disease susceptibility gene (GenBank L43964) was used as the query in a TBLASTN search against the *est* database. The goal was to identify cDNA clones from other organisms that may represent homologs of the human gene. (a) Portion of the hit list showing the 25 best hits. Each sequence is identified by GenBank accession number and a portion of the definition line. The reading frame and score of the best HSP are shown, together with the sum probability of a chance occurrence. The value in the last column gives the number of HSPs that were used in the sum probability calculation. At least 10 sequences from mouse and one from *Drosophila* may be seen on the hit list. (b) Match to the conceptual translation of the *Drosophila* EST sequence (GenBank AA390557). Two HSPs were found, each in a different reading frame. Identical residues are echoed to the central line, and plus (+) symbols indicate pairs of nonidentical amino acids with positive substitution scores.

finding of significant matches to repeat "warning sequences" that have been included in both GenBank and SWISS-PROT. These entries are consensus sequences (or translations thereof) for different subfamilies of human Alu repeats. However, with the large amount of human genomic sequence now present in the database, it is common to have many hits to individual repeats with scores greater than those for any consensus repeat. Consequently, hits to Alu-warning entries are less striking than when the database was smaller. Other indications of likely artifacts would be finding hits to many proteins that seem to have no functional relationship to one another or hits to genomic sequences from many different chromosomes. These patterns might also be seen if both query and database are contaminated with foreign sequences from the same source, for instance, cloning vectors.

Although it is always good practice to critically evaluate database search results and be suspicious of artifacts when the data don't make sense, a more proactive approach involves *masking* problematic sequences in the query before doing the search. The problem of repetitive elements is ably handled by the popular program RepeatMasker, which identifies, classifies, and masks several types of repetitive elements and simple repeats (A. F. A. Smit and P. Green, unpublished). A masking strategy strategy, which we will call "hard masking," is to replace subsequences with an ambiguity character ("N" for nucleotide sequences or "X" for proteins). Alternatively, a "soft-masking" approach, in which the resides are instead converted to lowercase letters, may be used with certain search programs. Because ambiguous residues are treated as mismatches (even when aligned to themselves), hard-masking effectively prohibits the identified repeats from making a positive contribution to the alignment score. Although hard masking is excellent for avoiding false hits, the fact that even the true alignments may be altered can present problems, particularly when alignment scores and lengths are used classify alignments. The solution to this dilemma is to use soft masking. Recent versions of the BLAST programs have an option that ignores regions of the query sequence that are lowercase when constructing the word dictionary. However, an alignment that is initiated in unique sequence may be extended through a repeat and would have the same alignment score as it would with unmasked sequence. With RepeatMasker, the `-xsmall` command-line option may be used for soft masking.

Both proteins and nucleic acids contain regions of biased composition, which can lead to confusing database search results. These *low-complexity regions* (LCRs) range from the obvious homopolymeric runs and short-period repeats to the more subtle cases where one or a few different residues may be overrepresented. Alignment of LCR-containing sequences is problematic because they do not fit the model of residue-by-residue sequence conservation. In some cases, the functionally relevant attributes may be only the periodicity or composition and not any specific sequence. Furthermore, methods for assessing the statistical significance of alignments are based on certain notions of randomness which LCRs do not obey. Consequently, many false positives may be observed in the output of a database search with an LCR-containing query sequence because the significance of matches can be overestimated (Altschul et al., 1994).

A program called seg has been developed to partition a protein sequence into segments of low and high compositional complexity (Wootton and Federhen, 1996; Wootton and Federhen, 1993). Using this program, it has been shown that more than half of the proteins in the database contain at least one LCR (Wootton, 1994; Wootton and Federhen, 1993). The evolutionary, functional, and structural properties of LCRs

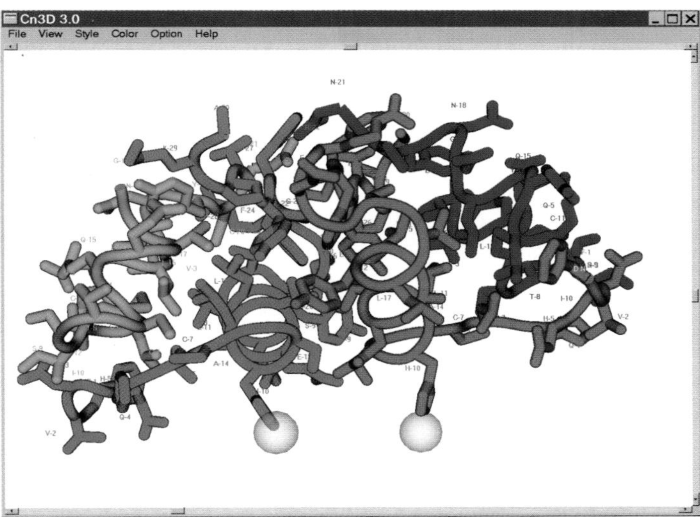

Figure 5.1. The insulin structure 3INS illustrated using Cn3D with OpenGL. Four chains are depicted in the crystallographic unit. This structure illustrates two of many bioinformatics bridges that must be spanned between sequence and structure databases, the lack of encoding of the active biological unit, and the lack of encoding of the relationship of the observed structure to the parent gene.

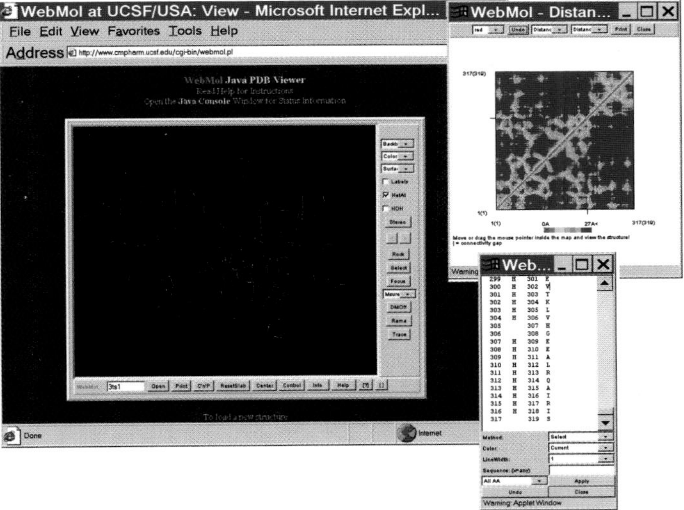

Figure 5.3. Testing a three-dimensional viewer for sequence numbering artifacts with the structure 3TS1 (Brick et al., 1989). WebMol, a Java applet, correctly indicates both the explicit and implicit sequences of the structure. Note the off-by-two difference in the numbering in the two columns of numbers in the inset window on the lower right. The actual sequence embedded in the PDB file is 419 residues long, but the COOH-terminal portion of the protein is lacking coordinates; it also has two missing residues.

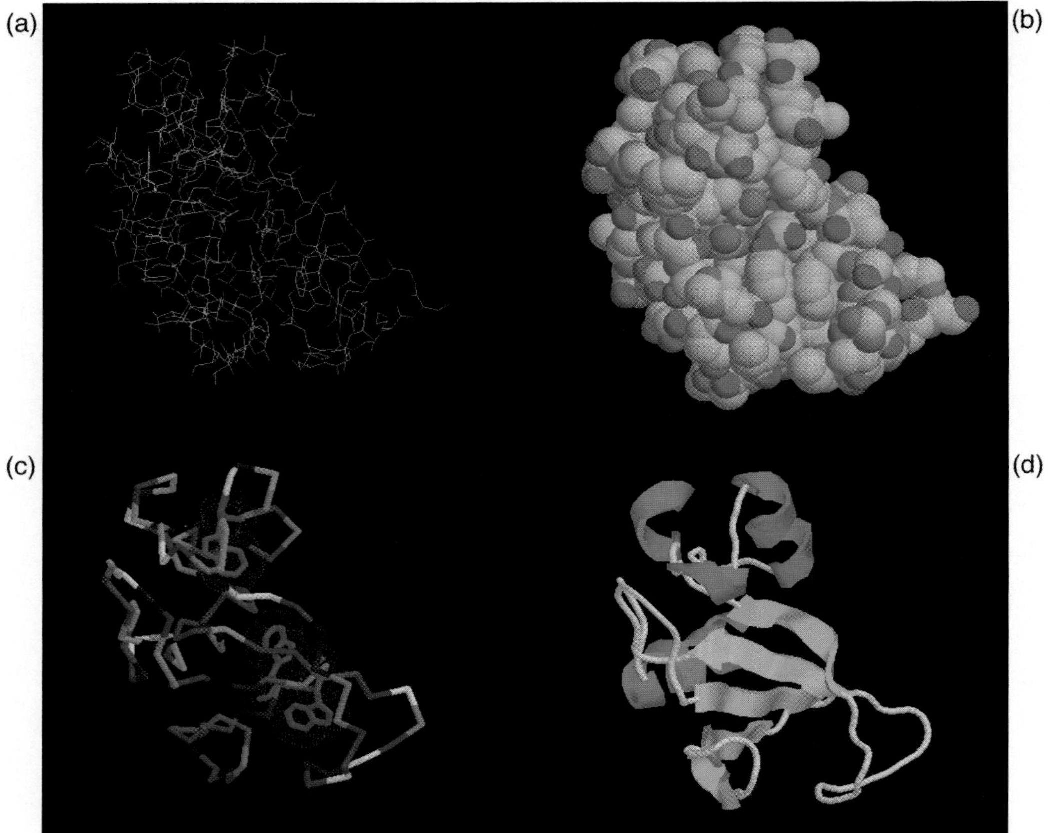

Figure 5.5. A constellation of viewing alternatives using RasMol with a portion of the barnase structure 1BN1 (Buckle et al., 1993). 1BN1 has three barnase molecules in the asymmetric unit. For this figure, the author edited the PDB file to remove two extra barnase molecules to make the images. Like most crystal structures, 1BN1 has no hydrogen locations. (a) Barnase in CPK coloring (element-based coloring) in a wire-frame representation. (b) Barnase in a space-filling representation. (c) Barnase in an α-carbon backbone representation, colored by residue type. The command line was used to select all the tryptophan residues, render them with "sticks," color them purple, and show a dot surface representation. (d) Barnase in a cartoon format showing secondary structure, α-helices in red; β-strands in yellow. Note that in all cases the default atom or residue coloring schemes used are at the discretion of the author of the software.

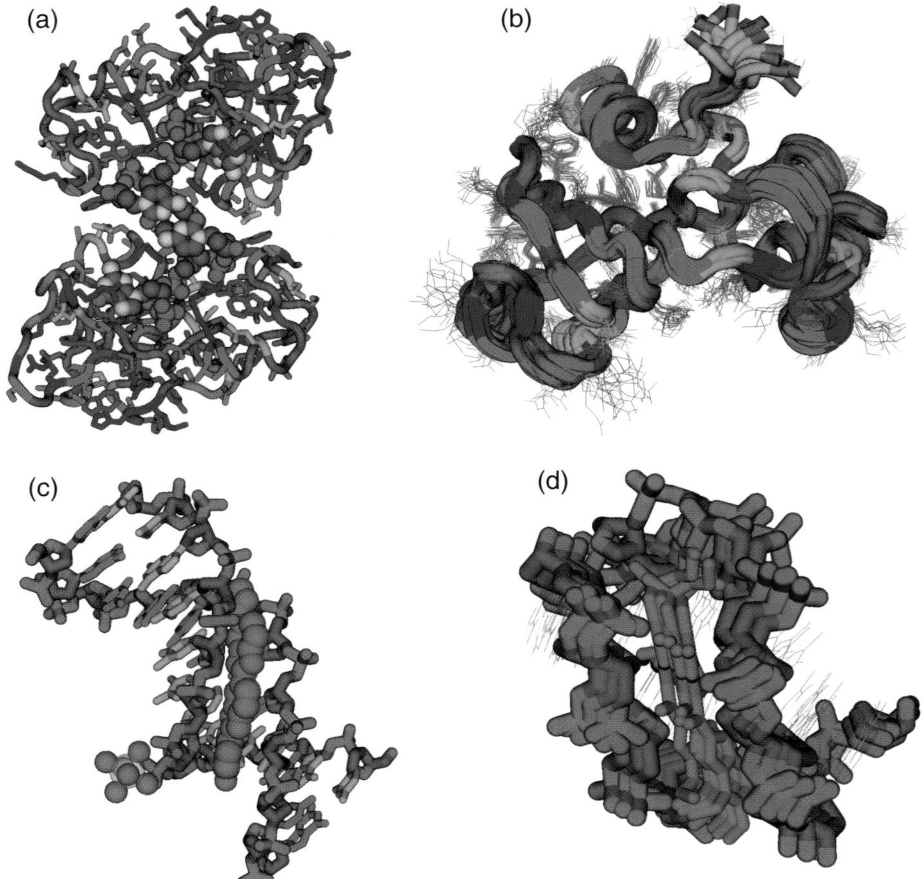

Figure 5.6. A comparison of three-dimensional structure data obtained by crystallography (left) and NMR methods (right), as seen in Cn3D. (a) The crystal structure 1BRN (Buckle and Fersht, 1994) has two barnase molecules in the asymmetric unit, although these are not dimers in solution. The image is rendered with an α-carbon backbone trace colored by secondary structure (green helices and yellow sheets), and the amino acid residues are shown with a wire-frame rendering, colored by residue type. (b) The NMR structure 1BNR (Bycroft et al., 1991) showing barnase in solution. Here, there are 20 different models in the ensemble of structures. The coloring and rendering are exactly as the crystal structure to its left. (c) The crystal structure 109D (Quintana et al., 1991) showing a complex between a minor-groove binding bis-benzimidazole drug and a DNA fragment. Note the phosphate ion in the lower left corner. (d) The NMR structure 107D showing four models of a complex between a different minor-groove binding compound (Duocarmycin A) and a different DNA fragment. It appears that the three-dimensional superposition of these ensembles is incorrectly shifted along the axis of the DNA, an error in PDB's processing of this particular file.

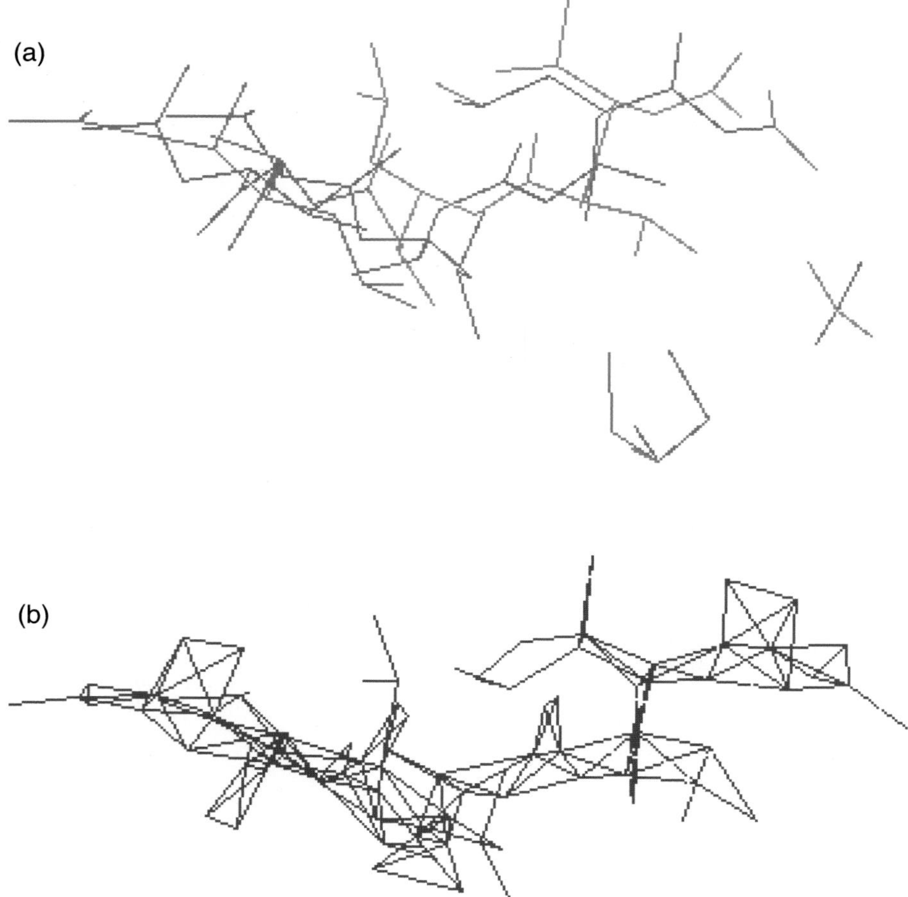

Figure 5.7. An example of crystallographic correlated disorder encoded in PDB files. This is chain C of the HIV protease structure 5HVP (Fitzgerald et al., 1990). This chain is in asymmetric binding site and can orient itself in two different directions. Therefore, it has a single chemical graph, but each atom can be in one of two different locations. (a) The correct bonding is shown with an MMDB-generated Kinemage file; magenta and red are the correlated disorder ensembles as originally recorded by the depositor, bonding calculated using standard-residue dictionary matching. (b) Bonding of the same chain in RasMol, wherein the disorder ensemble information is ignored, and all coordinates are displayed and all possible bonds are bonded together.

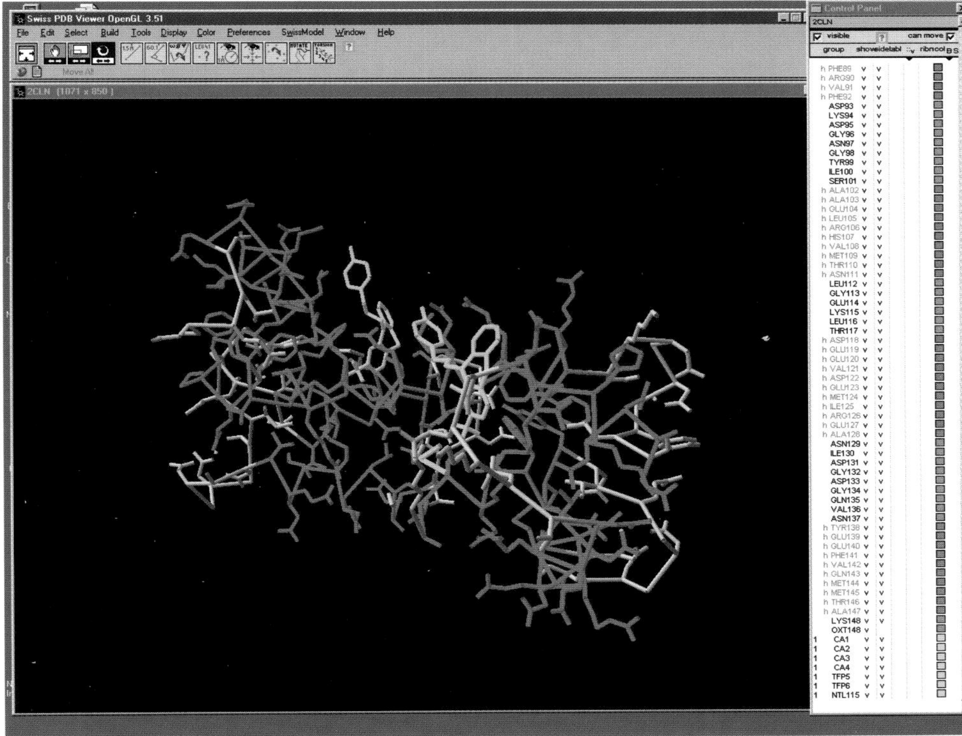

Figure 5.8. SwissPDB Viewer 3.51 with OpenGL, showing the calmodulin structure 2CLN. The binding of the inhibitor TFP is shown in yellow. The side panel allows great control over the rendering of the structure image, and menus provide a wealth of options and tools for structure superposition and modeling including mutagenesis and loop modeling, making it a complete structure modeling and analysis package.

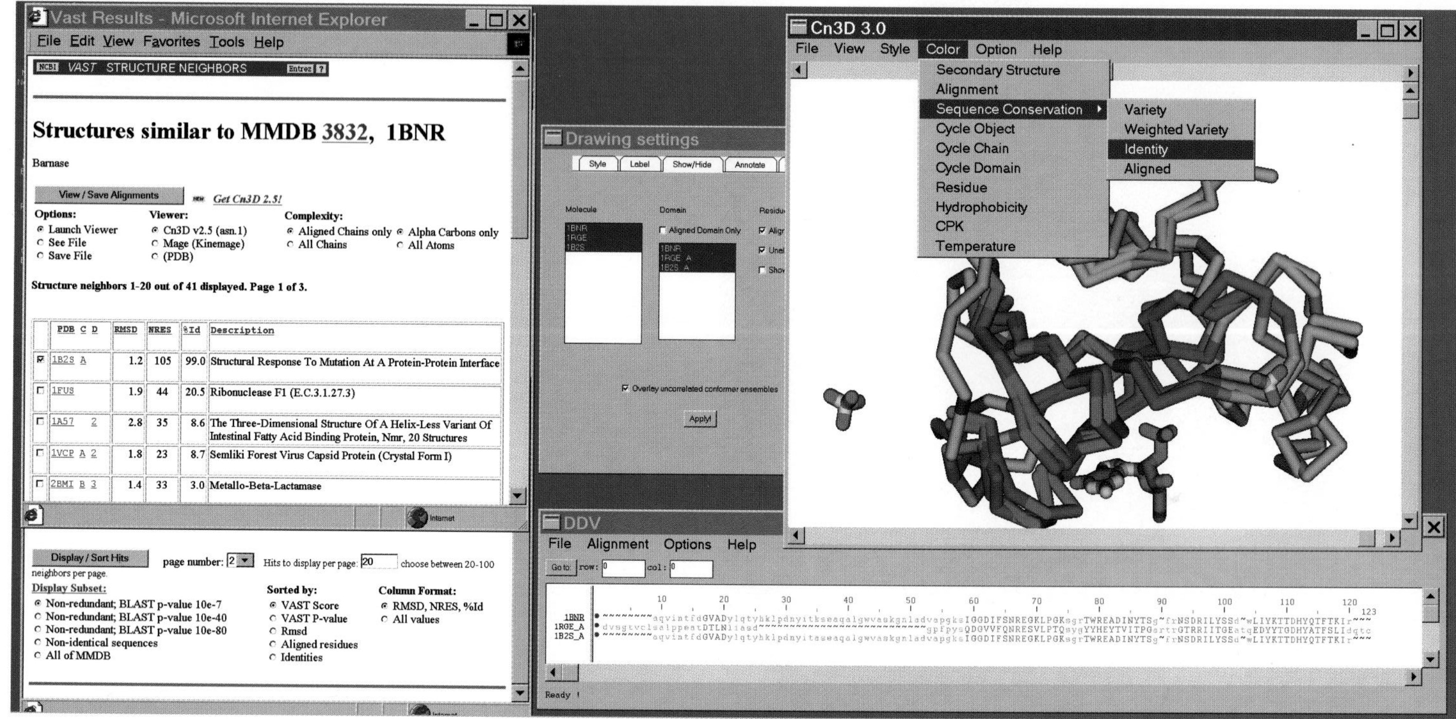

Figure 5.9. VAST structure neighbors of barnase. On the left is the query window obtained by clicking on the Structure Neighbors link from Figure 5.4. Structures to superposition are selected with the check boxes on the left, and Cn3D is launched from the top of the Web page. At the bottom left, controls that change the query are shown from the bottom of the VAST results page. The results shown here are selected as examples from a nonredundant set based on a BLAST probability of 10^{-7}, for the most concise display of hits that are not closely related to one another by sequence. The list may be sorted by a number of parameters, including RMSD from the query structure, number of identical residues, and the raw VAST score. More values can be displayed in the list as well. Cn3D is shown on the right, launched from the Web page with the structures 1RGE and 1B2S. Menu options show how Cn3D can highlight residues in the superposition (top right) and in the alignment (bottom right). The Cn3D drawing settings are shown in the top middle, where one can toggle structures on or off in the superposition window.

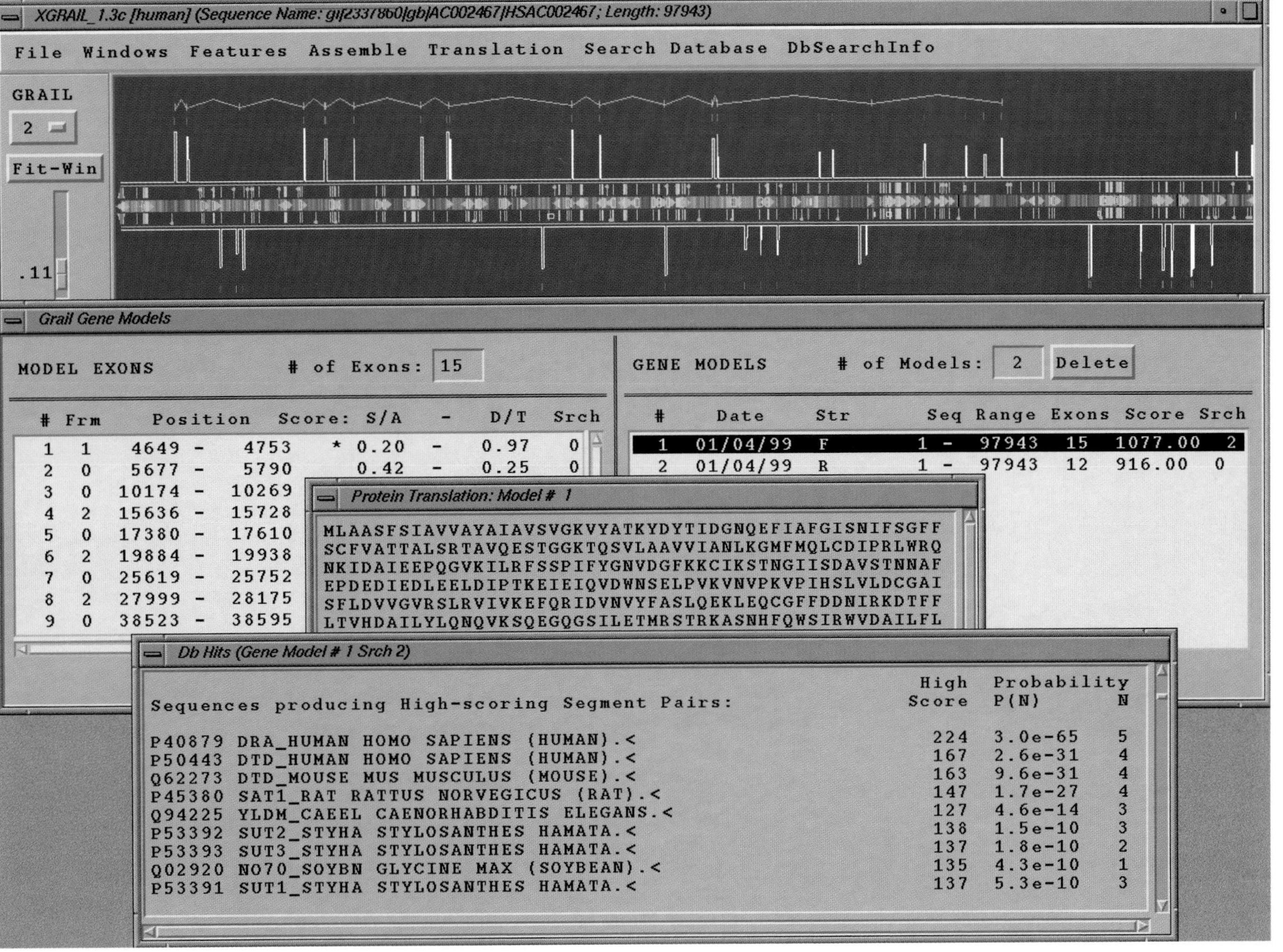

Figure 10.2. XGRAIL output using the human BAC clone RG364P16 from 7q31 as the query. The upper window shows the results of the prediction, with the histogram representing the probability that a given stretch of DNA is an exon. The various bars in the center represent features of the DNA (e.g., arrows represent repetitive DNA, and vertical bars represent repeat sequences). Exon and gene models, protein translations, and the results of a genQuest search using the protein translation are shown.

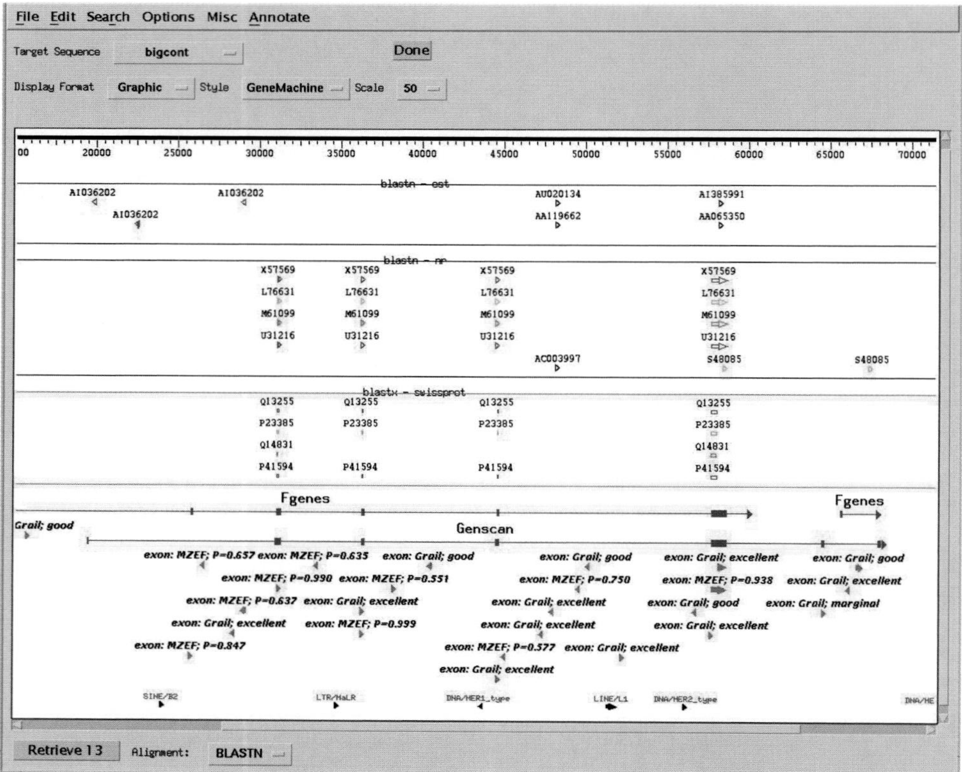

Figure 10.9. Annotated output from GeneMachine showing the results of multiple gene prediction program runs. NCBI Sequin is used at the viewer. The top of the output shows the results from various BLAST runs (BLASTN *vs.* dbEST, BLASTN vs. nr, and BLASTX vs. SWISS-PROT). Toward the bottom of the window are shown the results from the predictive methods (FGENES, GENSCAN, MZEF, and GRAIL 2). Annotations indicating the strength of the prediction are preserved and shown wherever possible within the viewer. Putative regions of high interest would be areas where hits from the BLAST runs line up with exon predictions from the gene prediction programs.

```
TARGET      1         QRRQ RTHFTSQQLQ QLEATFQRNR YPDMSTREEI AVWTNLTEAR
11FJL       0        KQRRS RTTFSASQLD ELERAFERTQ YPDIYTREEL AQRTNLTEAR
21FJL       1         QRRS RTTFSASQLD ELERAFERTQ YPDIYTREEL AQRTNLTEAR
11B72       1   ARTFDWMKVL RTNFTTRQLT ELEKEFHFNK YLSRARRVEI AATLELNETQ
22HDD       1          KRP RTAFSSEQLA RLKREFNENR YLTERRRQQL SSELGLNEAQ
12HOA       0        MRKRG RQTYTRYQTL ELEKEFHFNR YLTRRRRIEI AHALSLTERQ
                       .   *  ..  *    *   *     *     * .. .    * * .

TARGET
11FJL                          hhhhhh hhhhhhhhh    hhhhhhhh hhhhh hhhh
21FJL                          hhhhhh hhhhhhhhh    hhhhhhhh hhhhh hhhh
11B72                          hhhhhh hhhhhhhhh    hhhhhhhh hhhhh hhhh
22HDD                          hhhhhh hhhhhhhhh    hhhhhhhh hhhhh hhhh
12HOA                          hhhhhh hhhhhhhhh    hhhhhhhh hhhh  hhhh
```

```
ATOM      1  H1   GLN     1       9.226 107.177  13.966  1.00 99.00
ATOM      2  H2   GLN     1      10.769 107.671  13.751  1.00 99.00
ATOM      3  N    GLN     1       9.824 107.785  13.444  1.00 25.00
ATOM      4  H3   GLN     1       9.549 108.738  13.592  1.00 99.00
ATOM      5  CA   GLN     1       9.728 107.473  11.999  1.00 25.00
ATOM      6  CB   GLN     1       8.265 107.520  11.538  1.00 25.00
ATOM      7  CG   GLN     1       7.468 106.270  11.932  1.00 25.00
ATOM      8  CD   GLN     1       8.001 104.970  11.312  1.00 25.00
ATOM      9  OE1  GLN     1       8.748 104.928  10.343  1.00 25.00
ATOM     10  NE2  GLN     1       7.629 103.853  11.899  1.00 25.00
ATOM     11  HE21GLN     1       7.979 103.008  11.502  1.00 99.00
ATOM     12  HE22GLN     1       7.015 103.860  12.683  1.00 99.00
```

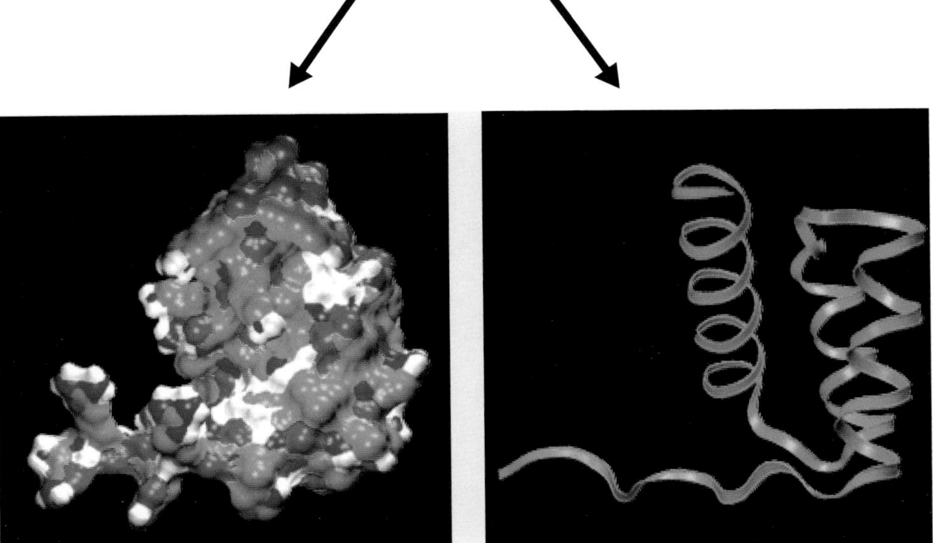

Figure 11.7. Molecular modeling using SWISS-MODEL. The input sequence for the structure prediction is the homeodomain region of human PITX2 protein. The output from SWISS-MODEL contains a text file containing a multiple sequence alignment, showing the alignment of the query against selected template structures from the Protein Data Bank (*top*). Also provided as part of the output is an atomic coordinate file for the target structure (*center*). In this example, the atomic coordinates of the target structure have been used to build a surface representation of the derived model using GRASP (*lower left*) and a ribbon representation of the derived model using RASMOL (*lower right*).

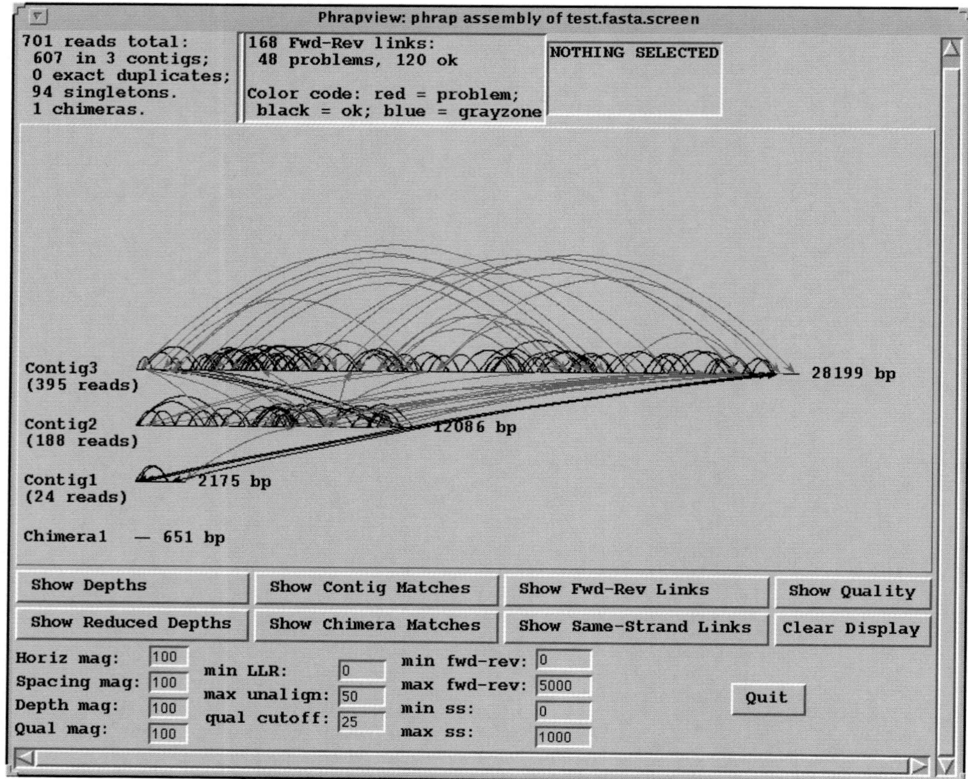

Figure 13.2. A screen dump from the program phrapview, showing a graphical display of the state of the data immediately after a run of the phrap assembly engine. See text for details.

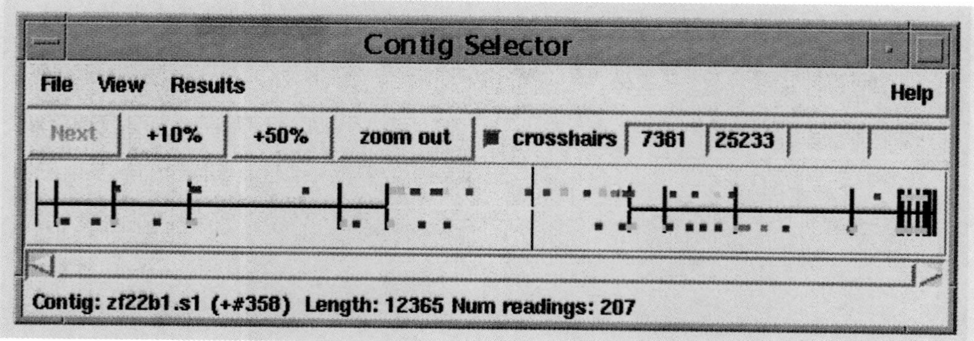

Figure 13.3. A screen dump of the gap4 Contig Selector, which gives an overview of the state of a sequencing project and provides a method for users to select contigs for processing. See text for details.

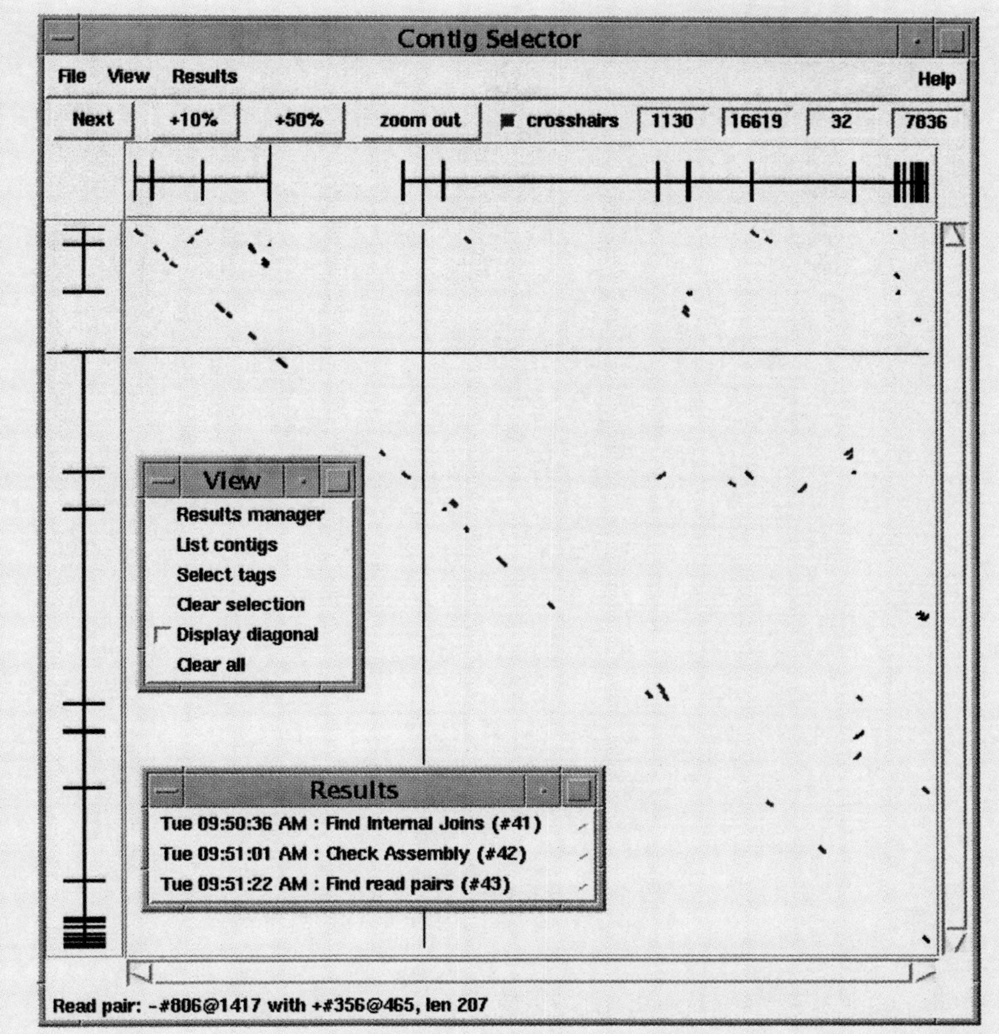

Figure 13.4. A screen dump of the gap4 Contig Comparator. This transformed version of the Contig Selector is used to display the results of analytical methods that give information about the relationships between contigs. For example, it can show sequence matches between contigs and the positions of read pairs that span contigs. See text for details.

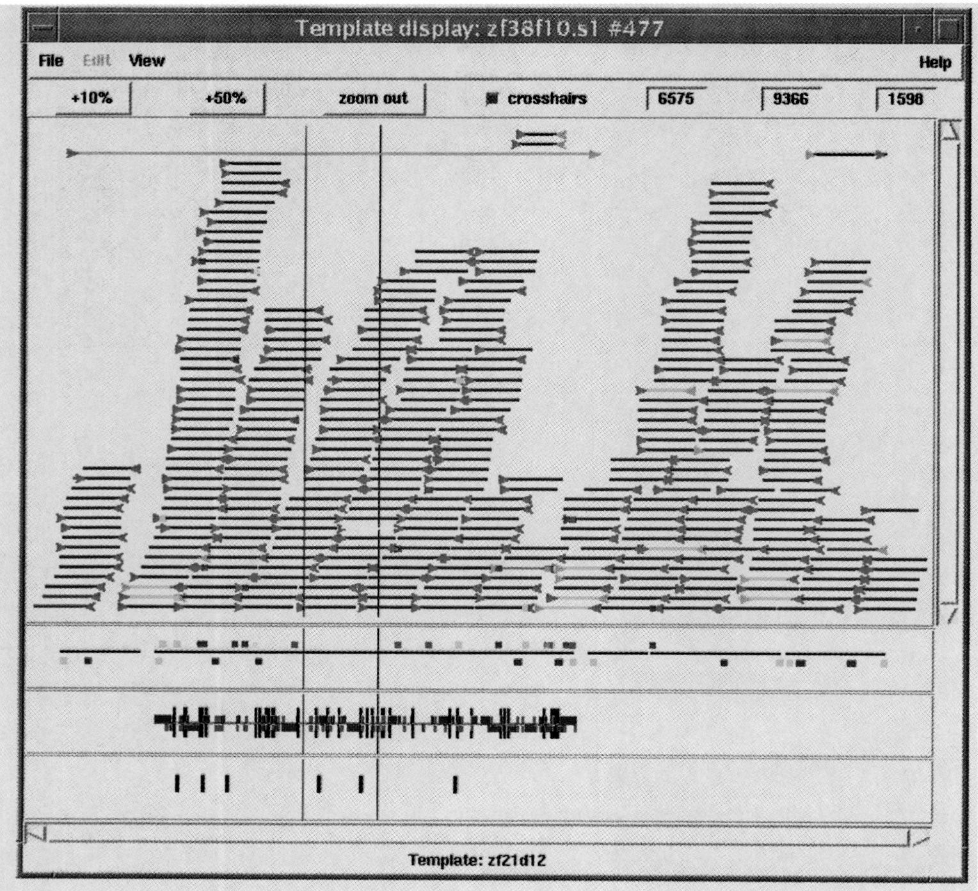

Figure 13.5. A screen dump of the gap4 Template Display, which shows the positions of DNA templates and the extent of readings derived from them. Color coding is used to distinguish between forward and reverse readings and to show consistent and inconsistent read pairs. See text for details.

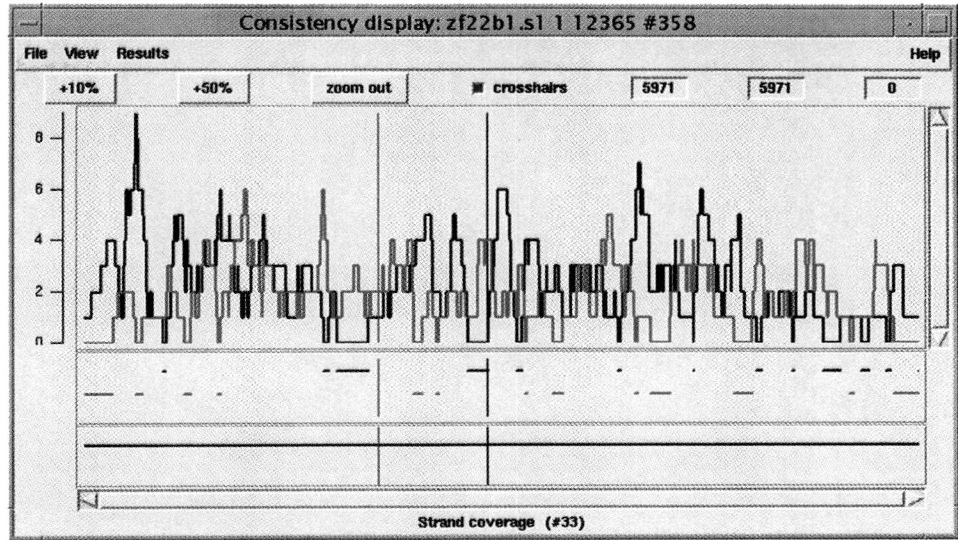

Figure 13.6. A screen dump of the gap4 Consistency Display. Here, it is being used to plot a histogram of the number of readings from each strand covering each position along a contig. Below that it is showing the segments with no data from one strand or the other. See text for details.

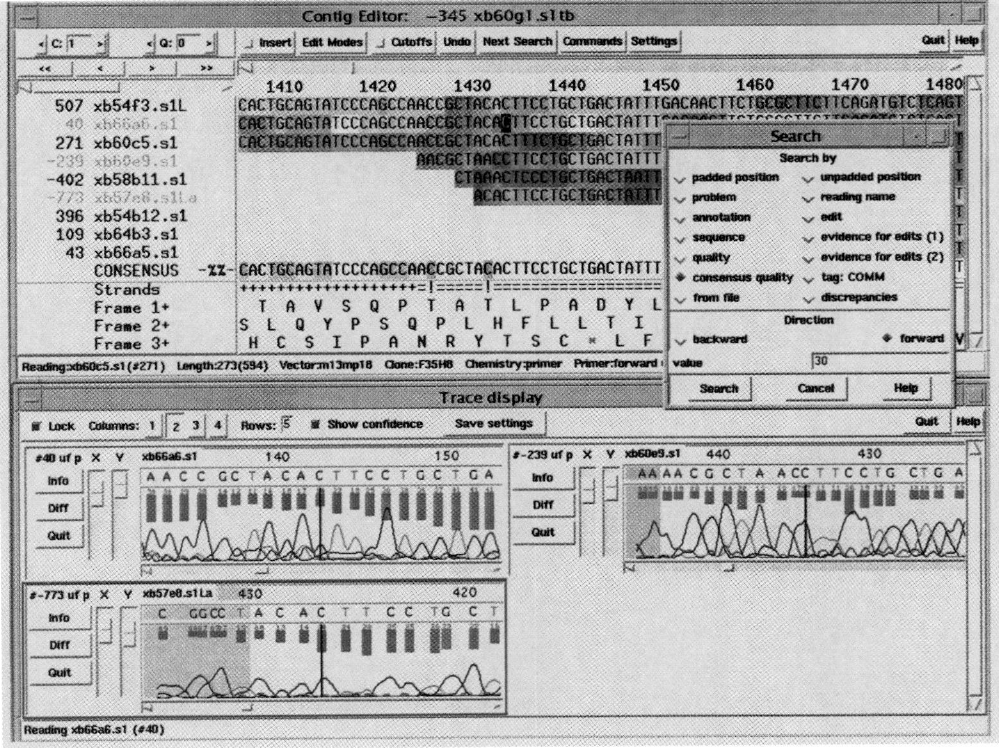

Figure 13.7. A screen dump of the gap4 Contig Editor and Trace display. See text for details.

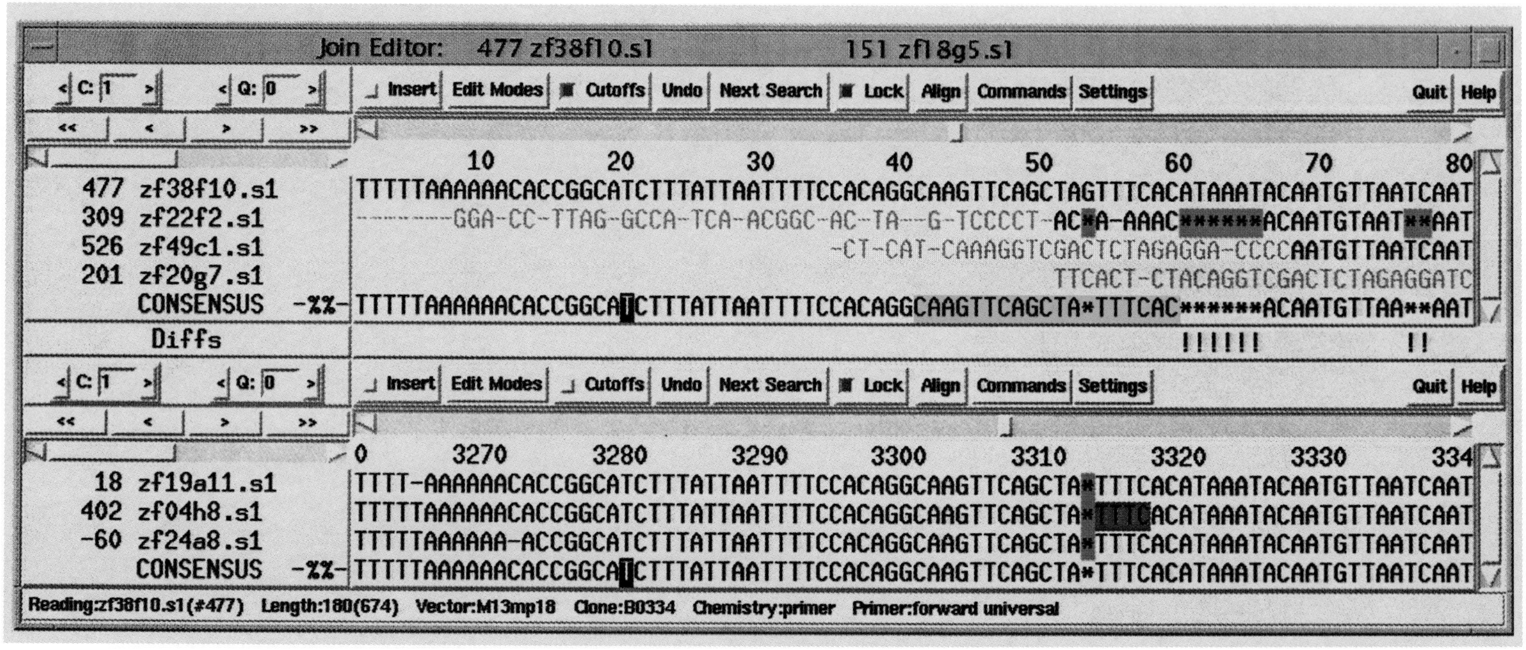

Figure 13.8. A screen dump of the gap4 Join Editor, which is used to align, edit, display traces, and join contigs. See text for details.

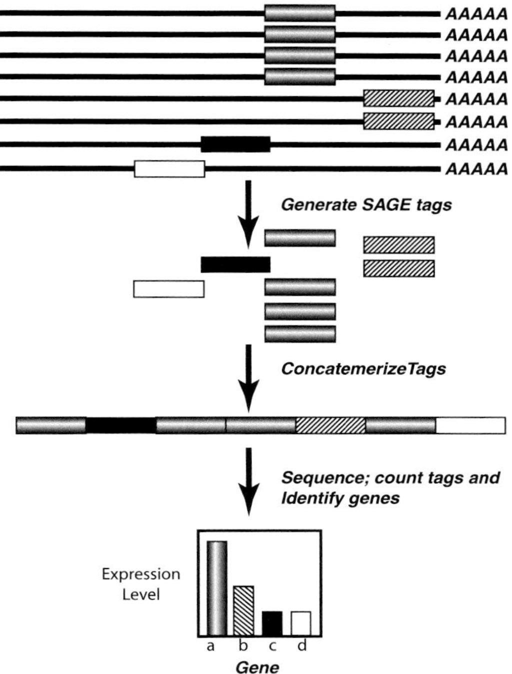

Figure 16.1. Serial Analysis of Gene Expression (SAGE) depends on the generation of a tag from the 3′ end of an mRNA. Tags are concatemerized and sequenced. These data are compared with a database of tags linked to individual transcripts to generate the frequency of each tag in the library, a measure of the expression level for that gene.

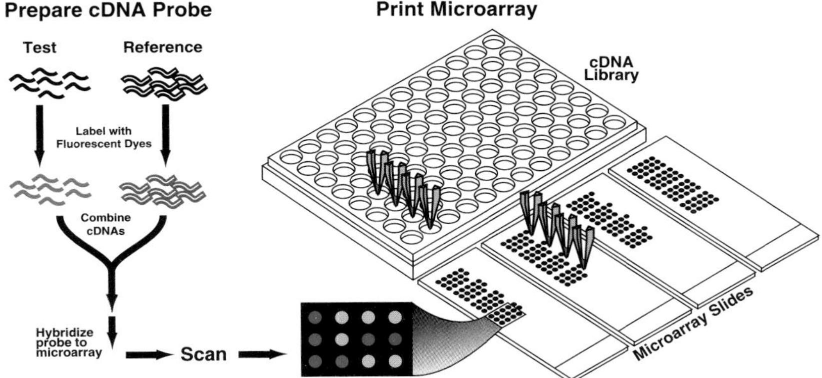

Figure 16.2. The process of microarray hybridization using printed DNA probes. A robotic printer deposits DNA in a regular array on a series of glass slides. After they are processed, the slides are hybridized to a mixture of two cDNA pools derived from test and reference samples that have been labeled with spectrally distinct fluorochromes. After stringency washes, the microarray is scanned in a laser-scanning device, and the image is processed to generate numerical data.

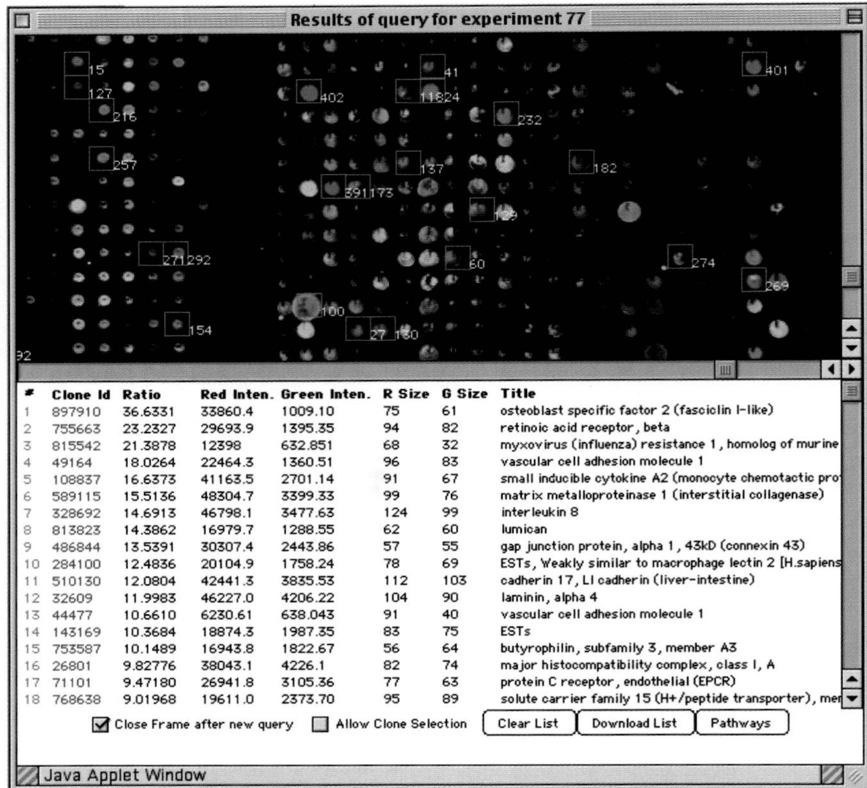

Results of query for experiment 77

#	Clone Id	Ratio	Red Inten.	Green Inten.	R Size	G Size	Title
1	897910	36.6331	33860.4	1009.10	75	61	osteoblast specific factor 2 (fasciclin I-like)
2	755663	23.2327	29693.9	1395.35	94	82	retinoic acid receptor, beta
3	815542	21.3878	12398	632.851	68	32	myxovirus (influenza) resistance 1, homolog of murine
4	49164	18.0264	22464.3	1360.51	96	83	vascular cell adhesion molecule 1
5	108837	16.6373	41163.5	2701.14	91	67	small inducible cytokine A2 (monocyte chemotactic pro
6	589115	15.5136	48304.7	3399.33	99	76	matrix metalloproteinase 1 (interstitial collagenase)
7	328692	14.6913	46798.1	3477.63	124	99	interleukin 8
8	813823	14.3862	16979.7	1288.55	62	60	lumican
9	486844	13.5391	30307.4	2443.86	57	55	gap junction protein, alpha 1, 43kD (connexin 43)
10	284100	12.4836	20104.9	1758.24	78	69	ESTs, Weakly similar to macrophage lectin 2 [H.sapiens
11	510130	12.0804	42441.3	3835.53	112	103	cadherin 17, LI cadherin (liver-intestine)
12	32609	11.9983	46227.0	4206.22	104	90	laminin, alpha 4
13	44477	10.6610	6230.61	638.043	91	40	vascular cell adhesion molecule 1
14	143169	10.3684	18874.3	1987.35	83	75	ESTs
15	753587	10.1489	16943.8	1822.67	56	64	butyrophilin, subfamily 3, member A3
16	26801	9.82776	38043.1	4226.1	82	74	major histocompatibility complex, class I, A
17	71101	9.47180	26941.8	3105.36	77	63	protein C receptor, endothelial (EPCR)
18	768638	9.01968	19611.0	2373.70	95	89	solute carrier family 15 (H+/peptide transporter), mer

Figure 16.9. In ArrayDB, query for outliers returns an image of the microarray, with outlying genes highlighted in the image and listed below the image, along with intensity data and clone identifiers.

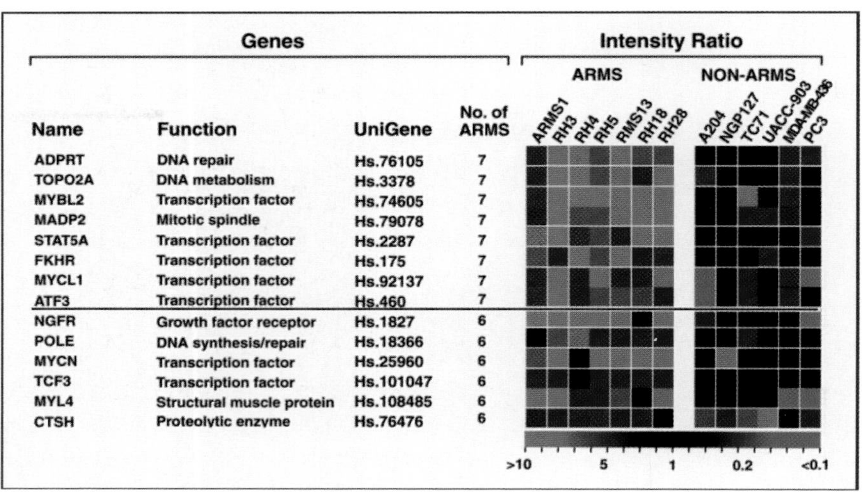

Name	Function	UniGene	No. of ARMS
ADPRT	DNA repair	Hs.76105	7
TOPO2A	DNA metabolism	Hs.3378	7
MYBL2	Transcription factor	Hs.74605	7
MADP2	Mitotic spindle	Hs.79078	7
STAT5A	Transcription factor	Hs.2287	7
FKHR	Transcription factor	Hs.175	7
MYCL1	Transcription factor	Hs.92137	7
ATF3	Transcription factor	Hs.460	7
NGFR	Growth factor receptor	Hs.1827	6
POLE	DNA synthesis/repair	Hs.18366	6
MYCN	Transcription factor	Hs.25960	6
TCF3	Transcription factor	Hs.101047	6
MYL4	Structural muscle protein	Hs.108485	6
CTSH	Proteolytic enzyme	Hs.76476	6

Figure 16.10. Display of microarray results retrieved from FileMaker Pro. This example illustrates the results of a query for genes upregulated in a series of cancer cell lines with ratios coded by a red to green color map (Khan et al., 1998).

are not well understood. Perhaps LCRs arise by such mechanisms as polymerase slippage, biased nucleotide substitution, or unequal crossing-over. In proteins, LCRs are likely to exist structurally as nonglobular regions. Regions that have been defined physicochemically as nonglobular are usually identified correctly using seg (Wootton, 1994). In DNA, there are many classes of satellite and microsatellite sequences that consist of many copies of a simple repeat unit.

The protein product of the human homolog of the *Drosophila* achaete-scute gene provides a good example of an LCR-containing protein. When analyzed with seg, two regions of low compositional complexity were identified. Figure 8.12a shows

a

```
>gi|1703441|sp|P50553|ASH1_HUMAN ACHAETE-SCUTE HOMOLOG 1

                                   1-11    MESSAKMESGG
agqqpqpqpqqpflppaacffataaaaaaa    12-72
aaaaaaqsaqqqqqqqqqqqqqqapqlrpa
                             a
                                  73-119   DGQPSGGGHKSAPKQVKRQRSSSPELMRCK
                                           RRLNFSGFGYSLPQQQP
                aavarrnerernrv   120-133
                                 134-238   KLVNLGFATLREHVPNGAANKKMSKVETLR
                                           SAVEYIRALQQLLDEHDAVSAAFQAGVLSP
                                           TISPNYSNDLNSMAGSPVSSYSSDEGSYDP
                                           LSPEEQELLDFTNWF
```

b

```
>gi|1703441|sp|P50553|ASH1_HUMAN ACHAETE-SCUTE HOMOLOG 1
MESSAKMESGGxxxxxxxxxxxxxxxxxxxxxxxxxxxxxxxxxxxxxxxxxxxxxxxxx
xxxxxxxxxxxxDGQPSGGGHKSAPKQVKRQRSSSPELMRCKRRLNFSGFGYSLPQQQPx
xxxxxxxxxxxxxKLVNLGFATLREHVPNGAANKKMSKVETLRSAVEYIRALQQLLDEHD
AVSAAFQAGVLSPTISPNYSNDLNSMAGSPVSSYSSDEGSYDPLSPEEQELLDFTNWF
```

c

```
>gi|540240 (U14590) achaete-scute homolog b [Danio rerio]
                Length = 195

Score =  193 bits (512), Expect = 7e-49
Identities = 107/155 (69%), Positives = 118/155 (76%)
Gaps = 8/155 (5%)

QUERY     86    KQVKRQRSSSPELMRCKRRLNFSGFGYSLPQQQPXXXXXXXXXXXXXXKLVNLGFATLRE 145
                K +KRQRSSSPEL+RCKRRL F+G GY++PQQQP              K VN+GF TLR+
540240    32    KVLKRQRSSSPELLRCKRRLTFNGLGYTIPQQQPMAVARRNERERNRVKQVNMGFQTLRQ 91

QUERY     146   HVPNGAANKKMSKVETLRSAVEYIRALQQLLDEHDAVSAAFQAGVLSPTISPNYSNDLNS 205
                HVPNGAANKKMSKVETLRSAVEYIRALQQLLDEHDAVSA  Q GV SP++S  YS
540240    92    HVPNGAANKKMSKVETLRSAVEYIRALQQLLDEHDAVSAVLQCGVPSPSVSNAYS----- 146

QUERY     206   MAG--SPVSSYSSDEGSYDPLSPEEQELLDFTNWF 238
                 AG  SP S+YSSDEGSY+ LS EEQELLDFT WF
540240    147   -AGPESPHSAYSSDEGSYEHLSSEEQELLDFTTWF 180
```

Figure 8.12. Identifying low-complexity regions with SEG. Analysis of the human achaete-scute protein (SWISS-PROT P50553) using seg reveals two regions of low compositional complexity. (a) Program output in the default "tree" format shows the low-complexity sequences in lower-case letters on the left and high-complexity in upper-case on the right. (b) Using the `-x` command-line switch, the seg program will generate a version of the sequence in which the low-complexity sequences have been masked. (c) For convenience, the BLAST programs can be instructed to perform the masking automatically. When a masked query sequence is used in a database search, some of the alignments may contain masked segments, as shown in this BLASTP output.

the default "tree" output, in which the low-complexity sequences are shown in lowercase letters on the left and the high-complexity sequences in uppercase on the right. The first region is a 61-residue segment containing homopolymeric tracts of glutamine and alanine. The second is a 14-residue segment with a bias toward arginine. Without filtering, many database sequences with biased regions involving these amino acids would be reported. Using a command-line option, seg can generate the masked version of the sequence for use as a search query (Figure 8.12b). Alternatively, filtering can be performed automatically by the BLAST programs through the use of optional parameters. Note that, in some implementations of BLAST, such as the Web version, filtering may be enabled by default.

POSITION-SPECIFIC SCORING MATRICES

In a standard substitution matrix, such as BLOSUM62, the substitution of one amino acid with another is associated with a single score—an obvious simplification given that the same amino acid may have different conservation patterns in one context than another in accordance with differing roles in biological function. Database searches can be tailored to find specific proteins families or domains through the use of substitution scores that reflect the substitution frequencies of each individual amino acid position in a domain. There is a large literature on the construction and application of these *position-specific scoring matrices* (PSSMs), which may also be called hidden Markov models (HMMs), motifs, or profiles (Bucher et al., 1996; Gribskov et al., 1987; Schneider et al., 1986; Staden, 1988; Tatusov et al., 1994). In its simplest form, a PSSM consists of a set of 20 substitution scores at each position along the motif—one for each of the amino acids. Amino acids that are commonly found at a particular position receive higher scores, whereas lower scores correspond to amino acids unlikely to appear at that position. It is also possible to assign scores to insertions and deletions in a position-specific manner.

A commonly used software package, HMMER (Eddy et al., 1995), contains a set of related programs for constructing and using PSSMs. Given a multiple alignment of several related proteins (e.g., one made using CLUSTAL W), the hmmbuild program may be used to calculate the position-specific scores and save it to a file (HMM file format). Using the hmmsearch program, the HMM file may be used as a query against a sequence database. Conversely, hmmpfam is used to compare a single query sequence against a database of PSSMs (HMMs). A comprehensive database of protein domains, Pfam (Bateman et al., 2000), is often used for this purpose.

The power of PSSMs in database searches can be further enhanced by iterative approaches in which the highest scoring matches in one search are incorporated into a PSSM used in successive searches. Position-Specific Iterated BLAST (PSI-BLAST) provides an automated facility for constructing, refining, and searching PSSMs within the context of a single program. Starting with a query sequence provided by the user, the process begins with a standard BLASTP search of a sequence database. Highly significant alignments found in this search are then used to construct a PSSM on-the-fly. Comparisons of the PSSM against the sequence database are performed using a variation of the word-based BLAST algorithm used for standard sequence comparisons. The process continues until no new matches are found or a specified limit on number of iterations is reached.

To demonstrate the improved sensitivity of the PSI-BLAST approach, the sequence of histidine triad (HIT) protein was used as a database search query. Simi-

```
                                                                         High    E
Sequences producing significant alignments:                              Score  Value
```

Pass 1:

```
sp|P49789|FHIT_HUMAN FRAGILE HISTIDINE TRIAD PROTEIN                      290   7e-79
sp|P49776|APH1_SCHPO BIS(5'-NUCLEOSYL)-TETRAPHOSPHATASE (ASYMME...        117   8e-27
sp|P49775|YD15_YEAST HYPOTHETICAL 24.8 KD HIT-LIKE PROTEIN               88.0   6e-18
sp|Q11066|YHIT_MYCTU HYPOTHETICAL 20.0 KD HIT-LIKE PROTEIN               52.7   3e-07
sp|Q04344|HIT_YEAST HIT1 PROTEIN (ORF U)                                 45.3   4e-05
```

Pass 2:

```
sp|P47378|YHIT_MYCGE HYPOTHETICAL 15.6 KD HIT-LIKE PROTEIN               70.5   1e-12
sp|P32083|YHIT_MYCHR HYPOTHETICAL 13.1 KD HIT-LIKE PROTEIN IN P...       59.0   3e-09
sp|P26724|YHIT_AZOBR HYPOTHETICAL 13.2 KD HIT-LIKE PROTEIN IN H...       57.6   9e-09
sp|P32084|YHIT_SYNP7 HYPOTHETICAL 12.4 KD HIT-LIKE PROTEIN IN P...       55.7   3e-08
sp|P53795|YHIT_CAEEL HYPOTHETICAL HIT-LIKE PROTEIN F21C3.3               54.3   9e-08
sp|P42856|ZB14_MAIZE 14 KD ZINC-BINDING PROTEIN (PROTEIN KINASE...       52.8   2e-07
sp|P42855|ZB14_BRAJU 14 KD ZINC-BINDING PROTEIN (PROTEIN KINASE...       50.2   1e-06
sp|P49774|YHIT_MYCLE HYPOTHETICAL 17.0 KD PROTEIN HIT-LIKE PROT...       49.5   2e-06
sp|P49773|IPK1_HUMAN PROTEIN KINASE C INHIBITOR 1 (PKCI-1)               49.1   3e-06
sp|P16436|IPK1_BOVIN PROTEIN KINASE C INHIBITOR 1 (PKCI-1) (17 ...       48.7   4e-06
sp|P44956|YCFF_HAEIN HYPOTHETICAL HIT-LIKE PROTEIN HI0961                47.3   1e-05
sp|P43424|GAL7_RAT GALACTOSE-1-PHOSPHATE URIDYLYLTRANSFERASE             41.0   8e-04
```

Pass 3:

```
sp|Q03249|GAL7_MOUSE GALACTOSE-1-PHOSPHATE URIDYLYLTRANSFERASE           87.2   1e-17
sp|P07902|GAL7_HUMAN GALACTOSE-1-PHOSPHATE URIDYLYLTRANSFERASE           79.8   2e-15
sp|P31764|GAL7_HAEIN GALACTOSE-1-PHOSPHATE URIDYLYLTRANSFERASE           64.7   6e-11
sp|P09148|GAL7_ECOLI GALACTOSE-1-PHOSPHATE URIDYLYLTRANSFERASE           62.5   3e-10
sp|P22714|GAL7_SALTY GALACTOSE-1-PHOSPHATE URIDYLYLTRANSFERASE           58.1   6e-09
sp|P09580|GAL7_KLULA GALACTOSE-1-PHOSPHATE URIDYLYLTRANSFERASE           48.5   4e-06
sp|P08431|GAL7_YEAST GALACTOSE-1-PHOSPHATE URIDYLYLTRANSFERASE           40.8   0.001
```

Pass 4:

```
sp|P40908|GAL7_CRYNE GALACTOSE-1-PHOSPHATE URIDYLYLTRANSFERASE           71.0   8e-13
sp|P13212|GAL7_STRLI GALACTOSE-1-PHOSPHATE URIDYLYLTRANSFERASE           57.0   1e-08
```

Figure 8.13. Increased sensitivity using PSI-BLAST. The human histidine triad (HIT) protein (SWISS-PROT P49789) was used as the query in a BLASTP search with the PSI-BLAST functionality enabled. Definition lines, scores, and *E* values are shown for all statistically significant matches newly identified in each iteration.

larity between HIT and galactose-1-phosphate uridylyltransferase (GalT) has recently been described based on superimposition of their three-dimensional structures (Holm and Sander, 1997). However, sequence similarity between these two proteins is extremely weak. With a standard (single-pass) BLASTP search, no significant hits to GalT sequences are observed. However, with a multipass search, new relationships are discovered at each iteration, as shown in Figure 8.13. The rat GalT protein is found in the second iteration and, after information from this alignment is incorporated into the profile, several additional homologs from other organisms are also identified.

SPLICED ALIGNMENTS

The identification of genes within long stretches of DNA sequence is a central problem for automatic annotation of complete genomes. For the very compact genomes of viruses and bacteria, this work amounts to little more than the enumeration of open reading frames. However, gene identification in eukaryotic genomes is significantly more challenging because of the larger amount of intergenic sequence and the fact that protein-coding regions may be interrupted by introns. The two fundamental strategies seek to identify genes using either the intrinsic signals in the DNA sequence (see Chapter 10) or alignments to mRNA and protein sequences.

At first glance, the problem of aligning mRNA and genomic sequences seems trivial—using a local alignment strategy each exon would come out as a separate locally optimal alignment separated by large 'deletions' in the mRNA sequence corresponding to the introns that have been spliced out. If the goal is merely to obtain a crude sense of how a gene is organized, a simple alignment is sufficient. However, for the purpose of genome annotation, it is important that all exons be found with precise endpoints or a correct protein translation cannot be obtained. The sim4 program is designed to address genome annotation needs by performing mRNA/genomic alignments rapidly and accurately (Florea et al., 1998). It begins with a BLAST-like search for finding the obvious exons—those with very high alignment scores—and follows this with search at lower stringency to identify any missed (usually short) exons. Splice donor and acceptor signals in the genomic sequences are used to adjust the exon boundaries (see Fig. 8.14). To avoid problems caused by tandemly repeated genes, an additional constraint is imposed to require that the order of the exons found in the genome match that implied by the mRNA. Other programs available for performing mRNA/genomic alignments are est_genome (Mott, 1997) and the est2gen program from the Wise package (Birney et al., 1996).

It should be noted that an mRNA sequence may align perfectly well to a pseudogene and that such an alignment may be difficult to distinguish from a functional gene. Certain features may be indicative of retropseudogenes, that is those resulting from integration of a reverse-transcribed mRNA. For example, a poly(A) tract found in the genomic sequence at the 3′ end of the gene is indicative of a pseudogene produced through an mRNA intermediate. Such an mRNA alignment will also lack the large gaps corresponding to introns, although this alone cannot be used to conclude that it is a retropseudogene because there many authentic genes that consist of a single exon. However, once all mRNA alignments have been generated for a genome, one strategy for retropseudogene identification involves looking for pairs of highly similar alignments in which one lacks and the other contains introns. Other types of pseudogenes may arise by gene duplication events followed by inactivation of one of the copies. Such cases can be very difficult to diagnose. Programs that perform mRNA/genomic alignments usually do not use any knowledge of reading frame; therefore, determining that a potential gene contains a frameshift will require subsequent analysis of the protein translations. In the case of genes predicted by alignment with ESTs only, the protein-coding sequence is not known so any interruptions of the reading frame will not be apparent. The possibility that an apparent frame shift may actually be a sequencing error in the genomic sequence should also be considered, particularly in the analysis of a working draft sequence. In many cases, determining whether an mRNA-predicted gene is a functional gene or a pseudogene must await experimental validation.

CONCLUSIONS

Sequence alignment and database searching are performed tens of thousands of times per day by scientists around the world and represent critical techniques that all mo-

```
seq1 = mrna (>gi|7661723|ref|NM_015372.1), 1247 bp
seq2 = genomic (>gi|1941922|emb|Z82248.1|HSN44A4), 40662 bp

(complement)
1-118  (15628-15745)   100% ->
119-318  (22863-23062)   100% ->
319-1247  (26529-27457)   100%

      0     .    :    .    :    .    :    .    :    .    :
      1 CCCCAGGCGTGGGAAGATGGAACCAGAACAATTCGAACGAGCAGAGCAAA
        ||||||||||||||||||||||||||||||||||||||||||||||||||
  15628 CCCCAGGCGTGGGAAGATGGAACCAGAACAATTCGAACGAGCAGAGCAAA

     50     .    :    .    :    .    :    .    :    .    :
     51 ACAGATCGGAATTGCAGACTTCAGGTCGTGGCAGAGAAAACCAGCTGAGA
        ||||||||||||||||||||||||||||||||||||||||||||||||||
  15678 ACAGATCGGAATTGCAGACTTCAGGTCGTGGCAGAGAAAACCAGCTGAGA

    100     .    :    .    :    .    :    .    :    .    :
    101 CAGGGCGCCACTTACTAG         CTCTGAAAGTCTAGGATATTTTG
        ||||||||||||||||||>>>...>>>|||||||||||||||||||||||
  15728 CAGGGCGCCACTTACTAGGTG...CAGCTCTGAAAGTCTAGGATATTTTG

    150     .    :    .    :    .    :    .    :    .    :
    142 CCACTGGAAGACCAGCAGACAATGTCATGACAACTCAAGAGGATACAACA
        ||||||||||||||||||||||||||||||||||||||||||||||||||
  22886 CCACTGGAAGACCAGCAGACAATGTCATGACAACTCAAGAGGATACAACA

    200     .    :    .    :    .    :    .    :    .    :
    192 GGGCTGCATCAAAAGACAAGTCTTTGGACCATGTCAAGACCTGGAGCGAA
        ||||||||||||||||||||||||||||||||||||||||||||||||||
  22936 GGGCTGCATCAAAAGACAAGTCTTTGGACCATGTCAAGACCTGGAGCGAA

    250     .    :    .    :    .    :    .    :    .    :
    242 GAAGGTAATGAACTCCTACTTCATAGCAGGCTGTGGGCCAGCAGTTTGCT
        ||||||||||||||||||||||||||||||||||||||||||||||||||
  22986 GAAGGTAATGAACTCCTACTTCATAGCAGGCTGTGGGCCAGCAGTTTGCT

    300     .    :    .    :    .    :    .    :    .    :
    292 ACTACGCTGTCTCTTGGTTAAGGCAAG         GTTTCAGTATCAAC
        |||||||||||||||||||||||||||>>>...>>>||||||||||||||
  23036 ACTACGCTGTCTCTTGGTTAAGGCAAGGTC...CAGGTTTCAGTATCAAC

    350     .    :    .    :    .    :    .    :    .    :
    333 CTGACTTCTTTTGGAAGGATCCCTTGGCCTCACGCTGGAGTGGGCACCTG
        ||||||||||||||||||||||||||||||||||||||||||||||||||
  26543 CTGACTTCTTTTGGAAGGATCCCTTGGCCTCACGCTGGAGTGGGCACCTG

    400     .    :    .    :    .    :    .    :    .    :
    383 CCCTAGCCCACAGAGCTGGATTTCTCCCTTTCTTCAATCACACAGGGAGC
        ||||||||||||||||||||||||||||||||||||||||||||||||||
  26593 CCCTAGCCCACAGAGCTGGATTTCTCCCTTTCTTCAATCACACAGGGAGC
```

ouput truncated for brevity

Figure 8.14. Spliced alignment. The sim4 program was used to align a novel human mRNA (RefSeq NM_015372) to the genomic sequence of a cosmid from chromosome 22 (EMBL Z82248). Three exons were identified on the complementary strand (the third one has been truncated for brevity). The ">>>" symbols indicate splice sites found at the exon/intron boundaries.

lecular biologists should be familiar with. It can be expected that these methods will continue to evolve to meet the challenges of an ever-increasing database size. This chapter has described some of the fundamental concepts involved, but it is useful to consult the documentation of the various programs for more detailed information. Researchers should have a basic understanding of how the programs work so that parameters can be intelligently selected. In addition, they should be aware of potential artifacts and know how to avoid them. Above all, it is important to apply the same powers of observation and critical evaluation that are used with any experimental method.

INTERNET RESOURCES FOR TOPICS PRESENTED IN CHAPTER 8

BLAST	*http://ncbi.nlm.nih.gov/BLAST/*
CLUSTAL W	*ftp://ftp.ebi.ac.uk/pub/software/*
dotter	*ftp://ftp.sanger.ac.uk/pub/dotter/*
FASTA, lalign	*ftp://ftp.virginia.edu/pub/fasta/*
hmmer	*http://hmmer.wustl.edu/*
RepeatMasker	*http://ftp.genome.washington.edu/RM/RepeatMasker.html*
seg	*ftp://ncbi.nlm.nih.gov/pub/seg/*
sim4	*http://globin.cse.psu.edu*
Wise package	*http://www.sanger.ac.uk/Software/Wise2/*

PROBLEM SET

1. What is the difference between a global and a local alignment strategy?
2. Calculate the score of the DNA sequence alignment shown below using the following scoring rules: +1 for a match, −2 for a mismatch, −3 for opening a gap, and −1 for each position in the gap.

```
GACTACGATCCGTATACGCACA--GGTTCAGAC
|||||||| |||||||||||||  |||||||||
GACTACGAGCCGTATACGCACACAGGTTCAGAC
```

3. If a match from a database search is reported to have a E-value of 0.0, should it be considered highly insignificant or highly significant?

REFERENCES

Altschul, S. F. (1991). Amino acid substitution matrices from an information theoretic perspective. *J. Mol. Biol.* 219, 555–565.

Altschul, S. F., Boguski, M. S., Gish, W., and Wootton, J. C. (1994). Issues in searching molecular sequence databases. *Nature Genet.* 6, 119–29.

Altschul, S. F., and Erickson, B. W. (1986). Locally optimal subalignments using nonlinear similarity functions. *Bull. Math. Biol.* 48, 633–660.

Altschul, S. F., and Erickson, B. W. (1985). Significance of nucleotide sequence alignments: A method for random sequence permutation that preserves dinucleotide and codon usage. *Mol. Biol. Evol.* 2, 526–538.

Altschul, S. F., and Gish, W. (1996). Local alignment statistics. *Methods Enzymol.* 266, 460–480.

Altschul, S. F., Gish, W., Miller, W., Myers, E. W., and Lipman, D. J. (1990). Basic local alignment search tool. *J. Mol. Biol.* 215, 403–410.

Baron, M., Norman, D. G., and Campbell, I. D. (1991). Protein modules. *Trends Biochem. Sci.* 16, 13–17.

Bateman, A., Birney, E., Durbin, R., Eddy, S. R., Howe, K. L., and Sonnhammer, E. L. (2000). The Pfam protein families database. *Nucleic Acids Res.* 28, 263–266.

Birney, E., Thompson, J. D., and Gibson, T. J. (1996). PairWise and SearchWise: finding the optimal alignment in a simultaneous comparison of a protein profile against all DNA translation frames. *Nucleic Acids Res.* 24, 2730–2739.

Bucher, P., Karplus, K., Moeri, N., and Hofmann, K. (1996). A flexible motif search technique based on generalized profiles. *Comput. Chem.* 20, 3–23.

Dayhoff, M. O., Schwartz, R. M., and Orcutt, B. C. (1978). A model of evolutionary change in proteins. In *Atlas of Protein Sequence and Structure*, M. O. Dayhoff, ed. (Washington: National Biomedical Research Foundation), p. 345–352.

Dembo, A., Karlin, S., and Zeitouni, O. (1984). Limit distribution of maximal non-aligned two-sequence segmental score. *Ann. Prob.* 22, 2022–2039.

Doolittle, R. F., Hunkapiller, M. W., Hood, L. E., Devare, S. G., Robbins, K. C., Aaronson, S. A., and Antoniades, H. N. (1983). Simian sarcoma virus onc gene, v-sis, is derived from the gene (or genes) encoding a platelet-derived growth factor. *Science* 221, 275–277.

Doolittle, R. J., and Bork, P. (1993). Evolutionarily mobile modules in proteins. *Sci. Am.* 269, 50–56.

Eddy, S. R., Mitchison, G., and Durbin, R. (1995). Maximum discrimination hidden Markov models of sequence consensus. *J. Comput. Biol.* 2, 9–23.

Fitch, W. M. (1969). Locating gaps in amino acid sequences to optimize the homology between two proteins. *Biochem. Genet.* 3, 99–108.

Fitch, W. M. (1983). Random sequences. *J. Mol. Biol.* 163, 171–176.

Florea, L., Hartzell, G., Zhang, Z., Rubin, G. M., and Miller, W. (1998). A computer program for aligning a cDNA sequence with a genomic DNA sequence. *Genome Res.* 8, 967–974.

Gibbs, A. J., and McIntyre, G. A. (1970). The diagram: a method for comparing sequences. Its use with amino acid and nucleotide sequences. *Eur. J. Biochem.* 16, 1–11.

Gonzalez, P., Hernandez-Calzadilla, C., Rao, P. V., Rodriguez, I. R., Zigler, J. S., Jr., and Borras, T. (1994). Comparative analysis of the zeta-crystallin/quinone reductase gene in guinea pig and mouse. *Mol. Biol. Evol.* 11, 305–315.

Gribskov, M., McLachlan, A. D., and Eisenberg, D. (1987). Profile analysis: detection of distantly related proteins. *Proc. Natl. Acad. Sci. USA* 84, 4355–4358.

Henikoff, S., and Henikoff, J. G. (1992). Amino acid substitution matrices from protein blocks. *Proc. Natl. Acad. Sci. USA* 89, 10915–10919.

Henikoff, S., and Henikoff, J. G. (1991). Automated assembly of protein blocks for database searching. *Nucleic Acids Res.* 19, 6565–6572.

Higgins, D. G., Thompson, J. D., and Gibson, T. J. (1996). Using CLUSTAL for multiple sequence alignments. *Methods Enzymol.* 266, 383–402.

Holm, L., and Sander, C. (1997). Enzyme HIT. *Trends Biochem. Sci.* 22, 16–117.

Huang, X., Hardison, R. C., and Miller, W. (1990). A space-efficient algorithm for local similarities. *Comput. Applic. Biosci.* 6, 373–381.

Karlin, S., and Altschul, S. F. (1990). Methods for assessing the statistical significance of molecular sequence features by using general scoring schemes. *Proc. Natl. Acad. Sci. USA* 87, 2264–2268.

Lipman, D. J., and Pearson, W. R. (1985). Rapid and sensitive protein similarity searches. *Science* 227, 1435–1441.

Mott, R. (1997). EST_GENOME: a program to align spliced DNA sequences to unspliced genomic DNA. *Comput. Appl. Biosci.* 13, 477–478.

Needleman, S. B., and Wunsch, C. (1970). A general method applicable to the search for similarities in the amino acid sequence of two proteins. *J. Mol. Biol.* 48, 443–453.

Patthy, L. (1991). Modular exchange principles in proteins. *Curr. Opin. Struct. Biol.* 1, 351–361.

Pearson, W. (2000). Flexible sequence similarity searching with the FASTA3 program package. *Methods Mol. Biol.* 132, 185–219.

Pearson, W. R. (1996). Effective protein sequence comparison. *Methods Enzymol.* 266, 227–258.

Pearson, W. R., and Lipman, D. J. (1988). Improved tools for biological sequence comparison. *Proc. Natl. Acad. Sci. USA* 85, 2444–2448.

Schneider, T. D., Stormo, G. D., Gold, L., and Ehrenfeucht, A. (1986). Information content of binding sites on nucleotide sequences. *J. Mol. Biol.* 188, 415–31.

Sellers, P. H. (1984). Pattern recognition in genetic sequences by mismatch density. *Bull. Math. Biol.* 46, 510–514.

Smith, T. F., and Waterman, M. S. (1981). Identification of common molecular subsequences. *J. Mol. Biol.* 147, 195–197.

Sonnhammer, E. L. L., and Durban, R. (1996). A dot-matrix program with dynamic threshold control suited for genomic DNA and protein sequence analysis. *Gene* 167, GC1–GC10.

Staden, R. (1988). Methods to define and locate patterns of motifs in sequences. *Comput. Appl. Biosci.* 4, 53–60.

Tatusov, R. L., Altschul, S. F., and Koonin, E. V. (1994). Detection of conserved segments in proteins: iterative scanning of sequence databases with alignment blocks. *Proc. Natl. Acad. Sci. USA* 91, 12091–12095.

Waterfield, M. D., Scrace, G. T., Whittle, N., Stroobant, P., Johnsson, A., Wasteson, A., Westermark, B., Heldin, C. H., Huang, J. S., and Deuel, T. F. (1983). Platelet-derived growth factor is structurally related to the putative transforming protein p28sis of simian sarcoma virus. *Nature* 304, 35–39.

Waterman, M. S., and Eggert, M. (1987). A new algorithm for best subsequence alignments with applications to tRNA-rRNA comparisons. *J. Mol. Biol.* 197, 723–728.

Waterman, M. S., and Vingron, M. (1994). Rapid and accurate estimates of statistical significance for sequence data base searches. *Proc. Natl. Acad. Sci. USA* 91, 4625–4628.

Wilbur, W. J., and Lipman, D. J. (1983). Rapid similarity searches of nucleic acid and protein data banks. *Proc. Natl. Acad. Sci. USA* 80, 726–730.

Wootton, J. C. (1994). Non-globular domains in protein sequences: automated segmentation using complexity measures. *Comput. Chem.* 18, 269–285.

Wootton, J. C., and Federhen, S. (1996). Analysis of compositionally biased regions in sequence databases. *Methods Enzymol.* 266, 554–571.

Wootton, J. C., and Federhen, S. (1993). Statistics of local complexity in amino acid sequences and sequence databases. *Comput. Chem.* 17, 149–163.

9

CREATION AND ANALYSIS OF PROTEIN MULTIPLE SEQUENCE ALIGNMENTS

Geoffrey J. Barton

European Molecular Biology Laboratory
European Bioinformatics Institute
Wellcome Trust Genome Campus
Hinxton, Cambridge
UK

INTRODUCTION

When a protein sequence is newly-determined, an important goal is to assign possible functions to the protein. The first computational step is to search for similarities with sequences that have previously been deposited in the DNA and protein sequence databases. If similar sequences are found, they may match the complete length of the new sequence or only to subregions of the sequence. If more than one similar sequence is found, then the next important step in the analysis is to multiply align all of the sequences. Multiple alignments are a key starting point for the prediction of protein secondary structure, residue accessibility, function, and the identification of residues important for specificity. Multiple alignments also provide the basis for the most sensitive sequence searching algorithms (cf. Gribskov et al., 1987; Barton and Sternberg, 1990; Attwood et al., 2000). Effective analysis of a well-constructed multiple alignment can provide important clues about which residues in the protein are important for function and which are important for stabilizing the secondary and tertiary structures of the protein. In addition, it is often also possible to make predictions about which residues confer specificity of function to subsets of the

Bioinformatics: A Practical Guide to the Analysis of Genes and Proteins
Edited by A. D. Baxevanis and B. F. F. Ouellette
ISBN 0-471-38390-2 (cloth), ISBN 0-471-383910 (paper)

sequences. In this chapter, some guidelines are provided toward the generation and analysis of protein multiple sequence alignments. This is not a comprehensive review of techniques; rather, it is a guide based on the software that have proven to be most useful in building alignments and using them to predict protein structure and function. A full summary of the software is available at the end of the chapter.

WHAT IS A MULTIPLE ALIGNMENT, AND WHY DO IT?

A protein sequence is represented by a string a of letters coding for the 20 different types of amino acid residues. A protein sequence alignment is created when the residues in one sequence are lined up with those in at least one other sequence. Optimal alignment of the two sequences will usually require the insertion of gaps in one or both sequences in order to find the best alignment. Alignment of two residues implies that those residues are performing similar roles in the two different proteins. This allows for information known about specific residues in one sequence to be potentially transferred to the residues aligned in the other. For example, if the active site residues of an enzyme have been characterized, alignment of these residues with similar residues in another sequence may suggest that the second sequence possesses similar catalytic activity to the first. The validity of such hypotheses depends on the overall similarity of the sequences, which in turn dictate the confidence with which an alignment can be generated. There are typically many millions of different possible alignments for any two sequences. The task is to find an alignment that is most likely to represent the chemical and biological similarities between the two proteins.

A *multiple sequence alignment* is simply an alignment that contains more than two sequences! Even if one is interested in the similarities between only two of the sequences in a set, it is always worth multiply-aligning all available sequences. The inclusion of these additional sequences in the multiple alignment will normally improve the accuracy of the alignment between the sequence pairs, as illustrated in Figure 9.1, as well as revealing patterns of conserved residues that would not have been obvious when only two sequences are directly studied. Although many programs exist that can generate a multiple alignment from unaligned sequences, extreme care must be taken when interpreting the results. An alignment may show perfect matching of a known active-site residue with an identical residue in a well-characterized protein family, but, if the alignment is incorrect, any inference about function will also be incorrect.

STRUCTURAL ALIGNMENT OR EVOLUTIONARY ALIGNMENT?

It is the precise arrangement of the amino acid side chains in the three-dimensional structure of the protein that dictates its function. Comparison of two or more protein three-dimensional structures will highlight which residues are in similar positions in space and hence likely to be performing similar functional roles. Such comparisons can be used to generate a sequence alignment from structure (e.g., see Russell and Barton, 1992). The *structural alignment* of two or more proteins is the gold standard against which sequence alignment algorithms are normally judged. This is because it is the structural alignment that most reliably aligns residues that are of functional importance. Unfortunately, structural alignments are only possible when the three-

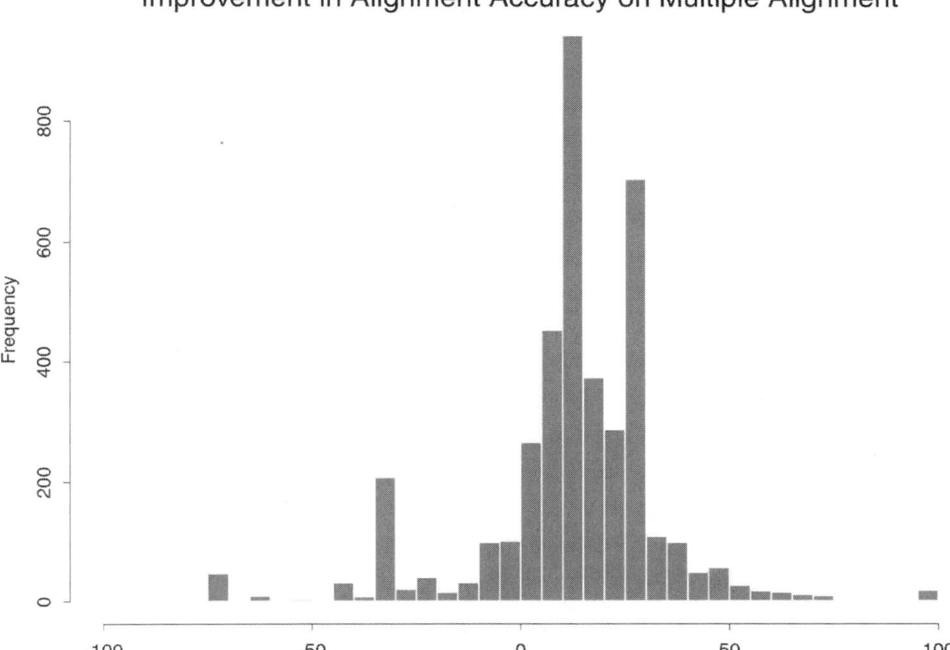

Figure 9.1. Histogram showing difference in accuracy between the same pairs of sequences aligned as a pair and as part of a larger multiple sequence alignment. On average, multiple alignments improve the overall alignment accuracy, which, in this example, is judged as the alignment obtained by comparison of the three-dimensional structures of the individual proteins rather than just their sequences (Russell and Barton, 1992).

dimensional structures of *all* the proteins to be aligned are known. This is not usually the case; therefore, the challenge for sequence alignment methods is to get as close as possible to the structural alignment without knowledge of structure. Although the structural alignment is the most important alignment for the prediction of function, it does not necessarily correspond to the *evolutionary alignment* implied by divergence from a common ancestor protein. Unfortunately, it is rarely possible to determine the evolutionary alignment of two divergent proteins with confidence because this would require knowledge of the precise history of substitutions, insertions, and deletions that have led to the creation of present-day proteins from their common ancestor.

HOW TO MULTIPLY ALIGN SEQUENCES

Automatic alignment programs such as CLUSTAL W (Thompson et al., 1994) will give good quality alignments for sequences that are more than 6σ similar (Barton and Sternberg, 1987). However, building good multiple alignments for sequences that are not trivially similar is a precise task even with the best available alignment tools. This section gives an overview of some of the steps to go through to make alignments that are good for structure/function predictions. This is *not* a universal recipe; in fact, there are very few universal recipes in bioinformatics in general. Each set of sequences presents its own biologically based problems, and only experience can guide the creation of high-quality alignments. Some collections of expertly created multiple alignments exist (described later), and these should always be consulted when studying sequences that are present there. The key steps in building a multiple alignment are as follows.

- Find the sequences to align by database searching or by other means.
- Locate the region(s) of each sequence to include in the alignment. *Do not* try to multiply align sequences that are substantially different in length. Most multiple alignment programs are designed to align sequences that are similar over their entire length; therefore, a necessary first step is to edit the sequences down to those regions that sequence database searches suggest are similar.
- Ideally, assess the similarities within the set of sequences by comparing them pairwise with randomizations. Select a subset of the sequences to align first that cluster above 6σ. Automatic alignment of such sequences are likely to be accurate (Barton and Sternberg, 1987). An alternative to doing randomization is to align only sequences that are similar to the query in a database search, say with an E-value of <1.
- Run the multiple alignment program.
- Manually inspect the alignment for problems. Pay particular attention to regions that appear to be speckled with gaps. Use an alignment visualization tool (e.g., ALSCRIPT/JalView, see below) to identify positions in the alignment that show conserved physicochemical properties across the complete alignment. If there are no such regions, then look at subsets of the sequences.
- Remove sequences that appear to disrupt the alignment seriously and then realign the remaining subset.
- After identifying key residues in the set of sequences that are straightforward to align, attempt to add the remaining sequences to the alignment so as to preserve the key features of the family.

Assessing Quality of Alignment

Multiple alignment programs will align *any* set of sequences. However, the fact that the program produces an alignment does not mean that the alignment has any biological meaning. Most programs will take unrelated protein sequences and align them just as easily as two genuinely related sequences. Even for related sequences, there is no guarantee that the resulting alignment is in any way meaningful. One way of assessing whether an alignment is meaningful is to perform a randomization or "Monte Carlo" test of significance. To do this, the two sequences are first aligned and the score (S) for the alignment is recorded. The sequences are then shuffled so that they maintain their length and amino acid composition but have a randomized order. The shuffled sequences are then compared again, and the score is recorded. The shuffling and realigning process is repeated a number of times (typically 100), and the mean and standard deviation (σ) for the scores are calculated. The Z-score provides an indication of the significance of the alignment. If $Z > 6$, then it is highly likely that the two sequences are alignable, and the alignment correctly relates the key functional and structural residues in the individual proteins to one another (Barton and Sternberg, 1987). Unfortunately, this can only be a rough guide. An alignment that gives a $Z < 6$ may be poor, and some alignments with low Z-scores are actually correct. This is simply a reflection of the fact that, during evolution, sequence similarity has diverged faster than structural or functional similarity. Z-scores are preferable to simple percent identities as a measure of similarity because it corrects for both compositional bias in the sequences as well as accounting for the varying lengths of sequences. The Z-score, therefore, gives an indication of the *overall* similarity between two sequences. Although it is a powerful measure, it does *not* help to locate parts of the sequence alignment that are incorrect. As a general rule, if the alignment is between two or more sequences that do indeed share a similar three-dimensional structure, then the majority of errors will be concentrated around regions where there are gaps (insertions/deletions).

Hierarchical Methods

The most accurate, practical methods for automatic multiple alignment are hierarchical methods. These work by first finding a guide tree and then following the guide tree to build the alignment. The process is summarized in Figure 9.2. First, all pairs of sequences in the set to be aligned are compared by a pairwise method of sequence comparison. This provides a set of pairwise similarity scores for the sequences that can be fed into a cluster analysis or tree calculating program. The tree is calculated to place more similar pairs of sequences closer together on the tree than sequences that are less similar. The multiple alignment is then built by starting with the pair of sequences that is most similar and aligning them and then aligning the next most similar pair, and so on. Pairs to be aligned need not be single sequences but can be alignments that have been generated earlier in the tree. If an alignment is compared with a sequence or another alignment, then gaps that exist in the alignment are preserved. There are many different variations of this basic multiple alignment technique. Because errors in alignment that occur early in the process can get locked in and propagated, some methods allow for realignment of the sequences after the initial alignment (e.g., Barton and Sternberg, 1987; Gotoh, 1996). Other refinements include using different similarity scoring matrices at different stages in building up the alignment (e.g., Thompson et al., 1994). Gaps (insertions/deletions) do not occur randomly in protein sequences.

Since a stable, properly-folded protein must be maintained, proteins with an insertion or deletion in the middle of a secondary structure (α-helix or β-strand) are usually selected against during the course of evolution. As a consequence, present-day proteins show a strong bias toward localizing insertions and deletions to loop regions that link the core secondary structures. This observation can be used to improve the accuracy of multiple sequence alignments when the secondary structure is known for one or more of the proteins in practice by making the penalty for inserting a gap higher when in secondary structure regions than when in loops (Barton and Sternberg, 1987; Jones, 1999. A further refinement is to bias where gaps are most likely to be inserted in the alignment by examining the growing alignment for regions that are most likely to accommodate gaps (Pascarella and Argos, 1992).

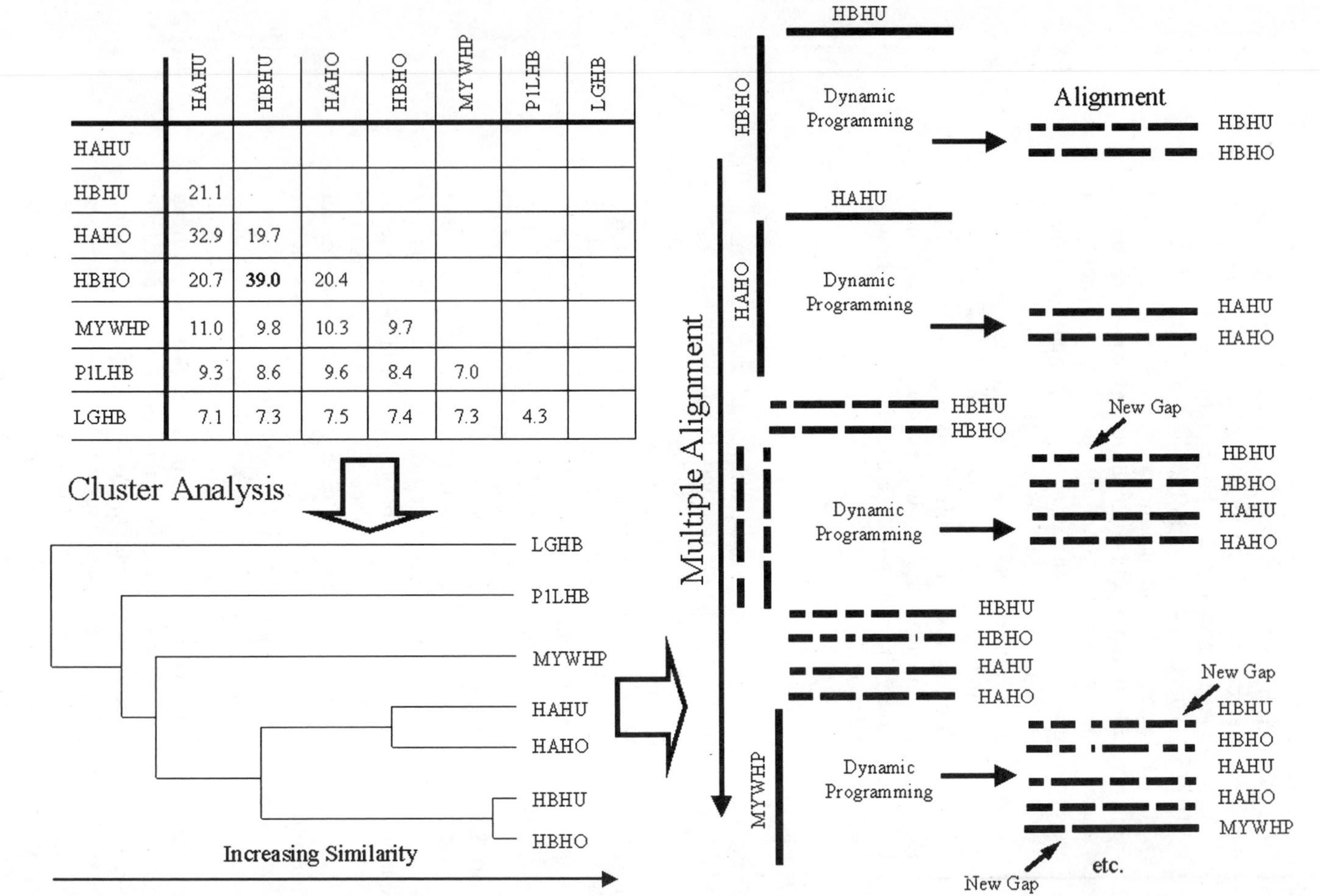

	HAHU	HBHU	HAHO	HBHO	MYWHP	P1LHB	LGHB
HAHU							
HBHU	21.1						
HAHO	32.9	19.7					
HBHO	20.7	**39.0**	20.4				
MYWHP	11.0	9.8	10.3	9.7			
P1LHB	9.3	8.6	9.6	8.4	7.0		
LGHB	7.1	7.3	7.5	7.4	7.3	4.3	

Figure 9.2. Illustration of the stages in hierarchical multiple alignment of seven sequences. The codes for these sequences are HAHU, HBHU, HAHO, HBHO, MYWHP, P1LHB, and LGHB. The table at the top left shows the pairwise *Z*-scores for comparison of each sequence pair. Higher numbers mean greater similarity (see text). Hierarchical cluster analysis of the *Z*-score table generates the dendrogram or tree shown at the bottom left. Items joined toward the right of the tree are more similar than those linked toward the left. Based on the tree, LGHB is least similar to the other sequences in the set, whereas HBHU and HBHO are the most similar pair (most similar to each other). The first four steps in building the multiple alignment are shown on the right. The first two steps are pairwise alignments. The third step is a comparison of profiles from the two alignments generated in *steps 1* and *2*. The fourth step adds a single sequence (MYWHP) to the alignment generated at *step 3*. Further sequences are added in a similar manner.

CLUSTAL W and Other Hierarchical Alignment Software

CLUSTAL W combines a good hierarchical method for multiple sequence alignment with an easy-to-use interface. The software is free, although a contribution to development costs is required when purchasing the program. CLUSTAL W runs on most computer platforms and incorporates many of the techniques described in the previous section. The program uses a series of different pair-score matrices, biases the location of gaps, and allows you to realign a set of aligned sequences to refine the alignment. CLUSTAL W can read a secondary structure "mask" and bias the positioning of gaps according to it; the program can also read two preexisting alignments and align them to each other or align a set of sequences to an existing alignment. CLUSTAL W also includes options to calculate neighbor-joining trees for use in inferring phylogeny. Although CLUSTAL W does not provide general tools for viewing these trees, the output is compatible with the PHYLIP package (Felsenstein, 1989) and the resultant trees can be viewed with that program. CLUSTAL W can read a variety of different common sequence formats and produce a range of different output formats. The manual for CLUSTAL W is clearly written and explains possible limitations of the alignment process. Although CLUSTAL W can be installed and run locally, users can also access it through a faster Web service via the EBI server by clicking the "Tools page". With the exception of manual editing and visualization, CLUSTAL W contains most of the tools that are needed to build and refine a multiple sequence alignment. When combined with JalView, as described below, the process of building and refining a multiple alignment is greatly simplified. Although CLUSTAL W is probably the most widely used multiple alignment program and for most purposes is adequate, other software exists having functionality not found in CLUSTAL W. For example, AMPS (Barton, 1990) provides a pairwise sequence comparison option with randomization, allowing *Z*-scores to be calculated. The program can also generate alignments without the need to calculate trees first. For large numbers of sequences, this can save a lot of time because it eliminates the need to perform all pairwise comparisons of the sequences. AMPS also has software to visualize trees, thus helping in the selection of sequences for alignment. However, the program has no simple menu interface; therefore, it is more difficult for the novice or occasional user to use.

More Rigorous Nonhierarchical Methods

Hierarchical methods do not guarantee finding the one mathematically optimal multiple alignment for an entire set of sequences. However, in practice, the mathematical optimum rarely makes any more biological sense than the alignment that is found by hierarchical methods. This is probably because a great deal of effort has gone into tuning the parameters used by CLUSTAL W and other hierarchical methods to produce alignments that are consistent with those that a human expert or three-dimensional structure comparison might produce. The widespread use of these techniques has also ensured that the parameters are appropriate for a wide range of alignment problems. More rigorous alignment methods that attempt to find the mathematically optimal alignment over a set of sequences (cf. Lipman et al., 1989) may be capable of giving better alignments, but, as shown in recent benchmark studies, they are, on average, no better than the hierarchical methods.

Multiple Alignment by PSI-BLAST

Multiple sequence alignments have long been used for more sensitive searches of protein sequence databases than is possible with a single sequence. The program PSI-BLAST (Altschul et al., 1997) has recently made these profile methods more easily available. As part of its search, PSI-BLAST generates a multiple alignment. However, this alignment is not like the alignments made by CLUSTAL W, AMPS, or other traditional multiple alignment tools. In a conventional multiple alignment, all sequences in the set have equal weight. As a consequence, a multiple alignment will normally be longer than any one of the individual sequences, since gaps will be inserted to optimize the alignment. In contrast, a PSI-BLAST multiple alignment is *always* exactly the length of the query sequence used in the search. If alignment of the query (or query profile) to a database sequence requires an insertion in the query, then the inserted region from the database sequence is simply discarded. The resulting alignment thus highlights the amino acids that may be aligned to each position in the query. Perhaps for this reason, PSI-BLAST multiple alignments and their associated frequency tables and profiles have proved very effective as input for programs that predict protein secondary structure (Jones, 1999; Cuff and Barton, 2000).

Multiple Protein Alignment From DNA Sequences

Although most DNA sequences will have translations represented in the EMBL-TrEMBL or NCBI-GenPept databases, this is not true of single-pass EST sequences. Because EST data are accumulating at an exponential pace, an automatic method of extracting useful protein information from ESTs has been developed. In brief, the ProtEST server (Cuff et al., 1999) searches EST collections and protein sequence databases with a protein query sequence. EST hits are assembled into species-specific contigs, and an error-tolerant alignment method is used to correct probable sequencing errors. Finally, any protein sequences found in the search are multiply aligned with the translations of the EST assemblies to produce a multiple protein sequence alignment. The JPred server (version 7.3) will generate a multiple protein sequence alignment when presented with a single protein sequence by searching the SWALL protein sequence database and building a multiple alignment. The JPred alignments are a good starting point for further analysis with more sensitive methods.

TOOLS TO ASSIST THE ANALYSIS OF MULTIPLE ALIGNMENTS

A multiple sequence alignment can potentially consist of several hundred sequences that are 500 or more amino acids long. With such a volume of data, it can be difficult to find key features and present the alignments in a form that can be analyzed by eye. In the past, the only option was to print out the alignment on many sheets of paper, stick these together, and then pore over the massive poster with colored highlighter pens. This sort of approach can still be useful, but it is rather inconvenient! Visualization of the alignment is an important scientific tool, either for analysis or for publication. Appropriate use of color can highlight positions that are either identical in all the aligned sequences or share common physicochemical properties. ALSCRIPT (Barton, 1993) is a program to assist in this process. ALSCRIPT takes a multiple sequence alignment and a file of commands and produces a file in

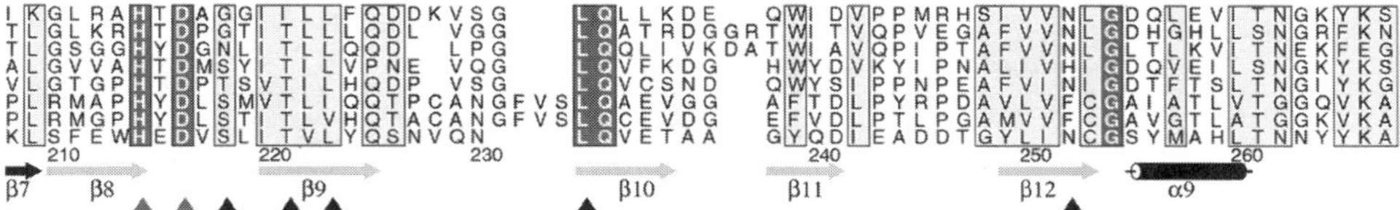

Figure 9.3. Example output from the program ALSCRIPT (Barton, 1993). Details can be found within the main text.

PostScript format suitable for printing out or viewing with a utility such as ghostview. Figure 9.3 illustrates a fragment of ALSCRIPT output (the full figure can be seen in color in Roach et al., 1995). In this example, identities across all sequences are shown in white on red and boxed, whereas positions with similar physicochemical properties are shown black on yellow and boxed. Residue numbering according to the bottom sequence is shown underneath the alignment. Green arrows illustrate the location of known β-strands, whereas α-helices are shown as black cylinders. Further symbols highlight specific positions in the alignment for easy cross-referencing to the text. ALSCRIPT is extremely flexible and has commands that permit control of font size and type, background coloring, and boxing down to the individual residue. The program will automatically split a large alignment over multiple pages, thus permitting alignments of any size to be visualized. However, this flexibility comes at a price. There is no point-and-click interface, and the program requires the user to be familiar with editing files and running programs from the command line. The ALSCRIPT distribution includes a comprehensive manual and example files that make the process of making a useful figure for your own data a little easier.

Subalignments—AMAS

ALSCRIPT provides a few commands for calculating residue conservation across a family of sequences and coloring the alignment accordingly. However, it is really intended as a display tool for multiple alignments rather than an analysis tool. In contrast, AMAS (Analysis of Multiply Aligned Sequences; Livingstone and Barton, 1993) is a program for studying the relationships between sequences in a multiple alignment to identify possible functional residues. AMAS automatically runs ALSCRIPT to provide one output that is a boxed, colored, and annotated multiple alignment.

Why might you want to run AMAS? A common question one faces is, "Which residues in a protein are important for its specificity?" AMAS can help identify these residues by highlighting similarities and differences between subgroups of sequences in a multiple alignment. For example, given a family of sequences that shows some variation, positions in a multiple alignment that are conserved across the entire family of sequences are likely to be important to stabilize the common fold of the protein or common functions. Positions that are conserved within a subset of the sequences, but different in the rest of the family, are likely to be those important to the specific function or specificity of that subset, and these positions can be easily identified using AMAS. There are a number of subtle types of differences that AMAS will search for, and these are summarized in Figure 9.4. To use AMAS, one must first have an idea of what subgroups of sequences exist in a multiple alignment of interest. One way to do this is to take a tree generated from the multiple alignment and

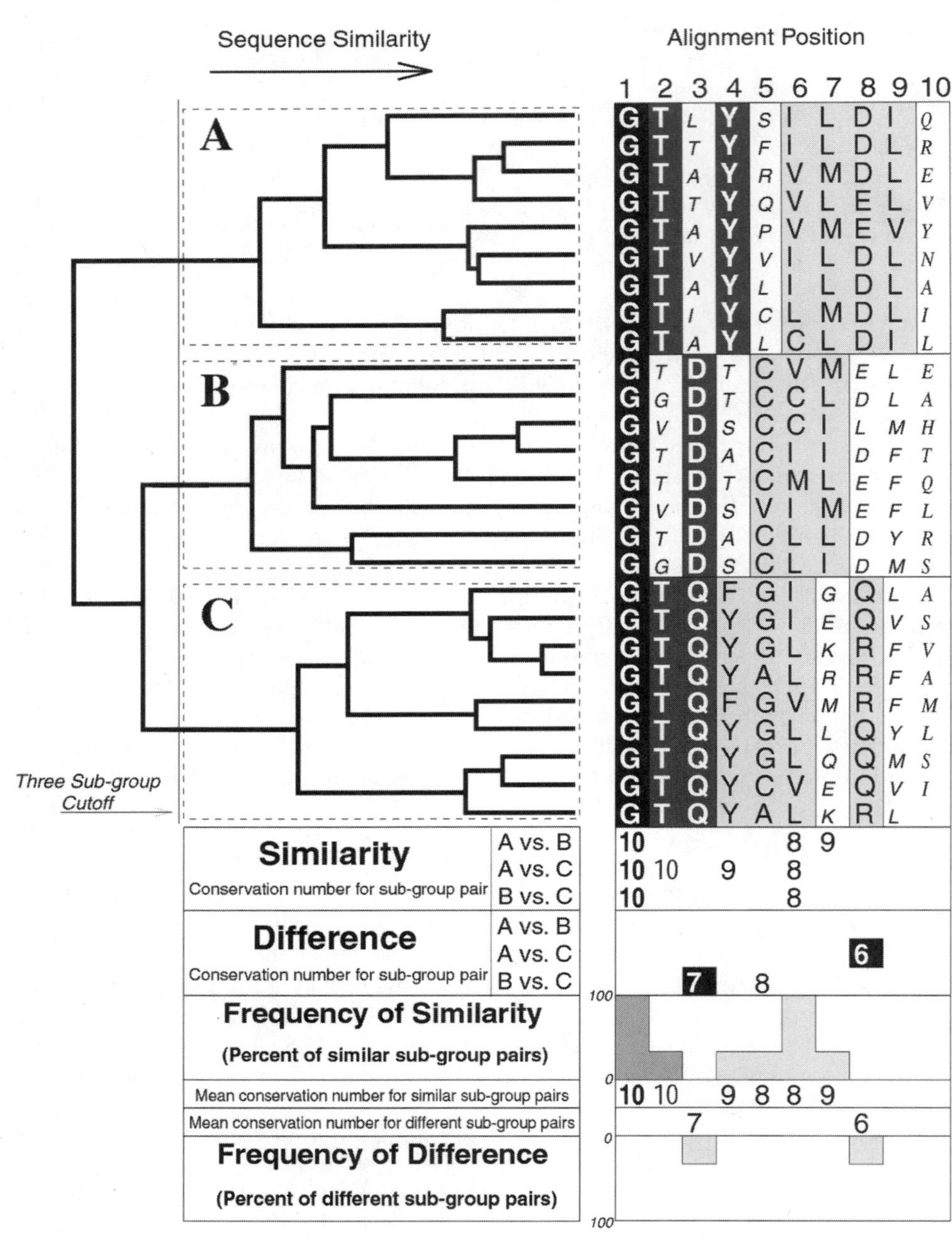

Sequence Similarity
Alignment Position
1 2 3 4 5 6 7 8 9 10
A
B
C
Three Sub-group Cutoff
Similarity
Conservation number for sub-group pair
A vs. B
A vs. C
B vs. C
Difference
Conservation number for sub-group pair
A vs. B
A vs. C
B vs. C
Frequency of Similarity
(Percent of similar sub-group pairs)
Mean conservation number for similar sub-group pairs
Mean conservation number for different sub-group pairs
Frequency of Difference
(Percent of different sub-group pairs)
KEY
Identical across all sequences
Identical within one sub-group
Conserved within one sub-group
Unconserved within one sub-group
Unconserved across all sequences

identify clusters of sequences at some similarity threshold. This is also illustrated in Figure 9.4, in which three groups have been selected on the basis of the tree shown at the top left. Alternatively, if one knows in advance that finding common features and differences between, for example, sequences 1–20 and 21–50 in a multiple alignment is important, one can specify these ranges explicitly. The output of AMAS is a detailed text summary of the analysis as well as a colored and shaded multiple sequence alignment. By default, AMAS searches for general features of amino acid physicochemical properties. However, this can be narrowed down just to a single feature of amino acids such as charge. An example of a charge analysis is shown in Figure 9.5 for repeats within the annexin supergene family of proteins (Barton et al., 1991). The analysis highlights a charge swap within two subgroups of the sequences, correctly predicting the presence of a salt bridge in the native folded protein (Huber et al., 1990). The AMAS program may either be downloaded and run locally, or a subset of its options can be accessed over the Web at a server hosted by EBI.

Secondary Structure Prediction and the Prediction of Buried Residues From Multiple Sequence Alignment

When aligning sequences, it is important to remember that the protein is a three-dimensional molecule and not just a string of letters. Predicting secondary structure either for the whole collection of sequences or subsets of the sequences can be used to help discover how the protein might fold locally and guide the alignment of more distantly related sequences. For example, it is common for proteins with similar topologies to have quite different sequences and be unalignable by an automatic alignment method (e.g., see Russell and Barton, 1994; *cf.* the SCOP database, see Murzin et al., 1995, Chapter 5). In these circumstances, the secondary structure may suggest which blocks of sequences should be equivalent. The prediction of secondary structure (α-helix and β-strand) is enhanced by around 6% when performed from a multiple alignment, compared with prediction from a single sequence (Cuff and

←

Figure 9.4. Stylized output from the program AMAS. The sequence alignment has been shaded to illustrate similarities within each subgroup of sequences. *Conservation numbers* (Livingstone and Barton, 1993; Zvelebil et al., 1987) run from 0 to 10 and provide a numerical measure of the similarity in physicochemical properties of each column in the alignment. Below the alignment, the lines "Similar Pairs" show the conservation values obtained when each pair of subgroups is combined and the combined conservation number is not less than a threshold. For example, at position 7, subgroups *A* and *B* combine with a conservation number of 9. The lines "Different Pairs" illustrate positions at which a combination of subgroups lowers the conservation number below the threshold. For example, at position 3, there is an identity in subgroup *B* and one in *C*, but, when the groups are combined, the identity is lost and the conservation drops below the threshold of 8 to 7. A summary of the similarities and differences is given as a frequency histogram. Each upward bar represents the proportion of subgroup pairs that preserve conservation, whereas each downward bar shows the percentage of differences. For example, at position 6, 3/3 pairs are conserved (100%), whereas at positions 3 and 8, 1/3 pairs show (33%) differences With a large alignment, the histogram can quickly draw the eye to regions that are highly conserved or to regions where there are differences in conserved physicochemical properties.

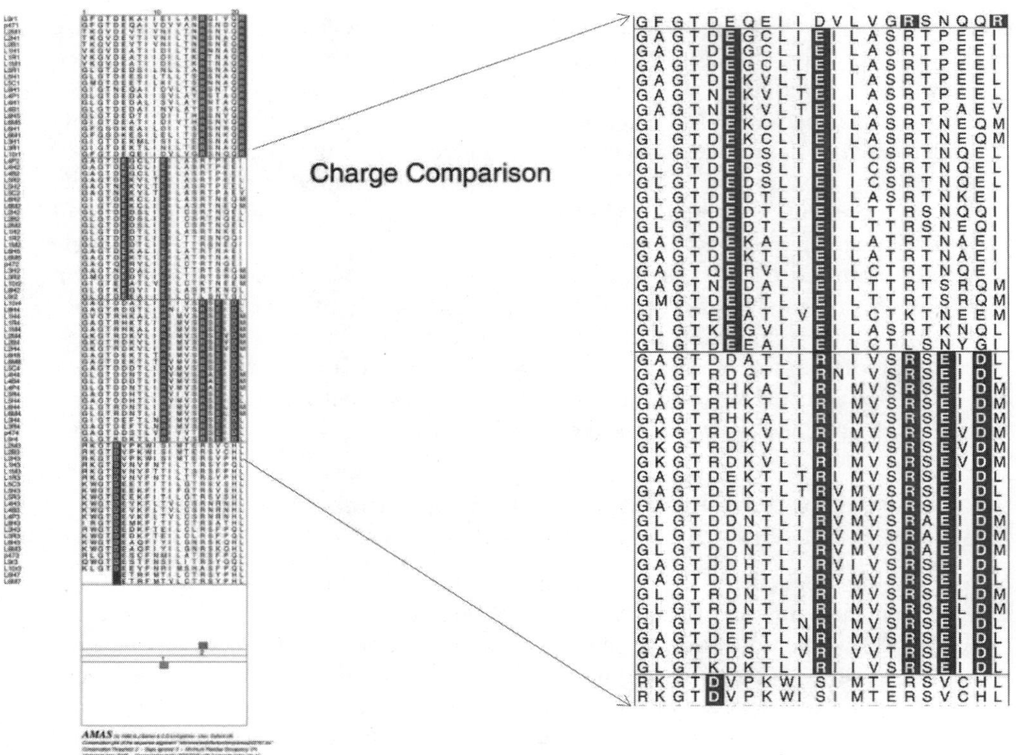

Figure 9.5. Illustration of an AMAS output used to find a charge pair in the annexins. There are four groups of sequences in the alignment. The highlighted positions highlight locations where the charge is conserved in each group of sequences yet different between groups. A change from glutamine to arginine is shown at position 1.

Barton 1999). The best current methods [PSIPRED (Jones, 1999) and JNET (Cuff and Barton, 2000)] give over 76% accuracy for the prediction of three states (α-helix, β-strand, and random coil) in rigorous testing. This high accuracy is possible because the prediction algorithms are able to locate regions in the sequences that show patterns of conserved physicochemical properties across the aligned family. These patterns are characteristic of particular secondary structure types and can often be seen by eye in a multiple sequence alignment, as summarized below:

- Short runs of conserved hydrophobic residues suggest a buried β-strand.
- i, $i + 2$, and $i + 4$ patterns of conserved hydrophobic amino acids suggest a surface β-strand, since the alternate residues in a strand point in the same direction. If the alternate residues all conserve similar physicochemical properties, then they are likely to form one face of a β-strand.
- i, $i + 3$, $i + 4$, and $i + 7$, and variations of that pattern, (e.g., i, $i + 4$, $i + 7$) of conserved residues suggest an α-helix with one surface facing the solvent.
- Insertions and deletions are normally only tolerated in regions not associated with the buried core of the protein. Thus, in a good multiple alignment, the location of indels suggests surface loops rather than α-helices or β-strands.

- Although glycine and proline may be found in all secondary structure types, a glycine or proline residue that is conserved across a family of sequences is a strong indicator of a loop.

Secondary structure prediction programs such as JNET (Cuff and Barton, 2000) and PHD (Rost and Sander, 1993) also exploit multiply aligned sequences to predict the likely exposure of each residue to solvent. Knowledge of solvent accessibility can help in the identification of residues key to stabilizing the fold of the protein as well as those that may be involved in binding. Both the JNET and PHD programs may be run from the JPred prediction server, whereas JNET may also be run from within JalView. [For further discussion of methods used to predict secondary structure, the reader is referred to Chapter 11.]

JalView

AMAS and ALSCRIPT are not interactive: they run a script or set of commands and produce a PostScript file, which can be viewed on-screen using a Postscript viewer or just printed out. Although this provides the maximum number of options and flexibility in its display, it is comparatively slow and sometimes difficult to learn. In addition, the programs require a separate program to be run to generate the multiple alignment for analysis. If the alignment requires modification or subsets of the alignment are needed, a difficult cycle of editing and realigning is often required. The program JalView overcomes most of these problems. JalView encapsulates many of the most useful features of AMAS and ALSCRIPT in an interactive, mouse-driven program that will run on most computers with a Java interpreter. The core of JalView is an interactive alignment editor. This allows an existing alignment to be read into the program and individual residues or blocks of residues to be moved around. A few mouse clicks permit the sequences to be subset into a separate copy of JalView. JalView can call CLUSTAL W (Thompson et al., 1994) either as a local copy on the same computer that is running JalView or the CLUSTAL W server at EBI. Thus, one can also read in a set of unaligned sequences, align them with CLUSTAL W, edit the alignment, and take subsets with great ease. Further functions of JalView will calculate a simple, neighbor-joining tree from a multiple alignment and allow an AMAS-style analysis to be performed on the subgroups of sequences. If the tertiary structure of one of the proteins in the set is available, then the three-dimensional structure may be viewed alongside the alignment in JalView. In addition, the JNET secondary structure prediction algorithm (Cuff and Barton, 2000) may be run on any subset of sequences in the alignment and the resulting prediction displayed along with the alignment. The JalView application is available for free download and, because it is written in Java, can also be run as an applet in a Web browser such as Netscape or Internet Explorer. Many alignment services such as the CLUSTAL W server at EBI and the Pfam server include JalView as an option to view the resulting multiple alignments. Figure 9.6 illustrates a typical JalView session with the alignment editing and tree windows open.

COLLECTIONS OF MULTIPLE ALIGNMENTS

This chapter has focused on methods and servers for building multiple protein sequence alignments. Although proteins that are clearly similar by the *Z*-score measure

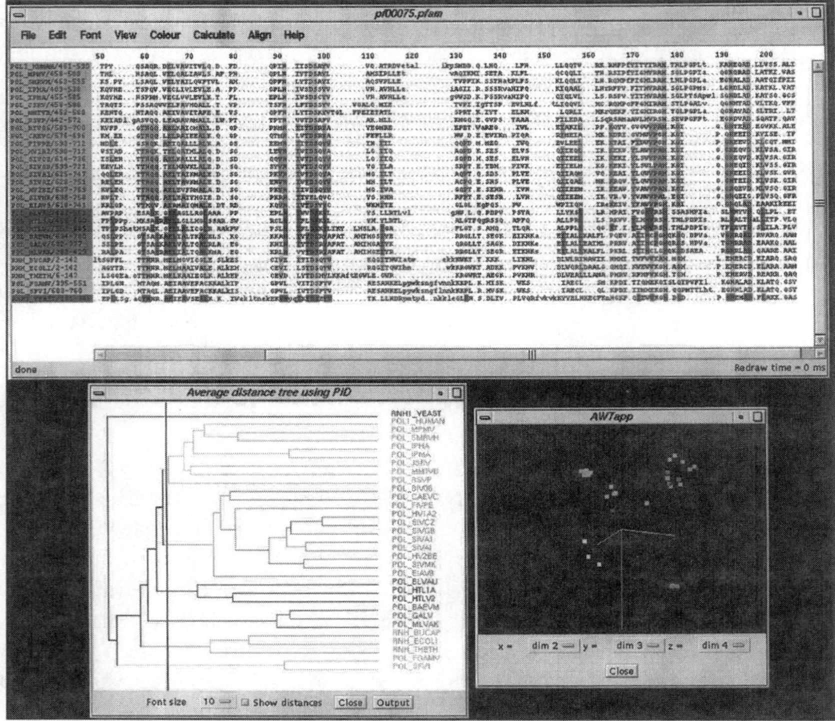

Figure 9.6. An example JalView alignment editing and analysis session. The top panel contains a multiple alignment, and the bottom left is the similarity tree resulting from that alignment. A vertical line on the tree has separated the sequences into subgroups, which have been colored to highlight conservation within each subgroup. The panel at the bottom right illustrates an alternative clustering method.

should be straightforward to align by the automatic methods discussed here, getting good alignments for proteins with more remote similarities can be a very time-consuming process. A number of groups have built collections of alignments using a combination of automation and expert curation [e.g., SMART (Schultz et al., 1998), Pfam (Bateman et al., 1999), and PRINTS (Attwood et al., 2000)], and these, together with the tools available at their Web sites, can provide an excellent starting point for further analyses.

INTERNET RESOURCES FOR TOPICS PRESENTED IN CHAPTER 9

CLUSTAL W	*ftp://ftp.ebi.ac.uk/pub/software*
AMAS	*http://barton.ebi.ac.uk/servers/amas.html*
JPred	*http://barton.ebi.ac.uk/servers/jpred.html*
ProtEST	*http://barton.ebi.ac.uk/servers/protest.html*
JalView	*http://barton.ebi.ac.uk/new/software.html*
AMPS	*http://barton.ebi.ac.uk/new/software.html*
European Bioinformatics Institute	*http://www.ebi.ac.uk*

PROBLEM SET

The following problems are based on the annexin supergene family, the same family used throughout the discussion in this chapter. This family contains a 100 amino acid residue unit that repeats either four, eight, or 16 times within each protein. The analysis required below will focus on the individual repeat units, rather than the organization of the repeat units within the full-length protein sequences.

The problems will require the use of CLUSTAL W and Jalview, which you may have to install (or have installed) on a UNIX- or Linux-based system to which you have access. The files referred to below are available on the book's Web site.

The file `ann_rep1.fa` contains the sequence of a single annexin domain. This sequence has been used as the query against the SWALL protein sequence database, using the program `scanps` to make the pairwise sequence comparisons. A partial listing of the results can be found in the file named `ann_rep1_frags.fa`.

Generation of a Multiple Sequence Alignment

1. Copy the file `ann_rep1_frags.fa` to a new directory.
2. Run CLUSTAL W on `ann_rep1_frags.fa`. Accept all defaults, and create an output file called `ann_rep1_frags.aln`.
3. Pass this output file to Jalview by typing `Jalview ann_rep1_frags.aln CLUSTAL`.
4. Select the fragment sequences by clicking on the ID code. Select Delete Selected Sequences from the Edit menu.
5. Save the modified alignment to a CLUSTAL-formatted file called `ann_rep1_frags_del1.aln`.
6. Select Average Distance Tree from the Calculate menu. A new window will now appear, and after a few moments, a tree (dendrogram) will be rendered within that window. There should be outliers at the very top of that tree, and these outliers will need to be eliminated.
7. Click on the tree to the left of where the outliers join the tree. A vertical line should now appear, and the outliers will be highlighted in a different color.
8. Return to the Alignment window and delete the outliers from the alignment, in the same way as was done in Step 4. Save the resulting alignment to a file named `ann_rep1_frags_del2.aln`.

This series of steps produces a "clean alignment" for inspection. Positions within the alignment can be colored in different ways to highlight certain features of the amino acids within the alignment. For example, selecting Conservation from the Calculate menu will shade each column on the basis of the relative amino acid conservation seen at that particular position in the alignment. By doing so, it immediately becomes apparent which parts of the protein may lie within regions of secondary structure. Examine the area around positions 60 to 70 of the alignment;

the pattern observed should be two conserved, two unconserved, and two conserved residues, a parttern that is characteristic of an alpha-helix.

Select Jnet from the Align menu. This will return a secondary structure prediction based on the alignment. Alternatively, the alignment file can be submitted to the JPRED2 server at EBI. In order to submit the alignment to the JPRED2 server, the alignment must first be saved in MSF format (`ann_rep1_frags_del2.msf`). Either of these methods should corroborate that there is an alpha-helical region in the area around residues 60–70.

By "cleaning" the alignment in this way, information about sequences (and sequences themselves) has been discarded. It is advisable to always save files at intermediate steps: the clean alignment will be relatively easy to interpret, but the results of the intermediate steps will have information about the parts of the alignment requiring more thought.

Subfamily Analysis

The following steps will allow a subfamily analysis to be performed on the annexin family. The input file is `ideal_annexins.als`.

1. Start Jalview and read in the alignment file by typing `ideal_annexins.blc BLC`.
2. Select Average Distance Tree from the Calculate menu. The resultant tree will have four clear clusters with one outlier. Click on the tree at an appropriate position to draw a vertical line and highlight the four clusters.
3. Return to the Alignment window. Select Conservation from the Calculate menu. The most highly-conserved positions within each subgroup of sequences will be colored the brightest. Examine the alignment, and identify the charge-pair shown as an example in this Chapter. Selecting either the Taylor or Zappo color schemes may help in identifying the desired region.
4. Submit the file `ideal_annexins.blc` to the AMAS Web server. On the Web page, paste the contents of `ideal_annexins.blc` into the Alignment window, then paste the contents of the file `ideal_annexins.grp` into the Sensible Groups window. The server should return results quickly, providing links to a number of output files. The Pretty Output file contains the PostScript alignment, which should be identical to `ideal_annexins_amas.ps` provided here.

REFERENCES

Altschul, S. F., Madden, T. L., Schaffer, A. A., Zhang, J., Zhang, Z., Miller, W., and Lipman, D. J. (1997). Gapped blast and psi-blast: a new generation of protein database search programs. *Nucl. Acids Res.* 25, 3389–3402.

Attwood, T. K., Croning, M. D. R., Flower, D. R., Lewis, A. P., Mabey, J. E., Scordis, P., Selley, J., and Wright. W. (2000). Prints-s: the database formerly known as prints. *Nucl. Acids Res.* 28, 225–227.

Barton, G. J. (1990). Protein multiple sequence alignment and flexible pattern matching. *Methods Enz.* 183, 403–428.

Barton, G. J. (1993). ALSCRIPT: A tool to format multiple sequence alignments. *Prot. Eng.* 6, 37–40.

Barton, G. J., Newman, R. H., Freemont, P. F., and Crumpton, M. J. (1991). Amino acid sequence analysis of the annexin super-gene family of proteins. *European J. Biochem.* 198, 749–760.

Barton, G. J., and Sternberg, M. J. E. Evaluation and improvements in the automatic alignment of protein sequences. (1987). *Prot. Eng.* 1, 89–94.

Barton, G. J., and Sternberg, M. J. E. (1987). A strategy for the rapid multiple alignment of protein sequences: Confidence levels from tertiary structure comparisons. *J. Mol. Biol.* 198, 327–337.

Barton, G. J., and Sternberg, M. J. E. (1990). Flexible protein sequence patternsa sensitive method to detect weak structural similarities. *J. Mol. Biol.* 212, 389–402.

Bateman, A., Birney, E., Durbin, R., Eddy, S. R., Finn, R. D., and Sonnhammer, E. L. L. Pfam 3.1: 1313 multiple alignments match the majority of proteins. *Nucl. Acids Res.* 27, 260–262.

Cuff, J. A., and Barton, G. J. (1999). Evaluation and improvement of multiple sequence methods for protein secondary structure prediction. *Proteins* 34, 508–519.

Cuff, J. A., and Barton, G. J. (2000). Application of multiple sequence alignment profiles to improve protein secondary structure prediction. *Proteins* 40, 502–511.

Cuff, J. A., Birney, E., Clamp, M. E., and Barton, G. J. (2000). ProtEST: Protein multiple sequence alignments from expressed sequence tags. *Bioinformatics* 6: 111–116.

Felsenstein, J. (1989). Phylip—phylogeny inference package (version 3.2). *Cladistics* 5, 164–166.

Murzin, A. G., Brenner, S. E., Hubbard, T., and Chothia, C. (1995). Scop: a structural classification of proteins database for the investigation of sequences and structures. *J. Mol. Biol.* 247, 536–540.

Gotoh, O. (1996). Significant improvement in accuracy of multiple protein sequence alignments by iterative refinement as assessed by reference to structural alignments. *J. Mol. Biol.* 264, 823–838.

Gribskov, M., McLachlan, A. D., and Eisenberg, D. (1987). Profile analysis: detection of distantly related proteins. *Proc. Nat. Acad. Sci.* USA 84, 4355–4358.

Huber, R., Romsich, J., and Paques, E.-P. (1990). The crystal and molecular structure of human annexin v, an anticoagulant protein that binds to calcium and membranes. *EMBO J.* 9, 3867–3874.

Jones, D. T. (1999). Protein secondary structure prediction based on position-specific scoring matrices. *J. Mol. Biol.* 17, 195–202.

Lesk, A. M., Levitt, M., and Chothia, C. (1986). Alignment of the amino acid sequences of distantly related proteins using variable gap penalties. *Prot. Eng.* 1, 77–78.

Lipman, D. J., Altschul, S. F., and Kececioglu, J. D. (1989). A tool for multiple sequence alignment. *Proc. Nat. Acad. Sci.* USA 86, 4412–4415.

Livingstone, C. D., and Barton, G. J. (1993). Protein sequence alignments: A strategy for the hierarchical analysis of residue conservation. *Comp. App. Biosci.* 9, 745–756.

Pascarella, S., and Argos, P. (1992). Analysis of insertions/deletions in protein structures. *J. Mol. Biol.* 224, 461–471.

Roach, P. L., Clifton, I. J., Fulop, V., Harlos, K., Barton, G. J., Hajdu, J., Andersson, I., Schofield, C. J., and Baldwin, J. E. (1995). Crystal structure of isopenicillin n synthase is the first from a new structural family of enzymes. *Nature* 375, 700–704.

Rost, B., and Sander, C. (1993). Prediction of protein secondary structure at better than 70% accuracy. *J. Mol. Biol.* 232, 584–599.

Russell, R. B., and Barton, G. J. (1992). Multiple protein sequence alignment from tertiary structure comparison: assignment of global and residue confidence levels. *Proteins* 14, 309–323.

Russell, R. B., and Barton, G. J. (1994). Structural features can be unconserved in proteins with similar folds. *J. Mol. Biol.* 244, 332–350.

Schultz, J., Milpetz, F., Bork, P., and Ponting, C. P. (1998). Smart, a simple modular architecture research tool: Identification of signalling domains. *Proc. Nat. Acad. Sci.* USA 95, 5857–5864.

Thompson, J. D., Higgins, D. G., and Gibson, T. J. (1994). Clustal W: improving the sensitivity of progressive multiple sequence alignment through sequence weighting, position-specific gap penalties and weight matrix choice. *Nucl. Acids Res.* 22, 4673–4680.

Zvelebil, M. J. J. M., Barton, G. J., Taylor, W. R., and Sternberg, M. J. E. (1987). Prediction of protein secondary structure and active sites using the alignment of homologous sequences. *J. Mol. Biol.* 195, 957–961.

10

PREDICTIVE METHODS USING DNA SEQUENCES

Andreas D. Baxevanis

Genome Technology Branch
National Human Genome Research Institute
National Institutes of Health
Bethesda, Maryland

With the announcement of the completion of a "working draft" of the sequence of the human genome in June 2000 and the Human Genome Project targeting the completion of sequencing in 2002, investigators will be faced with the challenge of developing a strategy by which they can deal with the oncoming flood of both unfinished and finished data, whether the data are generated in their own laboratories or at one of the major sequencing centers. These data undergo what can best be described as a maturation process, starting as single reads off of a sequencing machine, passing through a phase where the data become part of an assembled (yet incomplete) sequence contig, and finally ending up as part of a finished, completely assembled sequence with an error rate of less than one in 10,000 bases. Even before such sequencing data reach this highly polished state, investigators can begin to ask whether or not given stretches of sequence represent coding or noncoding regions. The ability to make such determinations is of great relevance in the context of systematic sequencing efforts, since all of the data being generated by these projects are, in essence, "anonymous" in nature—nothing is known about the coding potential of these stretches of DNA as they are being sequenced. As such, automated methods will become increasingly important in annotating the human and other genomes to increase the intrinsic value of these data as they are being deposited into the public databases.

In considering the problem of gene identification, it is important to briefly go over the basic biology underlying what will become, in essence, a mathematical

Bioinformatics: A Practical Guide to the Analysis of Genes and Proteins
Edited by A. D. Baxevanis and B. F. F. Ouellette
ISBN 0-471-38390-2 (cloth), ISBN 0-471-383910 (paper)

problem (Fig. 10.1). At the DNA level, upstream of a given gene, there are promoters and other regulatory elements that control the transcription of that gene. The gene itself is discontinuous, comprising both introns and exons. Once this stretch of DNA is transcribed into an RNA molecule, both ends of the RNA are modified, capping the 5′ end and placing a polyA signal at the 3′ end. The RNA molecule reaches maturity when the introns are spliced out, based on short consensus sequences found both at the intron-exon boundaries and within the introns themselves. Once splicing has occurred and the start and stop codons have been established, the mature mRNA is transported through a nuclear pore into the cytoplasm, at which point translation can take place.

Although the process of moving from DNA to protein is obviously more complex in eukaryotes than it is in prokaryotes, the mere fact that it can be described in its entirety in eukaryotes would lead one to believe that predictions can confidently be made as to the exact positions of introns and exons. Unfortunately, the signals that control the process of moving from the DNA level to the protein level are not very well defined, precluding their use as foolproof indicators of gene structure. For example, upward of 70% of the promoter regions contain a TATA box, but, because the remainder do not, the presence (or absence) of the TATA box in and of itself cannot be used to assess whether a region is a promoter. Similarly, during end modification, the polyA tail may be present or absent or may not contain the canonical

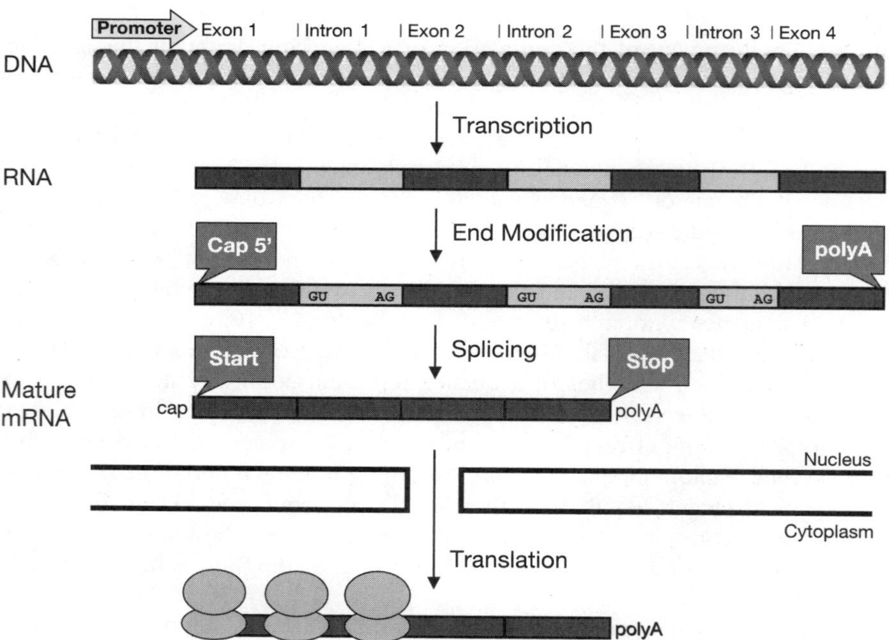

Figure 10.1. The central dogma. Proceeding from the DNA through the RNA to the protein level, various sequence features and modifications can be identified that can be used in the computational deduction of gene structure. These include the presence of promoter and regulatory regions, intron-exon boundaries, and both start and stop signals. Unfortunately, these signals are not always present, and, when they are present, they may not always be in the same form or context. The reader is referred to the text for greater detail.

AATAAA. Adding to these complications is the fact that an open reading frame is required *but is not sufficient* for judging a region as being an exon. Given these and other considerations, there is at present no straightforward method that will allow for 100% confidence in the prediction of an intron or an exon. Despite this, a combinatorial approach can be used, relying on a number of methods, to increase the confidence with which gene structure is predicted.

Briefly, gene-finding strategies can be grouped into three major categories. ***Content-based methods*** rely on the overall, bulk properties of a sequence in making a determination. Characteristics considered here include how often particular codons are used, the periodicity of repeats, and the compositional complexity of the sequence. Because different organisms use synonymous codons with different frequency, such clues can provide insight into determining regions that are more likely to be exons. In ***site-based methods***, the focus turns to the presence or absence of a specific sequence, pattern, or consensus. These methods are used to detect features such as donor and acceptor splice sites, binding sites for transcription factors, polyA tracts, and start and stop codons. Finally, ***comparative methods*** make determinations based on sequence homology. Here, translated sequences are subjected to database searches against protein sequences (cf. Chapter 8) to determine whether a previously characterized coding region corresponds to a region in the query sequence. Although this is conceptually the most straightforward of the methods, it is restrictive because most newly discovered genes do not have gene products that match anything in the protein databases. Also, the modular nature of proteins and the fact that there are only a limited number of protein motifs (Chothia and Lesk, 1986) make predicting anything more than just exonic regions in this way difficult. The reader is referred to a number of excellent reviews detailing the theoretical underpinnings of these various classes of methods (Claverie, 1997a; Claverie, 1997b; Guigó, 1997; Snyder and Stormo, 1997; Claverie, 1998; Rogic et al., 2001). Although many of the gene prediction methods belong strictly to one of these three classes of methods, most of the methods that will be discussed here use the strength of combining different classes of methods to optimize predictions.

With the complexity of the problem at hand and the various approaches described above for tackling the problem, it becomes important for investigators to gain an appreciation for when and how each particular method should be applied. A recurring theme in this chapter will be the fact that, *depending on the nature of the data, each method will perform differently*. Put another way, although one method may be best for human finished sequences, another may be better for unfinished sequences or for sequences from another organism. In this chapter, we will examine a number of the commonly used methods that are freely available in the public domain, focusing on their application to human sequence data; this will be followed by a general discussion of gene-finding strategy.

GRAIL

GRAIL, which stands for Gene Recognition and Analysis Internet Link (Uberbacher and Mural, 1991; Mural et al., 1992), is the elder statesman of the gene prediction techniques because it is among the first of the techniques developed in this area and enjoys widespread usage. As more and more has become known about gene structure in general and better Internet tools have become more widespread, GRAIL has continuously evolved to keep in step with the current state of the field.

There are two basic GRAIL versions that will be discussed in the context of this discussion. GRAIL 1 makes use of a neural network method to recognize coding potential in fixed-length (100 base) windows considering the sequence itself, without looking for additional features such as splice junctions or start and stop codons. An improved version of GRAIL 1 (called GRAIL 1a) expands on this method by considering regions immediately adjacent to regions deemed to have coding potential, resulting in better performance in both finding true exons and eliminating false positives. Either GRAIL 1 or GRAIL 1a would be appropriate in the context of searching for single exons. A further refinement led to a second version, called GRAIL 2, in which variable-length windows are used and contextual information (e.g., splice junctions, start and stop codons, polyA signals) is considered. Because GRAIL 2 makes its prediction by taking genomic context into account, it is appropriate for determining model gene structures.

In this chapter, the output of each of the methods discussed will be shown using the same set of input data as the query. The sequence that will be considered is that of a human BAC clone RG364P16 from 7q31, a clone established as part of the systematic sequencing of chromosome 7 (GenBank AC002467). By using the same example throughout, the strengths and weaknesses of each of the discussed methods can be highlighted. For purposes of this example, a client-server application called XGRAIL will be used. This software, which runs on the UNIX platform, allows for graphical output of GRAIL 1/1a/2 results, as shown in Figure 10.2. Because the DNA sequence in question is rather large and is apt to contain at least one gene, GRAIL 2 was selected as the method. The large, upper window presents an overview of the ~98 kb making up this clone, and the user can selectively turn on or off particular markings that identify features within the sequence (described in the figure legend). Of most importance in this view is the prediction of exons at the very top of the window, with the histogram representing the probability that a given region represents an exon. Information on each one of the predicted exons is shown in the Model Exons window, and the model exons can be assembled and shown as both Model Genes and as a Protein Translation. Only putative exons with acceptable probability values (as defined in the GRAIL algorithm) are included in the gene models. The protein translation can, in turn, be searched against the public databases to find sequence homologs using a program called genQuest (integrated into XGRAIL), and these are shown in the Db Hits window. In this case, the 15 exons in the first gene model (from the forward strand) are translated into a protein that shows significant sequence homology to a group of proteins putatively involved in anion transport (Everett et al., 1997).

Most recently, the authors of GRAIL have released GRAIL-EXP, which is based on GRAIL but uses additional information in making the predictions, including a database search of known complete and partial gene messages. The inclusion of this database search in deducing gene models has greatly improved the performance of the original GRAIL algorithm.

FGENEH/FGENES

FGENEH, developed by Victor Solovyev and colleagues, is a method that predicts internal exons by looking for structural features such as donor and acceptor splice

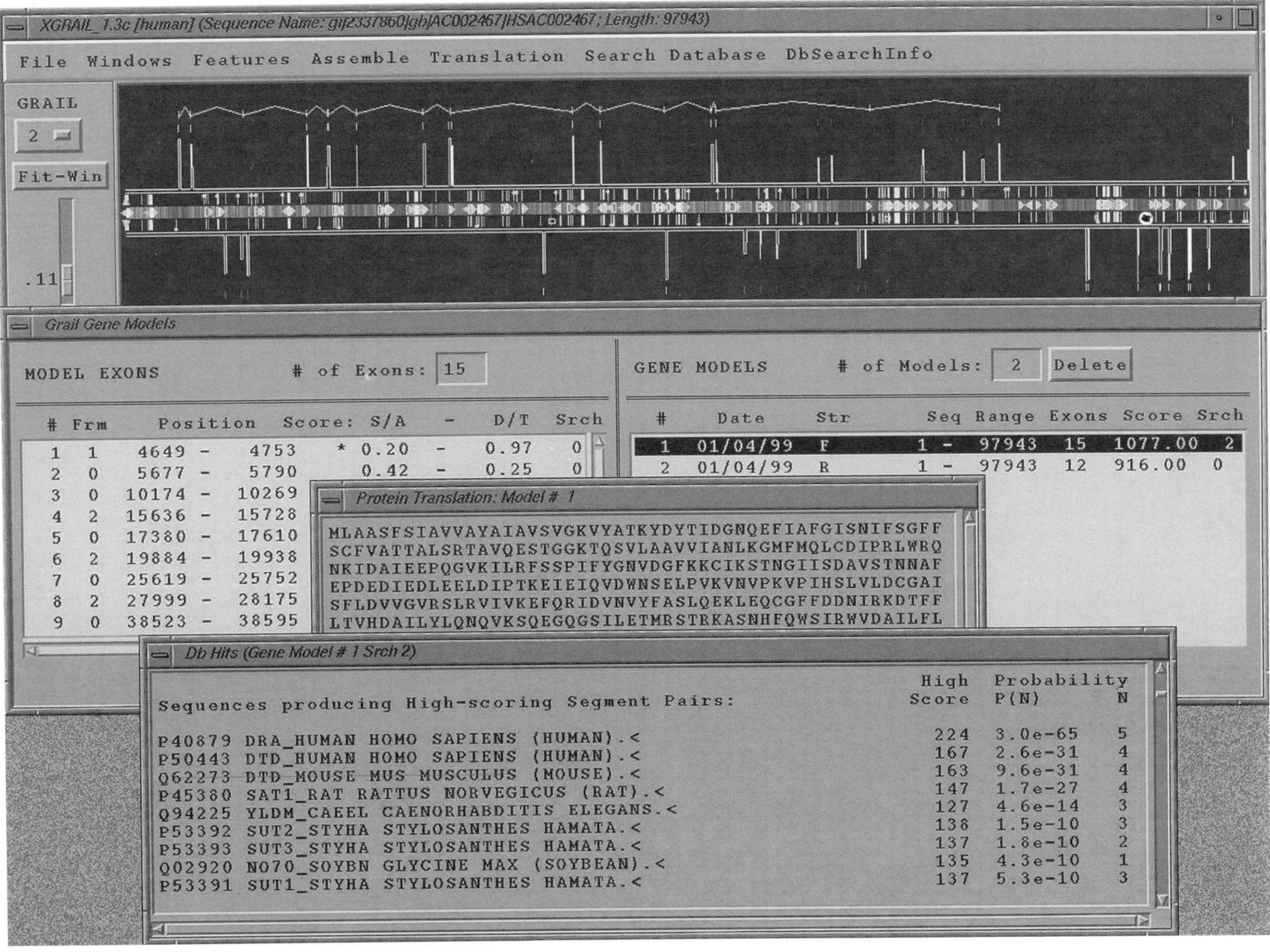

Figure 10.2. XGRAIL output using the human BAC clone RG364P16 from 7q31 as the query. The upper window shows the results of the prediction, with the histogram representing the probability that a given stretch of DNA is an exon. The various bars in the center represent features of the DNA (e.g., arrows represent repetitive DNA, and vertical bars represent repeat sequences). Exon and gene models, protein translations, and the results of a genQuest search using the protein translation are shown. (See color plate.)

sites, putative coding regions, and intronic regions both 5′ and 3′ to the putative exon (Solovyev et al., 1994a; Solovyev et al., 1994b; Solovyev et al., 1995). The method makes use of *linear discriminant analysis*, a mathematical technique that allows data from multiple experiments to be combined. Once the data are combined, a linear function is used to discriminate between two classes of events—here, whether a given stretch of DNA is or is not an exon. In FGENEH, results of the linear discriminant approach are then passed to a dynamic programming algorithm that determines how to best combine these predicted exons into a coherent gene model. An extension of FGENEH, called FGENES, can be used in cases when multiple genes are expected in a given stretch of DNA.

The Sanger Centre Web server provides a very simple front-end for performing FGENES. The query sequence (again, the BAC clone from 7q31) is pasted into the query box, an identifier is entered, and the search can then be performed. The results are returned in a tabular format, as shown in Figure 10.3. The total number of predicted genes and exons (2 and 33, respectively) is shown at the top of the output. The information for each gene (`G`) then follows. For each predicted exon, the strand (`Str`) is given, with + indicating the forward strand and − indicating the reverse. The Feature list in this particular case includes initial exons (`CDSf`), internal exons (`CDSi`), terminal exons (`CDSl`), and polyA regions (`PolA`). The nucleotide region for the predicted feature is then given as a range. In the current example, the features of the second predicted gene are shown in reverse order, since the prediction is based on the reverse strand. On the basis of the information in the table, predicted proteins are given at the bottom of the output in FASTA format. The definition line for each of the predicted proteins gives the range of nucleotide residues involved, as well as the total length of the protein and the direction (+/−) of the predicted gene.

MZEF

MZEF stands for "Michael Zhang's Exon Finder," after its author at the Cold Spring Harbor Laboratory. The predictions rely on a technique called *quadratic discriminant analysis* (Zhang, 1997). Imagine a case in which the results of two types of predictions are plotted against each other on a simple *XY* graph (for instance, splice site scores vs. exon length). If the relationship between these two sets of data is nonlinear or multivariate, the resulting graph will look like a swarm of points. Points lying in only a small part of this swarm will represent a "correct" prediction; to separate the correctly predicted points from the incorrectly predicted points in the swarm, a quadratic function is used, hence the name of the technique. In the case of MZEF, the measured variables include exon length, intron-exon and exon-intron transitions, branch sites, 3′ and 5′ splice sites, and exon, strand, and frame scores. MZEF is intended to predict internal coding exons and does not give any other information with respect to gene structure.

There are two implementations of MZEF currently available. The program can be downloaded from the CSHL FTP site for UNIX command-line use, or the program can be accessed through a Web front-end. The input is a single sequence, read in only one direction (either the forward *or* the reverse strand); to perform MZEF on both strands, the program must be run twice. Returning to the BAC clone from chromosome 7, MZEF predicts a total of 27 exons in the forward strand (Fig. 10.4). Focusing in on the first two columns of the table, the region of the prediction is

```
Number of predicted genes:   2 In +chain:   1 In -chain:   1
Number of predicted exons:  33 In +chain:  23 In -chain:  10
Positions of predicted genes and exons:
 G Str Feature  Start        End    Weight  ORF-start ORF-end

 1 +    1 CDSf   3413 -    3594    2.50    3413 -    3592
 1 +    2 CDSi   4606 -    4753    1.73    4607 -    4753
 1 +    3 CDSi   5677 -    5790    1.91    5677 -    5790
 1 +    4 CDSi   9956 -   10033    2.55    9956 -   10033
 1 +    5 CDSi  10174 -   10269    1.86   10174 -   10269
 1 +    6 CDSi  11486 -   11592    1.81   11486 -   11590
 1 +    7 CDSi  13595 -   13664    3.39   13596 -   13664
 1 +    8 CDSi  15636 -   15728    2.38   15636 -   15728
 1 +    9 CDSi  17380 -   17610    1.97   17380 -   17610
 1 +   10 CDSi  19884 -   19938    2.72   19884 -   19937
 1 +   11 CDSi  25607 -   25752    3.18   25609 -   25752
 1 +   12 CDSi  28092 -   28175    3.04   28092 -   28175
 1 +   13 CDSi  40915 -   40981    1.00   40915 -   40980
 1 +   14 CDSi  41081 -   41262    1.42   41083 -   41262
 1 +   15 CDSi  51053 -   51131    1.31   51053 -   51130
 1 +   16 CDSi  55392 -   55442    0.95   55394 -   55441
 1 +   17 CDSi  60609 -   60692    1.52   60611 -   60691
 1 +   18 CDSi  64433 -   64600    3.71   64435 -   64599
 1 +   19 CDSi  68964 -   69064    3.15   68966 -   69064
 1 +   20 CDSi  69448 -   69531    3.48   69448 -   69531
 1 +   21 CDSi  70971 -   71044    3.04   70971 -   71042
 1 +   22 CDSi  73696 -   74083    2.25   73697 -   74083
 1 +   23 CDSl  74150 -   74731    2.94   74150 -   74728
 1 +      PolA  75218              4.18

 2 -      PolA  82006              4.57
 2 -    1 CDSl  82727 -   82738    1.32   82730 -   82738
 2 -    2 CDSi  83132 -   83197    2.58   83132 -   83197
 2 -    3 CDSi  83319 -   83461    2.79   83319 -   83459
 2 -    4 CDSi  87607 -   87661    3.62   87608 -   87661
 2 -    5 CDSi  89473 -   89706    2.93   89473 -   89706
 2 -    6 CDSi  90330 -   90425    1.75   90330 -   90425
 2 -    7 CDSi  92005 -   92097    1.79   92005 -   92097
 2 -    8 CDSi  92190 -   92259    1.39   92190 -   92258
 2 -    9 CDSi  93728 -   93834    2.05   93730 -   93834
 2 -   10 CDSi  95221 -   95316    2.27   95221 -   95316

Predicted proteins:
>FGENES 1.5 AC002467      1 Multiexon gene    3413 -    74731
1087 a Ch+
MLSRPTVGSGFPTSCLSTDGVHSTVSLWGRMGYKEKRSLKINLTGRESKATRAENQTDLV
RFLPPELPPVSLFSEMLAASFSIAVVAYAIAVSVGKVYATKYDYTIDGNQEFIAFGISNI
FSGFFSCFVATTALSRTAVQESTGGKTQVAGIISAAIVMIAILALGKLLEPLQKSVLAAV
<remainder of output truncated>
```

Figure 10.3. FGENES output using the human BAC clone RG364P16 from 7q31 as the query. The columns, going from left to right, represent the gene number (G), strand (Str), feature (described in the main text), start and end points for the predicted exon, a scoring weight, and start and end points for corresponding open reading frames (ORF-start and ORF-end). Each predicted gene is shown as a separate block. The tables are followed by protein translations of any predicted gene products.

given as a range, followed by the probability that the prediction is correct (P). Predictions with $P > 0.5$ are considered correct and are included in the table. Immediately, one begins to see the difference in the predictions between methods. MZEF is again geared toward finding single exons; therefore, the exons are not shown in the context of a putative gene, as they are in GRAIL 2 or FGENES. However, the exons predicted by these methods are not the same, a point that we will return to later in this discussion.

```
Internal coding exons predicted by MZEF
Sequence_length: 97943 G+C_content: 0.391

 Coordinates     P     Fr1   Fr2   Fr3   Orf 3ss   Cds   5ss
 4606-4753    0.548 0.475 0.614 0.444 212 0.531 0.547 0.538
 5469-5543    0.557 0.588 0.461 0.600 121 0.499 0.594 0.622
 7353-7630    0.826 0.584 0.520 0.549 122 0.498 0.585 0.632
 10174-10269 0.546 0.605 0.443 0.442 122 0.517 0.552 0.515
 13595-13664 0.998 0.552 0.463 0.608 121 0.564 0.570 0.736
 15636-15728 0.534 0.444 0.432 0.544 221 0.488 0.500 0.636
 16654-16749 0.904 0.541 0.398 0.458 122 0.534 0.531 0.615
 17380-17610 0.940 0.614 0.470 0.442 122 0.518 0.569 0.594
 18736-18797 0.597 0.417 0.550 0.603 221 0.536 0.618 0.619
 19884-19938 0.866 0.434 0.406 0.537 221 0.550 0.504 0.657
 24126-24225 0.969 0.655 0.543 0.539 122 0.532 0.622 0.559
 25607-25752 0.977 0.551 0.452 0.466 122 0.530 0.542 0.647
 28107-28175 0.966 0.438 0.412 0.662 221 0.492 0.579 0.562
 37600-37687 0.605 0.328 0.610 0.434 212 0.515 0.549 0.586
 38297-38434 0.946 0.558 0.511 0.441 122 0.528 0.540 0.559
 50415-50823 0.632 0.557 0.451 0.470 122 0.543 0.533 0.519
 55133-55173 0.873 0.375 0.489 0.530 221 0.531 0.524 0.702
 57112-57175 0.518 0.562 0.424 0.469 122 0.514 0.530 0.618
 61089-61182 0.602 0.438 0.552 0.456 212 0.556 0.549 0.700
 64433-64600 0.980 0.614 0.552 0.505 122 0.517 0.599 0.606
 68964-69064 0.941 0.316 0.579 0.564 211 0.513 0.534 0.558
 69448-69531 0.997 0.565 0.444 0.364 122 0.536 0.523 0.705
 70971-71044 0.948 0.448 0.300 0.507 121 0.575 0.462 0.656
 73696-74083 0.968 0.487 0.594 0.498 212 0.552 0.574 0.536
 77911-77972 0.596 0.467 0.593 0.434 212 0.480 0.549 0.602
 80338-80413 0.944 0.467 0.464 0.590 221 0.507 0.555 0.662
 97197-97358 0.738 0.597 0.497 0.523 122 0.521 0.586 0.545
```

Figure 10.4. MZEF output using the human BAC clone RG364P16 from 7q31 as the query. The columns, going from left to right, give the location of the prediction as a range of included bases (`Coordinates`), the probability value (`P`), frame preference scores, an ORF indicator showing which reading frames are open, and scores for the 3′ splice site, coding regions, and 5′ splice site.

GENSCAN

GENSCAN, developed by Chris Burge and Sam Karlin (Burge and Karlin, 1997; Burge and Karlin, 1998), is designed to predict complete gene structures. As such, GENSCAN can identify introns, exons, promoter sites, and polyA signals, as do a number of the other gene identification algorithms. Like FGENES, GENSCAN does not expect the input sequence to represent one and only one gene or one and only one exon: it can accurately make predictions for sequences representing either partial genes or multiple genes separated by intergenic DNA. The ability to make these predictions accurately when a sequence is in a variety of contexts makes GENSCAN a particularly useful method for gene identification.

GENSCAN relies on what the author terms a "probabilistic model" of genomic sequence composition and gene structure. By looking for gene structure descriptions that match or are consistent with the query sequence, the algorithm can assign a probability as to the chance that a given stretch of sequence represents an exon, promoter, and so forth. The "optimal exons" are the ones with the highest probability and represent the part of the query sequence having the best chance of actually being an exon. The method will also predict "suboptimal exons," stretches of sequence having an acceptable probability value but one not as good as the optimal one. The authors of the method encourage users to examine both sets of predictions so that alternatively spliced regions of genes or other nonstandard gene structures are not missed.

With the use of the human BAC clone from 7q31 again, the query can be issued directly from the GENSCAN Web site, using Vertebrate as the organism, the default suboptimal cutoff, and Predicted Peptides Only as the print option. The results for this query are shown in Figure 10.5. The output indicates that there are three genes in this region, with the first gene having 11 exons, the second gene having 13 exons, and the third gene having 10 exons. The most important columns in the table are those labeled `Type` and `P`. The `Type` column indicates whether the prediction is for an initial exon (`Init`), an internal exon (`Intr`), a terminal exon (`Term`), a single-exon gene (`Sngl`), a promoter region (`Prom`), or a polyA signal (`PlyA`). The `P` column gives the probability that this prediction is actually correct. GENSCAN exons having a very high probability value ($P > 0.99$) are 97.7% accurate where the prediction matches a true, annotated exon. These high-probability predictions can be used in the rational design of PCR primers for cDNA amplification or for other purposes where extremely high confidence is necessary. GENSCAN exons that have probabilities in the range from 0.50 to 0.99 are deemed to be correct most of the time; the best-case accuracies for *P*-values over 0.90 is on the order of 88%. Any predictions below 0.50 should be discarded as unreliable, and those data are not given in the table. An alternative view of the data is shown in Figure 10.6. Here, both the optimal and suboptimal exons are shown, with the initial and terminal exons showing the direction in which the prediction is being made ($5' \rightarrow 3'$ or $3' \rightarrow 5'$). This view is particularly useful for large stretches of DNA, as the tables become harder to interpret when more and more exons are predicted.

By the time of this printing, a new program named GenomeScan will be available from the Burge laboratory at MIT. GenomeScan assigns a higher score to putative exons that overlap BLASTX hits than to comparable exons for which similarity evidence is lacking. Regions of higher similarity (according to BLASTX E-value, for example) are accorded more confidence than regions of lower similarity, since weak similarities sometimes do not represent homology. Thus, the predictions of GenomeScan tend to be consistent with all or almost all of the regions of high detected similarity but may sometimes ignore a region of weak similarity that either has weak intrinsic properties (e.g., poor splice signals) or is inconsistent with other extrinsic information. The accuracy of GenomeScan tends to be significantly higher than that of GENSCAN when a moderate or closely related protein sequence is available. An example of the improved accuracy of GenomeScan over GENSCAN, using the human BRCA1 gene as the query, is shown in Figure 10.7.

PROCRUSTES

Greek mythology heralds the story of Theseus, the king of Athens who underwent many trials and tribulations on his way to becoming a hero, along with Hercules. As if Amazons and the Minotaur were not enough, in the course of his travels, Theseus happened upon Procrustes, a bandit with a warped idea of hospitality. Procrustes, which means "he who stretches," would invite passersby into his home for a meal and a night's stay in his guest bed. The problem lay, quite literally, in the bed, in that Procrustes would make sure that his guests fit in the bed by stretching them out on a rack if they were too short or by chopping off their legs if they were too long.

```
Gn.Ex Type S .Begin ...End .Len Fr Ph I/Ac Do/T CodRg P.... Tscr..
----- ---- - ------ ------ ---- -- -- ---- ---- ----- ----- ------

 1.01 Init +   4697   4801  105  1  0   64   80   103 0.651   7.58
 1.02 Intr +   5725   5838  114  0  0   48   91   116 0.993   7.62
 1.03 Intr +  10004  10081   78  1  0   61   70    78 0.809   2.13
 1.04 Intr +  10222  10317   96  0  0   94   87   117 0.999  11.49
 1.05 Intr +  11534  11640  107  1  2  118   62    31 0.953   1.59
 1.06 Intr +  13643  13712   70  2  1   88  111    32 0.950   3.77
 1.07 Intr +  15684  15776   93  2  0   45   98    59 0.782   1.84
 1.08 Intr +  16702  16797   96  0  0   70  100    26 0.709   1.29
 1.09 Intr +  17428  17658  231  0  0   69   79   233 0.911  17.55
 1.10 Intr +  19932  19986   55  2  1   90   94    29 0.805   1.33
 1.11 Term +  25128  25375  248  1  2   48   48   167 0.867   3.67
 1.12 PlyA +  25382  25387    6                                1.05

 2.00 Prom +  26739  26778   40                               -7.05
 2.01 Init +  27929  28093  165  1  0   77   94    65 0.948   5.68
 2.02 Intr +  28140  28223   84  2  0   69   64   142 0.901   9.00
 2.03 Intr +  29931  30071  141  2  0  126   38    55 0.262   3.93
 2.04 Intr +  52002  52164  163  2  1   99   17   149 0.194   7.53
 2.05 Intr +  53036  53243  208  0  1   48   -2   191 0.028   3.31
 2.06 Intr +  58789  58968  180  1  0   82   35   127 0.411   4.86
 2.07 Intr +  59932  60222  291  1  0   69   20   255 0.369  12.13
 2.08 Intr +  63258  63277   20  0  2  102   86   -16 0.527  -5.06
 2.09 Intr +  64481  64648  168  0  0   47   86   162 0.939  10.90
 2.10 Intr +  69012  69112  101  1  2   56   75   115 0.967   5.91
 2.11 Intr +  69496  69579   84  0  0   25  115    57 0.615   1.20
 2.12 Intr +  71019  71092   74  2  2  105   90   -21 0.950  -2.91
 2.13 Term +  73744  74779 1036  1  1   85   44   805 0.960  66.40
 2.14 PlyA +  75266  75271    6                                1.05

 3.11 PlyA -  75947  75942    6                                1.05
 3.10 Term -  83049  82945  105  0  0   77   38    68 0.831  -1.87
 3.09 Intr -  83245  83180   66  1  0  113   94    43 0.948   5.58
 3.08 Intr -  83509  83367  143  2  2  108   69    88 0.995   8.05
 3.07 Intr -  87709  87655   55  1  1   50  115    63 0.988   2.83
 3.06 Intr -  89754  89539  216  0  0  110   42   182 0.727  13.58
 3.05 Intr -  90488  90378  111  2  0   25  100   169 0.499  11.46
 3.04 Intr -  92145  92053   93  0  0  109   59    52 0.893   3.64
 3.03 Intr -  92307  92238   70  2  1  101   67    38 0.955   1.27
 3.02 Intr -  93882  93776  107  0  2   70   68    84 0.640   2.69
 3.01 Intr -  95364  95269   96  0  0   68   75   106 0.661   6.59

Predicted peptide sequence(s):

>AC002467.seq|GENSCAN_predicted_peptide_1|430_aa
MLAASFSIAVVAYAIAVSVGKVYATKYDYTIDGNQEFIAFGISNIFSGFFSCFVATTALS
RTAVQESTGGKTQVAGIISAAIVMIAILALGKLLEPLQKSVLAAVVIANLKGMFMQLCDI
PRLWRQNKIDAVIWVFTCIVSIILGLDLGLLAGLIFGLLTVVLRVQFPSWNGLGSIPSTD
<remainer of output truncated>
```

Figure 10.5. GENSCAN output using the human BAC clone RG364P16 from 7q31 as the query. The columns, going from left to right, represent the gene and exon number (`Gn.Ex`), the type of prediction (`Type`), the strand on which the prediction was made (`S`, with + as the forward strand and − as the reverse), the beginning and endpoints for the prediction (`Begin` and `End`), the length of the prediction (`Len`), the reading frame of the prediction (`Fr`), several scoring columns, and the probability value (`P`). Each predicted gene is shown as a separate block; notice that the third gene has its exons listed in reverse order, reflecting that the prediction is on the reverse strand. The tables are followed by the protein translations for each of the three predicted genes.

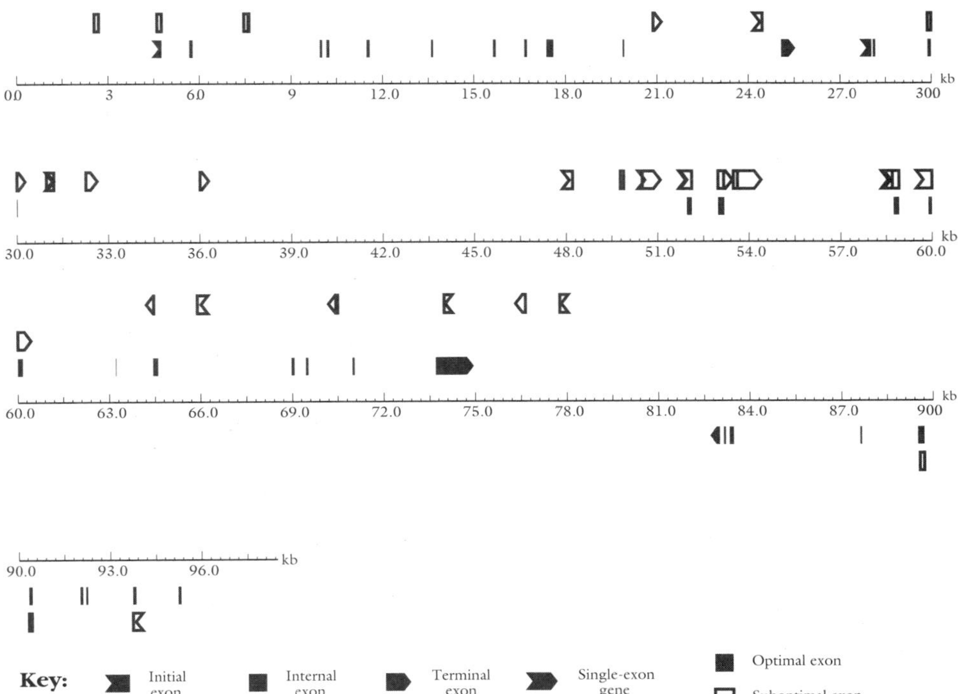

Figure 10.6. GENSCAN output in graphical form, using the human BAC clone RG364P16 from 7q31 as the query. Optimal and suboptimal exons are indicated, and the initial and terminal exons show the direction in which the prediction is being made ($5' \rightarrow 3'$ or $3' \rightarrow 5'$).

Theseus made short order of Procrustes by fitting him to his own bed, thereby sparing any other traveler the same fate. On the basis of this story, the phrase "bed of Procrustes" has come to convey the idea of forcing something to fit where it normally would not.

Living up to its namesake, PROCRUSTES takes genomic DNA sequences and "forces" them to fit into a pattern as defined by a related target protein (Gelfand et al., 1996). Unlike the other gene prediction methods that have been discussed, the algorithm does not use a DNA sequence *on its own* to look for content- or site-based signals. Instead, the algorithm requires that the user identify putative gene products *before* the prediction is made, so that the prediction represents the best fit of the given DNA sequence to its putative transcription product. The method uses a spliced alignment algorithm to sequentially explore all possible exon assemblies, looking for the best fit of predicted gene structure to candidate protein. If the candidate protein is known to arise from the query DNA sequence, correct gene structures can be predicted with an accuracy of 99% or better. By making use of candidate proteins in the course of the prediction, PROCRUSTES can take advantage of information known about this protein or related proteins in the public databases to better deter-

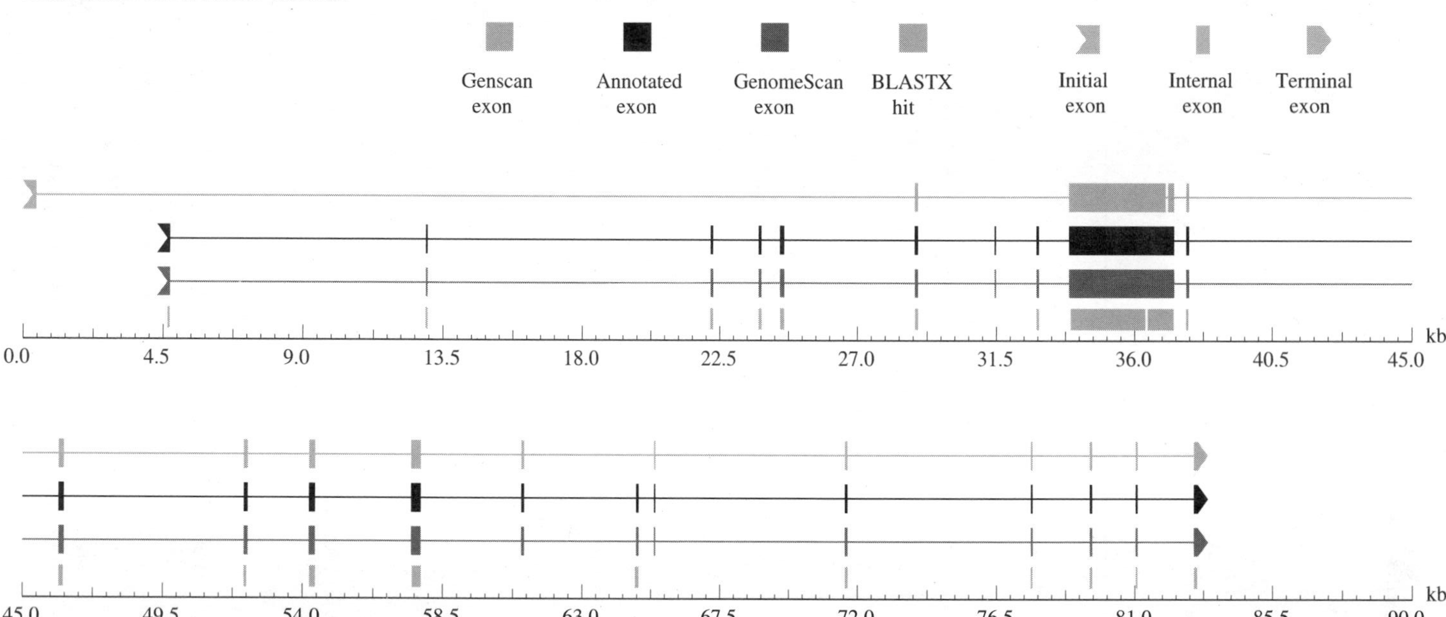

Figure 10.7. Comparison of GENSCAN with GenomeScan, using the human BRCA1 gene sequence as the query. The GENSCAN prediction (top line) is missing a number of the exons that appear in the annotation for the BRCA1 gene (second line; GenBank L78833), and the GENSCAN prediction is slightly longer than the actual gene at the 5′ end. The inclusion of BLASTX hit information (vertical bars closest to the scale) in GenomeScan produces a more complete and accurate prediction (third line).

mine the location of the introns and the exons in this gene. PROCRUSTES can handle cases where there are either partial or multiple genes in the query DNA sequence.

The input to PROCRUSTES is through a Web interface and is quite simple. The user needs to supply the nucleotide sequence *and* as many protein sequences as are relevant to this region. The supplied protein sequences will be treated as being similar, though not necessarily identical, to that encoded by the DNA sequence. Typical output from PROCRUSTES (not shown here) includes an aligned map of the predicted intron-exon structure for all target proteins, probability values, a list of exons with their starting and ending nucleotide positions, translations of the gene model (which may not be the same as the sequence of the initially supplied protein), and a "spliced alignment" showing any differences between the predicted protein and the target protein. The nature of the results makes PROCRUSTES a valuable method for refining results obtained by other methods, particularly in the context of positional candidate efforts.

GeneID

The current version of GeneID finds exons based on measures of coding potential (Guigó et al., 1992). The original version of this program was among the fastest in that it used a rule-based system to examine the putative exons and assemble them into the "most likely gene" for that sequence. GeneID uses position-weight matrices to assess whether or not a given stretch of sequence represents a splice site or a start or stop codon. Once this assessment is made, models of putative exons are built. On the basis of the sets of predicted exons that GeneID develops, a final refinement round is performed, yielding the most probable gene structure based on the input sequence.

The interface to GeneID is through a simple Web front-end, in which the user pastes in the DNA sequence and specifies whether the organism is either human or *Drosophila*. The user can specify whether predictions should be made only on the forward or reverse strand, and available output options include lists of putative acceptor sites, donor sites, and start and stop codons. Users can also limit output to only first exons, internal exons, terminal exons, or single genes, for specialized analyses. It is recommended that the user simply select All Exons to assure that all relevant information is returned.

GeneParser

GeneParser (Snyder and Stormo, 1993; Snyder and Stormo, 1997) uses a slightly different approach in identifying putative introns and exons. Instead of predetermining candidate regions of interest, GeneParser computes scores on all "subintervals" in a submitted sequence. Once each subinterval is scored, a neural network approach is used to determine whether each subinterval contains a first exon, internal exon, final exon, or intron. The individual predictions are then analyzed for the combination that represents the most likely gene. There is no Web front-end for this program, but the program itself is freely available for use on Sun, DEC, and SGI-based systems.

HMMgene

HMMgene predicts whole genes in any given DNA sequence using a hidden Markov model (HMM) method geared toward maximizing the probability of an accurate prediction (Krogh, 1997). The use of HMMs in this method helps to assess the confidence in any one prediction, enabling HMMgene to not only report the "best" prediction for the input sequence but alternative predictions on the same sequence as well. One of the strengths of this method is that, by returning multiple predictions on the same region, the user may be able to gain insight onto possible alternative splicings that may occur in a region containing a single gene.

The front-end for HMMgene requires an input sequence, with the organismal options being either human or *C. elegans*. An interesting addition is that the user can include known annotations, which could be from one of the public databases or based on experimental data that the investigator is privy to. Multiple sequences in FASTA format can be submitted as a single job to the server. Examples of sequence input format and resulting output are given in the documentation file at the HMMgene Web site.

HOW WELL DO THE METHODS WORK?

As we have already seen, different methods produce different types of results—in some cases, lists of putative exons are returned but these exons are not in a genomic context; in other cases, complete gene structures are predicted but possibly at a cost of less-reliable individual exon predictions. Looking at the absolute results for the 7q31 BAC clone, anywhere between one and three genes are predicted for the region, and those one to three genes have anywhere between 27 and 34 exons. In cases of similar exons, the boundaries of the exons are not always consistent. Which method is the "winner" in this particular case is not important; what is important is the variance in the results.

Returning to the cautionary note that different methods will perform better or worse, depending on the system being examined, it becomes important to be able to quantify the performance of each of these algorithms. Several studies have systematically examined the rigor of these methods using a variety of test data sets (Burset and Guigó, 1996; Claverie, 1997a; Snyder and Stormo, 1997, Rogic et al., 2001). Before discussing the results of these studies, it is necessary to define some terms.

For any given prediction, there are four possible outcomes: the detection of a true positive, true negative, false positive, or false negative (Fig. 10.8). Two measures of accuracy can be calculated based on the ratios of these occurrences: a *sensitivity* value, reflecting the fraction of actual coding regions that are correctly predicted as truly being coding regions, and a *specificity* value, reflecting the overall fraction of the prediction that is correct. In the best-case scenario, the methods will try to optimize the balance between sensitivity and specificity, to be able to find all of the true exons without becoming so sensitive as to start picking up an inordinate amount of false positives. An easier-to-understand measure that combines the sensitivity and specificity values is called the *correlation coefficient*. Like all correlation coefficients, its value can range from -1, meaning that the prediction is always wrong, through zero, to $+1$, meaning that the prediction is always right.

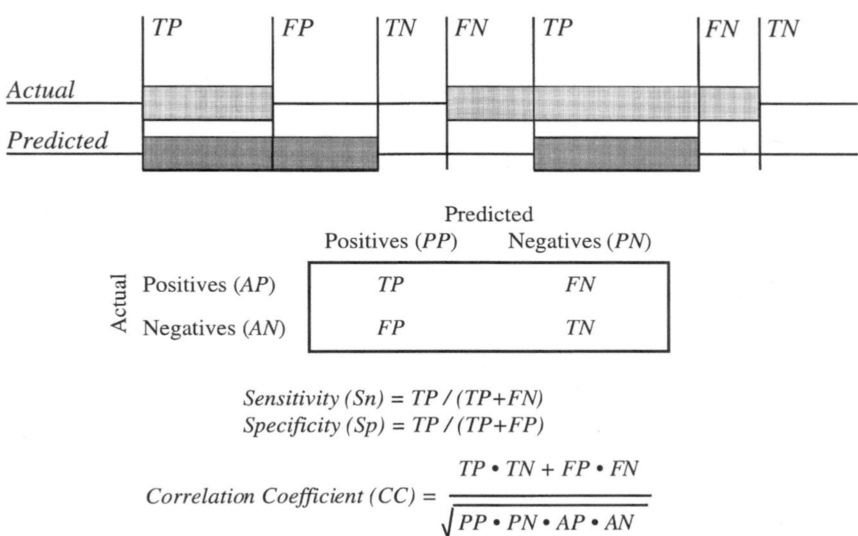

Figure 10.8. Sensitivity vs. specificity. In the upper portion, the four possible outcomes of a prediction are shown: a true positive (*TP*), a true negative (*TN*), a false positive (*FP*), and a false negative (*FN*). The matrix at the bottom shows how both sensitivity and specificity are determined from these four possible outcomes, giving a tangible measure of the effectiveness of any gene prediction method. (Figure adapted from Burset and Guigó, 1996; Snyder and Stormo, 1997.)

As a result of a Cold Spring Harbor Laboratory meeting on gene prediction,[1] a Web site called the "Banbury Cross" was created. The intent behind creating such a Web site was twofold: for groups actively involved in program development to post their methods for public use and for researchers actively deriving fully characterized, finished genomic sequence to submit such data for use as "benchmark" sequences. In this way, the meeting participants created an active forum for the dissemination of the most recent findings in the field of gene identification. Using these and other published studies, Jean-Michel Claverie at CNRS in Marseille compared the sensitivity and specificity of 14 different gene identification programs (Claverie, 1997, and references therein); PROCRUSTES was not one of the 14 considered, since the method varies substantially from that employed by other gene prediction programs. In examining data from these disparate sources, either the best performance found in an independent study *or* the worst performance reported by the authors of the method themselves was used in making the comparisons. On the basis of these comparisons, the best overall individual exon finder was deemed to be MZEF and the best gene structure prediction program was deemed to be GENSCAN. (By back-calculating as best as possible from the numbers reported in the Claverie paper, these two methods gave the highest correlation coefficients within their class, with $CC_{\mathrm{MZEF}} \sim 0.79$ and $CC_{\mathrm{GENSCAN}} \sim 0.86$.)

[1] Finding Genes: Computational Analysis of DNA Sequences. Cold Spring Harbor Laboratory, March 1997.

Because these gene-finding programs are undergoing a constant evolution, adding new features and incorporating new biological information, the idea of a comparative analysis of a number of representative algorithms was recently revisited (Rogic et al., 2001). One of the encouraging outcomes of this study was that these newer methods, as a whole, did a substantially better job in accurately predicting gene structures than their predecessors did. By using an independent data set containing 195 sequences from GenBank in which intron-exon boundaries have been annotated, GENSCAN and HMMgene appeared to perform the best, both having a correlation coefficient of 0.91. (Note the improvement of CC_{GENSCAN} from the time of the Burset and Guigó study to the time of the Rogic et al. study.)

STRATEGIES AND CONSIDERATIONS

Given these statistics, it can be concluded that both MZEF and GENSCAN are particularly suited for differentiating introns from exons at different stages in the maturation of sequence data. However, this should *not* be interpreted as a blanket recommendation to *only* use these two programs in gene identification. Remember that these results represent a compilation of findings from different sources, so keep in mind that the reported results may *not* have been derived from the same data set. It has already been stated numerous times that any given program can behave better or worse depending on the input sequences. It has also been demonstrated that the actual performance of these methods can be highly sensitive to G + C content. For example, Snyder and Stormo (1997) reported that GeneParser (Snyder and Stormo, 1993) and GRAIL2 (with assembly) performed best on test sets having high G + C content (as assessed by their respective *CC* values), whereas GeneID (Guigó et al., 1992) performed best on test sets having low G + C content. Interestingly, both GENSCAN and HMMgene were seen to perform "steadily," regardless of G + C content, in the Rogic study (Rogic et al., 2001).

There are several major drawbacks that most gene identification programs share that users need to be keenly aware of. Because most of these methods are "trained" on test data, they will work best in finding genes most similar to those in the training sets (that is, they will work best on things similar to what they have seen before). Often methods have an absolute requirement to predict both a discrete beginning and an end to a gene, meaning that these methods may miscall a region that consists of either a partial gene or multiple genes. The importance given to each individual factor in deciding whether a stretch of sequence is an intron or an exon can also influence outcomes, as the weighing of each criterion may be either biased or incorrect. Finally, there is the unusual case of genes that are transcribed but not translated (so-called "noncoding RNA genes"). One such gene, NTT (noncoding transcript in T cells), shows no exons or significant open reading frames, even though RT-PCR shows that NTT is transcribed as a polyadenlyated 17-kb mRNA (Liu et al., 1997). A similar protein, IPW, is involved in imprinting, and its expression is correlated to the incidence of Prader-Willi syndrome (Wevrick et al., 1996). Because hallmark features of gene structure are presumably absent from these genes, they cannot be reliably detected by any known method to date.

It begins to become evident that no one program provides the foolproof key to computational gene identification. The correct choice will depend on the nature of the data and where in the pathway of data maturation the data lie. On the basis of the studies described above, some starting points can be recommended. In the case of incompletely assembled sequence contigs (prefinished genome survey sequence), MZEF provides the best jumping-off point, since, for sequences of this length, one would expect no more than one exon. In the case of nearly finished or finished data, where much larger contigs provide a good deal of contextual information, GENSCAN or HMMgene would be an appropriate choice. In either case, users should supplement these predictions with results from *at least* one other predictive method, as consistency among methods can be used as a qualitative measure of the robustness of the results. Furthermore, utilization of comparative search methods, such as BLAST (Altschul et al., 1997) or FASTA (Pearson et al., 1997), should be considered an absolute requirement, with users targeting both dbEST and the protein databases for homology-based clues. PROCRUSTES again should be used when some information regarding the putative gene product is known, particularly when the cloning efforts are part of a positional candidate strategy.

A good example of the combinatorial approach is illustrated in the case of the gene for cerebral cavernous malformation (CCM1) located at 7q21–7q22; here, a combination of MZEF, GENSCAN, XGRAIL, and PowerBLAST (Zhang and Madden, 1997) was used in an integrated fashion in the prediction of gene structure (Kuehl et al., 1999). Another integrated approach to this approach lies in "workbenches" such as Genotator, which allow users to simultaneously run a number of prediction methods and homology searches, as well as providing the ability to annotate sequence features through a graphical user interface (Harris, 1997).

A combinatorial method developed at the National Human Genome Research Institute combines most of the methods described in this chapter into a single tool. This tool, named GeneMachine, allows users to query multiple exon and gene prediction programs in an automated fashion (Makalowska et al., 1999). A suite of Perl modules are used to run MZEF, GENSCAN, GRAIL2, FGENES, and BLAST. RepeatMasker and Sputnik are used to find repeats within the query sequence. Once GeneMachine is run, a file is written that can subsequently be opened using NCBI Sequin, in essence using Sequin as a workbench and graphical viewer. Using Sequin also has the advantage of presenting the results to the user in a familiar format—basically the same format that is used in Entrez for graphical views. The main feature of GeneMachine is that the process is fully automated; the user is only required to launch GeneMachine and then open the resulting file with NCBI Sequin. GeneMachine also does not require users to install local copies of the prediction programs, enabling users to pass-off to Web interfaces instead; although this reduces some of the overhead of maintaining the program, it does result in slower performance. Annotations can then be made to these results before submission to GenBank, thereby increasing the intrinsic value of these data. A sample of the output obtained using GeneMachine is shown in Figure 10.9, and more details on GeneMachine can be found on the NHGRI Web site.

The ultimate solution to the gene identification problem lies in the advancement of the Human Genome Project and other sequencing projects. As more and more gene structures are elucidated, this biological information can in turn be used to develop better methods, yielding more accurate predictions. Although the promise of such computational methods may not be completely fulfilled before the Human Genome Project reaches completion, the information learned from this effort will play a major role in facilitating similar efforts targeting other model genomes.

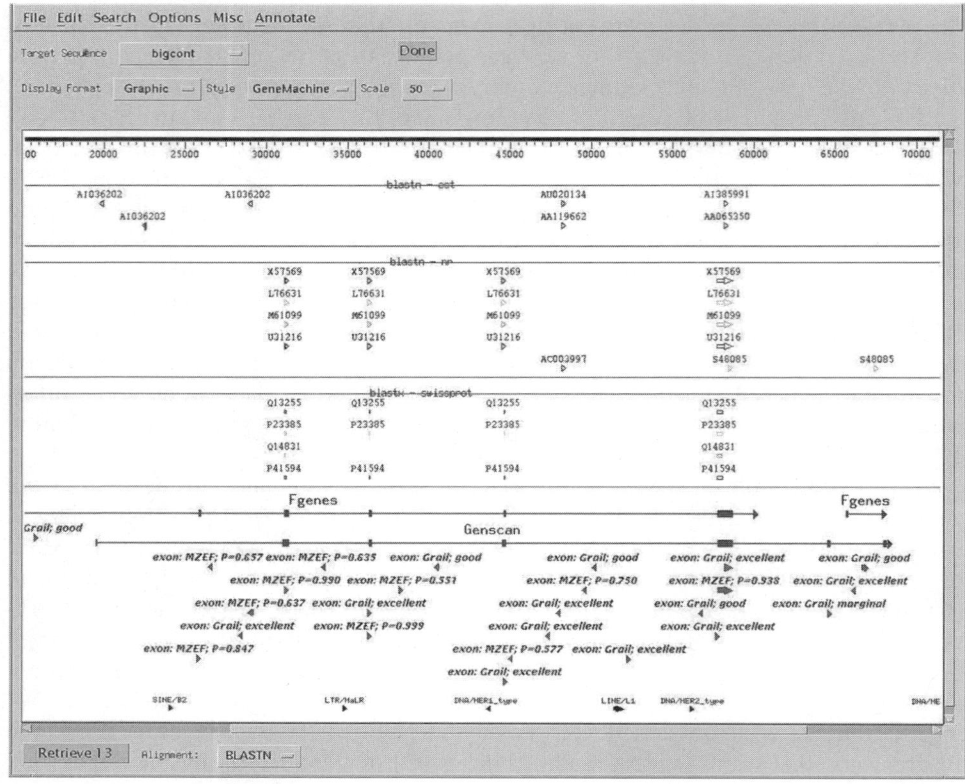

Figure 10.9. Annotated output from GeneMachine showing the results of multiple gene prediction program runs. NCBI Sequin is used at the viewer. The top of the output shows the results from various BLAST runs (BLASTN *vs.* dbEST, BLASTN vs. nr, and BLASTX vs. SWISS-PROT). Toward the bottom of the window are shown the results from the predictive methods (FGENES, GENSCAN, MZEF, and GRAIL 2). Annotations indicating the strength of the prediction are preserved and shown wherever possible within the viewer. Putative regions of high interest would be areas where hits from the BLAST runs line up with exon predictions from the gene prediction programs. (See color plate.)

INTERNET RESOURCES FOR TOPICS PRESENTED IN CHAPTER 10

Banbury Cross	*http://igs-server.cnrs-mrs.fr/igs/banbury*
FGENEH	*http://genomic.sanger.ac.uk/gf/gf.shtml*
GeneID	*http://www1.imim.es/geneid.html*
GeneMachine	*http://genome.nhgri.nih.gov/genemachine*
GeneParser	*http://beagle.colorado.edu/~eesnyder/GeneParser.htl*
GENSCAN	*http://genes.mit.edu/GENSCAN.html*
Genotator	*http://www.fruitfly.org/~nomi/genotator/*
GRAIL	*http://compbio.ornl.gov/tools/index.shtml*
GRAIL-EXP	*http://compbio.ornl.gov/grailexp/*
HMMgene	*http://www.cbs.dtu.dk/services/HMMgene/*
MZEF	*http://www.cshl.org/genefinder*

PROCRUSTES	*http://www-hto.usc.edu/software/procrustes*
RepeatMasker	*http://ftp.genome.washington.edu/RM/RepeatMasker.html*
Sputnik	*http://rast.abajian.com/sputnik/*

PROBLEM SET

An anonymous sequence from 18q requiring computational analysis is posted on the book's Web site (http://www.wiley.com/bioinformatics). To gain a better appreciation for the relative performance of the methods discussed in this chapter and how the results may vary between methods, use FGENES, GENSCAN, and HMMgene to answer each of the following questions.

1. How many exons are in the unknown sequence?
2. What are the start and stop points for each of these exons?
3. Which strand (forward or reverse) are the putative exons found on?
4. Are there any unique features present, like polyA tracts? Where are they located?
5. Can any protein translations be derived from the sequence? What is the length (in amino acids) of these translations?
6. For HMMgene only, can alternative translations be computed for this particular DNA sequence? If so, give the number of exons and the length of the coding region (CDS) for each possible alternative prediction. Note on which strand the alternative translations are found.

REFERENCES

Altschul, S. F., Madden, T. L., Schaffer, A. A., Zhang, J., Zhang, Z., Miller, W., and Lipman, D. J. (1997). Gapped BLAST and PSI-BLAST: a new generation of protein database search programs. *Nucleic Acids Res.* 25, 3389–3402.

Burge, C., and Karlin, S. (1997). Prediction of complete gene structures in human genomic DNA. *J. Mol. Biol.* 268, 78–94.

Burge, C. B., and Karlin, S. (1998). Finding the genes in genomic DNA. *Curr. Opin. Struct. Biol.* 8, 346–354.

Burset, M., and Guigó, R. (1996). Evaluation of gene structure prediction programs. *Genomics* 34, 353–367.

Chothia, C., and Lesk, A. M. (1986). The relation between the divergence of sequence and structure in proteins. *EMBO J* 5, 823–826.

Claverie, J. M. (1998). Computational methods for exon detection. *Mol. Biotechnol.* 10, 27–48.

Claverie, J. M. (1997a). Computational methods for the identification of genes in vertebrate genomic sequences. *Hum. Mol. Genet.* 6, 1735–1744.

Claverie, J. M. (1997b). Exon detection by similarity searches. *Methods Mol. Biol.* 68, 283–313.

Everett, L. A., Glaser, B., Beck, J. C., Idol, J. R., Buchs, A., Heyman, M., Adawi, F., Hazani, E., Nassir, E., Baxevanis, A. D., Sheffield, V. C., and Green, E. D. (1997). Pendred syndrome is caused by mutations in a putative sulphate transporter gene (PDS). *Nat. Genet.* 17, 411–422.

Gelfand, M. S., Mironov, A. A., and Pevzner, P. A. (1996). Gene recognition via spliced sequence alignment. *Proc. Natl. Acad. Sci. USA* 93, 9061–9066.

Guigó, R. (1997). Computational gene identification. *J. Mol. Med.* 75, 389–393.

Guigó, R., Knudsen, S., Drake, N., and Smith, T. (1992). Prediction of gene structure. *J. Mol. Biol.* 226, 141–157.

Harris, N. L. (1997). Genotator: a workbench for sequence annotation. *Genome Res.* 7, 754–762.

Krogh, A. (1997). Two methods for improving performance of an HMM and their application for gene finding. In *Proceedings of the Fifth International Conference on Intelligent Systems for Molecular Biology*, Gaasterland, T., Karp, P., Karplus, K., Ouzounis, C., Sander, C., and Valencia, A., eds. (AAAI Press, Menlo Park, CA), p. 179–186.

Kuehl, P., Weisemann, J., Touchman, J., Green, E., and Boguski, M. (1999). An effective approach for analyzing "prefinished genomic sequence data. *Genome Res.* 9, 189–194.

Liu, A. Y., Torchia, B. S., Migeon, B. R., and Siliciano, R. F. (1997). The human NTT gene: identification of a novel 17-kb noncoding nuclear RNA expressed in activated CD4+ T cells. *Genomics* 39, 171–184.

Makalowska, I., Ryan, J. F., and Baxevanis, A. D. (1999) GeneMachine: A Unified Solution for Performing Content-Based, Site-Based, and Comparative Gene Prediction Methods. 12th Cold Spring Harbor Meeting on Genome Mapping, Sequencing, and Biology, Cold Spring Harbor, NY.

Mural, R. J., Einstein, J. R., Guan, X., Mann, R. C., and Uberbacher, E. C. (1992). An artificial intelligence approach to DNA sequence feature recognition. *Trends Biotech.* 10, 67–69.

Pearson, W. R., Wood, T., Zhang, Z., and Miller, W. (1997). Comparison of DNA sequences with protein sequences. *Genomics* 46, 24–36.

Rogic, S., Mackworth, A., and Ouellette, B. F. F. (2001). Evaluation of Gene-Finding Programs. In press.

Snyder, E. E., and Stormo, G. D. (1993). Identification of coding regions in genomic DNA sequences: an application of dynamic programming and neural networks. *Nucleic Acids Res.* 21, 607–613.

Snyder, E. E., and Stormo, G. D. (1997). Identifying genes in genomic DNA sequences. In DNA and Protein Sequence Analysis, M. J. Bishop and C. J. Rawlings, eds. (New York: Oxford University Press), p. 209–224.

Solovyev, V. V., Salamov, A. A., and Lawrence, C. B. (1995). Identification of human gene structure using linear discriminant functions and dynamic programming. *Ismb* 3, 367–375.

Solovyev, V. V., Salamov, A. A., and Lawrence, C. B. (1994a). Predicting internal exons by oligonucleotide composition and discriminant analysis of spliceable open reading frames. *Nucleic Acids Res.* 22, 5156–5163.

Solovyev, V. V., Salamov, A. A., and Lawrence, C. B. (1994b). The prediction of human exons by oligonucleotide composition and discriminant analysis of spliceable open reading frames. *Ismb* 2, 354–362.

Uberbacher, E. C., and Mural, R. J. (1991). Locating protein-coding regions in human DNA sequences by a multiple sensor-neural network approach. *Proc. Natl. Acad. Sci. USA* 88, 11261–11265.

Wevrick, R., Kerns, J. A., and Francke, U. (1996). The IPW gene is imprinted and is not expressed in the Prader-Willi syndrome. *Acta Genet. Med. Gemellol.* 45, 191–197.

Zhang, J., and Madden, T. L. (1997). PowerBLAST: a new network BLAST application for interactive or automated sequence analysis and annotation. *Genome Res.* 7, 649–656.

11

PREDICTIVE METHODS USING PROTEIN SEQUENCES

Sharmila Banerjee-Basu

Genome Technology Branch
National Human Genome Research Institute
National Institutes of Health
Bethesda, Maryland

Andreas D. Baxevanis

Genome Technology Branch
National Human Genome Research Institute
National Institutes of Health
Bethesda, Maryland

The discussions of databases and information retrieval in earlier chapters of this book document the tremendous explosion in the amount of sequence information available in a variety of public databases. As we have already seen with nucleotide sequences, all protein sequences, whether determined directly or through the translation of an open reading frame in a nucleotide sequence, contain intrinsic information of value in determining their structure or function. Unfortunately, experiments aimed at extracting such information cannot keep pace with the rate at which raw sequence data are being produced. Techniques such as circular dichroism spectroscopy, optical rotatory dispersion, X-ray crystallography, and nuclear magnetic resonance are extremely powerful in determining structural features, but their execution requires many hours of highly skilled, technically demanding work. The gap in information becomes obvious in comparisons of the size of the protein sequence and structure databases; as of this writing, there were 87,143 protein entries (Release 39.0) in SWISS-PROT but only 12,624 structure entries (July, 2000) in PDB. Attempts to close the gap

Bioinformatics: A Practical Guide to the Analysis of Genes and Proteins
Edited by A. D. Baxevanis and B. F. F. Ouellette
ISBN 0-471-38390-2 (cloth), ISBN 0-471-383910 (paper)

center around theoretical approaches for structure and function prediction. These methods can provide insights as to the properties of a protein in the absence of biochemical data.

This chapter focuses on computational techniques that allow for biological discovery based on the protein sequence *itself* or on their comparison to protein families. Unlike nucleotide sequences, which are composed of four bases that are chemically rather similar (yet distinct), the alphabet of 20 amino acids found in proteins allows for much greater diversity of structure and function, primarily because the differences in the chemical makeup of these residues are more pronounced. Each residue can influence the overall physical properties of the protein because these amino acids are basic or acidic, hydrophobic or hydrophilic, and have straight chains, branched chains, or are aromatic. Thus, each residue has certain propensities to form structures of different types in the context of a protein domain. These properties, of course, are the basis for one of the central tenets of biochemistry: that *sequence specifies conformation* (Anfinsen et al., 1961).

The major precaution with respect to these or any other predictive techniques is that, regardless of the method, the results are *predictions*. Different methods, using different algorithms, may or may not produce different results, and it is important to understand *how* a particular predictive method works rather than just approaching the algorithm as a "black box": one method may be appropriate in a particular case but totally inappropriate in another. Even so, the potential for a powerful synergy exists: proper use of these techniques along with primary biochemical data can provide valuable insights into protein structure and function.

PROTEIN IDENTITY BASED ON COMPOSITION

The physical and chemical properties of each of the 20 amino acids are fairly well understood, and a number of useful computational tools have been developed for making predictions regarding the identification of unknown proteins based on these properties (and vice versa). Many of these tools are available through the ExPASy server at the Swiss Institute of Bioinformatics (Appel et al., 1994). The focus of the ExPASy tools is twofold: to assist in the analysis and identification of unknown proteins isolated through two-dimensional gel electrophoresis, as well as to predict basic physical properties of a known protein. These tools capitalize on the curated annotations in the SWISS-PROT database in making their predictions. Although calculations such as these are useful in electrophoretic analysis, they can be very valuable in any number of experimental areas, particularly in chromatographic and sedimentation studies. In this and the following section, tools in the ExPASy suite are identified, but the ensuing discussion also includes a number of useful programs made available by other groups. Internet resources related to these and other tools discussed in this chapter are listed at the end of the chapter.

AACompIdent and AACompSim (ExPASy)

Rather than using an amino acid sequence to search SWISS-PROT, AACompIdent uses the amino acid composition of an unknown protein to identify known proteins of the same composition (Wilkins et al., 1996). As inputs, the program requires the desired amino acid composition, the isoelectric point (pI) and molecular weight of the protein (if known), the appropriate taxonomic class, and any special keywords. In addition, the user must select from one of six amino acid "constellations," which influence how the analysis is performed; for example, certain constellations may combine residues like Asp/Asn (D/N) and Gln/Glu (Q/E) into Asx (B) and Glx (Z), or certain residues may be eliminated from the analysis altogether.

For each sequence in the database, the algorithm computes a score based on the difference in compositions between the sequence and the query composition. The results, returned by E-mail, are organized as three ranked lists:

- a list based on all proteins from the specified taxonomic class without taking pI or molecular weight into account;
- a list based on all proteins regardless of taxonomic class without taking pI or molecular weight into account; and
- a list based on the specified taxonomic class that does take pI and molecular weight into account.

Because the computed scores are a difference measure, a score of zero implies that there is exact correspondence between the query composition and that sequence entry.

AACompSim, a variant of AACompIdent, performs a similar type of analysis, but, rather than using an experimentally derived amino acid composition as the basis for searches, the sequence of a SWISS-PROT protein is used instead (Wilkins et al., 1996). A theoretical pI and molecular weight are computed before computation of the difference scores using Compute pI/MW (see below). It has been documented that amino acid composition across species boundaries is well conserved (Cordwell et al., 1995) and that, by considering amino acid composition, investigators can detect weak similarities between proteins whose sequence identity falls below 25% (Hobohm and Sander, 1995). Thus the consideration of composition in addition to the ability to perform "traditional" database searches may provide additional insight into the relationships between proteins.

PROPSEARCH

Along the same lines as AACompSim, PROPSEARCH uses the amino acid composition of a protein to detect weak relationships between proteins, and the authors have demonstrated that this technique can be used to easily discern members of the same protein family (Hobohm and Sander, 1995). However, this technique is more robust than AACompSim in that 144 different physical properties are used in performing the analysis, among which are molecular weight, the content of bulky residues, average hydrophobicity, and average charge. This collection of physical properties is called the query vector, and it is compared against the same type of vector precomputed for every sequence in the target databases (SWISS-PROT and PIR). Having this "database of vectors" calculated in advance vastly improves the processing time for a query.

The input to the PROPSEARCH Web server is just the query sequence, and an example of the program output is shown in Figure 11.1. Here, the sequence of human autoantigen NOR-90 was used as the input query. The results are ranked by a distance score, and this score represents the likelihood that the query sequence and new

```
Fragment search: OFF (POS1 and POS2 are begin and end of sequence)

Rank           ID   DIST  LEN2 POS1 POS2    pI DE
---------------------------------------------------------------------------------------------------
  1    >p1;s18193   0.00   727    1  727  5.33 autoantigen NOR-90 - human
  2    ubf1_human   1.36   764    1  764  5.62 NUCLEOLAR TRANSCRIPTION FACTOR 1 (UPSTREAM BINDING FACTOR 1) (UBF-1)
  3    ubf1_mouse   1.40   765    1  765  5.55 NUCLEOLAR TRANSCRIPTION FACTOR 1 (UPSTREAM BINDING FACTOR 1) (UBF-1).
  4      ubf1_rat   1.57   764    1  764  5.61 NUCLEOLAR TRANSCRIPTION FACTOR 1 (UPSTREAM BINDING FACTOR 1) (UBF-1).
  5    ubf1_xenla   3.95   677    1  677  5.79 NUCLEOLAR TRANSCRIPTION FACTOR 1 (UPSTREAM BINDING FACTOR-1) (UBF-1).
  6    ubf2_xenla   4.18   701    1  701  6.05 NUCLEOLAR TRANSCRIPTION FACTOR 2 (UPSTREAM BINDING FACTOR-2) (UBF-2).
  7    >p1;s57552   7.72   606    1  606  6.63 hypothetical protein YPR018w - yeast (Saccharomyces cerevisiae)
  8    >p1;i50463   8.49   772    1  772  5.71 protein kinase - chicken
  9    >p1;h54024   8.83   768    1  768  5.27 protein kinase (EC 2.7.1.37) cdc2-related PITSLRE alpha 2-3 - human
 10    >p1;b54024   8.87   777    1  777  5.27 protein kinase (EC 2.7.1.37) cdc2-related PITSLRE alpha 2-2 - human
 11    >p1;g54024   8.90   766    1  766  5.21 protein kinase (EC 2.7.1.37) cdc2-related PITSLRE beta 2-2 - human
 12    >p1;a55817   9.00   783    1  783  5.19 cyclin-dependent kinase p130-PITSLRE - mouse
 13    >p1;f54024   9.11   777    1  777  5.30 protein kinase (EC 2.7.1.37) cdc2-related PITSLRE beta 2-1 - human
 14    >p1;e54024   9.11   779    1  779  5.42 protein kinase (EC 2.7.1.37) cdc2-related PITSLRE alpha 2-1 - human
 15    yaa5_schpo   9.45   598    1  598  4.78 HYPOTHETICAL 69.5 KD PROTEIN C22G7.05 IN CHROMOSOME I.
 16    >p1;s62449   9.45   598    1  598  4.78 hypothetical protein SPAC22G7.05 - fission yeast (Schizosaccharomyces pombe)
 17    >f1;i58390   9.45   920    1  920  5.00 retinoblastoma binding protein 1 isoform I - human (fragment)
 18    >p1;s63193   9.58   590    1  590  6.15 hypothetical protein YNL227c - yeast (Saccharomyces cerevisiae)
 19    ynw7_yeast   9.58   590    1  590  6.15 HYPOTHETICAL 68.8 KD PROTEIN IN URE2-SSU72 INTERGENIC REGION.
 20    >p1;s49634   9.74   899    1  899  4.79 hypothetical protein YML093w - yeast (Saccharomyces cerevisiae)
 21    ymj3_yeast   9.74   899    1  899  4.79 HYPOTHETICAL 103.0 KD PROTEIN IN RAD10-PRS4 INTERGENIC REGION.
 22    radi_human   9.76   583    1  583  6.33 RADIXIN.
 23      radi_pig   9.81   583    1  583  6.21 RADIXIN (MOESIN B).
 24    >f1;i78883   9.83   866    1  866  4.77 retinoblastoma binding protein 1 isoform II - human (fragment)
 25    >p1;b42997   9.87   754    1  754  5.17 retinoblastoma-associated protein 2 - human
 26    >p1;a57467   9.91   647    1  647  5.74 RalBP1 - rat
```

Figure 11.1. Results of a PROPSEARCH database query based on amino acid composition. The input sequence used was that of the human autoantigen NOR-90. Explanatory material and a histogram of distance scores against the entire target database have been removed for brevity. The columns in the table give the rank of the hit based on the distance score, the SWISS-PROT or PIR identifier, the distance score, the length of the overlap between the query and subject, the positions of the overlap (from POS1 to POS2), the calculated pI, and the definition line for the found sequence.

sequences found through PROPSEARCH belong to the same family, thereby implying common function in most cases. A distance score of 10 or below indicates that there is a better than 87% chance that there is similarity between the two proteins. A score below 8.7 increases the reliability to 94%, and a score below 7.5 increases the reliability to 99.6%. Examination of the results showed NOR-90 to be similar to a number of nucleolar transcription factors, protein kinases, a retinoblastoma-binding protein, the actin-binding protein radixin, and RalBP1, a putative GTPase target. None of these hits would necessarily be expected, since the functions of these proteins are dissimilar; however, a good number of these are DNA-binding proteins, opening the possibility that a very similar domain is being used in alternative functional contexts. At the very least, a BLASTP search would be necessary to both verify the results and identify critical residues.

MOWSE

The Molecular Weight Search (MOWSE) algorithm capitalizes on information obtained through mass spectrometric (MS) techniques (Pappin et al., 1993). With the use of both the molecular weights of intact proteins and those resulting from digestion of the same proteins with specific proteases, an unknown protein can be unambiguously identified given the results of several experimental determinations. This approach substantially cuts down on experimental time, since the unknown protein does not have to be sequenced in whole or in part.

The MOWSE Web front end requires the molecular weight of the starting sequence and the reagent used, as well as the resultant masses and composition of the peptides generated by the reagent. A tolerance value may be specified, indicating the error allowed in the accuracy of the determined fragment masses. Calculations are based on information contained in the OWL nonredundant protein sequence database (Akrigg et al., 1988). Scoring is based on how often a fragment molecular weight occurs in proteins within a given range of molecular weights, and the output is returned as a ranked list of the top 30 scores, with the OWL entry name, matching peptide sequences, and other statistical information. Simulation studies produced an accuracy rate of 99% using five or fewer input peptide weights.

PHYSICAL PROPERTIES BASED ON SEQUENCE

Compute pI/MW and ProtParam (ExPASy)

Compute pI/MW is a tool that calculates the isoelectric point and molecular weight of an input sequence. Determination of pI is based on pK values, as described in an earlier study on protein migration in denaturing conditions at neutral to acidic pH (Bjellqvist et al., 1993). Because of this, the authors caution that pI values determined for *basic* proteins may not be accurate. Molecular weights are calculated by the addition of the average isotopic mass of each amino acid in the sequence plus that of one water molecule. The sequence can be furnished by the user in FASTA format, or a SWISS-PROT identifier or accession number can be specified. If a sequence is furnished, the tool automatically computes the pI and molecular weight for the entire length of the sequence. If a SWISS-PROT identifier is given, the definition and organism lines of the entry are shown, and the user may specify a range of amino acids so that the computation is done on a fragment rather than on the entire protein. ProtParam takes this process one step further. Based on the input sequence, ProtParam calculates the molecular weight, isoelectric point, overall amino acid composition, a theoretical extinction coefficient (Gill and von Hippel, 1989), aliphatic index (Ikai, 1980), the protein's grand average of hydrophobicity (GRAVY) value (Kyte and Doolittle, 1982), and other basic physicochemical parameters. Although this might seem to be a very simple program, one can begin to speculate about the cellular localization of the protein; for example, a basic protein with a high proportion of lysine and arginine residues may well be a DNA-binding protein.

PeptideMass (ExPASy)

Designed for use in peptide mapping experiments, PeptideMass determines the cleavage products of a protein after exposure to a given protease or chemical reagent (Wilkins et al., 1997). The enzymes and reagents available for cleavage through PeptideMass are trypsin, chymotrypsin, LysC, cyanogen bromide, ArgC, AspN, and GluC (bicarbonate or phosphate). Cysteines and methionines can be modified before the calculation of the molecular weight of the resultant peptides. By furnishing a SWISS-PROT identifier rather than pasting in a raw sequence, PeptideMass is able to use information within the SWISS-PROT annotation to improve the calculations, such as removing signal sequences or including known posttranslational modifications before cleavage. The results are returned in tabular format, giving a theoretical pI and molecular weight for the starting protein and then the mass, position, modified masses, information on variants from SWISS-PROT, and the sequence of the peptide fragments.

TGREASE

TGREASE calculates the hydrophobicity of a protein along its length (Kyte and Doolittle, 1982). Inherent in each of the 20 amino acids is its hydrophobicity: the relative propensity of the acid to bury itself in the core of a protein and away from surrounding water molecules. This tendency, coupled with steric and other considerations, influences how a protein ultimately folds into its final three-dimensional conformation. As such, TGREASE finds application in the determination of putative transmembrane sequences as well as the prediction of buried regions of globular proteins. TGREASE is part of the FASTA suite of programs available from the University of Virginia and runs as a stand-alone application that can be downloaded and run on either Macintosh or DOS-based computers.

The method relies on a hydropathy scale, in which each amino acid is assigned a score reflecting its relative hydrophobicity based on a number of physical characteristics (e.g., solubility, the free energy of transfer through a water-vapor phase transition, etc.). Amino acids with higher, positive scores are more hydrophobic; those with more negative scores are more hydrophilic. A moving average, or hydropathic index, is then calculated across the protein. The window length is adjustable, with a span of 7–11 residues recommended to minimize noise and maximize information content. The results are then plotted as hydropathic index versus residue number. The sequence for the human interleukin-8 receptor B was used to generate a TGREASE plot, as shown in Figure 11.2. Correspondence between the peaks and the actual location of the transmembrane segments, although not exact, is fairly good;

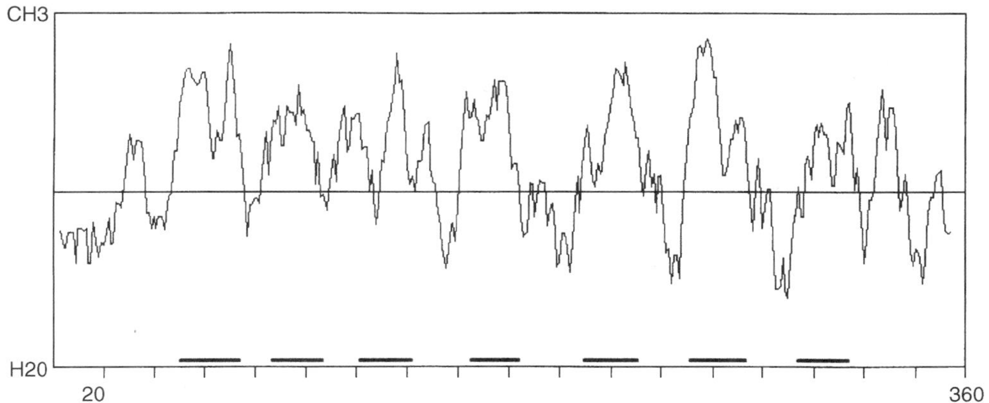

Figure 11.2. Results of a Kyte-Doolittle hydropathy determination using TGREASE. The input sequence was of the high affinity interleukin-8 receptor B from human. Default window lengths were used. The thick, horizontal bars across the bottom of the figure were added manually and represent the positions of the seven transmembrane regions of IL-8R-B, as given in the SWISS-PROT entry for this protein (P25025).

keep in mind that the method is predicting *all* hydrophobic regions, not just those located in transmembrane regions. The specific detection of transmembrane regions is discussed further below.

SAPS

The Statistical Analysis of Protein Sequences (SAPS) algorithm provides extensive statistical information for any given query sequence (Brendel et al., 1992). When a protein sequence is submitted via the SAPS Web interface, the server returns a large amount of physical and chemical information on the protein, based solely on what can be inferred from its sequence. The output begins with a composition analysis, with counts of amino acids by type. This is followed by a charge distribution analysis, including the locations of positively or negatively charged clusters, high-scoring charged and uncharged segments, and charge runs and patterns. The final sections present information on high-scoring hydrophobic and transmembrane segments, repetitive structures, and multiplets, as well as a periodicity analysis.

MOTIFS AND PATTERNS

In Chapter 8, the idea of direct sequence comparison was presented, where BLAST searches are performed to identify sequences in the public databases that are similar to a query sequence of interest. Often, this direct comparison may not yield any interesting results or may not yield any results at all. However, there may be very weak sequence determinants that are present that will allow the query sequence to be associated with a family of sequences. By the same token, a family of sequences can be used to identify new, distantly related members of the same protein family; an example of this is PSI-BLAST, discussed in Chapter 8.

Before discussing two of the methods that capitalize on such an approach, several terms have to be defined. The first is the concept of *profiles*. Profiles are, quite simply, a numerical representation of a multiple sequence alignment, much like the multiple sequence alignments derived from the methods discussed in Chapter 9. Imbedded within a multiple sequence alignment is intrinsic sequence information that represents the common characteristics of that particular collection of sequences, frequently a protein family. By using a profile, one is able to use these imbedded, common characteristics to find similarities between sequences with little or no absolute sequence identity, allowing for the identification and analysis of distantly related proteins. Profiles are constructed by taking a multiple sequence alignment representing a protein family and then asking a series of questions:

- What residues are seen at each position of the alignment?
- How often does a particular residue appear at each position of the alignment?
- Are there positions that show absolute conservation?
- Can gaps be introduced anywhere in the alignment?

Once those questions are answered, a *position-specific scoring table* (PSST) is constructed, and the numbers in the table now represent the multiple sequence alignment. The numbers within the PSST reflect the probability of any given amino acid occurring at each position. It also reflects the effect of a conservative or nonconservative substitution at each position in the alignment, much like a PAM or BLOSUM matrix does. This PSST can now be used for comparison against single sequences.

The second term requiring definition is *pattern* or *signature*. A signature also represents the common characteristics of a protein family (or a multiple sequence alignment) but does not contain any weighting information whatsoever—it simply provides a shorthand notation for what residues can be present at any given position. For example, the signature

```
[IV] - G - x - G - T -[LIVMF] - x(2) - [GS]
```

would be read as follows: the first position could contain *either* an isoleucine or a valine, the second position could contain only a glycine, and so on. An `x` means that any residue can appear at that position. The `x(2)` simply means that two positions can be occupied by any amino acid, the number just reflecting the length of the nonspecific run.

ProfileScan

Based on the classic Gribskov method of profile analysis (Gribskov et al., 1987, 1988), ProfileScan uses a method called pfscan to find similarities between a protein or nucleic acid query sequence and a profile library (Lüthy et al., 1994). In this case, three profile libraries are available for searching. First is PROSITE, an ExPASy database that catalogs biologically significant sites through the use of motif and sequence profiles and patterns (Hofmann, 1999). Second is Pfam, which is a collection of protein domain families that differ from most such collections in one important aspect: the initial alignment of the protein domains is done by hand, rather than by depending on automated methods. As such, Pfam contains slightly over 500 entries, but the entries are potentially of higher quality. The third profile set is referred to as the Gribskov collection.

Searches against any of these collections can be done through the ProfileScan Web page, which simply requires either an input sequence in plain text format, or an identifier such as a SWISS-PROT ID. The user can select the sensitivity of the search, returning either significant matches only or all matches, including borderline cases. To illustrate the output format, the sequence of a human heat-shock-induced protein was submitted to the server for searching against PROSITE profiles only.

```
normalized raw       from -    to Profile | Description
355.9801 41556 pos.     6 -  612 PF00012 | HSP70 Heat shock hsp70 proteins
```

Although the actual PROSITE entry returned is no great surprise, the output contains scores that are worth understanding. The raw score is the actual score calculated from the scoring matrix used during the search. The more informative number is the normalized or *N*-score. The *N*-score formally represents the number of matches one would expect in a database of given size. Basically, the larger the *N*-score the lower the probability that the hit occurred by chance. In the example, the *N*-score of 355 translates to 1.94×10^{-349} expected chance matches when normalized against SWISS-PROT—an extremely low probability of this being a false positive. The `from` and `to` numbers simply show the positions of the overlap between the query and the matching profile.

BLOCKS

The BLOCKS database utilizes the concept of blocks to identify a family of proteins, rather than relying on the individual sequences themselves (Henikoff and Henikoff, 1996). The idea of a block is derived from the more familiar notion of a motif, which usually refers to a conserved stretch of amino acids that confer a specific function or structure to a protein. When these individual motifs from proteins in the same family are aligned without introducing gaps, the result is a block, with the term "block" referring to the alignment, not the individual sequences themselves. Obviously, an individual protein can contain one or more blocks, corresponding to each of its functional or structural motifs.

The BLOCKS database itself is derived from the entries in PROSITE. When a BLOCKS search is performed using a sequence of interest, the query sequence is aligned against all the blocks in the database at all possible positions. For each alignment, a score is calculated using a position-specific scoring matrix, and results of the best matches are returned to the user. Searches can be performed optionally against the PRINTS database, which includes information on more than 300 families that do not have corresponding entries in the BLOCKS database. To ensure complete coverage, it is recommended that both databases be searched.

BLOCKS searches can be performed using the BLOCKS Web site at the Fred Hutchinson Cancer Research Center in Seattle. The Web site is straightforward, allowing both sequence-based and keyword-based searches to be performed. If a DNA sequence is used as the input, users can specify which genetic code to use and which strand to search. Regardless of whether the query is performed via a sequence or via keywords, a successful search will return the relevant block. An example is shown in Figure 11.3. In this entry (for a nuclear hormone receptor called a steroid finger),

```
ID     NUCLEAR_RECEPTOR; BLOCK
AC     BL00031A; distance from previous block=(4,603)
DE     Nuclear hormones receptors DNA-binding region proteins.
BL     CCR;    width=33; seqs=177; 99.5%=1562; strength=1584
CSR1_CAEEL|Q17370    (  11)  CAVCDDIATGKHYSVASCNGCKTFFRRALVNNR  41

FTFB_DROME|Q05192    ( 376)  CPICGDKISGFHYGIFSCESCKGFFKRTVQNRK  16

HR96_DROME|Q24143    (   7)  CAVCGDKALGYNFNAVTCESCKAFFRRNALAKK  59

NER_HUMAN|P55055     (  87)  CRVCGDKASGFHYNVLSCEGCKGFFRRSVVRGG  12

NHR2_CAEEL|Q10902    ( 105)  CMVCGDNSTGYHYGVQSCEGCKGFFRRSVHKNI  16

ODR7_CAEEL|P41933    ( 331)  QVCLSTHANGLHFGARTCAACAAFFRRTISDDK  85

TLL_DROME|P18102     (  34)  CKVCRDHSSGKHYGIYACDGCAGFFKRSIRRSR  27

YKC8_CAEEL|P41999    (  18)  CLVCSDISTGYHYGVPSCNGCKTFFRRTIMKNQ  20

YQN7_CAEEL|Q09528    (  33)  CLICGEPSTGKHYGIVACLGCKTFFRRAVVQRQ  24

YRG4_CAEEL|Q09587    (  97)  HVCSSPTANTLHFGGRSCKACAAFFRRSVSMSM 100

7UP1_DROME|P16375    ( 200)  CVVCGDKSSGKHYGQFTCEGCKSFFKRSVRRNL   6
7UP2_DROME|P16376    ( 200)  CVVCGDKSSGKHYGQFTCEGCKSFFKRSVRRNL   6
ARP1_HUMAN|P24468    (  79)  CVVCGDKSSGKHYGQFTCEGCKSFFKRSVRRNL   6
ARP1_MOUSE|P43135    (  79)  CVVCGDKSSGKHYGQFTCEGCKSFFKRSVRRNL   6
COT1_MOUSE|Q60632    (  85)  CVVCGDKSSGKHYGQFTCEGCKSFFKRSVRRNL   6
COTF_HUMAN|P10589    (  86)  CVVCGDKSSGKHYGQFTCEGCKSFFKRSVRRNL   6
EAR2_HUMAN|P10588    (  56)  CVVCGDKSSGKHYGVFTCEGCKSFFKRTIRRNL   5
EAR2_MOUSE|P43136    (  57)  CVVCGDKSSGKHYGVFTCEGCKSFFKRTIRRNL   5
HR78_DROME|Q24142    (  52)  CLVCGDRASGRHYGAISCEGCKGFFKRSIRKQL   5
TR2_HUMAN|P13056     ( 113)  CVVCGDKASGRHYGAVTCEGCKGFFKRSIRKNL   5
TR4_HUMAN|P49116     ( 117)  CVVCGDKASGRHYGAVSCEGCKGFFKRSVRKNL   5
TR4_MOUSE|P49117     ( 117)  CVVCGDKASGRHYGAVSCEGCKGFFKRSVRKNL   5
TR4_RAT|P55094       ( 117)  CVVCGDKASGRHYGAVSCEGCKGFFKRSVRKNL   5
```

Figure 11.3. Structure of a typical BLOCKS entry. This is part of the entry for one block associated with steroid fingers. The structure of the entry is discussed in the text.

the header lines marked ID, AC, and DE give, in order, a short description of the family represented by this block, the BLOCKS database accession number, and a longer description of the family. The BL line gives information regarding the original sequence motif that was used to construct this particular block. The `width` and `seqs` parameters show how wide the block is, in residues, and how many sequences are in the block, respectively. Some information then follows regarding the statistical validity and the strength of the construct. Finally, a list of sequences is presented, showing only the part of the sequence corresponding to this particular motif. Each line begins with the SWISS-PROT accession number for the sequence, the number of the first residue shown based on the entire sequence, the sequence itself, and a position-based sequence weight. These values are scaled, with 100 representing the sequence that is most distant from the group. Notice that there are blank lines between some of the sequences; parts of the overall alignment are clustered, and, in each cluster, 80% of the sequence residues are identical.

CDD

Recently, NCBI introduced a new search service aimed at identifying conserved domains within a protein sequence. The source database for these searches is called the Conserved Domain Database or CDD. This is a secondary database, with entries derived from both Pfam (described above) and SMART (Simple Modular Architecture Research Tool). SMART can be used to identify genetically mobile domains and analyze domain architectures and is discussed in greater detail within the context of comparative genomics in Chapter 15. The actual search is performed using reverse position-specific BLAST (RPS-BLAST), which uses the query sequence to search a database of precalculated PSSTs.

The CDD interface is simple, providing a box for the input sequence (alternatively, an accession number can be specified) and a pull-down menu for selecting the target database. If conserved domains are identified within the input sequence, a graphic is returned showing the position of each conserved domain, followed by the actual alignment of the query sequence to the target domain as generated by RPS-BLAST. In these alignments, the default view shows identical residues in red, whereas conservative substitutions are shown in blue; users can also select from a variety of representations, including the traditional BLAST-style alignment display. Hyperlinks are provided back to the source databases, providing more information on that particular domain. This "CD Summary" page gives the underlying source database information, references, the taxonomy spanned by this entry, and a sequence entry representative of the group. In the lower part of the page, the user can construct an alignment of sequences of interest from the group; alternatively, the user can allow the computer to select the top-ranked sequences or a subset of sequences that are most diverse within the group. If a three-dimensional structure corresponding to the CD is available, it can be viewed directly using Cn3D (see Chapter 5). Clicking on the CD link next to any of the entries on the CD Summary page will, in essence, start the whole process over again, using *that* sequence to perform a new RPS-BLAST search against CDD.

SECONDARY STRUCTURE AND FOLDING CLASSES

One of the first steps in the analysis of a newly discovered protein or gene product of unknown function is to perform a BLAST or other similar search against the public databases. However, such a search might not produce a match against a known protein; if there is a statistically significant hit, there may not be any information in the sequence record regarding the secondary structure of the protein, information that is very important in the rational design of biochemical experiments. In the absence of "known" information, there are methods available for predicting the ability of a sequence to form α-helices and β-strands. These methods rely on observations made from groups of proteins whose three-dimensional structure has been experimentally determined.

A brief review of secondary structure and folding classes is warranted before the techniques themselves are discussed. As already alluded to, a significant number of amino acids have hydrophobic side chains, whereas the main chain, or backbone, is hydrophilic. The required balance between these two seemingly opposing forces is accomplished through the formation of discrete secondary structural elements, first described by Linus Pauling and colleagues in 1951 (Pauling and Corey, 1951). An α-helix is a corkscrew-type structure with the main chain forming the backbone and the side chains of the amino acids projecting outward from the helix. The backbone is stabilized by the formation of hydrogen bonds between the CO group of each amino acid and the NH group of the residue four positions C-terminal ($n + 4$), creating a tight, rodlike structure. Some residues form α-helices better than others; alanine, glutamine, leucine, and methionine are commonly found in α-helices, whereas proline, glycine, tyrosine, and serine usually are not. Proline is commonly thought of as a helix breaker because its bulky ring structure disrupts the formation of $n + 4$ hydrogen bonds.

In contrast, the β-strand is a much more extended structure. Rather than hydrogen bonds forming within the secondary structural unit itself, stabilization occurs through bonding with one or more *adjacent* β-strands. The overall structure formed through the interaction of these individual β-strands is known as a *β-pleated sheet.* These sheets can be parallel or antiparallel, depending on the orientation of the N- and C-terminal ends of each component β-strand. A variant of the β-sheet is the *β-turn*; in this structure the polypeptide chain makes a sharp, hairpin bend, producing an antiparallel β-sheet in the process.

In 1976, Levitt and Chothia proposed a classification system based on the order of secondary structural elements within a protein (Levitt and Chothia, 1976). Quite simply, an α-structure is made up primarily from α-helices, and a β-structure is made up of primarily β-strands. Myoglobin is the classic example of a protein composed entirely of α-helices, falling into the α class of structures (Takano, 1977). Plastocyanin is a good example of the β class, where the hydrogen-bonding pattern between eight β-strands form a compact, barrel-like structure (Guss and Freeman, 1983). The combination class, α/β, is made up of primarily β-strands alternating with α-helices. Flavodoxin is a good example of an α/β-protein; its β-strands form a central β-sheet, which is surrounded by α-helices (Burnett et al., 1974).

Predictive methods aimed at extracting secondary structural information from the linear primary sequence make extensive use of neural networks, traditionally used for analysis of patterns and trends. Basically, a neural network provides computational processes the ability to "learn" in an attempt to approximate human learning versus following instructions blindly in a sequential manner. Every neural network has an *input layer* and an *output layer*. In the case of secondary structure prediction, the input layer would be information from the sequence itself, and the output layer would be the probabilities of whether a particular residue could form a particular structure. Between the input and output layers would be one or more *hidden layers* where the actual "learning" would take place. This is accomplished by providing a training data set for the network. Here, an appropriate training set would be all sequences for which three-dimensional structures have been deduced. The network can process this information to look for what are possibly weak relationships between an amino acid sequence and the structures they can form in a particular context. A more complete discussion of neural networks as applied to secondary structure prediction can be found in Kneller et al. (1990).

nnpredict

The nnpredict algorithm uses a two-layer, feed-forward neural network to assign the predicted type for each residue (Kneller et al., 1990). In making the predictions, the server uses a FASTA format file with the sequence in either one-letter or three-letter code, as well as the folding class of the protein (α, β, or α/β). Residues are classified as being within an α-helix (H), a β-strand (E), or neither (—). If no prediction can be made for a given residue, a question mark (?) is returned to indicate that an assignment cannot be made with confidence. If no information is available regarding the folding class, the prediction can be made without a folding class being specified; this is the default. For the best-case prediction, the accuracy rate of nnpredict is reported as being over 65%.

Sequences are submitted to nnpredict by either sending an E-mail message to *nnpredict@celeste.ucsf.edu* or by using the Web-based submission form. With the use of flavodoxin as an example, the format of the E-mail message would be as follows:

```
option: a/b
>flavodoxin - Anacystis nidulans
AKIGLFYGTQTGVTQTIAESIQQEFGGESIVDLNDIANADASDLNAYDYLIIGCPTWNVGELQSDWEGIY
DDLDSVNFQGKKVAYFGAGDQVGYSDNFQDAMGILEEKISSLGSQTVGYWPIEGYDFNESKAVRNNQFVG
LAIDEDNQPDLTKNRIKTWVSQLKSEFGL
```

The Option line specifies the folding class of the protein: n uses no folding class for the prediction, a specifies α, b specifies β, and a/b specifies α/β. Only one sequence may be submitted per E-mail message. The results returned by the server are shown in modified form in Figure 11.4.

PredictProtein

PredictProtein (Rost et al., 1994) uses a slightly different approach in making its predictions. First, the protein sequence is used as a query against SWISS-PROT to find similar sequences. When similar sequences are found, an algorithm called MaxHom is used to generate a profile-based multiple sequence alignment (Sander and Schneider, 1991). MaxHom uses an iterative method to construct the alignment: After the first search of SWISS-PROT, all found sequences are aligned against the query sequence and a profile is calculated for the alignment. The profile is then used to search SWISS-PROT again to locate new, matching sequences. The multiple alignment generated by MaxHom is subsequently fed into a neural network for prediction by one of a suite of methods collectively known as PHD (Rost, 1996). PHDsec, the method in this suite used for secondary structure prediction, not only assigns each residue to a secondary structure type, it provides statistics indicating the confidence of the prediction at each position in the sequence. The method produces an average accuracy of better than 72%; the best-case residue predictions have an accuracy rate of over 90%.

Sequences are submitted to PredictProtein either by sending an E-mail message or by using a Web front end. Several options are available for sequence submission; the query sequences can be submitted as single-letter amino acid code or by its SWISS-PROT identifier. In addition, a multiple sequence alignment in FASTA format or as a PIR alignment can also be submitted for secondary structure prediction.

The input message, sent to *predictprotein@embl-heidelberg.de*, takes the following form:

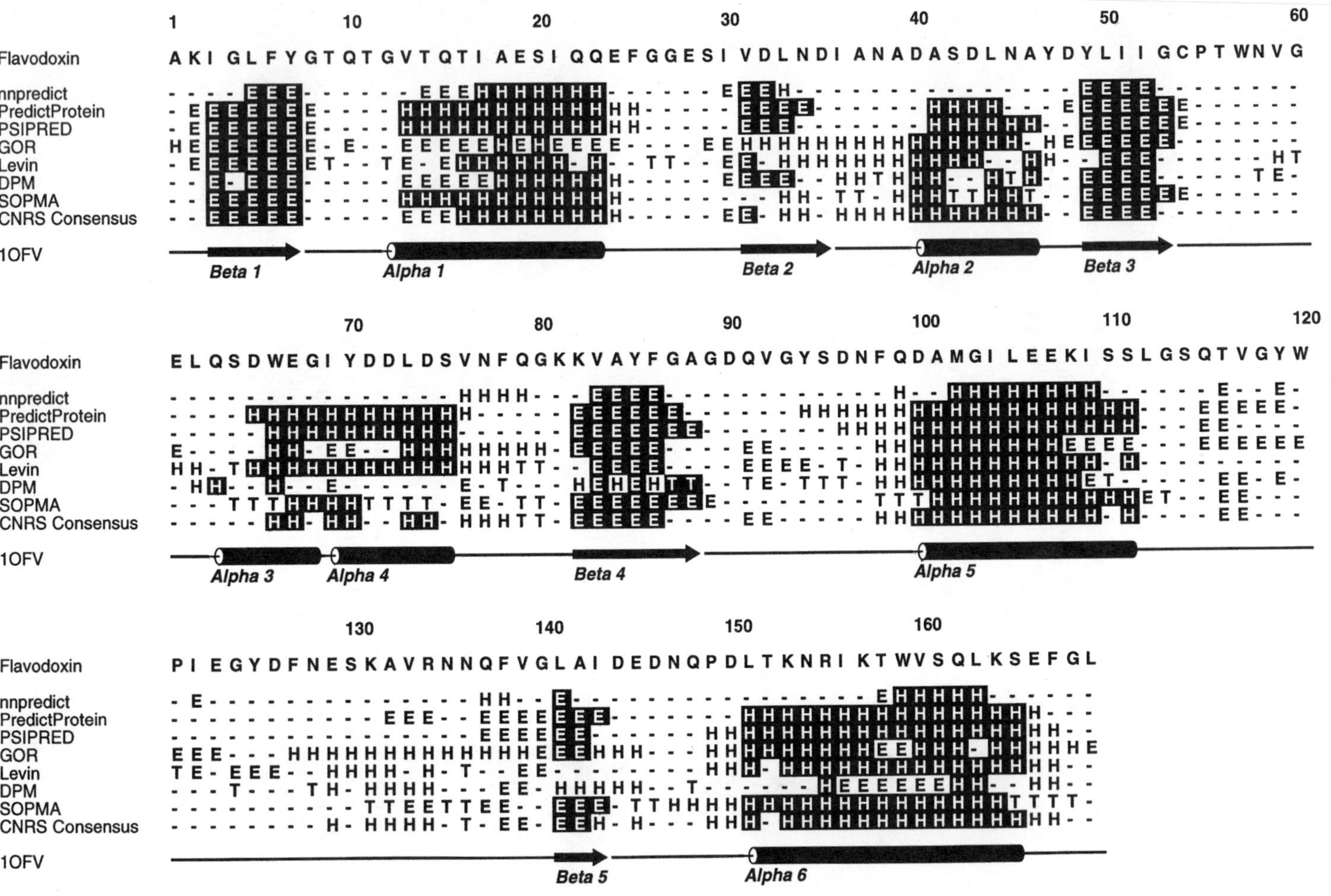

Figure 11.4. Comparison of secondary structure predictions by various methods. The sequence of flavodoxin was used as the query and is shown on the first line of the alignment. For each prediction, H denotes an α-helix, E a β-strand, and T a β-turn; all other positions are assumed to be random coil. Correctly assigned residues are shown in inverse type. The methods used are listed along the left side of the alignment and are described in the text. At the bottom of the figure is the secondary structure assignment given in the PDB file for flavodoxin (1OFV, Smith et al., 1983).

```
Joe Buzzcut
National Human Genome Research Institute, NIH
buzzcut@baldguys.org
do NOT align
# FASTA list     homeodomain proteins
>ANTP
---KRGRQTYTRYQTLELEKEFHFNRYLTRRRRIEIAHALSLTERQIKIWFQNRRMKWKK
>HDD
MDEKRPRTAFSSEQLARLKREFNENRYLTERRRQQLSSELGLNEAQIKIWFQNKRAKIKK
>DLX
-KIRKPRTIYSSLQLQALNHRFQQTQYLALPERAELAASLGLTQTQVKIWFQNKRSKFKK
>FTT
---RKRRVLFSQAQVYELERRFKQQKYLSAPEREHLASMIHLTPTQVKIWFQNHRYKMKR
>Pax6
--LQRNRTSFTQEQIEALEKEFERTHYPDVFARERLAAKIDLPEARIQVWFSNRRAKWRR
```

Above is an example of a FASTA-formatted multiple sequence alignment of homeodomain proteins submitted for secondary structure prediction. After the name, affiliation, and address lines, the # sign signals to the server that a sequence in one-letter code follows. The sequence format is essentially FASTA, except that blanks are not allowed. For this alignment, the phrase `do NOT align` before the line starting with # assures that the alignment will not be realigned. Nothing is allowed to follow the sequence. The output sent as an E-mail message is quite copious but contains a large amount of pertinent information. The results can also be retrieved from an ftp site by adding a qualifier `return no mail` in any line before the line starting with #. This might be a useful feature for those E-mail services that have difficulty handling very large output files. The format for the output file can be plain text or HTML files with or without PHD graphics.

The results of the MaxHom search are returned, complete with a multiple alignment that may be of use in further study, such as profile searches or phylogenetic studies. If the submitted sequence has a known homolog in PDB, the PDB identifiers are furnished. Information follows on the method itself and then the actual prediction will follow. In a recent release, the output can also be customized by specifying available options. Unlike nnpredict, PredictProtein returns a "reliability index of prediction" for each position ranging from 0 to 9, with 9 being the maximum confidence that a secondary structure assignment has been made correctly. The results returned by the server for this particular sequence, as compared with those obtained by other methods, are shown in modified form in Figure 11.4.

PREDATOR

The PREDATOR secondary structure prediction algorithm is based on recognition of potentially hydrogen-bonded residues in the amino acid sequence (Frishman and Argos, 1997). It uses database-derived statistics on residue-type occurrences in different classes of local hydrogen-bonded structures. The novel feature of this method is its reliance on local pairwise alignment of the sequence to be predicted between each related sequence. The input for this program can be a single sequence or a set of *unaligned*, related sequences. Sequences can be submitted to the PREDATOR server either by sending an E-mail message to *predator@embl-heidelberg.de* or by using a Web front end. The input sequences can be either FASTA, MSF, or CLUSTAL format. The mean prediction accuracy of PREDATOR in three structural states is 68% for a single sequence and 75% for a set of related sequences.

PSIPRED

The PSIPRED method, developed at the University of Warwick, UK, uses the knowledge inferred from PSI-BLAST (Altschul et al., 1997; cf. Chapter 8) searches of the input sequence to perform predictions. PSIPRED uses two feedforward neural networks to perform the analysis on the profile obtained from PSI-BLAST. Sequences can be submitted through a simple Web front end, in either single-letter raw format or in FASTA format. The results from the PSIPRED prediction are returned as a text file in an E-mail message. In addition, a link is also provided in the E-mail message to a graphical representation of the secondary structure prediction, visualized using a Java application called PSIPREDview. In this representation, the positions of the helices and strands are schematically represented above the target sequence. The average prediction accuracy for PSIPRED in three structural states is 76.5%, which is higher than any of the other methods described here.

SOPMA

The Protein Sequence Analysis server at the Centre National de la Recherche Scientifique (CNRS) in Lyons, France, takes a unique approach in making secondary structure predictions: rather than using a single method, it uses *five*, the predictions from which are subsequently used to come up with a "consensus prediction." The methods used are the Garnier–Gibrat–Robson (GOR) method (Garnier et al., 1996), the Levin homolog method (Levin et al., 1986), the double-prediction method (Deléage and Roux, 1987), the PHD method described above as part of PredictProtein, and the method of CNRS itself, called SOPMA (Geourjon and Déleage, 1995). Briefly, this self-optimized prediction method builds subdatabases of protein sequences with known secondary structures; each of the proteins in a subdatabase is then subjected to secondary structure prediction based on sequence similarity. The information from the subdatabases is then used to generate a prediction on the query sequence.

The method can be run by submitting just the sequence itself in single-letter format to *deleage@ibcp.fr*, using `SOPMA` as the subject of the mail message, or by using the SOPMA Web interface. The output from each of the component predictions, as well as the consensus, is shown in Figure 11.4.

Comparison of Methods

On the basis of Figure 11.4, it is immediately apparent that all the methods described above do a relatively good, but not perfect, job of predicting secondary structures. Where no other information is known, the best approach is to perform predictions using all the available algorithms and then to judge the validity of the predictions in comparison to one another. Flavodoxin was selected as the input query because it has a relatively intricate structure, falling into the α/β-folding class with its six α-helices and five β-sheets. Some assignments were consistently made by all methods; for example, all the methods detected $\beta1$, $\beta3$, $\beta4$, and $\alpha5$ fairly well. However, some methods missed some elements altogether (e.g., nnpredict with $\alpha2$, $\alpha3$, and $\alpha4$), and some predictions made no biological sense (e.g., the double-prediction method and $\beta4$, where helices, sheets, and turns alternate residue by residue). PredictProtein and PSIPRED, which both correctly found all the secondary structure elements and, in several places, identified structures of the correct length, *appear* to have made the best overall prediction. This is *not* to say that the other methods are not useful or not as good; undoubtedly, in some cases, another method would have emerged as having made a better prediction. This approach does not provide a fail-safe method of prediction, but it does reinforce the level of confidence resulting from these predictions.

A new Web-based server, Jpred, integrates six different structure prediction methods and returns a consensus prediction based on simple majority rule. The usefulness of this server is that it automatically generates the input and output requirements for all six prediction algorithms, which can be an important feature when handling large data sets. The input sequence for Jpred can be a single protein sequence in FASTA or PIR format, a set of unaligned sequences in PIR format, or a multiple sequence alignment in MSF or BLC format. In case of a single sequence, the server first generates a set of related sequences by searching the OWL database using the BLASTP algorithm. The sequence set is filtered using SCANPS and then pairwise-compared using AMPS. Finally, the sequence set is clustered using a 75% identity cutoff value to remove any bias in the sequence set, and the remaining sequences are aligned using CLUSTAL W. The Jpred server runs PHD (Rost and Sander, 1993), DSC (King and Sternberg, 1996), NNSSP (Salamov and Solovyev, 1995), PREDATOR (Frishman and Argos, 1997), ZPRED (Zvelebil et al., 1987), and MULPRED (Barton, 1988). The results from the Jpred server is returned as a text file in an E-mail message; a link is also provided to view the colored graphical representation in HTML or PostScript file format. The consensus prediction from the Jpred server has an accuracy of 72.9% in the three structural states.

SPECIALIZED STRUCTURES OR FEATURES

Just as the position of α-helices and β-sheets can be predicted with a relatively high degree of confidence, the presence of certain specialized structures or features, such as coiled coils and transmembrane regions, can be predicted. There are not as many methods for making such predictions as there are for secondary structures, primarily because the rules of folding that induce these structures are not completely understood. Despite this, when query sequences are searched against databases of known structures, the accuracy of prediction can be quite high.

Coiled Coils

The COILS algorithm runs a query sequence against a database of proteins known to have a coiled-coil structure (Lupas et al., 1991). The program also compares query sequences to a PDB subset containing globular sequences and on the basis of the differences in scoring between the PDB subset and the coiled coils database, determines the probability with which the input sequence can form a coiled coil. COILS can be downloaded for use with VAX/VMS or may more easily be used through a simple Web interface.

The program takes sequence data in GCG or FASTA format; one or more sequences can be submitted at once. In addition to the sequences, users may select one of two scoring matrices: MTK, based on the sequences of myosin, tropomyosin, and keratin, or MTIDK, based on myosin, tropomyosin, intermediate filaments types I–V, desmosomal proteins, and kinesins. The authors cite a trade-off between the scoring matrices, with MTK being better for detecting two-stranded structures and MTIDK being better for all other cases. Users may invoke an option that gives the same weight to the residues at the *a* and *d* positions of each coil (normally hydrophobic) as that given to the residues at the *b*, *c*, *e*, *f*, and *g* positions (normally hydrophilic). If the results of running COILS both weighted and unweighted are substantially different, it is likely that a false positive has been found. The authors caution that COILS is designed to detect solvent-exposed, left-handed coiled coils and that buried or right-handed coiled coils may not be detected. When a query is submitted to the Web server, a prediction graph showing the propensity toward the formation of a coiled coil along the length of the sequence is generated.

A slightly easier to interpret output comes from MacStripe, a Macintosh-based application that uses the Lupas COILS method to make its predictions (Knight, 1994). MacStripe takes an input file in FASTA, PIR, and other common file formats and, like COILS, produces a plot file containing a histogram of the probability of forming a coiled coil, along with bars showing the continuity of the heptad repeat pattern. The following portion of the statistics file generated by MacStripe uses the complete sequence of GCN4 as an example:

```
 89  89 L 5 a 0.760448 0.000047
 90  90 D 5 b 0.760448 0.000047
 91  91 D 5 c 0.760448 0.000047
 92  92 A 5 d 0.760448 0.000047
 93  93 V 5 e 0.760448 0.000047
 94  94 V 5 f 0.760448 0.000047
 95  95 E 5 g 0.760448 0.000047
 96  96 S 5 a 0.760448 0.000047
 97  97 F 5 b 0.760448 0.000047
 98  98 F 5 c 0.774300 0.000058
 99  99 S 5 d 0.812161 0.000101
100 100 S 5 e 0.812161 0.000101
101 101 S 5 f 0.812161 0.000101
102 102 T 5 g 0.812161 0.000101
```

The columns, from left to right, represent the residue number (shown twice), the amino acid, the heptad frame, the position of the residue within the heptad (`a-b-c-d-e-f-g`), the Lupas score, and the Lupas probability. In this case, from the fifth column, we can easily discern a heptad repeat pattern. Examination of the results for the entire GCN4 sequence shows that the heptad pattern is fairly well maintained but falls apart in certain areas. The statistics should not be ignored; however, the results are easier to interpret if the heptad pattern information is clearly presented. It is possible to get a similar type of output from COILS but not through the COILS Web server; instead, a C-based program must be installed on an appropriate Unix machine, a step that may be untenable for many users.

Transmembrane Regions

The Kyte-Doolittle TGREASE algorithm discussed above is very useful in detecting regions of high hydrophobicity, but, as such, it does not exclusively predict transmembrane regions because buried domains in soluble, globular proteins can also be primarily hydrophobic. We consider first a predictive method specifically for the prediction of transmembrane regions. This method, TMpred, relies on a database of transmembrane proteins called TMbase (Hofmann and Stoffel, 1993). TMbase, which is derived from SWISS-PROT, contains additional information on each sequence regarding the number of transmembrane domains they possess, the location of these domains, and the nature of the flanking sequences. TMpred uses this information in conjunction with several weight matrices in making its predictions.

The TMpred Web interface is very simple. The sequence, in one-letter code, is pasted into the query sequence box, and the user can specify the minimum and maximum lengths of the hydrophobic part of the transmembrane helix to be used in the analysis. The output has four sections: a list of possible transmembrane helices, a table of correspondences, suggested models for transmembrane topology, and a graphic representation of the same results. When the sequence of the G-protein-coupled receptor (P51684) served as the query, the following models were generated:

```
2 possible models considered, only significant TM-segments used

-----> STRONGLY preferred model: N-terminus outside
 7 strong transmembrane helices, total score : 14211
 # from   to length score orientation
 1   55   74 (20)    2707 o-i
 2   83  104 (22)    1914 i-o
 3  120  141 (22)    1451 o-i
 4  166  184 (19)    2170 i-o
 5  212  235 (24)    2530 o-i
 6  255  276 (22)    2140 i-o
 7  299  319 (21)    1299 o-i

-----> alternative model
 7 strong transmembrane helices, total score : 12079
 # from   to length score orientation
 1   47   69 (23)    2494 i-o
 2   84  104 (21)    1470 o-i
 3  123  141 (19)    1383 i-o
 4  166  185 (20)    1934 o-i
 5  219  236 (18)    2474 i-o
 6  252  274 (23)    1386 o-i
 7  303  319 (17)     938 i-o
```

Each of the proposed models indicates the starting and ending position of each segment, along with the relative orientation (inside-to-outside or outside-to-inside) of each segment. The authors appropriately caution that the models are based on the assumption that all transmembrane regions were found during the prediction. These models, then, should be considered in light of the raw data also generated by this method.

PHDtopology

One of the most useful methods for predicting transmembrane helices is PHDtopology, which is related to the PredictProtein secondary structure prediction method described above. Here, programs within the PHD suite are now used in an obviously different way to make a prediction on a membrane-bound rather than on a soluble protein. The method has reported accuracies that are nearly perfect: the accuracy of predicting a transmembrane helix is 92% and the accuracy for a loop is 96%, giving an overall two-state accuracy of 94.7%. One of the features of this program is that, in addition to predicting the putative transmembrane regions, it indicates the orientation of the loop regions with respect to the membrane.

As before, PHDtopology predictions can be made using either an E-mail server or a Web front end. If an E-mail server is used, the format is identical to that shown for PredictProtein above, except that the line `predict htm topology` must precede the line beginning with the pound sign. Regardless of submission method, results are returned by E-mail. An example of the output returned by PHDtopology is shown in Figure 11.5.

Signal Peptides

The Center for Biological Sequence Analysis at the Technical University of Denmark has developed SignalP, a powerful tool for the detection of signal peptides and their

```
Joe Buzzcut
National Human Genome Research Institute, NIH
buzzcut@nhgri.nih.gov
predict htm topology
# pendrin
MAAPGGRSEPPQLPEYSCSYMVSRPVYSELAFQQQHERRLQERKTLRESLAKCCSCSRKRAFGVLKTLVPILEWLPKYRV
KEWLLSDVISGVSTGLVATLQGMAYALLAAVPGYGLYSAFFPILTYFIFGTSRHISVGPFPVVSLMVGSVVLSMAP...
```

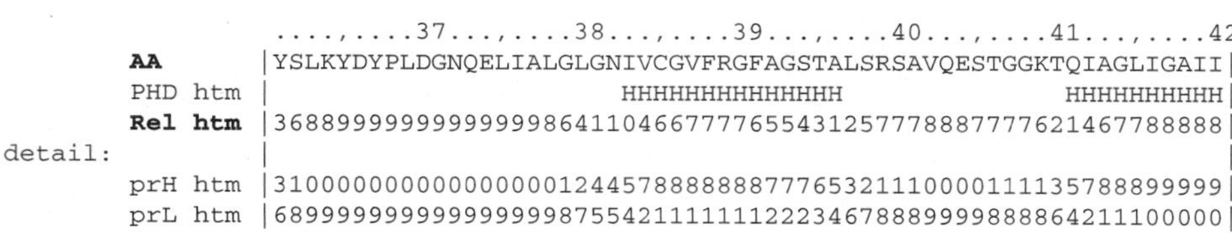

```
                    ....,....37...,....38...,....39...,....40...,....41...,....42
         AA        |YSLKYDYPLDGNQELIALGLGNIVCGVFRGFAGSTALSRSAVQESTGGKTQIAGLIGAII|
         PHD htm   |                  HHHHHHHHHHHH                    HHHHHHHH|
         Rel htm   |36889999999999999864110466777765543125777888777621467788888|
detail:            |                                                  |
         prH htm   |31000000000000000012445788888887776532111000011113578889999|
         prL htm   |68999999999999999987554211111112223467888999998888642111000|
                                   .
                                   .
                                   .
         PHDThtm   |iiiiiiiiiiiiiiiiTTTTTTTTTTTTTTToooooooooooooTTTTTTT|
```

Figure 11.5. Partial output from a PHDtopology prediction. The input sequence is pendrin, which is responsible for Pendred syndrome (Everett et al., 1998). The row labeled `AA` shows a portion of the input sequence, and the row labeled `Rel htm` gives the reliability index of prediction at each position of the protein; values range from 0 to 9, with 9 representing the maximum possible confidence for the assignment at that position. The last line, labeled `PHDThm`, contains one of three letters: a `T` represents a transmembrane region, whereas an `i` or `o` represents the orientation of the loop with respect to the membrane (inside or outside).

cleavage sites (Nielsen et al., 1997). The algorithm is neural-network based, using separate sets of Gram-negative prokaryotic, Gram-positive prokaryotic, and eukaryotic sequences with known signal sequences as the training sets. SignalP predicts secretory signal peptides and not those that are involved in intracellular signal transduction.

Using the Web interface, the sequence of the human insulin-like growth factor IB precursor (somatomedin C, P05019), whose cleavage site is known, was submitted to SignalP for analysis. The eukaryotic training set was used in the prediction, and the results of the analysis are as follows:

```
******************** SignalP predictions ********************
Using networks trained on euk data
>IGF-IB length = 195
# pos aa C S Y
  .


  .
 46 A 0.365 0.823 0.495
 47 T 0.450 0.654 0.577
 48 A 0.176 0.564 0.369
 49 G 0.925 0.205 0.855
 50 P 0.185 0.163 0.376
  .
  .
  .
< Is the sequence a signal peptide?
# Measure Position Value Cutoff Conclusion
 max. C 49 0.925 0.37 YES
 max. Y 49 0.855 0.34 YES
 max. S 37 0.973 0.88 YES
 mean S 1 - 48 0.550 0.48 YES
# Most likely cleavage site between pos. 48 and 49: ATA-GP
```

In the first part of the output, the column labeled `C` is a raw cleavage site score. The value of C is highest at the position C-terminal to the cleavage site. The column labeled `S` contains the signal peptide scores, which are high at all positions before the cleavage site and very low after the cleavage site. S is also low in the N-termini of nonsecretory proteins. Finally, the `Y` column gives the combined cleavage site score, a geometric average indicating when the C score is high and the point at which the S score shifts from high to low. The end of the output file asks the question, "Is the sequence a signal peptide?" On the basis of the statistics, the most likely cleavage site is deduced. On the basis of the SWISS-PROT entry for this protein, the mature chain begins at position 49, the same position predicted to be the most likely cleavage site by SignalP.

Nonglobular Regions

The use of the program SEG in the masking of low-complexity segments prior to database searches was discussed in Chapter 8. The same algorithm can also be used

```
                                  1-307  MAGAIASRMSFSSLKRKQPKTFTVRIVTMD
                                         AEMEFNCEMKWKGKDLFDLVCRTLGLRETW
                                         FFGLQYTIKDTVAWLKMDKKVLDHDVSKEE
                                         PVTFHFLAKFYPENAEEELVQEITQHLFFL
                                         QVKKQILDEKIYCPPEASVLLASYAVQAKY
                                         GDYDPSVHKRGFLAQEELLPKRVINLYQMT
                                         PEMWEERITAWYAEHRGRARDEAEMEYLKI
                                         AQDLEMYGVNYFAIRNKKGTELLLGVDALG
                                         LHIYDPENRLTPKISFPWNEIRNISYSDKE
                                         FTIKPLDKKIDVFKFNSSKLRVNKLILQLC
                                         IGNHDLF
mrrrkadslevqqmkaqareekarkqmerq  308-478
rlarekqmreeaertrdelerrllqmkeea
tmanealmrseetadllaekaqiteeeakl
laqkaaeaeqemqrikatairteeekrlme
qkvleaevlalkmaeeserrakeadqlkqd
        lqeareaerrakqklleiatk
                                479-496  PTYPPMNPIPAPLPPDIP
sfnligdslsfdfkdtdmkrlsmeiekekv  497-587
eymekskhlqeqlnelkteiealklkeret
aldilhnensdrggsskhntikkltlqsak
                             s
                                588-595  RVAFFEEL
```

Figure 11.6. Predicted nonglobular regions for the protein product of the neurofibromatosis type 2 gene (L11353) as deduced by SEG. The nonglobular regions are shown in the left-hand column in lowercase. Numbers denote residue positions for each block.

to detect putative nonglobular regions of protein sequences by altering the trigger window length W, the trigger complexity K_1, and extension complexity K_2. When the command `seg sequence.txt 45 3.4 3.75` is received, SEG will use a longer window length than the default of 12, thereby detecting long, nonglobular domains. An example of using SEG to detect nonglobular regions is shown in Figure 11.6.

TERTIARY STRUCTURE

By far the most complex and technically demanding predictive method based on protein sequence data has to do with structure prediction. The importance of being able to adequately and accurately predict structure based on sequence is rooted in the knowledge that, whereas sequence may specify conformation, the same conformation may be specified by multiple sequences. The ideas that structure is conserved to a much greater extent than sequence and that there is a limited number of backbone motifs (Chothia and Lesk, 1986; Chothia, 1992) indicate that similarities between proteins may not necessarily be detected through traditional, sequence-based methods only. Deducing the relationship between sequence and structure is at the root of the "protein-folding problem," and current research on the problem has been the focus of several reviews (Bryant and Altschul, 1995; Eisenhaber et al., 1995; Lemer et al., 1995).

The most robust of the structure prediction techniques is homology model building or "threading" (Bryant and Lawrence, 1993; Fetrow and Bryant, 1993; Jones and Thornton, 1996). The threading methods search for structures that have a similar fold without apparent sequence similarity. This method takes a query sequence whose structure is not known and threads it through the coordinates of a target protein whose structure has been solved, either by X-ray crystallography or NMR imaging. The sequence is moved position by position through the structure, subject to some predetermined physical constraints; for example, the lengths of secondary structure elements and loop regions may be either fixed or varying within a given range. For each placement of sequence against structure, pairwise and hydrophobic interactions between nonlocal residues are determined. These thermodynamic calculations are used to determine the most energetically favorable and conformationally stable alignment of the query sequence against the target structure. Programs such as this are computationally intensive, requiring, at a minimum, a powerful UNIX workstation; they also require knowledge of specialized computer languages. The threading methods are useful when the sequence-based structure prediction methods fail to identify a suitable template structure.

Although techniques such as threading are obviously very powerful, their current requirements in terms of both hardware and expertise may prove to be obstacles to most biologists. In an attempt to lower the height of the barrier, easy-to-use programs have been developed to give the average biologist a good first approximation for comparative protein modeling. (Numerous commercial protein structure analysis tools, such as WHAT-IF and LOOK, provide advanced capabilities, but this discussion is limited to Web-based freeware.)

The use of SWISS-MODEL, a program that performs automated sequence-structure comparisons (Peitsch, 1996), is a two-step process. The First Approach mode is used to determine whether a sequence can be modeled at all; when a sequence is submitted, SWISS-MODEL compares it with the crystallographic database (ExPdb), and modeling is attempted only if there is a homolog in ExPdb to the query protein. The template structures are selected if there is at least 25% sequence identity in a region more than 20 residues long. If the first approach finds one or more appropriate entries in ExPdb, atomic models are built and energy minimization is performed to generate the best model. The atomic coordinates for the model as well as the structural alignments are returned as an E-mail message. Those results can be resubmitted to SWISS-MODEL using its Optimize mode, which allows for alteration of the proposed structure based on other knowledge, such as biochemical information. An example of the output from SWISS-MODEL is shown in Figure 11.7.

Another automated protein fold recognition method, developed at UCLA, incorporates predicted secondary structural information on the probe sequence in addition to sequence-based matches to assign a probable protein fold to the query sequence. In this method, correct assignment of the fold depends on the ranked scores generated for the probe sequence, based on its compatibility with each of the structures in a library of target three-dimensional structures. The inclusion of the predicted secondary structure in the analysis improves fold assignment by about 25%. The input for this method is a single protein sequence submitted through a Web front end. A Web page containing the results is returned to the user, and the results are physically stored on the UCLA server for future reference.

The second approach compares structures with structures, in the same light as the vector alignment search tool (VAST) discussed in Chapter 5 does. The DALI algorithm looks for similar contact patterns between two proteins, performs an optimization, and returns the best set of structure alignment solutions for those proteins (Holm and Sander, 1993). The method is flexible in that gaps may be of any length,

```
TARGET       1           QRRQ RTHFTSQQLQ QLEATFQRNR YPDMSTREEI AVWTNLTEAR
11FJL        0          KQRRS RTTFSASQLD ELERAFERTQ YPDIYTREEL AQRTNLTEAR
21FJL        1           QRRS RTTFSASQLD ELERAFERTQ YPDIYTREEL AQRTNLTEAR
11B72        1     ARTFDWMKVL RTNFTTRQLT ELEKEFHFNK YLSRARRVEI AATLELNETQ
22HDD        1            KRP RTAFSSEQLA RLKREFNENR YLTERRRQQL SSELGLNEAQ
12HOA        0          MRKRG RQTYTRYQTL ELEKEFHFNR YLTRRRRIEI AHALSLTERQ
                              .    *  ..   *     *    *      *        * . .  .   * * .
TARGET
11FJL                               hhhhhh hhhhhhhhh      hhhhhhhh hhhhh hhhh
21FJL                               hhhhhh hhhhhhhhh      hhhhhhhh hhhhh hhhh
11B72                               hhhhhh hhhhhhhhh      hhhhhhhh hhhhh hhhh
22HDD                               hhhhhh hhhhhhhhh      hhhhhhhh hhhhh hhhh
12HOA                               hhhhhh hhhhhhhhh      hhhhhhhh hhhh  hhhh
```

```
ATOM      1  H1   GLN     1       9.226 107.177  13.966  1.00 99.00
ATOM      2  H2   GLN     1      10.769 107.671  13.751  1.00 99.00
ATOM      3  N    GLN     1       9.824 107.785  13.444  1.00 25.00
ATOM      4  H3   GLN     1       9.549 108.738  13.592  1.00 99.00
ATOM      5  CA   GLN     1       9.728 107.473  11.999  1.00 25.00
ATOM      6  CB   GLN     1       8.265 107.520  11.538  1.00 25.00
ATOM      7  CG   GLN     1       7.468 106.270  11.932  1.00 25.00
ATOM      8  CD   GLN     1       8.001 104.970  11.312  1.00 25.00
ATOM      9  OE1  GLN     1       8.748 104.928  10.343  1.00 25.00
ATOM     10  NE2  GLN     1       7.629 103.853  11.899  1.00 25.00
ATOM     11  HE21GLN     1       7.979 103.008  11.502  1.00 99.00
ATOM     12  HE22GLN     1       7.015 103.860  12.683  1.00 99.00
```

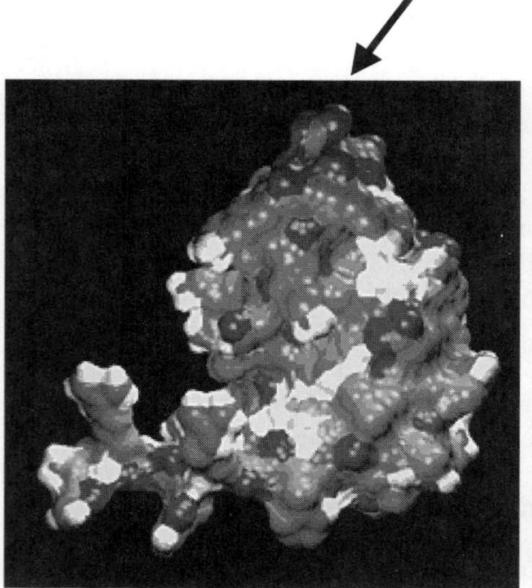

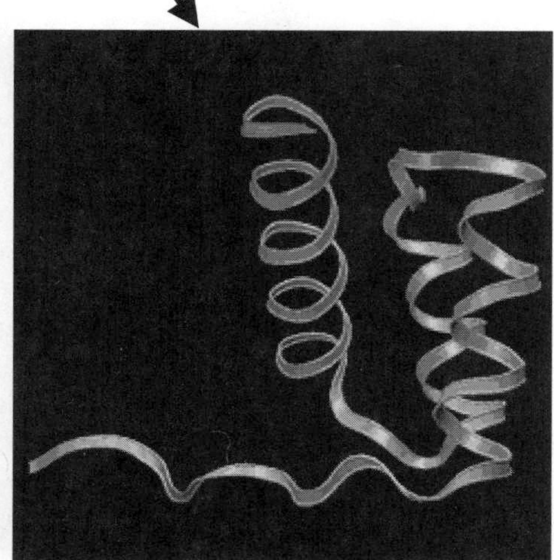

Figure 11.7. Molecular modeling using SWISS-MODEL. The input sequence for the structure prediction is the homeodomain region of human PITX2 protein. The output from SWISS-MODEL contains a text file containing a multiple sequence alignment, showing the alignment of the query against selected template structures from the Protein Data Bank (*top*). Also provided as part of the output is an atomic coordinate file for the target structure (*center*). In this example, the atomic coordinates of the target structure have been used to build a surface representation of the derived model using GRASP (*lower left*) and a ribbon representation of the derived model using RASMOL (*lower right*). (See color plate.)

and it allows for alternate connectivities between aligned segments, thereby facilitating identification of specific domains that are similar in two different proteins, even if the proteins as a whole are dissimilar. The DALI Web interface will perform the analysis on either two sets of coordinates already in PDB or by using a set of coordinates in PDB format submitted by the user. Alternatively, if both proteins of interest are present in PDB, their precomputed structural neighbors can be found by accessing the FSSP database of structurally aligned protein fold families (Holm and Sander, 1994), an "all-against-all" comparison of PDB entries.

The final method to be discussed here expands on the PHD secondary structure method discussed above. In the TOPITS method (Rost, 1995), a searchable database is created by translating the three-dimensional structure of proteins in PDB into one-dimensional "strings" of secondary structure. Then, the secondary structure and solvent accessibility of the query sequence is determined by the PHD method, with the results of this computation also being stored as a one-dimensional string. The query and target strings are then aligned by dynamic programming, to make the structure prediction. The results are returned as a ranked list, indicating the optimal alignment of the query sequence against the target structure, along with a probability estimate (Z-score) of the accuracy of the prediction.

The methods discussed here are fairly elementary, hence their speed in returning results and their ability to be adapted to a Web-style interface. Their level of performance is impressive in that they often can detect weak structural similarities between proteins. Although the protein-folding problem is nowhere near being solved, numerous protein folds can reliably be identified using intricate methods that are continuously being refined. Because different methods proved to have different strengths, it is always prudent to use a "consensus approach," similar to the approach used in the secondary structure prediction examples given earlier. The timing of these computational developments is quite exciting, inasmuch as concurrence with the imminent completion of the Human Genome Project will give investigators a powerful handle for predicting structure-function relationships as putative gene products are identified.

INTERNET RESOURCES FOR TOPICS PRESENTED IN CHAPTER 11

PREDICTION OF PHYSICAL PROPERTIES

Compute pI/MW	*http://www.expasy.ch/tools/pi_tool.html*
MOWSE	*http://srs.hgmp.mrc.ac.uk/cgi-bin/mowse*
PeptideMass	*http://www.expasy.ch/tools/peptide-mass.html*
TGREASE	*ftp://ftp.virginia.edu/pub/fasta/*
SAPS	*http://www.isrec.isb-sib.ch/software/SAPS_form.html*

PREDICTION OF PROTEIN IDENTITY BASED ON COMPOSITION

AACompIdent	*http://www.expasy.ch/tools/aacomp/*
AACompSim	*http://www.expasy.ch/tools/aacsim/*
PROPSEARCH	*http://www.embl-heidelberg.de/prs.html*

MOTIFS AND PATTERNS

BLOCKS	*http://blocks.fhcrc.org*
Pfam	*http://www.sanger.ac.uk/Software/Pfam/*
PRINTS	*http://www.bioinf.man.ac.uk/dbbrowser/PRINTS/PRINTS.html*

ProfileScan	*http://www.isrec.isb-sib.ch/software/PFSCAN_form.html*
PREDICTION OF SECONDARY STRUCTURE AND FOLDING CLASSES	
nnpredict	*http://www.cmpharm.ucsf.edu/~nomi/nnpredict.html*
PredictProtein	*http://www.embl-heidelberg.de/predictprotein/*
SOPMA	*http://pbil.ibcp.fr/*
Jpred	*http://jura.ebi.ac.uk:8888/*
PSIPRED	*http://insulin.brunel.ac.uk/psipred*
PREDATOR	*http://www.embl-heidelberg.de/predator/predator_info.html*
PREDICTION OF SPECIALIZED STRUCTURES OR FEATURES	
COILS	*http://www.ch.embnet.org/software/COILS_form.html*
MacStripe	*http://www.york.ac.uk/depts/biol/units/coils/mstr2.html*
PHDtopology	*http://www.embl-heidelberg.de/predictprotein*
SignalP	*http://www.cbs.dtu.dk/services/SignalP/*
TMpred	*http://www.isrec.isb-sib.ch/ftp-server/tmpred/www/TMPRED_form.html*
STRUCTURE PREDICTION	
DALI	*http://www2.ebi.ac.uk/dali/*
Bryant-Lawrence	*ftp://ncbi.nlm.nih.gov/pub/pkb/*
FSSP	*http://www2.ebi.ac.uk/dali/fssp/*
UCLA-DOE	*http://fold.doe-mbi.ucla.edu/Home*
SWISS-MODEL	*http://www.expasy.ch/swissmod/SWISS-MODEL.html*
TOPITS	*http://www.embl-heidelberg.de/predictprotein/*

PROBLEM SET

The sequence analyzed in the problem set in Chapter 10 yields at least one protein translation. Characterize this protein translation by answering the following questions.

1. Use ProtParam to determine the basic physicochemical properties of the unknown (leave the def line *out* when pasting the sequence into the query box).
- What is the molecular weight (in kilodaltons) and predicted isoelectric point (pI) for the protein?

2. Based on the pI and the distribution of charged residues, would this unknown possibly be involved in binding to DNA? Perform a BLASTP search on the unknown, using SWISS-PROT as the target database. Run BLASTP using pairwise as the Alignment View. *For each part of this question, consider the first protein in the hit list having a non-zero E-value.*
- What is the identity of this best, non-zero *E*-value hit, and what percent identity does the unknown share with this protein? For each alignment given, show the percent identity **and** the overall length of the alignment.
- Based on the BLASTP results *alone*, can any general observations be made regarding the putative function or cellular role of the unknown? *Do not just*

name the unknown—tell what you think the function of the unknown might be in the cell, based on ***all*** *of the significant hits in the BLASTP results.*

3. Does ProfileScan yield any additional information about the domain structure of this protein?
   - What types of domains were found? How many of each of these domains are present in the unknown?
   - Does the protein contain any low-complexity regions? If so, where?
   - Following the PDOC links to the right of the found domains, can any conclusions be made as to the cellular localization of this protein?

4. Does this protein have a putative signal sequence, based on SignalP? If so, what residues comprise the signal sequence? Is the result obtained from SignalP consistent with the BLASTP results and any associated GenBank entries?

5. Submit the sequence of the unknown to PHDtopology. On the basis of the results, draw a schematic of the protein, showing
   - the approximate location of any putative transmembrane helices and
   - the orientation of the N- and C-termini with respect to the membrane.

REFERENCES

Akrigg, D., Bleasby, A. J., Dix, N. I. M., Findlay, J. B. C., North, A. C. T., Parry-Smith, D., Wootton, J. C., Blundell, T. I., Gardner, S. P., Hayes, F., Sternberg, M. J. E., Thornton, J. M., Tickle, I. J., and Murray-Rust, P. (1988). A protein sequence/structure database. *Nature* 335, 745–746.

Altschul, S. F., Madden, T. L., Schaffer, A. A., Zhang, J., Zhang, Z., Miller, W. and Lipman, D. J. (1997) Gapped BLAST and PSI-BLAST: a new generation of protein database search programs. *Nucleic Acids Res.* 25, 3389–3402.

Anfinsen, C. B., Haber, E., Sela, M., and White, F. H. (1961). The kinetics of the formation of native ribonuclease during oxidation of the reduced polypeptide chain. *Proc. Natl. Acad. Sci. USA* 47, 1309–1314.

Appel, R. D., Bairoch, A., and Hochstrasser, D. F. (1994). A new generation of information retrieval tools for biologists: The example of the ExPASy WWW server. *Trends Biochem. Sci.* 19, 258–260.

Bjellqvist, B., Hughes, G., Pasquali, C., Paquet, N., Ravier, F., Sanchez, J.-C., Frutiger, S., and Hochstrasser, D. F. (1993). The focusing positions of polypeptides in immobilized pH gradients can be predicted from their amino acid sequence. *Electrophoresis* 14, 1023–1031.

Brendel, V., Bucher, P., Nourbakhsh, I., Blasidell, B. E., and Karlin, S. (1992). Methods and algorithms for statistical analysis of protein sequences. *Proc. Natl. Acad. Sci. USA* 89, 2002–2006.

Bryant, S. H., and Altschul, S. F. (1995). Statistics of sequence-structure threading. *Curr. Opin. Struct. Biol.* 5, 236–244.

Bryant, S. H., and Lawrence, C. E. (1993). An empirical energy function for threading protein sequence through the folding motif. *Proteins* 16, 92–112.

Burnett, R. M., Darling, G. D., Kendall, D. S., LeQuesne, M. E., Mayhew, S. G., Smith, W. W., and Ludwig, M. L. (1974). The structure of the oxidized form of clostridial flavodoxin at 1.9 Å resolution. *J. Biol. Chem.* 249, 4383–4392.

Chothia, C. (1992). One thousand families for the molecular biologist. *Nature* 357, 543–544.

Chothia, C., and Lesk, A. M. (1986). The relation between the divergence of sequence and structure in proteins. *EMBO J.* 5, 823–826.

Cordwell, S. J., Wilkins, M. R., Cerpa-Poljak, A., Gooley, A. A., Duncan, M., Williams, K. L., and Humphery-Smith, I. (1995). Cross-species identification of proteins separated by two-dimensional electrophoresis using matrix-assisted laser desorption ionization/time-of-flight mass spectrometry and amino acid composition. *Electrophoresis* 16, 438–443.

Deléage, G., and Roux, B. (1987). An algorithm for protein secondary structure based on class prediction. *Protein Eng.* 1, 289–294.

Eisenhaber, F., Persson, B., and Argos, P. (1995). Protein structure prediction: Recognition of primary, secondary, and tertiary structural features from amino acid sequence. *Crit. Rev. Biochem. Mol. Biol.* 30, 1–94.

Fetrow, J. S., and Bryant, S. H. (1993). New programs for protein tertiary structure prediction. *Bio/Technology* 11, 479–484.

Frishman, D. and Argos, P. (1997) Seventy-five percent accuracy in protein secondary structure prediction. *Proteins* 27, 329–335.

Garnier, J., Gibrat, J.-F., and Robson, B. (1996). GOR method for predicting protein secondary structure from amino acid sequence. *Methods Enzymol.* 266, 540–553.

Geourjon, C., and Déleage, G. (1995). SOPMA: Significant improvements in protein secondary structure prediction by consensus prediction from multiple alignments. *CABIOS* 11, 681–684.

Gill, S. C. and von Hippel, P. H. (1989) Calculation of protein extinction coefficients from amino acid sequence data. *Anal. Biochem.* 182, 319–326.

Gribskov, M., McLachlan, A. D. and Eisenberg, D. (1987) Profile analysis: detection of distantly related proteins. *Proc. Natl. Acad. Sci. USA* 84, 4355–4358.

Gribskov, M., Homyak, M., Edenfield, J. and Eisenberg, D. (1988) Profile scanning for three-dimensional structural patterns in protein sequences. *Comput. Appl. Biosci.* 4, 61–66.

Guss, J. M., and Freeman, H. C. (1983). Structure of oxidized poplarplastocyanin at 1.6 Å resolution. *J. Mol. Biol.* 169, 521–563.

Henikoff, J. G. and Henikoff, S. (1996) Using substitution probabilities to improve position-specific scoring matrices. *Comput. Appl. Biosci.* 12, 135–43.

Hobohm, U., and Sander, C. (1995). A sequence property approach to searching protein databases. *J. Mol. Biol.* 251, 390–399.

Hofmann, K., and Stoffel, W. (1993). TMbase: A database of membrane-spanning protein segments. *Biol. Chem. Hoppe-Seyler* 347, 166.

Hofmann, K., Bucher, P., Falquet, L., and Bairoch, A. (1999) The PROSITE database, its status in 1999. *Nucleic Acids Res.* 27, 215–219.

Holm, L., and Sander, C. (1993). Protein structure comparison by alignment of distance matrices. *J. Mol. Biol.* 233, 123–138.

Holm, L., and Sander, C. (1994). The FSSP database of structurally-aligned protein fold families. *Nucl. Acids Res.* 22, 3600–3609.

Ikai, A. (1980) Thermostability and aliphatic index of globular proteins. *J. Biochem. (Tokyo)* 88, 1895–1898.

Jones, D. T., and Thornton, J. M. (1996). Potential energy functions for threading. *Curr. Opin. Struct. Biol.* 6, 210–216.

King, R. D. and Sternberg, M. J. (1996) Identification and application of the concepts important for accurate and reliable protein secondary structure prediction. *Protein Sci.* 5, 2298–2310.

Kneller, D. G., Cohen, F. E., and Langridge, R. (1990). Improvements in protein secondary structure prediction by an enhanced neural network. *J. Mol. Biol.* 214, 171–182.

Knight, A. E. (1994). *The Diversity of Myosin-like Proteins* (Cambridge: Cambridge University Press).

Kyte, J., and Doolittle, R. F. (1982). A simple method for displaying the hydropathic character of a protein. *J. Mol. Biol.* 157, 105–132.

Lemer, C. M., Rooman, M. J., and Wodak, S. J. (1995). Protein structure prediction by threading methods: Evaluation of current techniques. *Proteins* 23, 337–355.

Levin, J. M., Robson, B., and Garnier, J. (1986). An algorithm for secondary structure determination in proteins based on sequence similarity. *FEBS Lett.* 205, 303–308.

Levitt, M., and Chothia, C. (1976). Structural patterns in globular proteins. *Nature* 261, 552–558.

Lupas, A., Van Dyke, M., and Stock, J. (1991). Predicting coiled coils from protein sequences. *Science* 252, 1162–1164.

Luthy, R., Xenarios, I. and Bucher, P. (1994) Improving the sensitivity of the sequence profile method. *Protein Sci.* 3, 139–146.

Mehta, P. K., Heringa, J., and Argos, P. (1995). A simple and fast approach to prediction of protein secondary structure from multiply aligned sequences with accuracy above 70%. *Protein Sci.* 4, 2517–2525.

Nielsen, H., Engelbrecht, J., Brunak, S., and von Heijne, G. (1997). Identification of prokaryotic and eukaryotic signal peptides and prediction of their cleavage sites. *Protein Eng.* 10, 1–6.

Pappin, D. J. C., Hojrup, P., and Bleasby, A. J. (1993). Rapid identification of proteins by peptide-mass fingerprinting. *Curr. Biol.* 3, 327–332.

Pauling, L., and Corey, R. B. (1951). The structure of proteins: Two hydrogen-bonded helical configurations of the polypeptide chain. *Proc. Natl. Acad. Sci. USA* 37, 205–211.

Peitsch, M. C. (1996). ProMod and SWISS-MODEL: Internet-based tools for automated comparative protein modeling. Biochem. *Soc. Trans.* 24, 274–279.

Persson, B., and Argos, P. (1994). Prediction of transmembrane segments in proteins utilizing multiple sequence alignments. *J. Mol. Biol.* 237, 182–192.

Rost, B. (1995). TOPITS: Threading one-dimensional predictions into three-dimensional structures. In *Third International Conference on Intelligent Systems for Molecular Biology*, C. Rawlings, D. Clark, R. Altman, L. Hunter, T. Lengauer, and S. Wodak, Eds. (Cambridge: AAAI Press), p. 314–321.

Rost, B. (1996). PHD: Predicting one-dimensional protein structure by profile-based neural networks. *Methods Enzymol.* 266, 525–539.

Rost, B. and Sander, C. (1993) Secondary structure prediction of all-helical proteins in two states. *Protein Eng.* 6, 831–836.

Rost, B., Sander, C., and Schneider, R. (1994). PHD: A mail server for protein secondary structure prediction. *CABIOS* 10, 53–60.

Salamov, A. A. and Solovyev, V. V. (1995) Prediction of protein secondary structure by combining nearest-neighbor algorithms and multiple sequence alignments *J. Mol. Biol.* 247, 11–15.

Sander, C., and Schneider, R. (1991). *Proteins* 9, 56–68.

Smith, W. W., Pattridge, K. A., Ludwig, M. L., Petsko, G. A., Tsernoglou, D., Tanaka, M., and Yasunobu, K. T. (1983). Structure of oxidized flavodoxin from *Anacystis nidulans. J. Mol. Biol.* 165, 737–755.

Takano, T. (1977). Structure of myoglobin refined at 2.0 Å. *J. Mol. Biol.* 110, 537–584.

Wilkins, M. R., Pasquali, C., Appel, R. D., Ou, K., Golaz, O., Sanchez, J.-C., Yan, J. X., Gooley, A. A., Hughes, G., Humphery-Smith, I., Williams, K. L., and Hochstrasser, D. F.

(1996). From proteins to proteomes: Large-scale protein identification by two-dimensional electrophoresis and amino acid analysis. *Bio/Techniques* 14, 61–65.

Wilkins, M. R., Lindskog, I., Gasteiger, E., Bairoch, A., Sanchez, J.-C., Hochstrasser, D. F., and Appel, R. D. (1997). Detailed peptide characterization using PeptideMass, a World Wide Web-accessible tool. *Electrophoresis* 18, 403–408.

Zvelebil, M. J., Barton, G. J., Taylor, W. R. and Sternberg, M. J. (1987) Prediction of protein secondary structure and active sites using the alignment of homologous sequences. *J. Mol. Biol.* 195, 957–961.

12

EXPRESSED SEQUENCE TAGS (ESTs)

Tyra G. Wolfsberg

Genome Technology Branch
National Human Genome Research Institute
National Institutes of Health
Bethesda, Maryland

David Landsman

Computational Biology Branch
National Center for Biotechnology Information
National Library of Medicine
National Institutes of Health
Bethesda, Maryland

The benefits arising from the rapid generation of large numbers of low-quality cDNA sequences were not universally recognized when the concept was originally proposed in the late 1980s. Proponents of this approach argued that these cDNA sequences would allow for the quick discovery of hundreds or thousands of novel protein-coding genes. Their critics countered that cDNA sequencing would miss important regulatory elements that could be found only in the genomic DNA. In the end, the cDNA sequencing advocates appear to have won. Since the original description of 609 Expressed Sequence Tags (ESTs) in 1991 (Adams et al., 1991), the growth of ESTs in the public databases has been dramatic. The number of ESTs in GenBank surpassed the number of non-EST records in mid-1995; as of June 2000, the 4.6 million EST records comprised 62% of the sequences in GenBank. Although the original ESTs were of human origin, NCBI's EST database (dbEST) now contains ESTs from over 250 organisms, including mouse, rat, *Caenorhabditis elegans*, and *Drosophila melanogaster*. In addition, several commercial establishments maintain privately funded, in-house collections of ESTs. ESTs are now widely used throughout

Bioinformatics: A Practical Guide to the Analysis of Genes and Proteins
Edited by A. D. Baxevanis and B. F. F. Ouellette
ISBN 0-471-38390-2 (cloth), ISBN 0-471-383910 (paper)

the genomics and molecular biology communities for gene discovery, mapping, polymorphism analysis, expression studies, and gene prediction.

WHAT IS AN EST?

An overview of an EST sequencing project is shown in Figure 12.1. In brief, a cDNA library is constructed from a tissue or cell line of interest. Individual clones are picked from the library, and one sequence is generated from each end of the cDNA insert. Thus, each clone normally has a 5′ and 3′ EST associated with it. The sequences average ~400 bases in length. Because the ESTs are short, they generally represent only fragments of genes, not complete coding sequences. Many sequencing centers have automated the process of EST generation, producing ESTs at a rapid rate. For example, at the time of this writing, the Genome Sequencing Center at Washington University was producing over 20,000 ESTs per week.

The ESTs that have been submitted to the public sequence databases to date were created from thousands of different cDNA libraries representing over 250 organisms. The libraries may be from whole organs, such as human brain, liver, lung, or skeletal muscle, specialized tissues or cells, such as cerebral cortex or epidermal keratinocyte, or cultured cell lines such as liver HepG2 or gastric carcinoma. Some libraries have been constructed to compare transcripts from different developmental stages, such as fetal versus infant human brain or embryonic 7-day versus neonatal 10-day rat heart ventricle. Others are used to highlight gene expression differences between normal and transformed tissue, such as normal colonic epithelium and colorectal carcinoma cell line. The libraries are constructed by isolating mRNA from the tissue or cell line of interest. The mRNA is then reverse-transcribed into cDNA, usually with an oligo(dT) primer, so that one end of the cDNA insert derives from the polyA tail at the end of the mRNA. The other end of the cDNA is normally

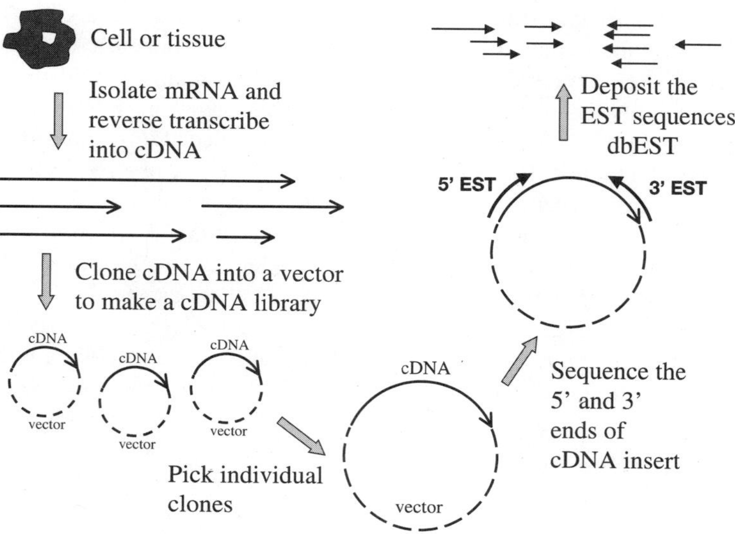

Figure 12.1. Overview of how ESTs are constructed.

within the coding sequence but may be in the 5′ untranslated region if the coding sequence is short. The resulting cDNA is cloned into a vector. In many libraries, the cDNA is cloned directionally. Some of the libraries are normalized to bring the frequency of occurrence of clones representing individual mRNA species into a narrow range (Bonaldo et al., 1996; Soares et al., 1994). Other libraries are constructed by a process of subtractive hybridization, in which a pool of mRNA sequences is removed from a library of interest, leaving behind sequences unique to that library (Bonaldo et al., 1996). For example, to construct a library for the study of bipolar disorder, researchers started with human frontal lobe cDNA from individuals with bipolar disorder, and subtracted out cDNA that hybridized to cDNA from mentally normal individuals (see *http://www.ncbi.nlm.nih.gov/dbEST/dbest_libs.html#lib1475*).

With the use of primers that hybridize to the vector sequence, the ends of the cDNA insert are sequenced. Automatic DNA sequencers generate most EST data. If the cDNA has been directionally cloned into the vector, the sequences can be classified as deriving from the 5′ or 3′ end of the clone. In most cases, both the 5′ and 3′ sequences are determined, but some EST projects have concentrated only on 5′ ESTs to maximize the amount of coding sequence determined. Because the sequence of each EST is generated only once, the sequences may (and often do) contain errors. Contaminating vector, mitochondrial, and bacterial sequences are routinely removed before the EST sequences are deposited into the public databases (Hillier et al., 1996). ESTs in the databases are identified by their clone number as well as their 5′ or 3′ orientation, if known.

The I.M.A.G.E. Consortium (Lennon et al., 1996) has picked individual clones from many of the libraries used for EST sequencing and arrayed them for easy distribution. These clones can be obtained royalty-free from I.M.A.G.E. Consortium distributors. As of the time of this writing, more than 3.8 million cDNA clones have been arrayed from 360 human and 108 mouse cDNA libraries; zebrafish and *Xenopus* clones have also been arrayed. I.M.A.G.E. Consortium sequences currently comprise more than half of the ESTs in GenBank. Most of the sequencing of I.M.A.G.E. clones is performed by the Genome Sequencing Center at Washington University/St. Louis. Merck sponsored human clone sequencing in 1995 and 1996; since then, the collaborative EST project has been sponsored by the National Cancer Institute as part of the Cancer Genome Anatomy Project. Sequencing by Washington University/St. Louis of mouse cDNAs is sponsored by the Howard Hughes Medical Institute. Sequence trace data from the ESTs sequenced by the Washington University/St. Louis projects are available online.

How to Access ESTs

ESTs are submitted to all three international sequence databases (GenBank, EMBL, and DDBJ), under the data-sharing agreement described in Chapter 2. Therefore, all ESTs can be accessed through all of these databases, regardless of where the sequence was originally submitted. The same ESTs are also available from the NCBI's dbEST, the database of Expressed Sequence Tags (Boguski et al., 1993). Instructions about how to submit EST sequences to GenBank are available online.

Like other sequences in GenBank, ESTs can be accessed through Entrez (see Chapter 7). Single ESTs are retrieved by accession or gi number. Advanced searches with multiple search terms can be limited to ESTs by selecting the `Properties` limit and entering `EST`. The two ESTs deriving from a particular I.M.A.G.E. clone can be retrieved by searching for "IMAGE:clone_number" (e.g., "IMAGE: 743313"). The Entrez version of the EST with accession AW592465 is shown in Figure 12.2. Various identifiers for the EST, including the accession number and GenBank gi, are shown in the top block. The CLONE INFO section specifies the number of the clone (2934602) and whether this EST derives from the 5′ or 3′ end of the clone (here, 3′). The nucleotide sequence is shown next, along with a note supplied by the submitter about where the high-quality sequence stops. The COMMENTS block tells how to order the clone from the I.M.A.G.E. Consortium. The last few sections present other information supplied by the submitter, including details about the cDNA library. Although many ESTs (especially 5′ ESTs) can be translated into a partial or sometimes full-length protein sequence, coding sequence features are not provided. Other views of the data, including a FASTA-formatted DNA sequence, can be selected from a pull-down at the top of the Entrez entry (not shown).

EST sequences are also available for BLAST searching. Because ESTs are nucleotide sequences, they can be retrieved only by using BLAST programs that search nucleotide databases (BLASTN for a nucleotide sequence query, TBLASTN for a protein sequence query, and TBLASTX for a translated nucleotide sequence query). Because they make up such a high proportion of sequences in GenBank, ESTs are not included in the BLAST *nr* database. To search against ESTs, select the *dbest* database or, for a specific organism, the *mouse ests*, *human ests*, or *other ests* database. Note that ESTs *are* also included in the *month* database, which contains *all* new or revised sequences released in the last 30 days.

Limitations of EST Data

Although ESTs are an excellent source of sequence data, these data are not of as high a quality as sequences determined by conventional means. Because EST sequences are generated in a single pass, they have a higher error rate than sequences that are verified by multiple sequencing runs, on the order of 3% (Boguski et al., 1993). In contrast, the standard for the human genome project is an error rate of <0.01% (Collins et al., 1998). ESTs may contain substitutions, deletions, or insertions compared with the parent mRNA sequence. The region of an EST between positions 100 and 300 may be the most accurate part of the sequence (Hillier et al., 1996).

Hillier et al. (1996) have performed a comprehensive analysis of potential EST artifacts. They found that ESTs may contain bacterial, mitochondrial, or vector sequence contamination. Most EST cDNA libraries are oligo(dT) primed, and the 3′ EST derives from the 3′ untranslated region of the gene. However, Hillier et al. found that 1.5% of oligo(dT)-primed 3′ ESTs do not align with the known 3′ end of the mRNA. These ESTs either represent nonspecific priming or indicate alternative splicing. cDNA for some libraries is synthesized with random primers, so the location of the 3′ EST is unknown. Another potential problem comes from inverted clones in directionally cloned libraries, in which the 5′ and 3′ EST are mislabeled. cDNA inserts may be inverted because of failures in the directional cloning procedure, or simply because of human error. Hillier et al. found that 6.25% of ESTs that match a known mRNA align in an inverted orientation. Chimeric clones, in which the 5′ EST matches one mRNA and the 3′ EST another mRNA, may arise either during library construction or sample handling. Hillier et al. found a chimera frequency of

```
IDENTIFIERS

dbEST Id:        4025315
EST name:        hf43a02.x1
GenBank Acc:     AW592465
GenBank gi:      7279647

CLONE INFO
Clone Id:        IMAGE:2934602 (3')
Source:          NCI
DNA type:        cDNA

PRIMERS
Sequencing:      -40UP from Gibco
PolyA Tail:      Unknown

SEQUENCE
                 TTTTTTTTTAAATTGCCAAGTGATTTTACTTCAAGATGACATCAGAATTGCTAAAAGGTG
                 ATGTAACCGTCAGAGTGACTATTGATTATAACTCCCAGTAAGTGTCAACGTGATTTTCTC
                 CATTGTGTGGGCTTCCATTAGTATTTACTCATTAGGTTCAGTAGTTTTCATTATTTTCTC
                 TTCCATAAATTCTATTGCTTGTGAAAAGCCACCAAAGAGAAGTGAAACCAGAAAAAGGAT
                 GCAACGAGTAAATATTAAAAGTAGTGCTCAGTTTATATTCGCAAGTGTGCTGGCTGTAAT
                 ACGATATTGTTTGTCAGGTGGAGGGCCACTATCTATACTACCTCCTTTTCCTCAGTTCAC
                 ATGTTGGTGGTTGCCACCCATGCAGACAGTGACAATGTTTTTTGTTGTTACATACTCCTT
                 TGTAATTGCATGTGTTAAGAACACACTCAAAATGCAGGTCTTGATAAGAAGGCAATTGTG
                 TTTAAGACAGTAGCTGCCTGGGCCACAGGTTGCACCATCCACTGACCGCCCCATTTCTGG
                 CAAGTCTGGACCCTGGTGTGGCTAATAACCAAGGCATTTATT
Quality:         High quality sequence stops at base: 356

Entry Created:   Mar 22 2000
Last Updated:    Mar 22 2000

COMMENTS
                 This clone is available royalty-free through LLNL ; contact
                 the IMAGE Consortium (info@image.llnl.gov) for further
                 information.

PUTATIVE ID      Assigned by submitter
                 TR:Q60815 Q60815 ADAM 4 PROTEIN PRECURSOR ;

LIBRARY
Lib Name:        Soares_NFL_T_GBC_S1
Organism:        Homo sapiens
Organ:           pooled
Lab host:        DH10B
Vector:          pT7T3D-Pac (Pharmacia) with a modified polylinker
R. Site 1:       Not I
R. Site 2:       Eco RI
Description:     Equal amounts of plasmid DNA from three normalized libraries
                 (fetal lung NbHL19W, testis NHT, and B-cell NCI_CGAP_GCB1)
                 were mixed, and ss circles were made in vitro. Following HAP
                 purification, this DNA was used as tracer in a subtractive
                 hybridization reaction. The driver was PCR-amplified cDNAs
                 from pools of 5,000 clones made from the same 3 libraries.
                 The pools consisted of I.M.A.G.E. clones 297480-302087,
                 682632-687239, 726408-728711, and 729096-731399. Subtraction
                 by Bento Soares and M. Fatima Bonaldo.

SUBMITTER
Name:            Robert Strausberg, Ph.D.
Tel:             (301) 496-1550
E-mail:          Robert_Strausberg@nih.gov

CITATIONS
Title:           National Cancer Institute, Cancer Genome Anatomy Project
                 (CGAP), Tumor Gene Index
Authors:         NCI-CGAP http://www.ncbi.nlm.nih.gov/ncicgap
Year:            1997
Status:          Unpublished

MAP DATA
```

Figure 12.2. The Entrez view of an EST, accession AI273896.

1%, but a separate study estimated the frequency at 11% (Wolfsberg and Landsman, 1997).

EST CLUSTERING

As of mid-2000, GenBank contained just under 1.9 million human EST records. Although original estimates of the number of genes in the human genome hovered around the 100,000 mark, predictions made based on experimental data and presented at the 2000 Cold Spring Harbor Genome meeting have drastically reduced the estimate to below 50,000. In any event, it is clear, even without doing any sequence comparisons, that these ESTs cannot each represent a unique sequence. Even with the process of library normalization, abundant transcripts are represented more frequently in dbEST than rare ones. For example, dbEST contains more than 200 ESTs for human alpha-fetoprotein alone. A number of efforts are geared at simplifying this abundance of DNA sequences by grouping together records that likely derive from the same gene. Other resources, including those for mapping and gene discovery, can then make use of this condensed set of gene-based clusters, rather than the expansive and relatively unorganized collection of all ESTs and other mRNA sequences.

UniGene

The UniGene resource, developed at NCBI, clusters ESTs and other mRNA sequences, along with coding sequences (CDSs) annotated on genomic DNA, into subsets of related sequences (Boguski and Schuler, 1995; Wagner, L. et al., unpublished observations). In most cases, each cluster is made up of sequences produced by a single gene, including alternatively spliced transcripts (Fig. 12.3). However, some genes may be represented by more than one cluster. The clusters are organism specific and are currently available for human, mouse, rat, zebrafish, and cattle. They are built in several stages, using an automatic process based on special sequence comparison algorithms. First, the nucleotide sequences are searched for contaminants, such as mitochondrial, ribosomal, and vector sequence, repetitive elements, and low-complexity sequences. After a sequence is screened, it must contain at least 100 bases to be a candidate for entry into UniGene. mRNA and genomic DNA are clustered first into gene links. A second sequence comparison links ESTs to each other and to the gene links. At this stage, all clusters are "anchored," and contain either a sequence with a polyadenylation site or two ESTs labeled as coming from the 3′ end of a clone. Clone-based edges are added by linking the 5′ and 3′ ESTs that derive from the same clone. In some cases, this linking may merge clusters identified at a previous stage. Finally, unanchored ESTs and gene clusters of size 1 (which may represent rare transcripts) are compared with other UniGene clusters at lower stringency. The UniGene build is updated weekly, and the sequences that make up a cluster may change. Thus, it is not safe to refer to a UniGene cluster by its cluster identifier; instead, one should use the GenBank accession numbers of the sequences in the cluster. A summary of the UniGene build procedure is shown in Figure 12.4a. Additional information about the UniGene build is available online.

As of July 2000, the human subset of UniGene contained 1.7 million sequences in 82,000 clusters; 98% of these clustered sequences were ESTs, and the remaining

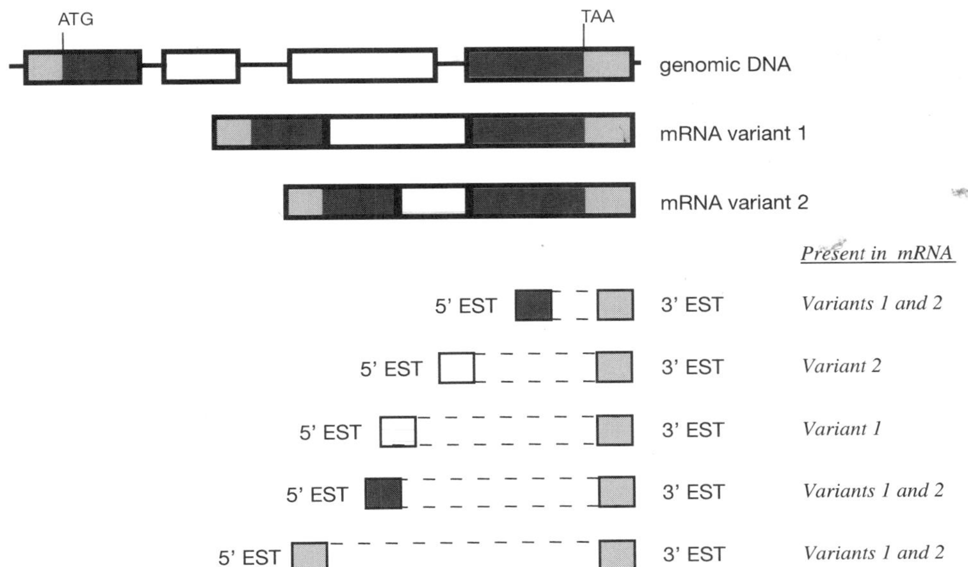

Figure 12.3. Sequences in a UniGene cluster. This cluster contains a genomic DNA sequence with an annotated coding sequence (CDS), two alternatively-spliced mRNA sequences, and 10 ESTs from five clones that derive from the mRNA sequences.

2% were from mRNAs or CDSs annotated on genomic DNA. These human clusters could represent fragments of up to ~82,000 unique human genes, implying that many human genes are now represented in a UniGene cluster. (This number is undoubtedly an overestimate of the number of genes in the human genome, as some genes may be represented by more than one cluster.) Only 1.4% of clusters totally lack ESTs, implying that most human genes are represented by at least one EST. Conversely, it appears that the majority of human genes have been identified *only* by ESTs; only 16% of clusters contain either an mRNA or a CDS annotated on a genomic DNA. Because fewer ESTs are available for mouse, rat, and zebrafish, the UniGene clusters are not as representative of the unique genes in the genome. Mouse UniGene contains 895,000 sequences in 88,000 clusters, and rat UniGene contains 170,000 sequences in 37,000 clusters.

A new UniGene resource, HomoloGene, includes curated and calculated orthologs and homologs for genes from human, mouse, rat, and zebrafish. Calculated orthologs and homologs are the result of nucleotide sequence comparisons between all UniGene clusters for each pair of organisms. Homologs are identified as the best match between a UniGene cluster in one organism and a cluster in a second organism. When two sequences in different organisms are best matches to one another (a reciprocal best match), the UniGene clusters corresponding to the pair of sequences are considered putative orthologs. A special symbol indicates that UniGene clusters in three or more organisms share a mutually consistent ortholog relationship. The calculated orthologs and homologs are considered putative, since they are based only on sequence comparisons. Curated orthologs are provided by the Mouse Genome Database (MGD) at the Jackson Laboratory and the Zebrafish Information Database (ZFIN) at the University of Oregon and can also be obtained from the scientific

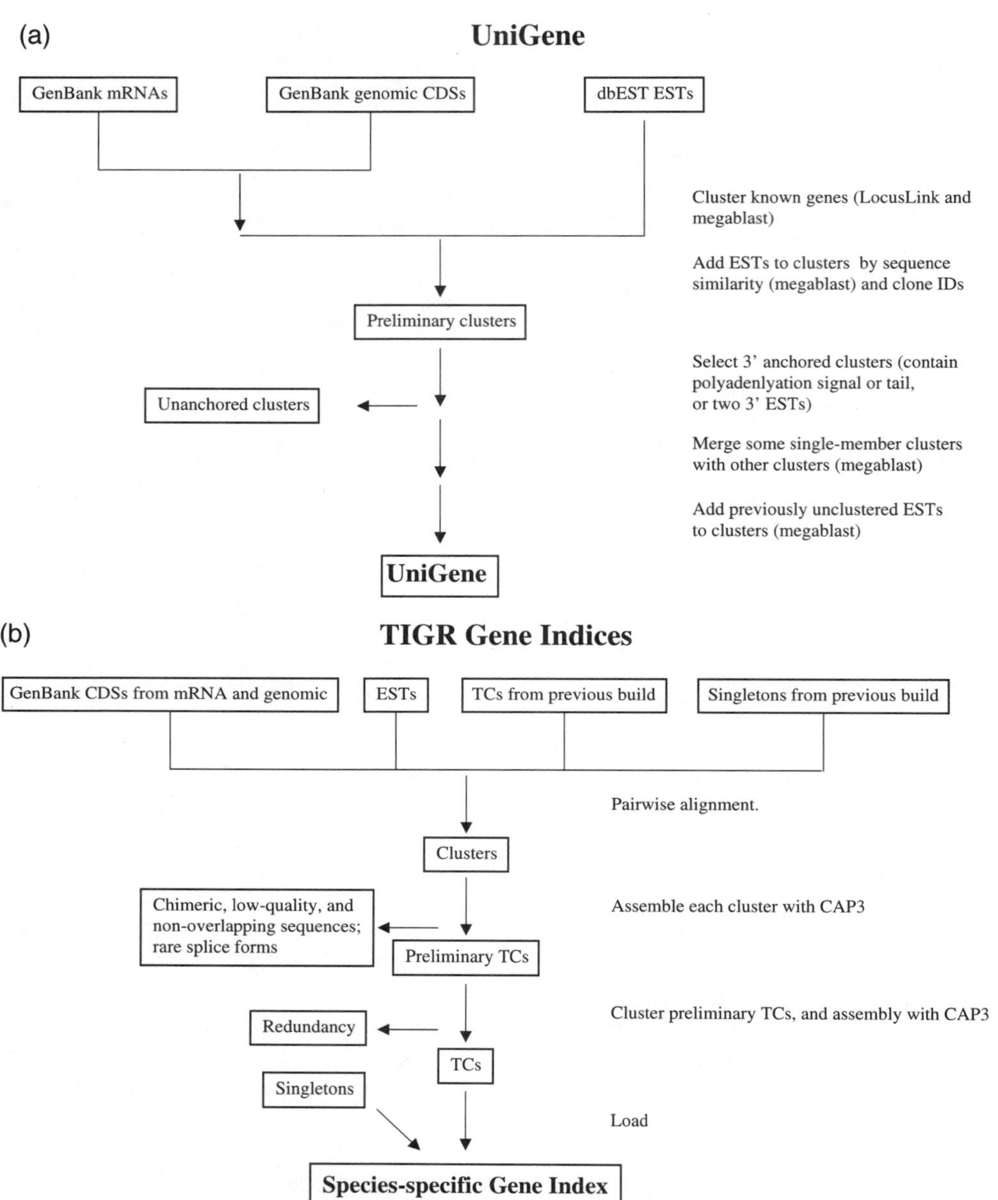

Figure 12.4. Schematics for clustering of ESTs. All three methods prescreen ESTs for contaminating sequence. (a) UniGene. Most sequence analysis is done with MegaBLAST (Zhang et al., 2000), a fast version of BLAST. The minimum alignment length is 70 nucleotides, and an alignment must extend over at least 70% of the alignable region in the first two steps or 55% of the alignable region in the last two steps. (b) TIGR Gene Indices. Sequences are clustered if they share a minimum of 95% identity over a 40 nucleotide region, with fewer than 20 nucleotides of mismatched sequence at either end. Sequences are assembled with CAP3 (Huang and Madan, 1999). (c) STACK. Sequences are clustered if they share 96% identity over 150 nucleotides. Clustering is done with d2_cluster (Burke et al., 1999) and aligned with PHRAP (Green, 1996).

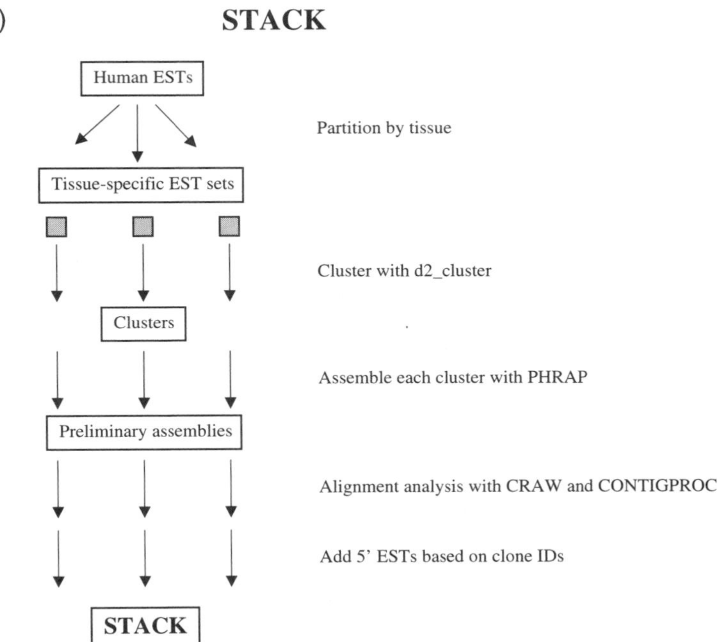

Figure 12.4. Continued

literature. Direct links to HomoloGene are provided for UniGene clusters that have a candidate ortholog or homolog.

Queries to UniGene are entered into a text box on any of the UniGene pages. Query terms can be, for example, the UniGene identifier, a gene name, a text term that is found somewhere in the UniGene record, or the accession number of an EST or gene sequence in the cluster. For example, the cluster entitled "A disintegrin and metalloprotease domain 10" that contains the sequence for human ADAM10 can be retrieved by entering `ADAM10`, `disintegrin`, `AF009615` (the GenBank accession number of ADAM10), or `H69859` (the GenBank accession number of an EST in the cluster). Enter multiple terms to get a list of entries containing all terms. To query a specific part of the UniGene record, use the @ symbol. For example, `@gene(symbol)` looks for genes with the name of the symbol enclosed in the parentheses, `@chr(num)` searches for entries that map to chromosome num, `@lib(id)` returns entries in a cDNA library identified by id, and `@pid(id)` selects entries associated with a GenBank protein identifier id.

The query results page contains a list of all UniGene clusters that match the query. Each cluster is identified by an identifier, a description, and a gene symbol, if available. Cluster identifiers are prefixed with Hs for *Homo sapiens*, Rn for *Rattus norvegicus*, Mm for *Mus musculus*, or Dn for *Danio rerio*. The descriptions of UniGene clusters are taken from LocusLink, if available, or from the title of a sequence in the cluster. The UniGene report page for each cluster links to data from other NCBI resources (Fig. 12.5). At the top of the page are links to LocusLink, which provides descriptive information about genetic loci (Pruitt et al., 2000), OMIM, a catalog of human genes and genetic disorders, and HomoloGene. Next are

Hs.172028 *Homo sapiens* **ADAM10**

A disintegrin and metalloprotease domain 10

SEE ALSO

LocusLink: 102
OMIM: 602192
HomoloGene: Hs.172028

SELECTED MODEL ORGANISM PROTEIN SIMILARITIES
organism, protein and percent identity and length of aligned region

H. sapiens :	PID:g2393947 - ADAM10	**100 % / 747 aa**
M. musculus :	PID:g2282608 - kuzbanian	**95 % / 747 aa**
R. norvegicus :	PIR:S52477 - disintegrin	**97 % / 543 aa**
D. melanogaster :	PID:g1531633 - kuzbanian	**46 % / 584 aa**
C. elegans :	PID:g3875352 - similar to Zinc-binding metalloprotease domain	**43 % / 715 aa**

MAPPING INFORMATION

Chromosome:	15
Cytogenetic Position:	15q22
Gene Map 98:	Marker H69847 , Interval D15S117-D15S159
Gene Map 98:	Marker stSG15641 , Interval D15S117-D15S159

EXPRESSION INFORMATION

cDNA sources:	Brain, Breast, CNS, Foreskin, Heart, Kidney, Liver, Lung, Lymph, Muscle, Ovary, Pancreas, Pooled, Thyroid, Tonsil, Uterus, Whole embryo, breast_normal, colon, genitourinary tract, head_neck, kidney, lung
SAGE :	Gene to Tag mapping

mRNA/GENE SEQUENCES (3)

AF009615	Homo sapiens ADAM10 (ADAM10) mRNA, complete cds	P
Z48579	H.sapiens mRNA for disintegrin-metalloprotease (partial)	P
NM_001110	Homo sapiens disintegrin and metalloprotease domain 10 (ADAM10) mRNA	

EST SEQUENCES (10 of 122)[Show all ESTs]

AI680390	cDNA clone IMAGE:2264325 Uterus	3' read	3.4 kb	C
AI188417	cDNA clone IMAGE:1723156	3' read	2.2 kb	A C
AA812209	cDNA clone IMAGE:1338085 Tonsil		2.0 kb	A C
AI302180	cDNA clone IMAGE:1902444 Kidney	3' read	1.9 kb	A C
AI742324	cDNA clone IMAGE:2368543 Pooled	3' read	1.8 kb	C
AI273896	cDNA clone IMAGE:1964283 Ovary	3' read	1.7 kb	C
AI472608	cDNA clone IMAGE:2149114 Pooled	3' read	1.7 kb	C
AI368402	cDNA clone IMAGE:2011372 Brain	3' read	1.6 kb	C
H69450	cDNA clone IMAGE:212359	5' read	1.5 kb	P
H69859	cDNA clone IMAGE:212359	3' read	1.5 kb	

Key to Symbols

P Has similarity to known **P**roteins (after translation)
A Contains a poly-**A**denylation signal
S Contains a mapped **S**equence-tagged site (STS)
C Clone source is a **C**GAP library

Figure 12.5. UniGene cluster for ADAM10. The contents of the cluster are subject to change as additional sequences are submitted to the public databases.

listed similarities between the translations of DNA sequences in the cluster and protein sequences from model organisms, including human, mouse, rat, fruit fly, and worm. The subsequent section describes relevant mapping information. It is followed by "expression information," which lists the tissues from which the ESTs in the cluster have been created, along with links to the SAGE database (see below). Sequences making up the cluster are listed next, along with a link to download these sequences.

It is important to note that clusters that contain ESTs only (i.e., no mRNAs or annotated CDSs) will be missing some of these fields, such as LocusLink, OMIM, and mRNA/Gene links. UniGene titles for such clusters, such as "EST, weakly similar to ORF2 contains a reverse transcriptase domain [H. sapiens]," are derived from the title of a characterized protein with which the translated EST sequence aligns. The cluster title might be as simple as "EST" if the ESTs share no significant similarity with characterized proteins.

TIGR GENE INDICES

The TIGR Gene Indices represent another effort to consolidate EST and other annotated gene sequences (Quackenbush et al., 2000). A significant difference between the Gene Indices and UniGene is that the Gene Indices are assemblies of ESTs and other gene sequences rather than clusters (Figure 12.4b). The assemblies tend to represent one transcript, so alternatively spliced products are grouped separately. Furthermore, the process generates a single consensus sequence per assembly.

A Gene Index is maintained for 14 organisms, including human, mouse, rat, *Drosophila*, zebrafish, *Arabidopsis*, and several crop plants. Gene Indices are created from publicly available GenBank and dbEST sequences by clustering ESTs with the DNA sequences encoding the coding sequences annotated on DNA and mRNA sequences. The elements of a cluster are assembled with other EST sequences into tentative consensus sequences (TCs, or THCs for human). TCs are updated as sequence flow into the public databases. The TIGR databases, as of mid-2000, contain 85,000 THCs, 43,000 TCs from mouse, and 18,000 TCs from rat. These numbers are somewhat lower than the numbers of UniGene clusters, probably due to different methods used for clustering. The TIGR Gene Indices, like UniGene, can be queried with text searches. A BLAST interface, for BLASTN and TBLASTN, is also available. A related project at TIGR is to identify orthologous genes between human, mouse, and rat using the TCs. The TIGR Orthologous Gene Alignment (TOGA) database represents the ortholog sets.

STACK

The STACK resource at the South African National Bioinformatics Institute (SANBI) uses a third method, a combination of clustering and assembly, to group related ESTs into clusters (Burke et al., 1999; Miller et al., 1999). At this time, STACK clusters are available only for human ESTs (Fig. 12.4c). STACK clusters consolidate ESTs into a smaller number of groups than does UniGene. Unlike UniGene or the TIGR Gene Indices, ESTs in STACK are separated by tissue type before being clustered. BLAST queries can be performed on STACK clusters.

ESTs AND GENE DISCOVERY

ESTs have been widely used for gene discovery (Boguski et al., 1994). Because ESTs outnumber other nucleotide sequences in GenBank, researchers hunting for a novel gene are much more likely to find it in dbEST than in the rest of GenBank. ESTs are not included in the BLAST *nr* database, and sequence similarity searches for ESTs must target the est database. Gene discovery methods using ESTs include, for example, hunting for new members of gene families in the same species (paralogs), for functionally equivalent genes in other species (orthologs), or even for alternatively spliced forms of known genes.

Gene discovery using dbEST is very rapid, requiring only a few minutes for the BLAST search. For example, ADAM 10, also known as "kuzbanian," is a well-studied gene with known orthologs in human, mouse, rat, cow, frog, pig, fruit fly, and worm. A TBLASTN search of *dbest* with the human protein sequence quickly reveals not only many of these known genes but also an additional likely ortholog in zebrafish. Discovering alternatively spliced transcripts among ESTs is more problematic. For one, it is difficult to determine if the new sequences are due to alternative splicing or to the presence of contaminating genomic DNA sequence in EST libraries (Wolfsberg and Landsman, 1997). An analysis of the TIGR Human Gene Index using a spliced alignment algorithm provided evidence that up to 35% of human genes may undergo alternative splicing and that the majority of these events occur in the 5′ untranslated regions (Mironov et al., 1999).

The uses of ESTs extend beyond mammals. For example, until recently, the public databases contained little sequence data from *Toxoplasma gondii*, a disease-causing protozoan parasite of human. A large scale project generated 7,000 5′ ESTs, representing ~4,000 unique sequences, from *T. gondii* (Ajioka et al., 1998). Comparisons between the ESTs and sequences in public databases identified potential functions for 500 novel *T. gondii* genes. Some ESTs are phylogenetically restricted to *T. gondii* and other members of the Apicomplexa phylum.

THE HUMAN GENE MAP

ESTs are also being used to create gene maps by the use of sequence-tagged sites (STSs), short stretches of unique sequence identified by polymerase chain reaction (PCR) assays. An international consortium agreed to coordinate a mapping effort for the human genome, using UniGene clusters to represent individual human genes. In the initial effort, the gene-based STSs were mapped relative to two radiation hybrid (RH) maps and one YAC panel; in subsequent work, the STSs have been mapped only to the two RH panels, the Genebridge4 (GB4) and Stanford G3. STSs were generated from the 3′ untranslated regions of UniGene clusters. GeneMap '96 reported the mapping of 16,000 gene-based STSs (Schuler et al., 1996), and GeneMap '98 nearly doubled that number to 30,000 (Deloukas et al., 1998). Thus, current maps detail the position of up to one-half of all human protein-coding genes. The gene map is updated as new STSs are mapped. GeneMap '98 is described in more detail in Chapter 6.

Information about the map location of individual ESTs is provided by UniGene. If an STS exists for an EST in the cluster, the map position of that STS is indicated in the record. This data may confirm what is already known about the location of a mapped human disease gene, but, for the majority of cases involving yet unmapped ESTs, GeneMap '98 provides novel information. Because the STS markers correspond to individual genes, the project also shows the density of genes on each human chromosome. For example, chromosomes 1, 17, and 22 have a higher than expected gene density, and chromosomes 4, 13, 18, and X have a lower than expected density.

GENE PREDICTION IN GENOMIC DNA

Another use of ESTs is to predict or refine computational predictions of the location of genes in genomic DNA. With the appropriate use of sequence alignment parameters, up to 90% of genes annotated on human genomic DNA are also detected by ESTs (Bailey et al., 1998). ESTs can complement other algorithms used for gene prediction because they may do a better job at predicting alternative splicing and 3′ untranslated regions. A study is underway to reannotate the *C. elegans* genome using EST sequences (Kohara, Y., unpublished observations). With the use of the acembly program, 126,000 ESTs were aligned to 98 Mb of genomic DNA. The genes predicted by EST clones were compared with those predicted by the *C. elegans* genomic sequencing consortium, which were constructed using GeneFinder with hand editing (The *C. elegans* Sequencing Consortium, 1998). The following points are noteworthy:

1. In about half the cases, the computationally predicted genes were identical to the EST alignments; 25% of the genes were predicted with less accuracy, and the remaining 25% were predicted poorly. In some cases, 5′ sequences from ESTs showed that the gene predictions were either too long or too short.
2. Comparisons of the EST sequences to the genomic sequence confirm that the error rate of the worm genome sequence is less than 1 mistake per 10,000 nucleotides. Instances where many ESTs share identical sequence may indicate errors in genomic sequence. Alternatively, these differences could be sequence polymorphisms.
3. About 30% of the ESTs exist in alternatively-spliced forms. Many of these alternative splices are not annotated on the genomic sequence.
4. Computational methods may predict separate genes, whereas EST analysis shows that these segments are actually exons of a single gene. Conversely, the computational method may predict exons in cases that should be separate genes.

ESTs have also been used in genome sequencing projects to make estimates about gene expression along the chromosome. In the complete sequences of *Arabidopsis thaliana* chromosomes 2 (Lin et al., 1999) and 4 (Mayer et al., 1999), about one-third of the computationally predicted genes have an EST match. Histograms plotting the EST distribution along the chromosomes predict that some genes are highly expressed, at least within the tissues from which EST libraries were constructed. On chromosome 4, 75% of the matching ESTs aligned with only 6% of the genes, implying that these genes are transcribed at high rates.

ESTs AND SEQUENCE POLYMORPHISMS

Single nucleotide polymorphisms (SNPs) can help to associate sequence variations with heritable phenotypes, facilitate studies in population and evolutionary biology, and aid in positional cloning and physical mapping. On average, SNPs occur every 500–1,000 nucleotides in human DNA. Gene-associated SNPs are found in untranslated regions as well as coding sequences (cSNPs). Because ESTs are sequenced redundantly from libraries prepared from different individuals, they seem an ideal source of polymorphic data. Indeed, a number of recent studies demonstrate that analysis of aligned EST sequences can lead to SNP discovery (Buetow et al., 1999; Garg et al., 1999; Marth et al., 1999; Picoult-Newberg et al., 1999). All rely on alignment of EST sequences, identification of sequence differences, and a method to distinguish real polymorphisms from base calling (sequencing) errors and other artifacts. The public database for SNPs, dbSNP, is maintained at the NCBI (Sherry et al., 1999; Smigielski et al., 2000). dbSNP accepts submissions not only of single nucleotide polymorphisms but of other polymorphisms, such as short deletions and inserts, microsatellites, and polymorphic insertion elements like retrotransposons. As of mid-July 2000, dbSNP contains data from 600,000 SNPs. dbSNP is integrated with other NCBI resources such as GenBank, PubMed, genome sequences, and LocusLink.

ASSESSING LEVELS OF GENE EXPRESSION USING ESTs

Because ESTs are generated by random sequencing of clones from many different libraries, they appear, at least at first glance, to be a good source of data source for studies of gene expression levels. However, any conclusions about transcript levels must be made very carefully. Many libraries are normalized or generated by subtractive hybridization. Both of these processes change the relative representation of cDNAs. Normalization results in abundant messages being seen less frequently and rare messages more frequently, whereas subtractive hybridization removes entire pools of transcripts from the library. Although libraries made by these processes can provide very general ideas about which genes are expressed at higher levels, detailed analysis is not possible.

CGAP

A subset of the EST libraries was constructed for the purpose of gene expression profiling, and these libraries were not normalized or created by subtractive hybridization. Many of these libraries were constructed by the Cancer Genome Anatomy Project (CGAP), an NCI initiative that is working to decipher the molecular anatomy of the cancer cell (Wheeler et al., 2000). CGAP has developed libraries from normal, precancerous, and cancerous cell types. Comparing the genes expressed in these three tissue types can lead to predictions about the genes involved in cancer progression. ESTs from the CGAP project are submitted to dbEST and are available in UniGene. CGAP has developed online tools to compare computationally gene expression levels between libraries. Digital Differential Display (DDD) uses a statistical test to calculate the number of times sequences from different libraries are assigned to a particular UniGene cluster (Krizman et al., 1999). By selecting pools of libraries, users can compare gene expression levels between tissues (e.g., liver, lung, muscle, and spleen) or cancer stages (e.g., normal vs. premalignant vs. cancerous prostate tissue). The DDD results show the genes that are expressed at different levels in the selected pools (i.e., the UniGene clusters in which the number of ESTs from one set of libraries is significantly different from the number of ESTs from another set of libraries). The results are presented as an easily interpreted graphic similar to a Northern blot and also as text. A detailed explanation of how to perform a DDD experiment, including a worked example, is provided on the CGAP Web site. The CGAP xProfiler compares gene expression levels between two pools of libraries by listing the genes that are expressed either in both library pools or in one pool but not the other. The calculations are also based on the tissue distribution of ESTs in UniGene clusters.

SAGE

Serial analysis of gene expression (SAGE) is an experimental technique used for quantitative, high throughput gene expression analysis (Velculescu et al., 1995). SAGE involves the isolation of short unique sequence tags from a specific location within each transcript. These sequence tags are concatenated, cloned, and sequenced. The frequency of particular transcripts within the starting cell population is reflected by the number of times the associated sequence tag is encountered within the sequence population. The SAGEmap database is a repository for some of this SAGE data, and tools that allow gene expression analysis are also available on the SAGEmap Web site (Lal et al., 1999). The virtual Northern predicts the SAGE tag in a user-supplied mRNA sequence and calculates the distribution of the tag in the SAGE libraries, thus providing a virtual picture of the expression pattern of the mRNA. In the SAGE xProfiler, the user selects two pools of libraries, such as colon cancer versus normal colon. The tool calculates which SAGE tags are more abundant in one pool or the other. The SAGE tags are mapped to UniGene clusters to provide biological context for the results.

Microarrays

High-density oligonucleotide and cDNA microarrays are a relatively new technique being used to monitor gene expression on a genome-wide scale. The technique uses the same principles of nucleic acid hybridization as do Northern and Southern blots but on a much larger scale. Thousands of gene-specific probes are arrayed on a small matrix, such as a glass slide or microchip, and this matrix is probed with labeled nucleic acid synthesized from a tissue type, developmental stage, or other condition of interest. The expression profiles of thousands of genes under that condition can thus be assayed simultaneously. The array probes can be derived from oligonucleotides or cDNAs. In many cases, the probes for cDNA arrays are 3' ESTs (Duggan et al., 1999). For human expression analysis, UniGene clusters can be used as a source of additional information about the ESTs on the array. Microarray technologies and the bioinformatics challenges surrounding them are discussed in Chapter 16.

INTERNET RESOURCES FOR TOPICS PRESENTED IN CHAPTER 12

dbEST home page	*http://www.ncbi.nlm.nih.gov/dbEST/*
List of dbEST libraries	*http://www.ncbi.nlm.nih.gov/dbEST/libs_byorg.html*
dbEST summary by organism	*http://www.ncbi.nlm.nih.gov/dbEST/dbEST_summary.html*
How to submit ESTs to dbEST	*http://www.ncbi.nlm.nih.gov/dbEST/how_to_submit.html*
EST Projects at Washington University	*http://genome.wustl.edu/gsc/est/navest.pl*
The I.M.A.G.E. Consortium	*http://image.llnl.gov/*
UniGene	*http://www.ncbi.nlm.nih.gov/UniGene/*
The UniGene build procedure	*http://www.ncbi.nlm.nih.gov/UniGene/build.html*
UniGene query engine	*http://www.ncbi.nlm.nih.gov/UniGene/query.cgi*
HomoloGene	*http://www.ncbi.nlm.nih.gov/HomoloGene/*
STACK	*http://www.sanbi.ac.za/Dbases.html*
TIGR Gene Indices	*http://www.tigr.org/tdb/tgi.html*
TIGR Orthologous Gene Alignment database	*http://www.tigr.org/tdb/toga/toga.html*
GeneMap '98	*http://www.ncbi.nlm.nih.gov/genemap/*
dbSNP	*http://www.ncbi.nlm.nih.gov/SNP/*
Cancer Genome Anatomy Project (CGAP)	*http://www.ncbi.nlm.nih.gov/ncicgap/*
CGAP Digital Differential Display (DDD)	*http://www.ncbi.nlm.nih.gov/CGAP/info/ddd.cgi*
CGAP xProfiler	*http://www.ncbi.nlm.nih.gov/CGAP/hTGI/xprof/cgapxpsetup.cgi*
Serial Analysis of Gene Expression (SAGE)	*http://www.ncbi.nlm.nih.gov/SAGE/*
SAGE virtual Northern	*http://www.ncbi.nlm.nih.gov/SAGE/sagevn.cgi*
SAGE xProfiler	*http://www.ncbi.nlm.nih.gov/SAGE/sagexpsetup.cgi*

PROBLEM SET

You have been studying the histone deacetylase gene, RPD3, in the yeast *Saccharomyces cerevisiae*. You are moving to a lab that works on zebrafish, and you would like to continue your work on this gene. You wonder how difficult it will it be to clone the zebrafish ortholog of RPD3.

1. What is the GenBank accession number of the first listed RPD3 protein sequence from *Saccharomyces cerevisiae*?

2. Do the public sequence databases already contain any zebrafish proteins that are likely orthologs of RPD3?

a. What type of sequence comparison search should you perform?
b. To interpret the search results of your sequence comparison, you will need to know the scientific name for zebrafish. What is the scientific name?
c. Are there any zebrafish protein orthologs of yeast RPD3?

3. You remember that the EST database is an excellent source of sequence data.
a. What type of sequence comparison should you perform to find EST hits to the yeast protein sequence?
b. Are there any zebrafish EST hits to this yeast protein sequence?

4. Do the five top scoring ESTs belong to the same UniGene cluster?

5. What is the GenBank accession number of the human sequence that matches this UniGene cluster?

6. What cDNA clone does the top scoring EST hit come from?

7. Is this EST from the or 3′ end of the cDNA clone?

8. From which cDNA library was this clone sequenced?

9. Is the EST that comes from the opposite end of this cDNA clone also a member of this UniGene cluster?

10. Does this EST also align with the yeast RPD3 protein sequence? Why or why not?

11. Is the top-scoring zebrafish EST also present in the TIGR Zebrafish Gene Index?

12. Is the EST that comes from the opposite end of the cDNA clone also in this TIGR TC?

13. Are the sequences in the UniGene cluster and the TIGR TC basically the same?

14. How does the TIGR consensus sequence for the 5′ EST TC compare with that produced by UniGene?

15. Is the top-scoring zebrafish EST hit to the yeast RPD3 protein present in STACK?

16. Based on what you have learned, how would you get a cDNA clone of the zebrafish RPD3 gene?

REFERENCES

Adams, M. D., Kelley, J. M., Gocayne, J. D., Dubnick, M., Polymeropoulos, M. H., Xiao, H., Merril, C. R., Wu, A., Olde, B., Moreno, R. F., Kerlavage, A. R., McCombie, W. R., and Venter, J. C. (1991). Complementary DNA sequencing: expressed sequence tags and human genome project. *Science* 252: 1651–1656.

Ajioka, J. W., Boothroyd, J. C., Brunk, B. P., Hehl, A., Hillier, L., Manger, I. D., Marra, M., Overton, G. C., Roos, D. S., Wan, K. L., Waterston, R., and Sibley, L. D. (1998). Gene discovery by EST sequencing in *Toxoplasma gondii* reveals sequences restricted to the Apicomplexa. *Genome Res* 8: 18–28.

Bailey, L. C., Jr., Searls, D. B., and Overton, G. C. (1998). Analysis of EST-driven gene annotation in human genomic sequence. *Genome Res* 8: 362–376.

Boguski, M. S., Lowe, T. M., and Tolstoshev, C. M. (1993). dbEST—database for "expressed sequence tags." *Nat Genet* 4: 332–333.

Boguski, M. S., and Schuler, G. D. (1995). ESTablishing a human transcript map. *Nat Genet* 10: 369–371.

Boguski, M. S., Tolstoshev, C. M., and Bassett, D. E., Jr. (1994). Gene discovery in dbEST. *Science* 265: 1993–1994.

Bonaldo, M. F., Lennon, G., and Soares, M. B. (1996). Normalization and subtraction: two approaches to facilitate gene discovery. *Genome Res* 6: 791–806.

Buetow, K. H., Edmonson, M. N., and Cassidy, A. B. (1999). Reliable identification of large numbers of candidate SNPs from public EST data. *Nat Genet* 21: 323–325.

Burke, J., Davison, D., and Hide, W. (1999). d2_cluster: A Validated Method for Clustering EST and Full-Length cDNA Sequences. *Genome Res* 9: 1135–1142.

Collins, F. S., Patrinos, A., Jordan, E., Chakravarti, A., Gesteland, R., and Walters, L. (1998). New goals for the U.S. Human Genome Project: 1998–2003. *Science* 282: 682–689.

Deloukas, P., Schuler, G. D., Gyapay, G., Beasley, E. M., Soderlund, C., Rodriguez-Tome, P., Hui, L., Matise, T. C., McKusick, K. B., Beckmann, J. S., Bentolila, S., Bihoreau, M., Birren, B. B., Browne, J., Butler, A., Castle, A. B., Chiannilkulchai, N., Clee, C., Day, P. J., Dehejia, A., Dibling, T., Drouot, N., Duprat, S., Fizames, C., Bentley, D. R., and et al. (1998). A physical map of 30,000 human genes. *Science* 282: 744–746.

Duggan, D. J., Bittner, M., Chen, Y., Meltzer, P., and Trent, J. M. (1999). Expression profiling using cDNA microarrays. *Nat Genet* 21: 10–14.

Garg, K., Green, P., and Nickerson, D. A. (1999). Identification of candidate coding region single nucleotide polymorphisms in 165 human genes using assembled expressed sequence tags. *Genome Res* 9: 1087–1092.

Green, P. (1996). http://www.genome.washington.edu/uwgc/analysistools/phrap.htm.

Hillier, L. D., Lennon, G., Becker, M., Bonaldo, M. F., Chiapelli, B., Chissoe, S., Dietrich, N., DuBuque, T., Favello, A., Gish, W., Hawkins, M., Hultman, M., Kucaba, T., Lacy, M., Le, M., Le, N., Mardis, E., Moore, B., Morris, M., Parsons, J., Prange, C., Rifkin, L., Rohlfing, T., Schellenberg, K., Marra, M., and et al. (1996). Generation and analysis of 280,000 human expressed sequence tags. *Genome Res* 6: 807–828.

Huang, X., and Madan, A. (1999). CAP3: A DNA sequence assembly program. *Genome Res* 9: 868–877.

Krizman, D. B., Wagner, L., Lash, A., Strausberg, R. L., and Emmert-Buck, M. R. (1999). The Cancer Genome Anatomy Project: EST Sequencing and the Genetics of Cancer Progression. *Neoplasia* 1: 101–106.

Lal, A., Lash, A. E., Altschul, S. F., Velculescu, V., Zhang, L., McLendon, R. E., Marra, M. A., Prange, C., Morin, P. J., Polyak, K., Papadopoulos, N., Vogelstein, B., Kinzler, K. W., Strausberg, R. L., and Riggins, G. J. (1999). A public database for gene expression in human cancers. *Cancer Res* 59: 5403–5407.

Lennon, G., Auffray, C., Polymeropoulos, M., and Soares, M. B. (1996). The I.M.A.G.E. Consortium: an integrated molecular analysis of genomes and their expression. *Genomics* 33: 151–152.

Lin, X., Kaul, S., Rounsley, S., Shea, T. P., Benito, M. I., Town, C. D., Fujii, C. Y., Mason, T., Bowman, C. L., Barnstead, M., Feldblyum, T. V., Buell, C. R., Ketchum, K. A., Lee, J., Ronning, C. M., Koo, H. L., Moffat, K. S., Cronin, L. A., Shen, M., Pai, G., Van Aken, S., Umayam, L., Tallon, L. J., Gill, J. E., and Venter, J. C. (1999). Sequence and analysis of chromosome 2 of the plant *Arabidopsis thaliana. Nature* 402: 761–768.

Marth, G. T., Korf, I., Yandell, M. D., Yeh, R. T., Gu, Z., Zakeri, H., Stitziel, N. O., Hillier, L., Kwok, P. Y., and Gish, W. R. (1999). A general approach to single-nucleotide polymorphism discovery. *Nat Genet* 23: 452–456.

Mayer, K., Schuller, C., Wambutt, R., Murphy, G., Volckaert, G., Pohl, T., Dusterhoft, A., Stiekema, W., Entian, K. D., Terryn, N., Harris, B., Ansorge, W., Brandt, P., Grivell, L., Rieger, M., Weichselgartner, M., de Simone, V., Obermaier, B., Mache, R., Muller, M., Kreis, M., Delseny, M., Puigdomenech, P., Watson, M., and McCombie, W. R. (1999). Sequence and analysis of chromosome 4 of the plant *Arabidopsis thaliana*. *Nature* 402: 769–777.

Miller, R. T., Christoffels, A. G., Gopalakrishnan, C., Burke, J., Ptitsyn, A. A., Broveak, T. R., and Hide, W. A. (1999). A comprehensive approach to clustering of expressed human gene sequence: The sequence tag alignment and consensus knowledge base. *Genome Res* 9: 1143–1155.

Mironov, A. A., Fickett, J. W., and Gelfand, M. S. (1999). Frequent alternative splicing of human genes. *Genome Res* 9: 1288–1293.

Picoult-Newberg, L., Ideker, T. E., Pohl, M. G., Taylor, S. L., Donaldson, M. A., Nickerson, D. A., and Boyce-Jacino, M. (1999). Mining SNPs from EST databases. *Genome Res* 9: 167–174.

Pruitt, K. D., Katz, K. S., Sicotte, H., and Maglott, D. R. (2000). Introducing RefSeq and LocusLink: curated human genome resources at the NCBI. *Trends Genet* 16: 44–47.

Quackenbush, J., Liang, F., Holt, I., Pertea, G., and Upton, J. (2000). The TIGR Gene Indices: reconstruction and representation of expressed gene sequences. *Nucleic Acids Res* 28: 141–145.

Schuler, G. D., Boguski, M. S., Stewart, E. A., Stein, L. D., Gyapay, G., Rice, K., White, R. E., Rodriguez-Tome, P., Aggarwal, A., Bajorek, E., Bentolila, S., Birren, B. B., Butler, A., Castle, A. B., Chiannilkulchai, N., Chu, A., Clee, C., Cowles, S., Day, P. J., Dibling, T., Drouot, N., Dunham, I., Duprat, S., East, C., Hudson, T. J., and et al. (1996). A gene map of the human genome. *Science* 274: 540–546.

Sherry, S. T., Ward, M., and Sirotkin, K. (1999). dbSNP-database for single nucleotide polymorphisms and other classes of minor genetic variation. *Genome Res* 9: 677–679.

Smigielski, E. M., Sirotkin, K., Ward, M., and Sherry, S. T. (2000). dbSNP: a database of single nucleotide polymorphisms. *Nucleic Acids Res* 28:352–355.

Soares, M. B., Bonaldo, M. F., Jelene, P., Su, L., Lawton, L., and Efstratiadis, A. (1994). Construction and characterization of a normalized cDNA library. *Proc Natl Acad Sci USA* 91: 9228–9232.

The *C. elegans* Sequencing Consortium. (1998). Genome sequence of the nematode *C. elegans*: a platform for investigating biology. *Science* 282: 2012–2018.

Velculescu, V. E., Zhang, L., Vogelstein, B., and Kinzler, K. W. (1995). Serial analysis of gene expression. *Science* 270: 484–487.

Wheeler, D. L., Chappey, C., Lash, A. E., Leipe, D. D., Madden, T. L., Schuler, G. D., Tatusova, T. A., and Rapp, B. A. (2000). Database resources of the National Center for Biotechnology Information. *Nucleic Acids Res* 28: 10–14.

Wolfsberg, T. G., and Landsman, D. (1997). A comparison of expressed sequence tags (ESTs) to human genomic sequences. *Nucleic Acids Res* 25: 1626–1632.

Zhang, Z., Schwartz, S., Wagner, L., and Miller, W. (2000). A greedy algorithm for aligning DNA sequences. *J. Comput. Biol.* 7: 203–214.

13

SEQUENCE ASSEMBLY AND FINISHING METHODS

Rodger Staden

MRC Laboratory of Molecular Biology
Cambridge, United Kingdom

David P. Judge

Department of Biochemistry
University of Cambridge
Cambridge, United Kingdom

James K. Bonfield

MRC Laboratory of Molecular Biology
Cambridge, United Kingdom

That genome centers are able to produce DNA sequence data at very high speed using robotics and sequencing instruments should not be taken to imply that sequencing is always straightforward or routine. The nature of genomes and the limitations of mapping and sequencing techniques still make such projects challenging and provide scope for improved computing methods. To understand the requirements of sequence assembly software, it is necessary to know a little about genome sequences and sequencing techniques. After a general introduction to the task of sequencing and assembly and an illustration of the reality of routine sequencing operations, we describe how the Staden package can be employed in effectively assembling sequence data. Other packages will include their own, often similar, methods, to deal with the problems.

Bioinformatics: A Practical Guide to the Analysis of Genes and Proteins
Edited by A. D. Baxevanis and B. F. F. Ouellette
ISBN 0-471-38390-2 (cloth), ISBN 0-471-383910 (paper)

The large-scale organization and local composition of DNA affects the difficulty and complication of determining its sequence. Genomes of higher organisms vary considerably in their relative GC-to-AT content and in the number and types of repetitive elements they contain. For example, within its components of known utility, the human genome consists of genes for proteins which occur in single or few copies, multigene families scattered throughout the chromosomes, and gene clusters in a variety of arrangements, including multiple copies of genes for proteins and RNAs needed in high abundance. A large proportion of the remainder of the genome consists of various types of repetitive elements including LINEs and SINEs (Hutchinson et al., 1989) of which Alu sequences are the most widely known. Many other simple elements are also repeated, sometimes hundreds of times.

The Sanger dideoxy sequencing technique (Sanger et al., 1977), which is employed by all the major public genome sequencing projects, uses DNA polymerase to synthesize a complementary copy of the target DNA. The inclusion of dideoxy nucleotides (which terminate the growing DNA chains) in the reaction mixture ensures that a proportion of all the fragments so produced stop after each base, hence creating a complete set of nested sequences. The polymerase extends, in the 5′ to 3′ direction, a short segment of DNA (known as a *primer*) that has to be annealed to the 5′ end of the target sequence to initiate the process. Either the primers or the dideoxy "terminators" are fluorescently labeled. The fragments are run through a sieving material such as a gel, which separates them according to their length. The fluorescence of the separated fragments is measured, and the resulting traces are analyzed to produce the base calls. The results of individual experiments are known as "gel readings" or "reads." If the sequence is 100 bases in length, this requires separating fragments having length differing by only 1%, and this effectively limits the size of sequence that can be obtained by a single experiment to around 1,000 bases. Most sequence runs obtain approximately 500 bases.

There are many potential sources of problems when preparing the DNA "template" for sequencing, performing the sequencing reactions, and loading and running the sequencing instrument; all of these factors influence the reliability of the data obtained. The characteristics of individual segments of DNA also influence the difficulty of obtaining reliable sequence, dictating whether special techniques need to be applied.

At the level of the individual gel reading, potential sources of problems include (1) "compressions," in which secondary structure in the DNA fragments causes them to move anomalously in the gel so that more than one size of fragment may migrate to the same position; (2) "stops," where the polymerase has a tendency to dissociate and hence, in dye primer chemistry, produce a large band on the gel; (3) regions of extreme composition, including high GC or GT composition and homopolymer regions; and (4) repetitive DNA. These problems all affect the accuracy and reliability of the data obtained from these regions. At the level of joining the readings together in the correct order, the various types of repeats listed above are the major difficulty (especially if the problem is compounded by poor quality data).

As mentioned above, the primer sequence (of around 20 bases) must be complementary to the sequence at the 5′ end of the target DNA, and the sequence from each experiment is limited to around 1,000 bases. These two factors have led to the widespread use of the so-called "shotgun" strategy of DNA sequencing in which the target DNA is randomly broken into overlapping fragments of around 2,000 bases and cloned into a "sequencing vector." This sidesteps the problem of having to know and synthesize the sequence at the 5′ end of each segment of the target DNA because they now all have a common sequence, namely, that just to the 5′ side of the cloning site in the vector. The use of cloning techniques also provides a way of producing sufficient pure quantities of the DNA. However, this strategy creates the problem of not knowing the order of the fragments obtained from each sequencing experiment. An alternative method of producing random samples from along the target DNA while retaining the ability of using the same primer for each experiment is to use transposons. One further and important experimental strategy is to determine the sequence from both ends of each of the cloned fragments. Together, these "forward" and "reverse" readings, known as "read pairs," give data from opposite strands of the DNA and provide information about the relative positions and orientations of the pairs of readings from the same fragment or template. The recent introduction of capillary electrophoresis instruments has made the production of forward and reverse readings more useful because this technology removes the problem of gel lanes being confused and, hence, with readings being assigned the wrong clone name (and subsequent incorrect assignment of read pairs).

THE USE OF BASE CALL ACCURACY ESTIMATES OR CONFIDENCE VALUES

When the idea of using numerical estimates of base calling was put forward (Dear and Staden, 1992; Bonfield and Staden, 1995), it was expected that these methods would be supplied as part of the instrument manufacturers' base-calling software. Instead, the first useable numerical values were produced by the program phred, which was devised by an academic group (Ewing, and Green, 1998). This was a very important step forward and has had a major impact on genome projects.

Through the program phrap, phred-derived confidence values have been used to improve the quality of sequence assembly. Through the Staden Package, confidence values are also extensively used during the finishing stages of sequencing projects. Here, much of the tedious and time-consuming trace checking is obviated by the software; the confidence values are used to decide if human expertise is required to adjudicate between conflicting base calls. The result is that the majority of conflicts need never be brought to the user's attention, greatly reducing thetime required to check and edit a contig (Bonfield and Staden, 1995). When the phred-style confidence values became available, the Staden program was adjusted to work on the decibel scale that was then defined. Green's group have written a contig editor known as consed (Gordon et al., 1998), which has some similarities to the one in gap4 (Bonfield et al., 1995) described later in this chapter.

Phred produces a confidence value that defines the probability that the base call is correct. The values were calculated by analyzing the traces from which the readings are derived: peak spacing over seven peaks, peak height ratios over seven and three peaks, and the number of peaks to the nearest uncalled base. The confidence value is given by the formula

$$\text{C-value} = -10 \times \log_{10} (\text{probability of error})$$

A confidence value of 10 corresponds to an error rate of 1/10, 20 to 1/100, 30 to 1/1000, and so on.

Another program that does base calling and produces confidence values is TraceTuner. Also, the program ATQA calculates confidence values and produces probabilities for insertions and deletions but does not recall the bases.

THE REQUIREMENTS FOR ASSEMBLY SOFTWARE

The task that sequence assembly software needs to accomplish is to infer the original sequence from the evidence of the readings and ideally to give, for each base, a probability that it is correct. It might be thought that this is simply a matter of comparing and aligning all the readings in a fully automated process, but this is not the case. The combination of limited reading length, reading errors, and repeats means that this ideal of automation is still some way off; therefore, interactive software tools are essential for helping to solve the many problems that confront sequencing staffs worldwide. As should be clear from the section above, there is ample opportunity for joining readings in the wrong order or wrong orientation, for missing out or duplicating regions, as well as for making minor base assignment errors, insertions, or deletions. Tools are required to check for these contingencies, sometimes in concert with extra experiments such as restriction enzyme digests. At present, the sequencing process is often talked of as consisting of two parts, namely, assembly and finishing, but in practice there is considerable overlap between the two. Assembly is the process of attempting to order and align the readings, and finishing is the task of checking and editing the assembled data. This includes performing new sequencing experiments to fill gaps or to cover segments where the data is poor and adjudicating between conflicting readings when editing.

GLOBAL ASSEMBLY

The global sequence assembly problem can be divided into three steps:

1. Find all possible overlaps between readings by comparing each one, in both orientations, to all the others;
2. from the list of overlaps, produce the best layout of the readings;
3. from the alignment of the readings in the final layout, derive a censensus sequence.

Step 1 is usually performed in two stages. First, a rapid comparison is performed to find all pairs of reading that share an exact match of, for example, 14 consecutive bases. Second, those that contain these matches are aligned using dynamic programming methods. The alignments that satisfy some preset criteria are "stored" in a graph, in which the vertices represent the readings and the edges represent the overlaps. Several different algorithms have been published that can analyze and prune these graphs to produce a consistent left-to-right ordering, orientation, and positioning for the readings. The resulting layout of the readings usually still requires multiple sequence alignment, as it is based only on individual pairwise alignments, each of which may conflict with others that they overlap. Once this has been done, a consensus can be derived. Elegant descriptions of the assembly problem and particular algorithmic solutions can be found in Kececioglu and Myers (1995) and Myers (1995).

Working programs usually include a number of important and effective extra methods. All the readings can be prescreened to see if they contain the sequences of known repeats. Those that do can be set aside or treated in other special ways. For example, the segments containing repeat elements can be ignored during the search for an exact initial match but then used during the alignment phase. The layout can be checked and altered to be consistent with known read-pair data. The quality of the alignments can be scored by using the confidence values of the bases, and these scores can be used when the overlap graph is analyzed to produce the layout.

There are several widely used global assembly engines. Those that are currently available free of charge to the academic community include phrap (Green, unpublished), FAKII (Myers et al., 1996), CAP3 (Huang, 1996), the TIGR Assembler (Sutton et al., 1995) and gap4 (Bonfield et al., 1995). Those available commercially include Sequencher and DNASTAR.

After the global assembly engines have done their best with the initial shotgun data, the readings will be arranged into overlapping sets, and it is part of the job of the "finishing" process to complete the project by obtaining readings to fill the gaps. At the end of the project, there will be only one overlapping set, and it should cover the whole of the target sequence. Although its usage has now been expanded to include any set of overlapping clones, even those for which the sequence is unknown, the word "contig" was originally defined to mean a set of sequence readings that overlap (Staden, 1982). A consensus sequence can be calculated for each contig. New readings obtained for the finishing process can be compared against the consensus sequences. If the reading overlaps the end of one contig, it can be extended; if it overlaps two, these can be joined. This more limited assembly problem can be performed by the global engines or more quickly by simpler algorithms that use the consensus sequence.

FILE FORMATS

Raw sequence assembly data consist of traces, base calls, and confidence values, and most programs store these in SCF-format files (Dear and Staden, 1992). In addition to this, processing programs need data about how the readings were obtained—which sequencing vector they were cloned into, which primer was used, which template they were derived from, and which chemistry was used. A variety of methods are in use to manage this data. Many groups store their readings in FASTA-format files and thus have to use so-called "naming conventions" or "naming schemes" in which the entry name is used to encode data about readings. Others use relational databases, such as ABI's BioLIMS system. The Staden package uses Experiment files. The format is identical to EMBL/SWISS-PROT entries in that two-character record-type identifiers begin each line, but the record types are extended to include the necessary data about sequence readings. This simple text format is easily parsed. A set of C programs for reading and writing SCF and Experiment file formats is available on the Staden package Web site. Assemblies describe readings, chemistries, contigs, alignments and edits, and, for this, a "gap4 database" is used for each sequencing project.

The remainder of this chapter illustrates the use of the Staden package as it is applied to sequence assembly data, but readers should be aware that other packages include similar features. Because the examples here use a specific package, we have not gone into the details of how to perform particular tasks but instead give a flavor of the possible operations by describing some of the components of the individual programs.

PREPARING READINGS FOR ASSEMBLY

Data must be passed through several preassembly steps before being entered into the gap4 database, usually via a program called pregap4 (Bonfield and Staden, unpublished), which can operate on batches of traces. The possible steps in this process are shown in Figure 13.1.

Usually, trace files that are in proprietary format, such as those produced by ABI sequencers, are first converted to SCF files. Confidence values are added to the SCF file, and then its Experiment file is initialized with copies of the base calls and confidence values. The trace file is not needed again until the assembly is edited, and, because these programs can uncompress them on the fly, they are now compressed. The trace files are not altered but are kept as archival data so that it is always possible to check the original base calls and traces. Any changes to the data before assembly are made to the copy of the sequence in the Experiment file. It is recommended that no changes are made to the data until readings can be viewed aligned with others. The Experiment file is augmented to include data about how the readings were obtained—which sequencing vector they were cloned into, which primer was used, which template they were derived from, and which chemistry was used. This information can be obtained from a variety of sources. Next, the readings are analyzed to mark segments that are of low confidence, which can aid some assembly engines. With the use of the information in the Experiment file, the reading is searched for the presence of sequencing vectors at each end. A similar search is then done for cloning vectors (for example, for a BAC or cosmid vector). A final check is then performed for missed vectors or other possible contaminant sequences such as *E. coli* or yeast. All of these searches write their results back into the Experiment file, ready for use by later programs in the pipeline. For the assembly stage, pregap4 can use phrap, CAP3, FAKII, or gap4; at the time of this writing, we believe that phrap is the most effective method. Only gap4 can read and write to gap4 databases; therefore, if an external assembly engine is used, the resulting assembly is copied into the gap4 database in the "enter assembly" step. When processing has finished, pregap4 will produce a report containing information from each module and the final list of "passed" and "failed" sequences. From this stage on, all changes are made to the copies of the data in the gap4 database and the Experiment files are no longer required. Pregap4 provides the interfaces to configure all of the above operations, to save the configuration for future use, and to perform the actual work.

Phrapview

Before other components of the gap4 package are described, a brief introduction to the phrapview program is warranted. Phrapview is distributed along with the phrap assembly engine and is a graphical viewer for phrap assemblies. It is intended to

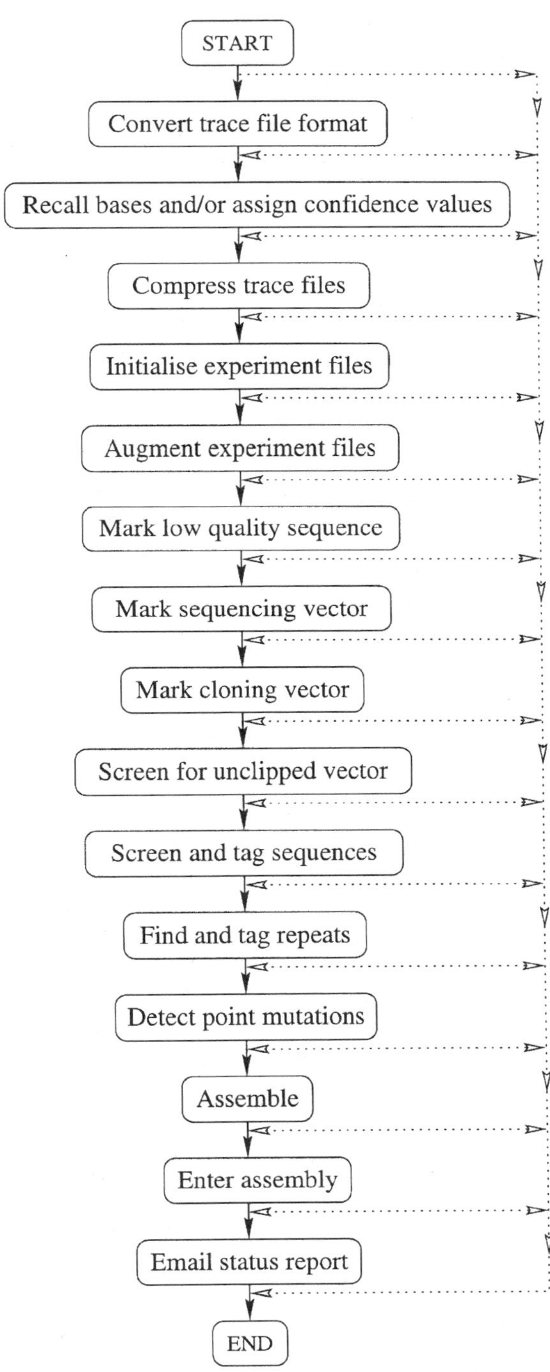

Figure 13.1. A schematic of the tasks performed by the program pregap4 when processing a batch of sequence trace files. See text for details.

provide a "global' view of the assembly, complementing the individual base and trace view provided by consed. This global view focuses on information pertaining to possible incorrectness, incompleteness, or nonuniqueness of the phrap-generated assembly. Phrapview displays depth of coverage, forward-reverse read pairs, significant pairwise matches involving reads in different locations in the assembly, and chimeric reads.

The input to phrapview is a `.view` file, which is produced by running phrap with the View option. Note that phrapview does not perform any of the analyses itself; rather, it provides a way of displaying a file that contains an already completed analysis of the project. A screen dump for a typical phrapview display of a 40-kb cosmid sequencing project (still in three pieces) is shown in Figure 13.2. In this display, color-coded lines are drawn between read pairs, where red indicates a problem and black indicates "OK." Here, 48 problems and 120 OKs are reported. Read pairs will only be indicated properly if the read-naming convention assumed by phrap is used. The other types of data mentioned above can also be displayed in the same window.

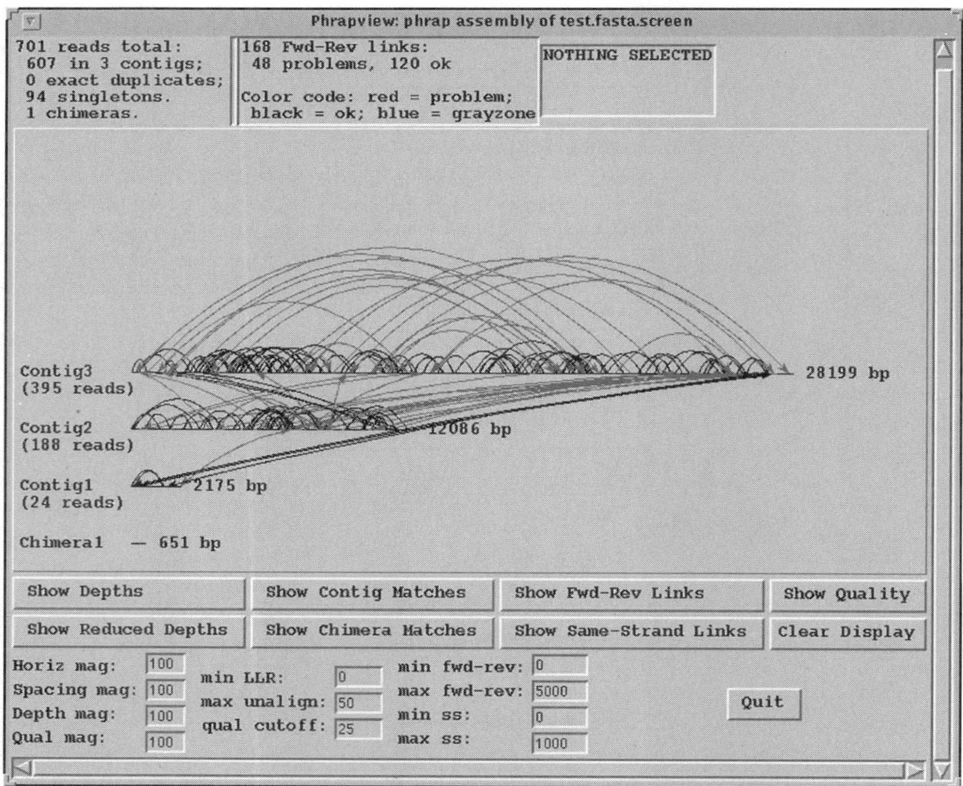

Figure 13.2. A screen dump from the program phrapview, showing a graphical display of the state of the data immediately after a run of the phrap assembly engine. See text for details. (See color plate.)

INTRODUCTION TO GAP4

Gap4 is an interactive program used for working on data from sequencing projects. It contains a comprehensive set of functions, many of which present their results graphically. Others, such as the Experiment Suggestion functions, produce textual output ready for parsing by external programs. One of its important components, used by many of the other functions, is the consensus algorithm. The gap4 database does not store the consensus sequence; rather, it is calculated whenever it is needed. When appropriate, it can be calculated separately for each strand, and, in the Contig Editor and Contig Joining Editor, it is instantly updated for each edit made. When phred-style confidence values are available, the algorithm uses them with strand and chemistry data to calculate a confidence value for each base in the consensus. At the end of a project, the algorithm can produce a FASTA-format file or an Experiment file containing the consensus and its confidence values. Preprocessing programs used by pregap4 and routines within gap4 can add annotations to readings (for example, the position of an Alu segment or a custom primer). Throughout the text, these annotations are referred to as "tags."

The gap4 top-level window is divided into an Output Window for showing textual results, an Error Window for displaying error messages, and, at the top, an array of menus for selecting the main functions. The contents of the two text windows can be searched, edited, and saved. Each set of results is preceded by a header containing the time and date when it was generated. Most of gap4's tools are used within other windows, which are invoked when required or when an analytic function producing graphical output is selected. The most important windows are the Contig Selector, Contig Comparator, Template Display, Consistency Plot, Restriction Map, Contig Editor, Contig Joining Editor, and Trace Display. All of the graphical displays and the Contig Editors can be scrolled in register. The base of the graphical display windows usually contains an Information Line that shows short textual data about results or items touched by the mouse cursor. The next sections give short descriptions of the windows in order of their resolution.

THE CONTIG SELECTOR

When an assembly database is opened, gap4 will bring up the Contig Selector, which is used to display, select and reorder contigs. In the Contig Selector, all contigs are shown as colinear horizontal lines separated by short vertical lines. The length of the horizontal lines is proportional to the length of the contigs, and their left-to-right order represents the current ordering of the contigs. The contig order can be changed by using the mouse to drag the lines representing the contigs. The Contig Selector can also be used to select contigs for processing. For example, clicking with the right mouse button on the line representing a contig will invoke a menu containing the commands that can be performed on that contig. As the mouse is moved over a contig, it is highlighted and the contig name and length are displayed in the Information Line.

A sample Contig Selector is shown in Figure 13.3. Along the top are three menus: File, View, and Results. Below this are buttons for zooming, activating the crosshair, and showing its position. The leftmost button, labeled Next, is used to sequentially step through sets of results. For example, if the user employs the Find

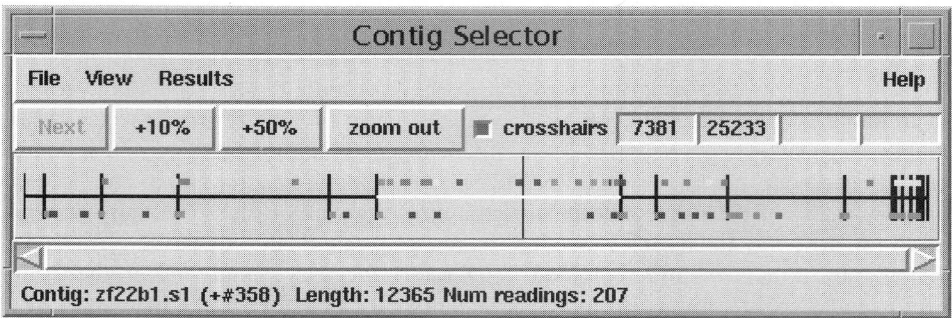

Figure 13.3. A screen dump of the gap4 Contig Selector, which gives an overview of the state of a sequencing project and provides a method for users to select contigs for processing. See text for details. (See color plate.)

Internal Joins option (see below), which finds possible overlaps between contigs, the results are automatically sorted into descending order of overlap quality, and the Next button can be used to process them in that order. Each time the user clicks on the Next button, the Contig Joining Editor will be invoked to show the next overlap, enabling the user to examine the match and edit the two contigs and make the join. Below this row of buttons is the schematic of the contigs. The small boxes around the contig lines show the positions of tags on the readings and the consensus sequence.

THE CONTIG COMPARATOR

Gap4 commands such as Find Internal Joins, Find Repeats, Check Assembly, and Find Read Pairs automatically transform the Contig Selector to produce the Contig Comparator. For this transformation, a copy of the Contig Selector is added at right angles to the original window to create a two-dimensional rectangular surface on which to display the results of comparing or checking contigs. It is therefore equivalent to the "dot plot" display commonly used for comparing pairs of sequences (cf. Chapter 8).

As mentioned above, Find Internal Joins compares the ends of contigs to see if there are possible overlaps. Find Repeats is similar, but, unlike Find Internal Joins, it does not require the found matches to continue to the ends of contigs. Check Assembly compares every reading with the segment of the consensus it overlaps to see how well they align. Those that align poorly are plotted along the main diagonal of the Contig Comparator. Find Read Pairs plots the positions of consistent read pairs that may indicate the order of contigs. Each of the functions plots its results as diagonal lines of different colors. Lines parallel to the main diagonal represent contigs that are in the correct orientation relative to one another. Those perpendicular to the main diagonal show results for which one contig would need to be reversed before the pair could be joined. The manual contig-dragging procedure mentioned above can be used to change the relative positions of contigs. As the contigs are dragged, the plotted results will automatically be moved to their corresponding new positions. Because this plot can simultaneously show the results of independent types of search, users are able to determine whether different analyses produce corroborating or conflicting evidence for the ordering of readings or contigs. The plotted results can be used to invoke a subset of commands by the use of pop-up menus. For example, if the user clicks the right mouse button over a result from Find Internal Joins, a menu containing Invoke Join Editor and Invoke Contig Editors will pop up. If the user selects Invoke Join Editor, the Join Editor will be started with the two contigs aligned at the match position contained in the result. If required, one of the contigs will be complemented to allow their alignment.

A typical display from the Contig Comparator is shown in Figure 13.4. Although they cannot be distinguished without their usual color coding, this screen dump includes results for Find Internal Joins (black), Check Assembly (green), and Find Read Pairs (blue). Superimposed on the bottom left corner are the View menu and the Results Manager menu. The crosshairs show the positions for a pair of contigs. The vertical line continues into the Contig Selector part of the display, and the position represented by the horizontal line is also duplicated there.

THE TEMPLATE DISPLAY

The next level of resolution is the Template Display, which can show schematic plots of readings, templates, tags, restriction enzyme sites, and the consensus quality. It can show one or several contigs and uses color coding to distinguish reading, primer, and template types. It is often used to view the results of the Order Contigs option, which uses read-pair data to find the most likely left-to-right order of the contigs. These data may contain errors due to misnaming of readings (and hence which templates they were obtained from); thus, it is useful to view the results and, if required, modify the order. Again, this can be done by dragging the lines representing the contigs (but this time in the Template Display rather than in the Contig Selector). If the contigs are dragged, the plot is immediately redrawn to reflect the new ordering. Figure 13.5 shows a Template Display containing the data for several contigs after they have been ordered by Order Contigs.

Under the menus and buttons, the largest section of the display contains stacks of lines and arrows representing the readings and templates for the contigs (shown by single nonoverlapping lines surrounded by tags directly below). Beneath this is a representation of the consensus quality for one contig and below that its restriction map. Note that the high depth of coverage seen here is template coverage. An alternative plot in which only the readings (and not their templates) are drawn would show much lower coverage. The template and reading sections of the display are in two parts. The top part contains the templates that have been sequenced from both ends but that are in some way inconsistent—for example, given the current relative positions of their readings, they may have a length that is larger or greater than that expected or the two readings may point in opposite directions. In this screen dump, there are four inconsistent templates, three within contigs and one spanning a pair of contigs. Color coding is used to distinguish between different types of inconsistencies and whether the inconsistencies involve readings within or between contigs. The rest of the data (mostly for templates sequenced from only one end) is plotted below the data for the inconsistent templates. Templates with only one reading are shown in dark blue. Superimposed on the lines representing the templates are forward readings, shown as light blue arrows, and reverse readings, shown as orange arrows.

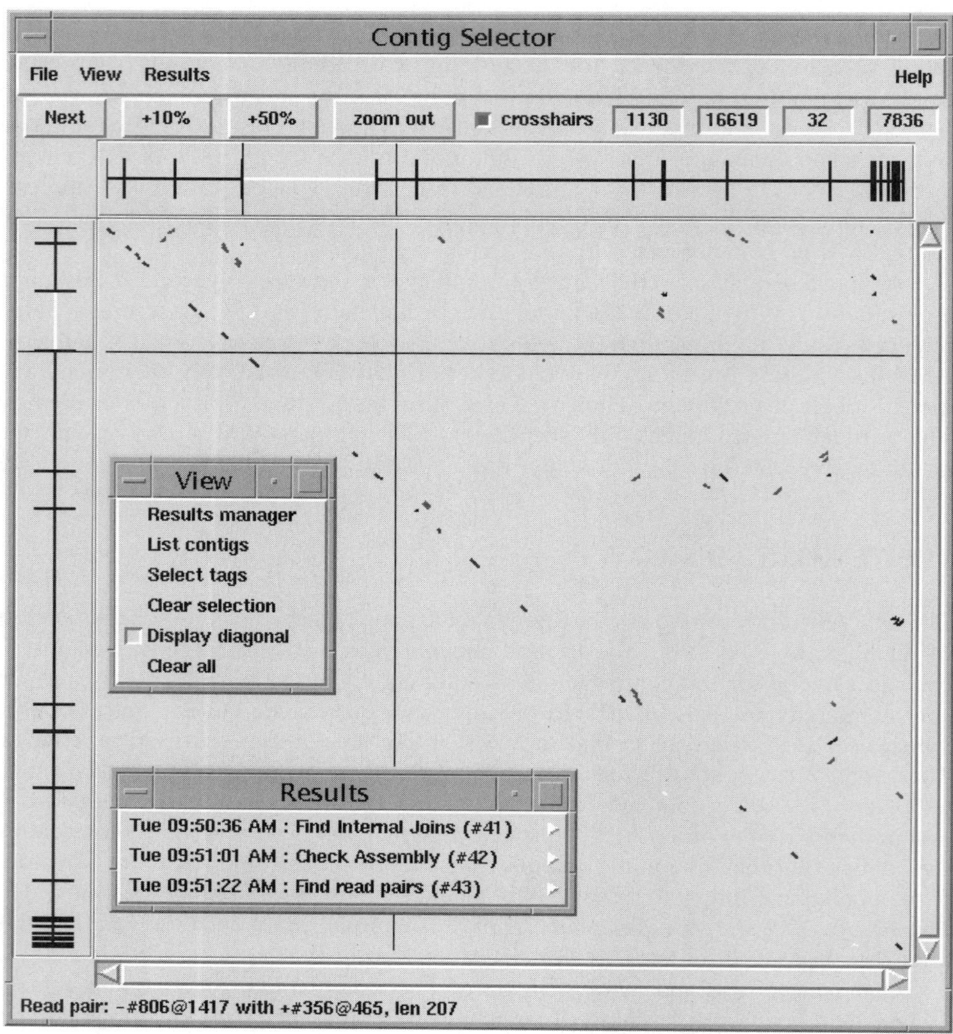

Figure 13.4. A screen dump of the gap4 Contig Comparator. This transformed version of the Contig Selector is used to display the results of analytical methods that give information about the relationships between contigs. For example, it can show sequence matches between contigs and the positions of read pairs that span contigs. See text for details. (See color plate.)

Templates in bright yellow have been sequenced from both ends, are consistent, and span a pair of contigs (and thus show the relative orientation and separation of the contigs). All data are scaled: the templates with one reading are given their expected length (here, 2,000 bases), templates within single contigs and with readings from both ends are shown with their actual length, and templates spanning contigs are shown with a length calculated from the positions of their readings. The screen dump contains two vertical lines that extend the full height of the plot. One is the crosshair, and the other shows the position of the Contig Editor editing cursor. Not only does

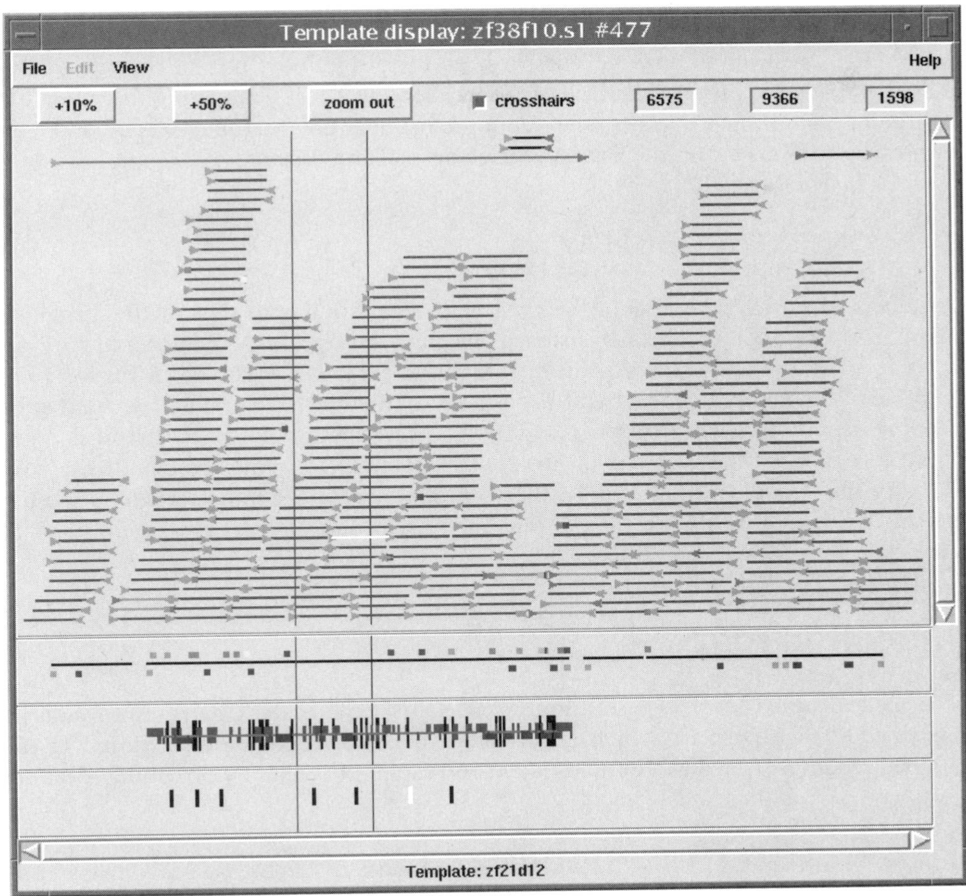

Figure 13.5. A screen dump of the gap4 Template Display, which shows the positions of DNA templates and the extent of readings derived from them. Color coding is used to distinguish between forward and reverse readings and to show consistent and inconsistent read pairs. See text for details. (See color plate.)

this vertical line show the editing cursor position but, by using the mouse, it can be used to scroll the editor.

In the plot, below the line representing the longest contig, the quality of its consensus is illustrated by color and line height. Here, the "quality" of the consensus is calculated using a simple algorithm and is mostly useful for data that has no confidence values. If a consensus base in a contig has good data from both strands and the consensus, when calculated separately for each of the two strands, is in agreement, then the plot for that position will show nothing. However, if the two strands disagree, or there are good data for only one strand, lines will be plotted of a color and height that signifies the problem. An alternative plot for data with confidence values, which shows the confidence for the consensus, is also available.

The position of a crosshair is shown in the two leftmost boxes in the top right-hand corner. The leftmost shows the distance in bases between the crosshair and the start of the contig underneath the crosshair. The middle box shows the distance between the crosshair and the start of the first contig. The bottom plot is of the positions of restriction enzyme cut sites, and the rightmost box at the top of the display shows the distance between two selected cut sites. Comparing the predicted restriction pattern and that obtained from the original physical map clone can be an important final check on the overall correctness of the assembly.

THE CONSISTENCY DISPLAY

An alternative way of viewing the data at a scale similar to that of the Template Display is provided by the Consistency Display, which can plot histograms of the reading coverage and the read-pair coverage. An example is shown in Figure 13.6. The blocky histogram shows two lines (one red, one black) depicting the number of readings covering each position in a contig. The coverage for each strand is summarized below by the two sets of horizontal lines; where a line appears, there is no data for the strand. In the example, there are no data for the minus strand at the left end of the contig and none for the plus strand just left of the right-hand cursor or crosshair.

THE CONTIG EDITOR

The most detailed level of resolution provided by gap4 is the Contig Editor and its associated Trace Display, which are used for the final checking and editing of the aligned readings. It makes this an efficient and rapid procedure by providing a variety

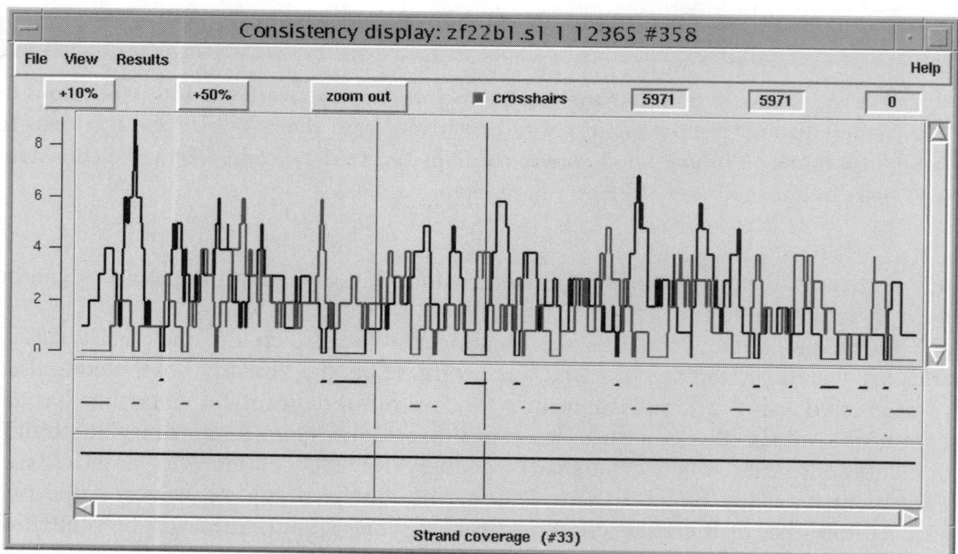

Figure 13.6. A screen dump of the gap4 Consistency Display. Here, it is being used to plot a histogram of the number of readings from each strand covering each position along a contig. Below that it is showing the segments with no data from one strand or the other. See text for details. (See color plate.)

of search methods and by automatically displaying the trace data for any problems found. Users can alter bases or their confidence values to resolve errors, and the consensus sequence is instantly updated. Editing should be performed only to obtain the correct consensus at the required level of confidence. The less editing done, the quicker the project will be completed, and the more original base calls visible in the Contig Editor, the easier it will be to check if any doubts arise in the future. The consensus calculation is the same as that used to produce the final consensus to be sent to the sequence library, and it will remove all padding characters that appear in the Contig Editor consensus. Note that the numbering shown along the top of the Contig Editor window can be set in two ways: positions can include or exclude consensus padding characters.

Figure 13.7 shows a screen dump from the Contig Editor that contains segments of aligned readings, their consensus, and a three-phase translation. Below this is the Trace Display. The Search dialogue is obscuring the right-hand side of the Editor window. The main components of the editor are the controls at the top, reading names on the left, sequences to their right, and status lines at the bottom. To the left of each reading name is the reading number, which is negative for readings that have been reversed and complemented. The reliability of the individual base calls is shown using gray scale—the lighter the background, the higher the confidence. Below the reading data is their consensus sequence, with its reliability again shown using gray

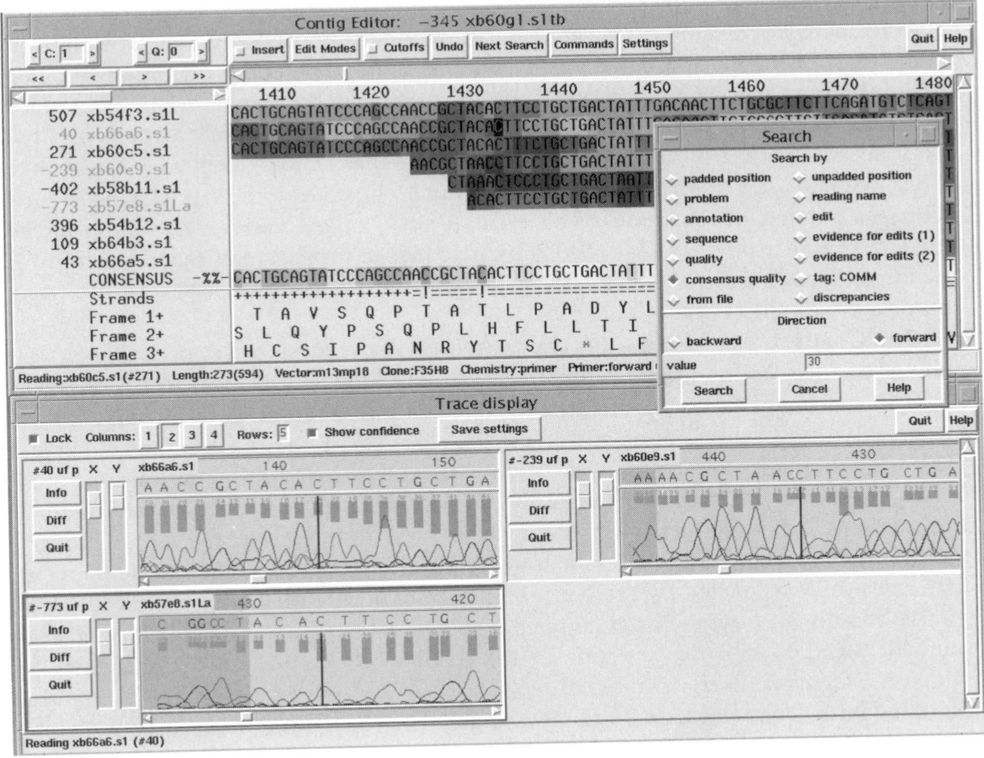

Figure 13.7. A screen dump of the gap4 Contig Editor and Trace display. See text for details. (See color plate.)

scale. The first of the status lines, labeled "Strands," shows a summary of strand coverage. The left end of the segment of sequence being displayed is covered only by readings from one strand of the DNA (+), but the rest contains data from both strands (=). Two positions are marked with an exclamation mark (!), indicating that, when the consensus is calculated separately for each strand, they disagree.

Along the top of the editor window is a row of command buttons and menus. The rightmost pair of buttons provide help and exit. To their left are two menus, Settings and Commands. To the left of this is a button that initially displays the search dialogue (shown); by pressing it again, the selected search will be performed. Further left is the Undo button; each time the user clicks on this box, the program reverses the previous edit command. The next button, labeled Cutoffs, is used to toggle between showing or hiding the reading data of poor quality or are vector sequences. The next button to the left is the Edit Modes menu, which allows users to select which editing commands are enabled. The next command toggles between insert and replace and thus governs the effect of typing in the edit window. The Information Line at the bottom of the window can show information about readings, annotations, and base calls, depending on what is under the mouse cursor.

With the use of the Settings menu, the Contig Editor can be configured to display disagreements and edits. Disagreements between the consensus and individual base calls are shown in dark green. Edits are shown with a light green background, replacements/insertions are shown in pink, deletions are shown in red, and confidence value changes are shown in purple. As can be seen from the contents of the Search dialogue superimposed on the Contig Editor, there are many types of search that can be performed. For example, the cursor can explicitly be moved to any position in the contig, either ignoring or counting padding characters. Other operations include finding the next non-ACGT character in the consensus; the next edit, the next place where the confidence of the consensus falls below some user-defined value, and the next place where an edit has been performed, but there is no evidence for the change in the original data (i.e., the current base does not appear in any of the original readings covering the position). The type of search used varies depending on the user's role. For example, a user finishing a sequence project may search mainly for low-confidence consensus sequences, whereas someone in a supervisory role may go through and check every edit made.

The Contig Editor can display the traces for any reading or set of readings. The number of rows and columns of traces displayed can be set by the user. The traces scroll in register with one another, as well as with the cursor in the Contig Editor. Conversely, the Contig Editor cursor can be scrolled by the trace cursor. A typical view containing three traces is shown in the screen dump of the Contig Editor (Figure 13.7). These are the best two traces from each strand plus a trace from a reading that contains a disagreement with the consensus. With the use of the Settings menu, the program can be configured to automatically bring up this combination of traces for each problem located by the Next Search option. The histogram or vertical bars plotted top-down show the original confidence values for each base call. The reading number, together with the direction of the reading (+ or −) and the chemistry by which it was determined, is given at the top left of each subwindow. There are three buttons (Info, Diff, and Quit) arranged vertically, with X and Y scale bars to their right. The Info button produces a window containing notes about the trace. The Diff button is mostly used for mutation detection and causes a pair of traces to be subtracted from one another and the result plotted, hence revealing their differences.

THE CONTIG JOINING EDITOR

The Join Editor looks like a pair of Contig Editors stacked one above the other with a strip in between, labeled `Diffs` for displaying the disagreements between the two consensus sequences. The other two main differences are the Align and Lock buttons. The former performs an alignment between the two consensus sequences and inserts padding characters accordingly, and the latter toggles between scrolling the contigs separately or in unison. Each contig can be edited separately, and it is important that, before a join is confirmed, the alignment is checked over its full extent. The Contig Join Editor is usually invoked by clicking on a Find Internal Joins or Find Repeats result in the Contig Comparator, in which case the two contigs will appear aligned at the match found by these searches. The few differences between the Join Editor and the Contig Editor can be seen in Figure 13.8. This figure shows the right-hand end of one contig in the lower editor and the left-hand end of another in the top editor. The Cutoff or Hidden data are displayed for the right-hand contig but not for the left-hand contig.

DISASSEMBLING READINGS

Pregap4 and gap4 use lists of readings and contigs for processing batches of data. Obviously, the input to pregap4 is usually a list of trace files. Within gap4, users can create these lists in a number of ways. For example, when using the Contig Editor, users can create a list of readings to be disassembled. These are automatically sent to the gap4 Disassemble Readings function when the editor is exited. Readings can either be removed from the database or from the contigs they were in. If they are removed only from the contigs, they each start new contigs of their own; therefore, if the Find Internal Joins function is applied to all the contigs in the database, it will reveal any overlaps they may have elsewhere. If they are removed from the database, the list can be used by the assembly function to reassemble them. During its disassembly, if a reading is the only one covering part of a contig, the contig will be automatically broken in two. Gap4 also has a specific function for this purpose that will break a contig at the start of a given reading.

EXPERIMENT SUGGESTION AND AUTOMATION

Much of the foregoing has been to illustrate the types of tools that are available for solving difficult problems interactively, but ideally one would prefer to automate as much of the work as possible. Genome sequencing center staff have put many hours into developing their own in-house procedures for automating their work. The current release of gap4 enables this approach in two ways. The first is by use of its Experiment Suggestion functions, and the second is by use of the gap4 scripting language.

Using its consensus algorithms, gap4 can analyze assembly databases to find segments of sequence that require further readings, either to resolve disagreements, to fill the requirement for data from both strands of the DNA, or to extend contigs to try join them to others. It can then suggest the experiments to perform and the templates to use because it has all the necessary information in its assembly database. However, in the current version of gap4, the types of experiments are limited, and

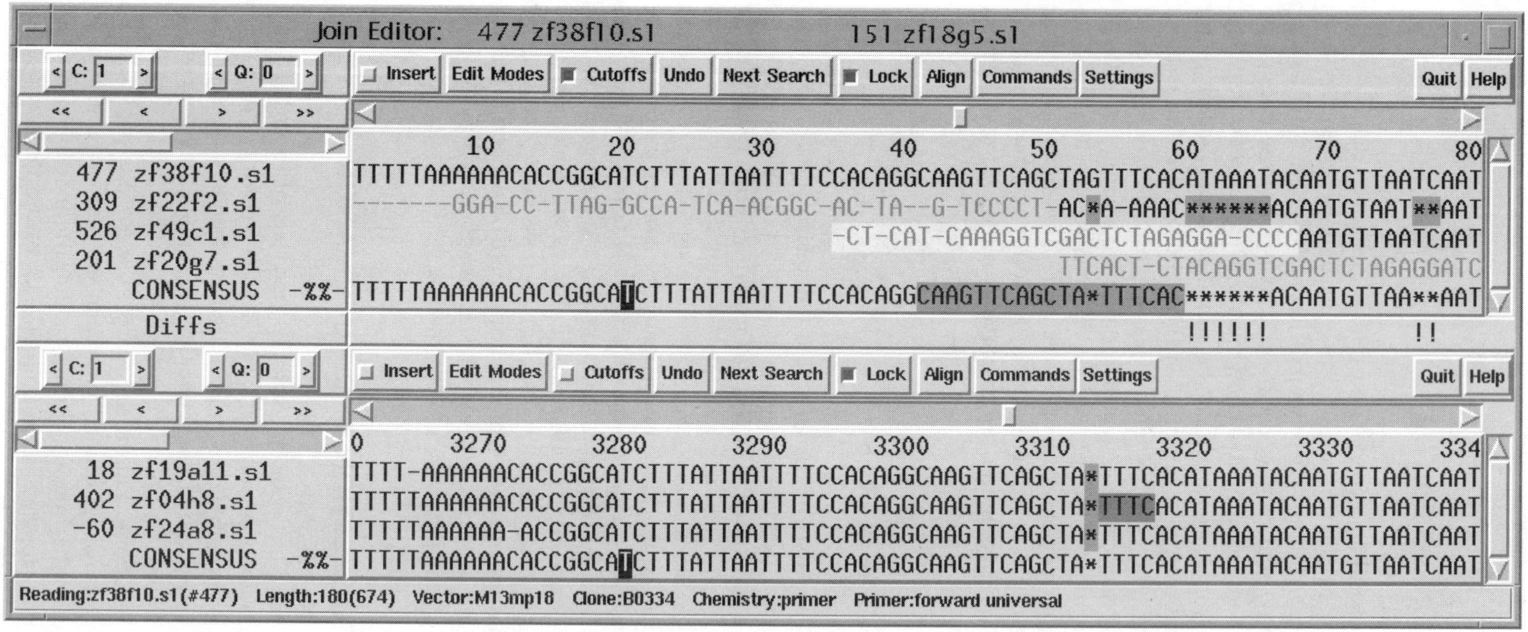

Figure 13.8. A screen dump of the gap4 Join Editor, which is used to align, edit, display traces, and join contigs. See text for details. (See color plate.)

it has no knowledge of the very latest techniques or the ones that may be available in any particular laboratory. For large laboratories using gap4, at present, it is better to use its algorithms to analyze the database and to use the results of that analysis with a set of external routines customized to local methods. As outlined below, this is enabled by the design of gap4.

An entirely new set of gap4 finishing functions is currently being tested at the Sanger Center. In addition to the consensus algorithm, these new functions employ an analysis of reading and template depthas well as a knowledge of sequencing chemistry to design sets of potential problem-solving experiments. These experiments are geared towards satisfying the finishing criteria specified by the user.

Tcl (Ousterhout, 1994) is a portable and extensible scripting language written in the C language. The programs in the Staden package are written in the form of additional commands understood by the Tcl interpreter. Each of gap4's algorithms can therefore be used in Tcl scripts. This means that users can write scripts to perform analysis of gap4 databases (e.g., apply a consensus calculation or read-pair analysis algorithm), and these scripts could also be linked to local laboratory information management systems and robotics. A further type of automation is furnished by the fact that the gap4 Contig Editor can be driven by an external file of commands.

CONCLUDING REMARKS

We have tried to give insight into the types of tasks performed during routine sequence assembly projects and used our own software as an illustration. We recommend interested readers to look at the Web sites of some of the main sequencing centers to find out more about their finishing criteria and the software they have developed to aid their projects. At the time of this writing, much of the effort of publicly funded genome centers was devoted to producing a low-coverage sequence of the human genome; therefore, the emphasis was on assembly. When that stage is completed, the low-coverage data will need to be finished, which will require more detailed attention of the types we have attempted to describe here.

INTERNET RESOURCES FOR TOPICS PRESENTED IN CHAPTER 13

ATQA	*http://www.wagner.com*
CAP3	*http://www.cs.mtu.edu.faculty/Huang.html*
Consed	*http://bozeman.genome.washington.edu/consed/consed.html*
DNASTAR	*http://www.dnastar.com*
FAKII	*http://www.cs.arizona.edu/faktory*
Phrap	*http://bozeman.genome.washington.edu/phrap.docs/phrap.html*
Phred	*http://bozeman.genome.washington.edu/phrap.docs/phred.html*
Gap4	*http://mrc-lmb.cam.ac.uk/pubseq/index.html*
Sequencher	*http://www.genecodes.com*
Staden package	*http://www.mrc-lmb.cam.ac.uk/pubseq/index.html*
TIGR Assembler	*http://www.tigr.org/softlab*
TraceTuner	*http://www.paracel.com/html/tracetuner.html*

PROBLEM SET

Data and lecture notes for a short course in the use of pregap4 and gap4 are available on the book's Web site (http://www.wiley.com/bioinformatics). This information is derived from a two-day practical course, but the notes are written such that users can perform the exercises in an autotutorial fashion. The exercises revolve around a series of ABI trace files and expose the user to format conversion, experiment file creation, vector clipping, quality clipping, contaminant screening, assembly, join finding, read-pair analysis, finishing experiments, use of confidence values, use of the contig editor and trace display, and consensus calculation. The notes are available in both PostScript and Microsoft Word format, and the data can be used with either the UNIX or Microsoft Windows versions of the programs.

REFERENCES

Bonfield, J. K., Smith, K. F., and Staden, R. (1995). A new DNA sequence assembly program. *Nucleic Acids Res.* 23, 4992–4999.

Bonfield, J. K., and Staden, R. (1996). Experiment files and their application during large-scale sequencing projects. *DNA Sequence* 6, 109–117.

Bonfield, J. K., and Staden, R. (1995).The application of numerical estimates of base calling accuracy to DNA sequencing projects. *Nucleic Acids Res.* 23, 1406–1410.

Dear, S., and Staden, R. (1992). A standard file format for data from DNA sequencing instruments. *DNA Sequence* 3, 107–110.

Ewing, B., and Green, P. (1998). Base-calling of automated sequencer traces using phred. II. Error probabilities. *Genome Res.* 8, 186–194.

Gordon, D., Abajian, C., and Green, P. (1998). Consed: A graphical tool for sequence finishing. *Genome Res.* 8, 195–202.

Huang, X. (1996). *Genomics* 33, 21.

Hutchinson, C. A., et al (1989). In *Mobile DNA*, D. E. Berg and M. M. Howe, Eds. (American Society of Microbiology, Washington, DC). p. 593–618.

Kececioglu, J., and Myers, E. (1995). Combinatorial algorithms for DNA sequence assembly. *Algorithmica* 13, 7–51.

Myers, E. (1995). Toward simplifying and accurately formulating fragment assembly. *J. Comput. Biol.* 2, 275–290.

Myers, E. W., Jain, M., and Larson, S. (1996). Internal report. University of Arizona.

Ousterhout, J. K. (1994). Tcl and the TK toolkit. (Addison-Wesley, Reading, MA).

Sanger, F., Nicklen, S., and Coulson, A. R. (1977). DNA sequencing with chain terminator inhibitors. *Proc. Natl. Acad. Sci. USA* 74, 5463–5467.

Staden, R. (1982) Automation of the computer handling of gel reading data produced by the shotgun method of DNA sequencing. *Nucleic Acids Res.* 10, 4731–4751.

Sutton G., White O., Adams M., and Kerlavage A. (1995). TIGR Assembler: A new tool for assembling large shotgun sequencing projects. *Genome Sci. Technol.* 1, 9–19.

14

PHYLOGENETIC ANALYSIS

Fiona S. L. Brinkman

Department of Microbiology and Immunology
University of British Columbia
Vancouver, British Columbia, Canada

Detlef D. Leipe

National Center for Biotechnology Information
National Library of Medicine
National Institutes of Health
Bethesda, Maryland

Phylogenetics is the study of evolutionary relationships. Phylogenetic analysis is the means of *inferring* or estimating these relationships. The evolutionary history inferred from phylogenetic analysis is usually depicted as branching, treelike diagrams that represent an estimated pedigree of the inherited relationships among molecules ("gene trees"), organisms, or both. Phylogenetics is sometimes called cladistics because the word "clade," a set of descendants from a single ancestor, is derived from the Greek word for branch. However, cladistics is a particular method of hypothesizing about evolutionary relationships.

The basic tenet behind cladistics is that members of a group or clade share a common evolutionary history and are more related to each other than to members of another group. A given group is recognized by sharing unique features that were not present in distant ancestors. These shared, *derived* characteristics can be anything that can be observed and described—from two organisms having developed a spine to two sequences having developed a mutation at a certain base pair of a gene. Usually, cladistic analysis is performed by comparing multiple characteristics or "characters" at once, either multiple phenotypic characters or multiple base pairs or amino acids in a sequence.

Bioinformatics: A Practical Guide to the Analysis of Genes and Proteins
Edited by A. D. Baxevanis and B. F. F. Ouellette
ISBN 0-471-38390-2 (cloth), ISBN 0-471-383910 (paper)

- There are three basic assumptions in cladistics: Any group of organisms is related by descent from a common ancestor (fundamental tenet of evolutionary theory).
- There is a bifurcating pattern of cladogenesis. This assumption is controversial.
- Change in characteristics occurs in lineages over time. This is a necessary condition for cladistics to work.

The resulting relationships from cladistic analysis are most commonly represented by a phylogenetic tree:

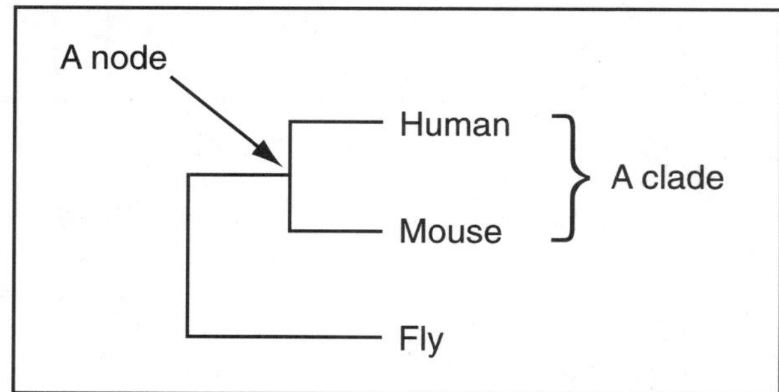

Even with this simple tree, a number of terms that are used frequently in phylogenetic analysis can be introduced:

- A **clade** is a monophyletic taxon. Clades are groups of organisms or genes that include the most recent common ancestor of all of its members and all of the descendants of that most recent common ancestor. Clade is derived from the Greek word "klados," meaning branch or twig.
- A **taxon** is any named group of organisms but not necessarily a clade.
- In some analyses, **branch** lengths correspond to divergence (e.g., in the above example, mouse is slightly more related to fly than human is to fly).
- A **node** is a bifurcating branch point.

Macromolecules, especially sequences, have surpassed morphological and other organismal characters as the most popular form of data for phylogenetic or cladistic analysis. It is this molecular phylogenetic analysis that we will introduce here.

It is unrealistic to believe that an all-purpose phylogenetic analysis recipe can be delineated (Hillis et al., 1993). Although numerous phylogenetic algorithms, procedures, and computer programs have been devised, their reliability and practicality are, in all cases, dependent on the structure and size of the data. The merits and pitfalls of various methods are the subject of often acrimonious debates in taxonomic and phylogenetic journals. Some of these debates are summarized in a series of useful reviews of phylogenetics (Saitou, 1996; Li, 1997; Swofford et al., 1996). An especially concise introduction to molecular phylogenetics is provided by Hillis et al. (1993).

The danger of generating incorrect results is inherently greater in computational phylogenetics than in many other fields of science. The events yielding a phylogeny happened in the past and can only be inferred or estimated (with a few exceptions, see Hillis et al., 1994). Despite the well-documented limitations of available phylogenetic procedures, current biological literature is replete with examples of conclusions derived from the results of analyses in which data had been simply run through one or another phylogeny program. Occasionally, the limiting factor in phylogenetic analysis is not so much the computational method used; more often than not, the limiting factor is the users' understanding of what the method is actually doing with the data.

This brief guide to phylogenetic analysis has several objectives. First, a conceptual approach that describes some of the most important principles underlying the most widely and easily applied methods of phylogenetic analyses of biological sequences and their interpretation will be introduced. The aim is to show that practical phylogenetic analysis should be conceived as a search for a correct model, as much as a search for the correct tree. In this context, some of the particular models assumed by various popular methods and how these models might affect analysis of particular data sets will be discussed. Finally, some examples of the application of particular methods to the inferences of evolutionary history are provided.

Note that the principles for DNA analysis will be initially discussed, although most also apply to protein sequences (except where further description of protein sequences is indicated). As there is a growing interest in the analysis of protein sequences, the reader is directed to further descriptions of protein-specific problems, as reviewed by Felsenstein (1996).

FUNDAMENTAL ELEMENTS OF PHYLOGENETIC MODELS

Phylogenetic tree-building methods presume particular evolutionary models. For a given data set, these models can be violated because of occurrences such as the transfer of genetic material between organisms. Thus, when interpreting a given analysis, one should always consider the model used and its assumptions and entertain other possible explanations for the observed results. As an example, consider the tree in Figure 14.1. An investigation of organismal relationships in the tree suggests the eukaryote 1 is more related to the bacteria than to the other eukaryotes. Because the vast majority of other cladistic analyses, including those based on morphological features, suggest that eukaryote 1 is more related to the other eukaryotes than to bacteria, we suspect that for this analysis the assumptions of a bifurcating pattern of evolution are incorrect. We suspect that horizontal gene transfer from an ancestor of the bacteria 1, 2, and 3 to the ancestor of eukaryote 1 occurred because this would most simply explain the results.

Models inherent in phylogenetics methods make additional "default" assumptions:

1. The sequence is correct and originates from the specified source.
2. The sequences are homologous (i.e., are all descended in some way from a shared ancestral sequence).
3. Each position in a sequence alignment is homologous with every other in that alignment.
4. Each of the multiple sequences included in a common analysis has a common phylogenetic history with the others (e.g., there are no mixtures of nuclear and organellar sequences).

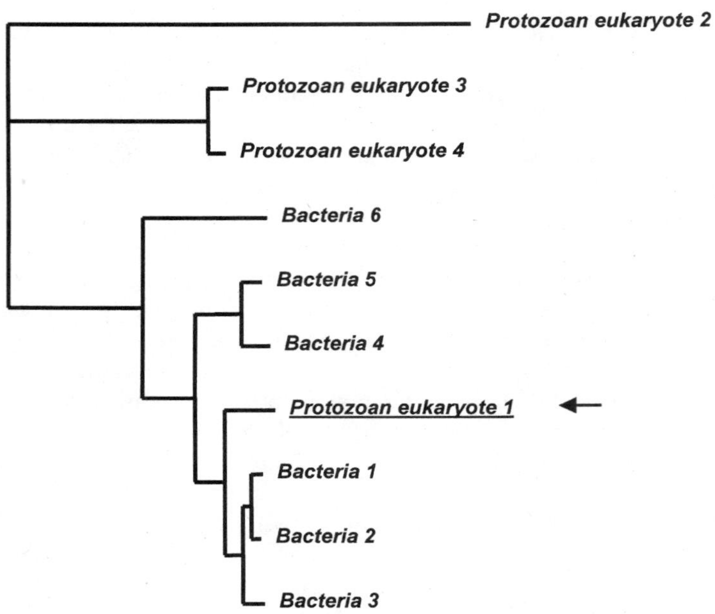

Figure 14.1. Example of a phylogenetic tree based on genes that do not match organismal phylogeny, suggesting horizontal gene transfer has occurred. The ancestor of protozoan eukaryote 1 (underlined and marked with an arrow) appears to have obtained the gene from the ancestor of Bacteria 1, 2, and 3, as this is the simplest explanation for the results. This unexpected result is not without precedent: there have been a number of reported phylogenetic analyses that suggest that protozoa have taken up genes from bacteria, most likely from bacteria that they have ingested.

5. The sampling of taxa is adequate to resolve the problem of interest.
6. Sequence variation among the samples is representative of the broader group of interest.
7. The sequence variability in the sample contains phylogenetic signal adequate to resolve the problem of interest.

There are additional assumptions that are defaults in some methods but can be at least partially corrected for in others:

1. The sequences in the sample evolved according to a single stochastic process.
2. All positions in the sequence evolved according to the same stochastic process.
3. Each position in the sequence evolved independently.

Errors in published phylogenetic analyses can often be attributed to violations of one or more of the foregoing assumptions. Every sequence data set must be evaluated against these assumptions, with other possible explanations for the observed results considered.

TREE INTERPRETATION—THE IMPORTANCE OF IDENTIFYING PARALOGS AND ORTHOLOGS

As more genomes are sequenced, we are becoming more interested in learning about protein or gene evolution (i.e., investigating gene phylogeny, rather than organismal phylogeny). This can aid our understanding of the function of proteins and genes.

Studies of protein and gene evolution involve the comparison of *homologs*—sequences that have common origins but may or may not have common activity. Sequences that share an arbitrary, threshold level of similarity determined by alignment of matching bases are termed *homologous*. They are inherited from a common ancestor that possessed similar structure, although the structure of the ancestor may be difficult to determine because it has been modified through descent.

Homologs are most commonly either orthologs, paralogs, or xenologs.

- *Orthologs* are homologs produced by speciation. They represent genes derived from a common ancestor that diverged due to divergence of the organisms they are associated with. *They tend to have similar function.*
- *Paralogs* are homologs produced by gene duplication. They represent genes derived from a common ancestral gene that duplicated within an organism and then subsequently diverged. *They tend to have different functions.*
- *Xenologs* are homologs resulting from horizontal gene transfer between two organisms. The determination of whether a gene of interest was recently transferred into the current host by horizontal gene transfer is often difficult. Occasionally, the %(G + C) content may be so vastly different from the average gene in the current host that a conclusion of external origin is nearly inescapable, however often it is unclear whether a gene has horizontal origins. Function of xenologs can be variable depending on how significant the change in context was for the horizontally moving gene; however, in general, the function tends to be similar.

An example of how the identification of orthologs and paralogs can be used to aid prediction of protein function is illustrated in Figure 14.2.

PHYLOGENETIC DATA ANALYSIS: THE FOUR STEPS

A straightforward phylogenetic analysis consists of four steps:

1. Alignment (both building the data model and extracting a phylogenetic dataset)
2. Determining the substitution model
3. Tree building
4. Tree evaluation

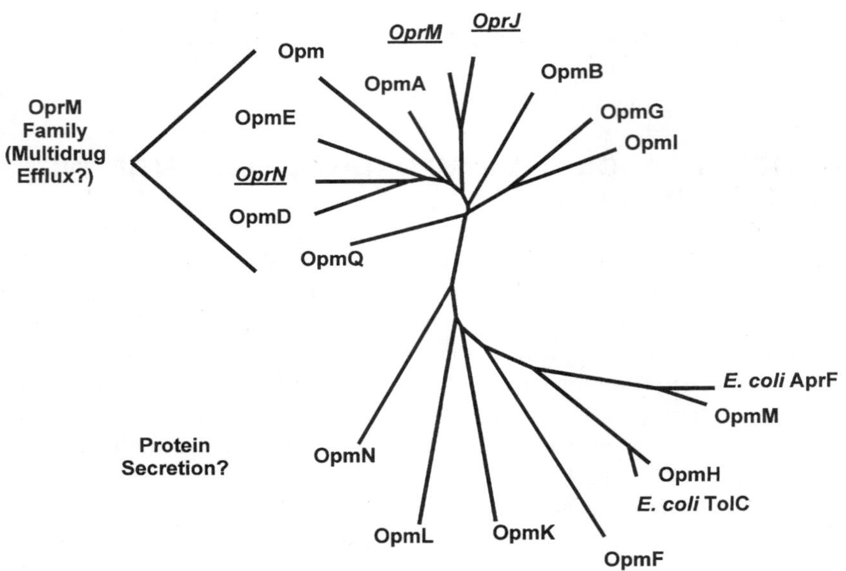

Figure 14.2. Insight into protein function from an investigation of paralogs and orthologs —an example. *Pseudomonas aeruginosa*, a bacteria that is one of the top three causes of opportunistic infections, is noted for its antimicrobial resistance and resistance to detergents. Three homologous outer membrane proteins, OprJ, OpM, and OprN, have been identified as playing a role in this antimicrobial resistance, by pumping different antimicrobials out of the cell as they entered. When the genome of this bacterium was sequenced, it was found that there were no less than 14 homologs of the genes encoding these three proteins (given names starting with "Opm"). Phylogenetic analysis of these protein sequences, using the neighbor joining distance method within the PHYLIP 5.3 package, showed that this 17-member family was divided into two clades, one containing all three genes with roles in antimicrobial efflux pumps (underlined italics). Two members of the other clade were found to share highest similarity with proteins AprF and TolC from another organism, *E. coli*. AprF and TolC are both involved in secreting proteins. This analysis allowed us to hypothesize that the clade containing OprM, OprJ, and OprN, nicknamed the OprM family, comprises a series of paralogous genes involved in efflux of different antimicrobials or antimicrobial-like compounds. The other cluster with homologs to AprF and TolC may be a functionally rated group of paralogs involved in secretion of proteins (of which OpmM appears to be the ortholog of AprF and OpmH is likely the ortholog of TolC). Currently, efforts are expanding to characterize *P. aeruginosa* with mutations in these genes to evaluate their ability to efflux antimicrobials. This phylogenetic analysis allows us to prioritize the analysis of genes in this extended family, analyzing the OprM family genes first as they are more likely to have the functions of interest. This tree was drawn using Treeview.

Each step is critical for the analysis and should be handled accordingly. For example, trees are only as good as the alignment they are based on. When performing a phylogenetic analysis, it often insightful to build trees based on different modifications of the alignment to see how the alignment proposed influences the resulting tree.

ALIGNMENT: BUILDING THE DATA MODEL

Phylogenetic sequence data usually consist of multiple sequence alignments; the individual, aligned-base positions are commonly referred to as "sites." These sites are equivalent to "characters" in theoretical phylogenetic discussions, and the actual base (or gap) occupying a site is the "character state."

Multiple alignment methods are reviewed in Chapter 9. This chapter reviews similar alignment methods in the context of phylogenetic analysis. Aligned sequence positions subjected to phylogenetic analysis represent a priori phylogenetic conclusions because the sites themselves (not the actual bases) are effectively assumed to be genealogically related, or homologous. Sites at which one is confident of homology and that contain changes in character states useful for the given phylogenetic analysis are often referred to as "informative sites."

Steps in building the alignment include selection of the alignment procedure(s) and extraction of a phylogenetic data set from the alignment. The latter procedure requires determination of how ambiguously aligned regions and insertion/deletions (referred to as *indels*, or gaps) will be treated in the tree-building procedure.

A typical alignment procedure involves the application of a program such as CLUSTAL W, followed by manual alignment editing and submission to a tree-building program. This procedure should be performed with the following questions and considerations in mind:

How much computer dependence? Fully computational multiple alignment is sometimes advocated on the grounds that manual editing is inexplicit and/or unobjective (Gatesy et al., 1993). Usually, however, manual alignment editing is advocated (e.g., Thompson et al., 1994) because alignment algorithms and programs are not optimally adapted for phylogenetic alignment (see Fig. 14.3).

Phylogenetic criteria preferred. Some computational multiple alignment methods align sequences strictly based on the order they receive them (the input order) without any consideration of their relationship. However, many current methods (e.g., CLUSTAL W, PileUp, ALIGN in ProPack) align according to an explicitly phylogenetic criterion (a "guide tree"). These guide trees are generated on the basis of initial pairwise sequence alignments. SAM (Hughey et al., 1996) and MACAW (Lawrence et al., 1993) are examples of multiple alignment programs that do not explicitly invoke phylogenetic criteria, although it is possible to manipulate parameters in these programs to mimic phylogenetic processes. Theory holds that more closely related sequences should be aligned first and then the resulting groups of sequences, which may be less related to one another but still have a common ancestor, should share the same ancestral indels. This means that they could then be more accurately aligned.

The guide tree from CLUSTAL W (Fig. 14.4) is formatted as a PHYLIP tree file and can be imported in various tree-drawing programs. Some programs are designed to simultaneously (recursively) optimize an alignment and a phylogenetic tree (e.g., TreeAlign and MALIGN). In theory, an optimal simultaneous solution or set of solutions to an alignment/phylogeny problem exists, but the hazard of the recursive approach lies in the possibility of funneling the analysis toward a wrong or

A

		116	122-144	155
1	ARABI	GCGCCC	---CAAGCCTTCT-GGCCG----	AGGGCACGTCT
2	LYCOP	C	---GAAGCCATTT-GGCCG----	A..........
3	triti	C	---GAGGCCACTC-GGCCG----	A.......C..
4	LACTU	C	---GAAGCCATCC-GGCTG----	A.......C..
5	SILEN	C	---GAAGC--TTC-GGCTG----	A..........
6	vicia	C	---GATGCCATTA-GGTTG----	A..........
7	CANEL	C	---GAGGCCACTA-GGCTG----	A....T..C..
8	potam	C	---TAAGCTTCCG-GGCCG----	A.....A.C..
9	ephed	.-...C	---GAAGCC--TC-CGCCA----	A..........
10	gnetu	.-...C	TCCG-AGCC--TA-GGCCG----	A..........
11	PINUS	C	---GAGGCC--TC-GGTCG----	A..........
12	PICEA	C	---GAGNCC--TC-GGTCG----	A..........
13	TAXUS	.GC..G	---GAG-C---TC-GGCCG----	A..........
14	marsi	.-...C	---GAGGC---TC-GTCCG----	A..........
15	osmun	.-...C	---GCGGC---TC-GTCCA----	A.....T.C..
16	mnium	C	---GAGGC---TC-GTCCG----	A.....TT..C
17	CHLAM	....TC	---GAGGC--TTC-GGCCA----	A.A...T....
18	SPERM	....TC	---GAGGC--TTC-GGCCG----	A.A...T..T.
19	TETRA	....TC	---GAGGC--CTC-GGCCA----	A.A.....C..
20	CHLOR	....TC	---GAGAC--CTC-GGTCA----	A.A...T....
21	CLADO	....TC	---AAGTC--TAC-GGACT----	T.A...T....
22	HETER	C	TTT--GGT-ATT----CCGA---	A..-....C..
23	VOLVA	....TC	TTT--GGCCATT----CCGA---	A.A...T.C..
24	SCLER	C	CTT--GGT-ATT----CCGG---	G.....T.C..
25	sacch	C	CTT--GGT-ATT----CCAG---	G.....T.C..
26	BIPOL	C	TTT--GGT-ATT----CCAA---	A.....T.C..
27	GLOMU	....TC	CCT--GGT-ATT----CCGG---	G.A.T.T.C..
28	CYANI	....TT	TC--AGGAGAATTTTATTTTCCT	G.A........
29	SARCO	....TC	GC---GGTAA-TC--------CT	GCA.--T....
30	PHYTO	..A.TT	CCG--GGTTAGTC---CTG----	G.A.T.T.C..
31	SCYTO	...-TT	CCG--GGATATGC---CTG----	G.A...T.CT.
32	crypt	.---CT	CC---AGC--TGA---CT-----	------T...A
33	PRORO	....TT	TCG--GGATATCC---CTG----	AA....T.C..

B

		116	122-144	155	[122′-141′]
1	ARABI	GCGCCC	???CAAGCCTTCT?GGCCG????	AGGGCACGTCT	????????????????????
2	LYCOP	GCGCCC	???GAAGCCATTT?GGCCG????	AGGGCACGTCT	????????????????????
3	triti	GCGCCC	???GAGGCCACTC?GGCCG????	AGGGCACGCCT	????????????????????
4	LACTU	GCGCCC	???GAAGCCATCC?GGCTG????	AGGGCACGCCT	????????????????????
5	SILEN	GCGCCC	???GAAGC?-TTC?GGCTG????	AGGGCACGTCT	????????????????????
6	vicia	GCGCCC	???GATGCCATTA?GGTTG????	AGGGCACGTCT	????????????????????
7	CANEL	GCGCCC	???GAGGCCACTA?GGCTG????	AGGGCTCGCCT	????????????????????
8	potam	GCGCCC	???TAAGCTTCCG?GGCCG????	AGGGCAAGCCT	????????????????????
9	ephed	G-GCCC	???GAAGC?-?TC?CGCCA????	AGGGCACGTCT	????????????????????
10	gnetu	G-GCCC	TCCG??GC?-?TA?GGCCG????	AGGGCACGTCT	????????????????????
11	PINUS	GCGCCC	???GAGGC?-?TC?GGTCG????	AGGGCACGTCT	????????????????????
12	PICEA	GCGCCC	???GAG?C?-?TC?GGTCG????	AGGGCACGTCT	????????????????????
13	TAXUS	GGCCCG	???GAG-C?-?TC?GGCCG????	AGGGCACGTCT	????????????????????
14	marsi	G-GCCC	???GAGGC??-TC?GTCCG????	AGGGCACGTCT	????????????????????
15	osmun	G-GCCC	???GCGGC??-TC?GTCCA????	AGGGCATGCCT	????????????????????
16	mnium	GCGCCC	???GAGGC??-TC?GTCCG????	AGGGCATTTCC	????????????????????
17	CHLAM	GCGCTC	???GAGGC?-TTC?GGCCA????	AGAGCATGTCT	????????????????????
18	SPERM	GCGCTC	???GAGGC?-TTC?GGCCG????	AGAGCATGTTT	????????????????????
19	TETRA	GCGCTC	???GAGGC?-CTC?GGCCA????	AGAGCACGCCT	????????????????????
20	CHLOR	GCGCTC	???GAGAC?-CTC?GGTCA????	AGAGCATGTCT	????????????????????
21	CLADO	GCGCTC	???AAGTC?-TAC?GGACT????	TGAGCATGTCT	????????????????????
22	HETER	GCGCCC	???????????????????????	AGG-CACGCCT	TTT??GGT-ATT????CCGA
23	VOLVA	GCGCTC	???????????????????????	AGAGCATGCCT	TTT??GGCCATT????CCGA
24	SCLER	GCGCCC	???????????????????????	GGGGCATGCCT	CTT??GGT-ATT????CCGG
25	sacch	GCGCCC	???????????????????????	GGGGCATGCCT	CTT??GGT-ATT????CCAG
26	BIPOL	GCGCCC	???????????????????????	AGGGCATGCCT	TTT??GGT-ATT????CCAA
27	GLOMU	GCACTC	???????????????????????	GGAGTATGCCT	CCT??GGT-ATT????CCGG
28	CYANI	GCGCTT	???????????????????????	GGAGCACGTCT	????????????????????
29	SARCO	GCGCTC	???????????????????????	GCAG-?TGTCT	????????????????????
30	PHYTO	GCACTT	???????????????????????	GGAGTATGCCT	????????????????????
31	SCYTO	GCG-TT	???????????????????????	GGAGCATGCTT	????????????????????
32	crypt	G??-CT	???????????????????????	??????TGTCA	????????????????????
33	PRORO	GCGCTT	???????????????????????	AAGGCATGCCT	????????????????????

incomplete solution (Thorne and Kishino, 1992). Thus, when the tree-building analysis based on the alignment is followed, one should consider whether other evolutionary relationships might be favored using a slightly modified alignment.

Alignment Parameter Estimation. The most important parameters in an alignment method are those that determine the placement of indels or gaps in an alignment of length-variable sequences. Alignment parameters should vary dynamically with evolutionary divergence (Thompson et al., 1994), such that base mismatches are more likely as the sequences become more divergent. Alignment parameters should also be adjusted to prevent closely related, overrepresented sequences from adversely influencing the alignment of underrepresented sequences (Thompson et al., 1994, Hughey et al., 1996). This is accomplished by downweighting the alignment score contribution of closely related sequences. These dynamic parameter adjustments are both implemented in CLUSTAL W, whereas sequence weighting is implemented in SAM.

Which Alignment Procedure is Best for Phylogenetic Analysis? The short answer is "the method that is closest to understanding the evolutionary relationships

Figure 14.3. Alignment modification for phylogenetic analysis. (A) Alignment showing a length-variable region (boxed) of 5.8S rDNA for the taxa in the guide tree of Figure 14.4 Taxa 1–8 are angiosperms; 9 and 10, gnetophytes; 11–13, conifers; 14 and 15, ferns; 16, moss; 17–21, green algae; 22–27, fungi; and 28–33, protists. The alignment positions correspond to those published elsewhere (Hershkovitz and Lewis, 1996). Each sequence is unique in the shaded region. Taxa represented in the Figure 14.4 tree having the same sequence as any shown here were omitted for brevity. Note that taxa grouped in the guide tree (based on the entire sequence) appear to form alignment groups in the length-variable region. On a pairwise basis, alternative alignments of some of the distantly related taxa seem plausible. For example, if moved two spaces to the left, the TAC in the center of the CLADO sequence might appear to align better with YAY in several angiosperms than the YYC in other green algae. Sufficient sampling, however, shows that YAY is not universal in the angiosperms, and the guide tree supports the present alignment, which allows no length variability in green algae. In the absence of sufficient sampling, a guide tree, or other prior phylogenetic evidence, no such conclusion could be drawn. Note also that the taxa of the green plant lineage (1–21) do not align well with the fungi and protists. The variability in the shaded region and the divergences indicated in the guide tree suggest that there is no true alignment between these distantly related groups, that the alignment indicated is arbitrary, and that the actual bases are not likely homologous. (B) The same alignment, modified as follows for phylogenetic analysis: (1) the fungi and protists are rescored as "missing" for all positions in the shaded region, where alignment with the green plant lineage is ambiguous; (2) the length-variable regions of the fungi were appended to the end of the alignment because these sequences are alignable among fungi and include phylogenetically useful variation; and (3) multiple-position gaps were rescored as one gap position and the rest missing, so that, in MP analysis, multiposition gaps are not counted as several independent deletions. The length-variable region of protists was not appended to the end of the alignment because both the alignment and the guide tree indicate that the original alignment is arbitrary.

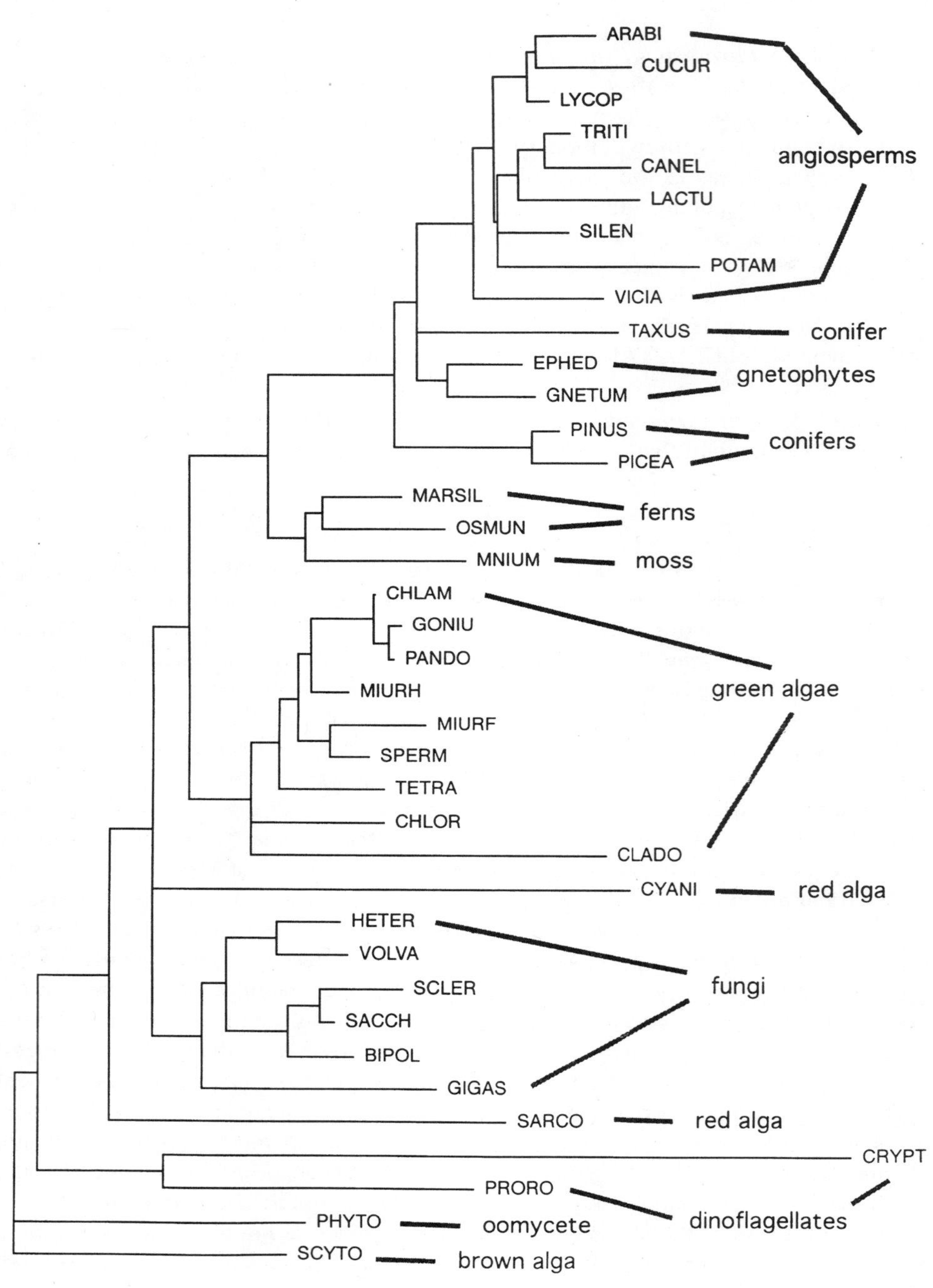
ARABI
CUCUR
LYCOP
TRITI
CANEL
LACTU
SILEN
POTAM
VICIA
angiosperms
TAXUS
conifer
EPHED
GNETUM
gnetophytes
PINUS
PICEA
conifers
MARSIL
OSMUN
ferns
MNIUM
moss
CHLAM
GONIU
PANDO
MIURH
MIURF
SPERM
TETRA
CHLOR
CLADO
green algae
CYANI
red alga
HETER
VOLVA
SCLER
SACCH
BIPOL
GIGAS
fungi
SARCO
red alga
CRYPT
PRORO
dinoflagellates
PHYTO
oomycete
SCYTO
brown alga

between the sequences being examined." Unless the actual phylogenetic relationships are known beforehand, there is no clear way to determine which alignment procedure is best for a given phylogenetic analysis. In general, it is inadvisable to simply subject a computer-generated alignment to a tree-building procedure because the latter is blind to errors in the former. This caution especially applies to tree-building programs included in alignment packages (e.g., CLUSTAL W and TREE in ProPack) because the tree-building methods in these programs are not rigorous (Feng and Doolittle, 1996). However, as long as the entire alignment is scrutinized in view of independent phylogenetic evidence, methods such as CLUSTAL W that utilize some degree of phylogenetic criteria are among the best currently available.

Mathematical optimization and analysis of structures. Some alignment programs (e.g., MACAW, SAM) optimize according to a statistical model, but the relationship of these statistics to phylogenetic models is not yet clear. No methods are yet available for determining whether one multiple alignment is significantly better than another based on a phylogenetic model.

Aligning according to secondary or tertiary sequence structure is considered phylogenetically more reliable than sequence-based alignment because confidence in homology assessment is greater when comparisons are made to complex structures rather than to simple characters (primary sequence). However, there does not appear to be any way to computationally facilitate phylogeny-based structural multiple alignment. Hopefully, new insights into these areas will be developed in the near future.

ALIGNMENT: EXTRACTION OF A PHYLOGENETIC DATA SET

In alignments that include length variation, the phylogenetic data set is usually not identical to the alignment. Even in alignments of length-invariable sequences, the data set can be different—for example, when only first and second codon positions are to be analyzed to avoid the strong G + C bias in the third codon position from affecting the final results.

<

Figure 14.4. CLUSTAL guide tree for 5.8S rDNA sequences of selected plants, fungi, and protists. The taxa and sequences corresponding to the acronyms are described elsewhere (Hershkovitz and Lewis, 1996). The tree is a neighbor-joining (distance) resolution of pairwise sequence similarities determined by pairwise alignment according to specified (in this case, default) gap penalties in CLUSTAL. Similarity is calculated as the proportion of pairwise shared bases, ignoring gap positions in either sequence. The tree can be generated as an end product or as a preliminary step in a multiple-alignment procedure. Either way, it is saved to a PHYLIP-formatted tree file. For the multiple-alignment procedure, the guide tree topology determines the sequence input order (outermost clusters are aligned first) and the branch lengths determine the sequence weights. This tree includes (see Hershkovitz and Lewis, 1996) several groupings that contradict broader evidence (e.g., polyphyly of conifers and red algae; monophyly of ferns plus moss). Such inaccuracies potentially mislead the multiple alignment. This tree was drawn and printed using the tree-drawing feature in the Macintosh version of PAUP.

In the case of length-variable sequences, the degree of difference between an alignment and the phylogenetic data set is determined mainly by how alignment ambiguities and indels are treated. The most extreme way to treat indels is to remove from the analysis all sites that include gaps (cf. Swofford et al., 1996). This approach has the advantage of permitting all the variation in the sequences to be described in terms of the substitution model, without the need for an ad hoc model to account for indels. The disadvantage of this approach is that phylogenetic signals contained within the indel regions are discarded.

Maximum parsimony (MP; see below) is the only method that permits for the incorporation of alignable gaps as characters. These can be included in either of two ways: as an additional character state (a ''fifth'' nucleotide base or ''twenty-first'' amino acid) or as a set of characters independent of base substitutions. The first approach is not tenable for gaps occupying more than one site, for these will be counted as independent character state changes. The latter approach is useful for analyzing an alignment in which subsets of sequences contain perfectly aligned gaps. A set of gap characters can be appended to the aligned sequence data set, or the gaps can be scored ''in place'' by using the extra base approach but scoring only one of the gap positions in a sequence as a gap and the remainder as missing. These approaches can be implemented using PAUP.

For some alignments, procedures that ignore all gap scores or all sites including gap scores are less than ideal. However, there is not yet any program that allows one to ignore individual sites in individual sequences. When alignment might be unambiguous within groups of sequences but ambiguous among them, alignment ''surgery'' is warranted to ensure that unambiguous information relevant to groups of sequences can be retained and ambiguous information removed.

An example of alignment surgery is given in Figure 14.3. In gapped regions, one should determine whether alternative alignments seem reasonably plausible and, just as important, whether they might bias the tree-building analysis. When alignment ambiguities are resolved manually, phylogenetic relations, substitution processes, and base composition should be considered. It is perfectly reasonable at this stage to resolve ambiguities in favor of phylogenetic evidence and in some cases to delete ambiguous regions in the alignment. The advantage of this latter approach is that unambiguous information relevant to particular sequences can be retained over ambiguous data. The disadvantage is that parsimony and likelihood tree-building methods can interpret the ''missing'' information as zero divergence.

In summary, the following points should be considered when constructing a multiple sequence alignment for a phylogenetic analysis:

- The alignment step in phylogenetic analysis is one of the most important because it produces the data set on which models of evolution are used.
- It is not uncommon to edit the alignment, deleting unambiguously aligned regions and inserting or deleting gaps to more accurately reflect probable evolutionary processes that led to the divergence between sequences.
- It is useful to perform phylogenetic analyses based on a series of slightly modified alignments to determine how ambiguous regions in the alignment affect the results and what aspects of the results one may have more or less confidence in.

DETERMINING THE SUBSTITUTION MODEL

The substitution model should be given the same emphasis as alignment and tree building. As implied in the preceding section, the substitution model influences both alignment and tree building; hence, a recursive approach is warranted. At the present time, two elements of the substitution model can be computationally assessed for nucleotide data but not for amino acid or codon data. One element is the model of substitution between particular bases; the other is the relative rate of overall substitution among different sites in the sequence. Simple computational procedures have not been developed for assessing more complex variables (e.g., site- or lineage-specific substitution models). An overview of substitution models is presented below.

Models of Substitution Rates Between Bases

In general, substitutions are more frequent between bases that are biochemically more similar. In the case of DNA, the four types of transition (A $\rightarrow$ G, G $\rightarrow$ A, C $\rightarrow$ T, T $\rightarrow$ C) are usually more frequent than the eight types of transversion (A $\rightarrow$ C, A $\rightarrow$ T, C $\rightarrow$ G, G $\rightarrow$ T, and the reverse). Such biases will affect the estimated divergence between two sequences.

Specification of the relative rates of substitution among particular residues usually takes the form of a square matrix; the number of rows/columns is four in the case of bases, 20 in the case of amino acids (e.g., in PAM and BLOSUM matrices), and 61 in the case of codons (excluding stop codons). The off-diagonal elements of the matrix correspond to the relative costs of going from one base to another. The diagonal elements represent the cost of having the same base in different sequences.

The cost schedule can be fixed a priori to ensure that the tree-building method will tally an exact cost for each substitution incurred. Fixed-cost matrices are character-state weight matrices and are applied in maximum parsimony (MP) tree building (Fig. 14.5). When such weights are applied, the method is referred to as "weighted parsimony." For distance and maximum likelihood (ML) tree building, the costs can be derived from instantaneous rate matrices representing ML estimators of the probability that a particular type of substitution will occur (Fig. 14.6). Although application of the MP weight matrix is just simple arithmetic, application of the distance and ML rate matrices can involve complex algebra. To avoid the blind application of possibly inappropriate methods, practitioners are advised to familiarize themselves with the relevant underlying theory (see Li, 1997; Swofford et al., 1996).

Character-state weight matrices have usually been estimated more or less by eye, but they can also be derived from a rate matrix. For example, if it is presumed that each of the two transitions occurs at double the frequency of each transversion, a weight matrix can simply specify, for example, that the cost of A-G is 1 and the cost of A-T is 2 (Fig. 14.5). (The parsimony method dictates that the diagonal elements of the matrix, or the cost of having the same base in different sequences, be zero. This proves to be a shortcoming of parsimony; this will be discussed further below.) In the subsequent tree-building step, this set of assumptions will minimize the overall number of transversions and tend to cluster sequences differing mainly by transitions.

A simplified substitution rate matrix used in ML and distance phylogenetic analysis is presented in Figure 14.6. The matrix is analogous to that presented in Figure

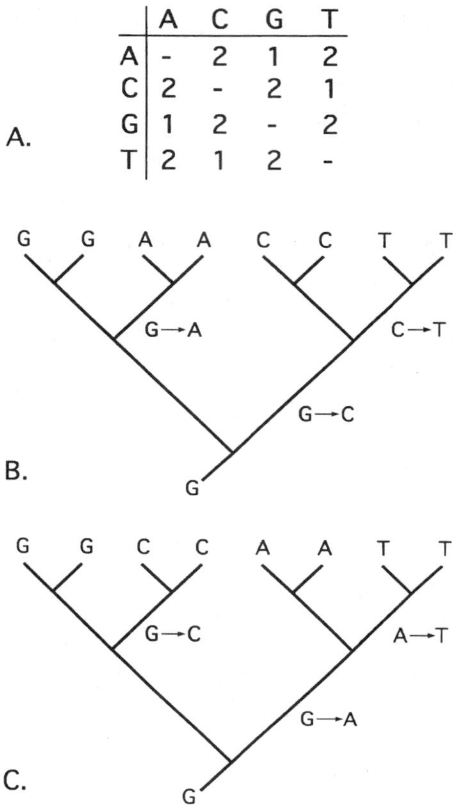

Figure 14.5. Character weight matrix and application in MP phylogenetic analysis. (A) Matrix indicating that a transversion substitution costs twice that of a transition. Because, according to MP bases shared between two sequences cannot ever have changed, diagonal elements of the matrix are ignored. (B, C) Two phylogenetic resolutions and reconstructions of the evolution of a hypothetical pattern of aligned bases at a particular site in eight sequences. With unweighted MP, both reconstructions (among several others) have the same cost (three steps); hence, they are equally acceptable. With the weight matrix in (A), the reconstruction of (B) requires four steps, and the reconstruction of (C) requires five. Thus, the first reconstruction (B) and others requiring four steps are preferred.

14.5, but the actual computation of divergence involves more complex algebra and cannot be determined by simply counting steps between bases.

The paralinear or "log-det" transformation corrects for nonstationarity (see Swofford et al., 1996). In this method, which is applicable only to distance tree building, the numbers of raw substitutions of each type and in each direction are tallied for each sequence pair in a four-by-four matrix as shown in Figure 14.7. Each matrix has an algebraic determinant, the log of which becomes a factor in estimating sequence divergence, hence the name "log-det." Pairwise comparisons of sequences having various and assorted patterns of base frequencies will yield a variety of matrix patterns, giving a variety of determinant values. Thus, each estimated pairwise distance will be affected by the determinant particular to each pair, which effectively

$$
\begin{array}{c|cccc}
 & A & C & G & T \\
A & -(a_1+a_2+a_3) & a_1 & a_2 & a_3 \\
C & a_4 & -(a_4+a_5+a_6) & a_5 & a_6 \\
G & a_7 & a_8 & -(a_7+a_8+a_9) & a_9 \\
T & a_{10} & a_{11} & a_{12} & -(a_{10}+a_{11}+a_{12})
\end{array}
$$

Figure 14.6. Simplified substitution rate matrix used in ML and distance phylogenetic analysis. The off-diagonal values a_n represent a product of an instantaneous rate of change, a relative rate between the different substitutions, and the frequency of the target base. In practice, the forward rates (upper triangular values) are presumed to equal the reverse rates (corresponding lower triangular values). The diagonal elements are nonzero, which effectively accounts for the possibility that more divergent sequences are more likely to share the same base by chance. In the simplest model of sequence evolution (the Jukes-Cantor model), all values of a are the same: all substitution types and base frequencies are presumed equal.

allows the substitution model to be different for each, varying along different branches of a phylogenetic tree. Log-det is especially sensitive to among-site rate heterogeneity (see below), since base frequency bias can exist only in sites that are subject to variation.

Models of Among-Site Substitution Rate Heterogeneity

In addition to variation in substitution patterns, variation in substitution rates among different sites in a sequence has been shown to profoundly affect the results of tree building (Swofford et al., 1996). The most obvious example of among-site rate variation, or heterogeneity, is that evident among the three codon positions in a coding

		Sclerotinium sclerotiorum				
		A	C	G	T	total
Spinacia oleracea	A	340	6	13	4	363
	C	10	229	6	36	281
	G	25	8	229	12	372
	T	5	22	6	312	345
	total	380	265	352	364	1361

Figure 14.7. Pairwise sequence comparison. The table compares 1,361 sites of 18S rDNA aligned between spinach (*Spinacia oleracea*) and a rust fungus (*Sclerotinium sclerotiorum*). The rows indicate the distribution of bases in the fungus aligned to particular bases in spinach. The columns indicate the reverse. The diagonal values are the number of site-wise identities between the sequences. Note the AT bias in the fungus: 83 (10 + 36 + 25 + 12) sites that are G or C in spinach are A or T in the fungus. In contrast, only 47 sites (6 + 22+ 13 + 6) that are G or C in the fungus are A or T in spinach. This bias is muted in simple comparison of base frequencies in the two sequences (the totals) because most sites are the same in both sequences and are probably mutationally constrained. Note also the obviously larger number of transition (13 + 36 + 25 + 22 = 96) versus transversion (6 + 4 + 10 + 6+ 8 + 12 + 5 + 6 = 57) substitutions and that C-T transitions account for 58/153 total differences. The data shown can be generated using the PAUP or MEGA programs.

sequence. Due to the degeneracy of the genetic code, changes in the third codon position can more frequently occur without affecting the ultimately encoded protein sequence. Therefore, this third codon position tends to be much more variable than the first two. For this reason, many phylogenetic analyses of coding sequences exclude the third codon position. In some cases, however, rate variation patterns are more subtle (e.g., those corresponding to conserved regions of proteins or rRNA).

Approaches to the estimation of substitution rate heterogeneity are the nonparametric models (Yang et al., 1996), the invariants model, and the gamma distribution models (Swofford et al., 1996). The nonparametric approach derives categories of relative rates for particular sites. This approach can be used with MP tree building simply by weighting particular sites according to relative mutation frequency, although such weighting tends to require prior knowledge of the true tree. The approach is also applicable to ML tree building, but it is considered computationally impractical (Yang et al., 1996). The invariants approach estimates a proportion of sites that are not free to vary. The remaining sites are presumed to vary with equal probability. The gamma approach assigns a substitution probability to sites by assuming that, for a given sequence, the probabilities vary according to a gamma distribution. The shape of the gamma distribution, as described by the shape parameter , describes the distribution of substitution probabilities among sites in a sequence (Swofford et al., 1996, p. 444, Fig. 13; cf. Li, 1997, p. 76, Fig. 3.10; note that the scales differ). In a combined approach, it can be presumed that a proportion of sites are invariant and that the remainder varies according to a gamma distribution.

In practice, gamma correction can be continuous, discrete, or "autodiscrete" (Yang et al., 1996). "Continuous gamma" means that sites are assigned to a change probability along a continuous curve. At present, this approach is computationally impractical in most cases. The discrete gamma approximation assigns sites to a specified number of categories that approximate the shape of the gamma curve. The autodiscrete model assumes that adjacent sites have correlated rates of change. Groups of sites are assigned to categories, and sites within a category can be assumed to have either constant or heterogeneous rates.

Various rate heterogeneity corrections are implemented in several tree-building programs. For nucleotide data, PAUP 4.0 implements both invariants and discrete gamma models for separate or combined use with time-reversible distance and likelihood tree-building methods and invariants in conjunction with the log-det distance method (see below). For nucleotide, amino acid, and codon data, PAML implements continuous, discrete, and autodiscrete models. For nucleotide and amino acid data, PHYLIP implements a discrete gamma model.

Models of Substitution Rates Between Amino Acids

The most widely used models of amino acid substitution include distance-based methods, which are based on matrixes such as PAM and BLOSUM. Again, such matrices are described further in other chapters in this book. Briefly, Dayhoff's PAM 001 matrix (Dayhoff, 1979) is an empirical model that scales probabilities of change from one amino acid to another in terms of an expected 1% change between two amino acid sequences. This matrix is used to make a transition probability matrix that allows prediction of the probability of changing from one amino acid to another and also predicts equilibrium amino acid composition. Phylogenetic distances are calculated with the assumption that the probabilities in the matrix are correct. The distance that is computed is scaled in units of expected fraction of amino acids changed. Kimura's distance is another method used in PROTDIST, one of the PHYLIP family of programs (mentioned further below), and is a rough distance formula for approximating PAM distance by simply measuring the fraction of amino acids that differ between two sequences and computing the distance by a set formula (see Kimura, 1983). This is a more rapid method, but it has some obvious limitations. It does not take into account which amino acids differ or what amino acids are changed, so some information is lost. The distance measure is represented as the fraction of amino acids differing; this is also the case with PAM distances. If the fraction of amino acids differing gets larger than 0.8541, the distance becomes infinite.

Although PROTDIST is one of the most widely used programs providing substitution models for calculating protein distances, others that are faster and make use of additional matrices such as BLOSUM are now more widely-used (e.g., PUZZLE).

The model used in parsimony (not a distance-based method) insists that any amino acid changes be consistent with the genetic code so that, for example, lysine is allowed to substitute to methionine but not to proline. However, changes between two amino acids via a third are allowed *and* are counted as two changes if each of the two replacements is individually allowed. This sometimes allows changes that, at first sight, one would think should be outlawed. Thus, phenylalanine can be changed to glutamine via leucine in two steps total. Genetic code translation tables show that there is a leucine codon one step away from a phenylalanine codon; there is also a leucine codon one step away from a glutamine codon. These leucine codons, however, are not identical. It actually takes three base substitutions to get from either of the phenylalanine codons (UUU and UUC) to either of the glutamine codons (CAA or CAG). Why, then, does this program count only two? The answer is that recent DNA sequence comparisons seem to show that synonymous changes (changes in the nucleotide sequence of a codon region that do not change what amino acids are encoded by that region) are considerably faster and easier than ones that change the amino acid outright. We are assuming that, in effect, synonymous changes occur so much more readily that they need not be counted. Thus, in the chain of changes UUU (Phe) $\rightarrow$ CUU (Leu) $\rightarrow$ CUA (Leu) $\rightarrow$ CAA (Glu), the middle one is not counted because it does not actually change the amino acid (leucine).

Which Substitution Model to Use?

Although any of the parameters in a substitution model might prove critical for a given data set, the best model is not always the one with the most parameters. To the contrary, the fewer the parameters, the better. This is because every parameter estimate has an associated variance. As additional parametric dimensions are introduced, the overall variance increases, sometimes prohibitively (see Li, 1997, p. 84, Table 4.1). For a given DNA sequence comparison, a two-parameter model will require that the summed base differences be sorted into two categories and into six for a six-parameter model. Obviously, the number of sites sampled in each of the six categories would be much smaller (and perhaps too small) to give a reliable estimate.

A good strategy for substitution model specification for DNA sequences is the "describe tree" feature in PAUP, which uses likelihood to simultaneously estimate the six reversible substitution rates, the -shape parameter of the gamma distribution, and the proportion of invariant sites. These parameters can be estimated by means of equal or specified base frequencies. Usually, any reasonable phylogenetic tree (e.g., an easily generated neighbor-joining tree) is suitable for this procedure because parameter estimates are apparently influenced predominantly by the character pattern rather than by the tree topology (Swofford et al., 1996). This estimation procedure is not overly time consuming for up to 50 sequences. If there will be more sequences or less time, the test tree can be selectively pruned to reduce the number of taxa while retaining the overall phylogenetic range and structure. From the estimated substitution parameters, one can determine whether a simpler model is justified (e.g., whether the six substitution categories can be reduced to two) by comparing likelihood scores estimated for this tree using more or fewer parameters. Parameters for and the proportion of invariant sites sometimes can substitute for each other, so one should compare likelihoods with each estimated alone versus both together. Note that, unlike MP and ME, the ML scores derived using different parameter values are directly comparable (Swofford et al., 1996).

In the case of protein-coding DNA sequences, it is sometimes obvious that, depending on the divergence of the samples, the useful variation is essentially either in the first and second codon positions, with the third positions randomized across the data set, or in the third position, with the first and second positions invariant. The procedure above will correct for this rate heterogeneity, although removing the "useless" sites may permit a more precise estimate of rate heterogeneity in the remaining sites.

For protein sequences, the model used is often dependent on the degree of sequence similarity. For more divergent sequences, the BLOSUM matrices are often better, whereas the PAM matrix is suited for more highly similar sequences. Both parsimony and distance matrix methods (mentioned further below) have benefits and disadvantages, and their use depends on one's philosophy about protein sequence changes: Is it better to retain information about each character when determining a tree (i.e., through parsimony) or to derive distance measures to base the tree (i.e., using a distance matrix)? Is a matrix based on empirical data a more accurate reflection of evolutionary change than a matrix based on generated theories about sequence change? Again, although cladistic analysis can be a powerful method for investigating evolutionary relationships, keep in mind that there is no one clear method that is better than the other. Each has its own benefits and disadvantages that differ depending on the type of analyses performed and the philosophy of the investigator.

TREE-BUILDING METHODS

Tree-building methods implemented in available software are discussed in detail in the literature (Saitou, 1996; Swofford et al., 1996; Li, 1997) and described on the Internet. This section briefly describes some of the most popular methods. Tree-building methods can be sorted into distance-based vs. character-based methods. Much of the discussion in molecular phylogenetics dwells on the utility of distance- and character-based methods (e.g., Saitou, 1996; Li, 1997). Distance methods compute pairwise distances according to some measure and then discard the actual data, using only the fixed distances to derive trees. Character-based methods derive trees that optimize the distribution of the actual data patterns for each character. Pairwise distances are, therefore, not fixed, as they are determined by the tree topology. The most commonly applied distance-based methods include neighbor-joining and the Fitch-Margoliash method, and the most common character-based methods include maximum parsimony and maximum likelihood.

Distance-Based Methods

Distance-based methods use the amount of dissimilarity (the distance) between two aligned sequences to derive trees. A distance method would reconstruct the true tree if all genetic divergence events were accurately recorded in the sequence (Swofford et al., 1996). However, divergence encounters an upper limit as sequences become mutationally saturated. After one sequence of a diverging pair has mutated at a particular site, subsequent mutations in either sequence cannot render the sites any more "different." In fact, subsequent mutations can make them again equal (for example, if a valine mutates to an isoleucine, which mutates back to a valine). Therefore, most distance-based methods correct for such "unseen" substitutions. In practice, application of the rate matrix effectively presumes that some proportion of observed pairwise base identities actually represents multiple mutations and that this proportion increases with increasing overall sequence divergence. Some programs implement, at least optionally, calculation of uncorrected distances, whereas, for example, the MEGA program (Kumar et al., 1994) implements only uncorrected distances for codon and amino acid data. Unless overall divergences are very low, the latter approach is virtually guaranteed to give inaccurate results.

Pairwise distance is calculated using maximum-likelihood estimators of substitution rates. The most popular distance tree-building programs have a limited number of substitution models, but PAUP 4.0 implements a number of models, including the actual model estimated from the data using maximum likelihood, as well as the log-det distance method.

Distance methods are much less computationally intensive than maximum likelihood but can employ the same models of sequence evolution. This is their biggest advantage. The disadvantage is that the actual character data are discarded. The most commonly applied distance-based methods are the unweighted pair group method with arithmetic mean (UPGMA), neighbor joining (NJ), and methods that optimize the additivity of a distance tree, including the minimum evolution (ME) method. Several methods are available in more than one phylogenetics software package but not all implementations allow the same parameter specifications and/or tree optimization features (e.g., branch swapping; see below).

Unweighted Pair Group Method with Arithmetic Mean (UPGMA). UPGMA is a clustering or phenetic algorithm—it joins tree branches based on the criterion of greatest similarity among pairs and averages of joined pairs. It is not strictly an evolutionary distance method (Li, 1997). UPGMA is expected to generate an accurate topology with true branch lengths only when the divergence is according to a molecular clock (ultrametric; Swofford et al., 1996) or approximately equal to raw sequence dissimilarity. As mentioned earlier, these conditions are rarely met in practice.

Neighbor Joining (NJ). The neighbor-joining algorithm is commonly applied with distance tree building, regardless of the optimization criterion. The fully resolved tree is "decomposed" from a fully unresolved "star" tree by successively

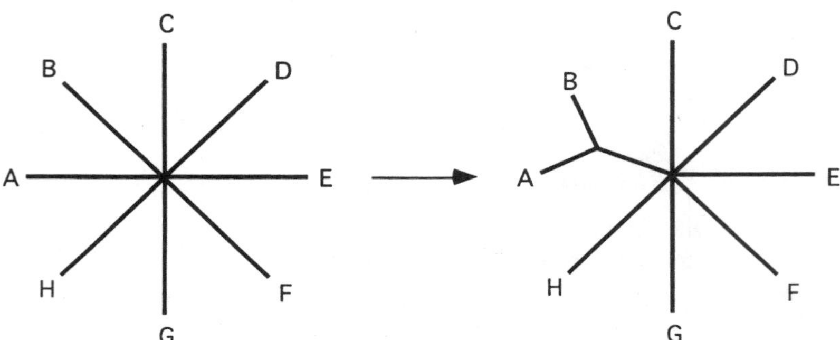

Figure 14.8. Star decomposition. This is how tree-building algorithms such as neighbor-joining work. The most similar terminals are joined, and a branch is inserted between them and the remainder of the star. Subsequently, the new branch is consolidated so that its value is a mean of the two original values, yielding a star tree with n-1 terminals. The process is repeated until only one terminal remains.

inserting branches between a pair of closest (actually, most isolated) neighbors and the remaining terminals in the tree (Fig. 14.8). The closest neighbor pair is then consolidated, effectively reforming a star tree, and the process is repeated. The method is comparatively rapid.

Fitch-Margoliash (FM). The Fitch-Margoliash (FM) method seeks to maximize the fit of the observed pairwise distances to a tree by minimizing the squared deviation of all possible observed distances relative to all possible path lengths on the tree (Felsenstein, 1997). There are several variations that differ in how the error is weighted. The variance estimates are not completely independent because errors in all the internal tree branches are counted at least twice (Rzhetsky and Nei, 1992).

Minimum Evolution (ME). Minimum evolution seeks to find the shortest tree that is consistent with the path lengths measured in a manner similar to FM; that is, ME works by minimizing the squared deviation of observed to tree-based distances (Rzhetsky and Nei, 1992; Swofford et al., 1996; Felsenstein, 1997). Unlike FM, ME does not use all possible pairwise distances and all possible associated tree path lengths. Rather, it fixes the location of internal tree nodes based on the distance to external nodes and then optimizes the internal branch length according to the minimum measured error between these "observed" points. It thus purports to eliminate the nonindependence of FM measurements.

Which Distance-Based Tree-Building Procedure Is Best? ME and FM appear to be the best procedures, and they perform nearly identically in simulation studies (Huelsenbeck, 1995). ME is becoming more widely implemented in computer programs, including METREE (Rzhetsky and Nei, 1994) and PAUP. For protein data, the FM procedure in PHYLIP offers the greatest range of substitution models but no correction for among-site rate heterogeneity. The MEGA (Kumar et al., 1994) and METREE packages include a gamma correction for proteins, but only in conjunction with a raw ("*p*-distance") divergence model (no distance or bias correction), which is unreliable except for small divergences (Rzhetsky and Nei, 1994). MEGA also computes separate distances for synonymous and nonsynonymous sites, but this method is valid only in the absence of substitution or base frequency bias and when there is no correction for among-site rate heterogeneity. Thus, for most data sets, using the nucleotide data under a more realistic model might be preferable to MEGA's methods.

Simulation studies indicate that UPGMA performs poorly over a broad range of tree shape space (Huelsenbeck, 1995). The use of this method is not recommended; it is mentioned here only because its application seems to persist, as evidenced by UPGMA gene trees appearing in publications (Huelsenbeck, 1995).

NJ is clearly the fastest procedure and generally yields a tree close to the ME tree. (Rzhetsky and Nei, 1992; Li, 1997). However, it yields only one tree. Depending on the structure of the data, numerous different trees might be as good or significantly better than the NJ tree (Swofford et al., 1996).

Character-Based Methods

The character-based methods have little in common with each other, besides the use of the character data at all steps in the analysis. This allows the assessment of the reliability of each base position in an alignment on the basis of all other base positions.

Maximum Parsimony (MP). Maximum parsimony is an optimization criterion that adheres to the principle that the best explanation of the data is the simplest, which in turn is the one requiring the fewest ad hoc assumptions. In practical terms, the MP tree is the shortest—the one with the fewest changes—which, by definition, is also the one with the fewest parallel changes. There are several variants of MP that differ with regard to the permitted directionality of character state change (Swofford et al., 1996).

To accommodate substitution bias, MP is amenable to weighting; for example, the transformation of a transversion can be weighted relative to a transition (see above). The easiest way to do this is to create a weighting step matrix in which the weights are the reciprocal of the rates estimated using ML as described above. However, step-matrix weighting can greatly slow MP computation.

The MP method performs poorly when there is substantial among-site rate heterogeneity (Huelsenbeck, 1995). There are few good fixes for this problem. One approach is to modify the data set to include only sites that exhibit little or no heterogeneity as determined by likelihood estimation (see above). Another approach is to recursively reweight positions according to their propensity to change as observed in preliminary trees. This "successive approximations" approach is automatically facilitated in PAUP, but it is prone to error to the degree that the preliminary trees are incorrect.

MP analyses tend to yield numerous (and sometimes many thousands of) trees that have the same score. Because each is held to be as optimal as any other, only groupings present in the strict consensus of all trees are considered to be supported by the data. The reason that distance and ML tree methods tend to arrive at a single best tree is that their calculations involve division and decimals, whereas MP merely counts discrete steps. For a given data set, a strict consensus of all ME or ML trees that are not significantly worse than optimal probably would yield resolution more or less comparable to the MP consensus. Unfortunately, whereas MP users conventionally present strict consensus (and sometimes consensus of trees one or two steps worse), ME and ML users typically do not.

Simulation studies have shown that MP performs no better than ME and worse than ML when the amount of sequence evolution since lineages diverged is much greater than the amount of divergence that occurred between lineage splits (i.e., in a tree with very long terminal branches and short internal internodes) (Huelsenbeck, 1995). This condition produces "long branch attraction"—the long branches become artificially connected because the number of nonhomologous similarities the sequences have accumulated exceeds the number of homologous similarities they have retained with their true closest relatives (Swofford et al., 1996). Character weighting improves the performance of MP under these conditions (Huelsenbeck, 1995).

Maximum Likelihood (ML). ML turns the phylogenetic problem inside out. ML searches for the evolutionary model, including the tree itself, that has the highest likelihood of producing the observed data.

In practice, ML is derived for each base position in an alignment. The likelihood is calculated in terms of the probability that the pattern of variation at a site would be produced by a particular substitution process, given a particular tree and the overall observed base frequencies. The likelihood becomes the sum of the probabilities of each possible reconstruction of substitutions under a particular substitution process. The likelihoods for all the sites are multiplied to give an overall "likelihood of the tree" (i.e., the probability of the data given the tree and the substitution process). As one can imagine, for one particular tree, the likelihood of the data is low at some sites and high at others. For a "good" tree, many sites will have higher likelihood, so the product of likelihoods is high. For a "poor" tree, the reverse will be true.

The substitution model should be optimized to fit the observed data. For example, if there is a transition bias, evident by an inordinate number of sites that include only purines or pyrimidines, the likelihood of the data under a model that assumes no bias will never be as good as one that does. Likewise, if a substantial proportion of the sites are occupied by a single base and another substantial proportion have equal base frequencies, the likelihood of the data under a model that assumes that all sites evolve equally will be less than that of a model that allows rate heterogeneity. Modifying the substitution parameters, however, modifies the likelihood of the data associated with particular trees. Thus, the tree yielding the highest likelihood under one substitution model might yield much lower likelihood under another.

Because ML uses great amounts of computational time, it is usually impractical to perform a complete search that simultaneously optimizes the substitution model and the tree for a given data set. An economical, heuristic approach is recommended (Adachi and Hasegawa, 1996; Swofford et al., 1996). Perhaps the best time saver in this regard is preliminary ML estimation of the substitution model (as can be performed using PAUP). This procedure can be applied iteratively, searching for better ML trees, then reestimating the parameters, and then searching for better trees.

As algorithms, computers, and phylogenetic understanding have improved, the ML criterion has become more popular for molecular phylogenetic analysis. In simulation studies, ML has consistently outperformed ME and MP when the data analysis proceeds according to the same model that generates the data (Huelsenbeck, 1995). ML will always be the most computationally intensive method of all, however, so there will always be situations in which it is not practical.

DISTANCE, PARSIMONY, AND MAXIMUM LIKELIHOOD: WHAT'S THE DIFFERENCE?

Distance matrix methods simply count the number of differences between two sequences. This number is referred to as the evolutionary distance, and its exact size depends on the evolutionary model used. The actual tree is then computed from the matrix of distance values by running a clustering algorithm that starts with the most similar sequences (i.e., those that have the shortest distance between them) or by trying to minimize the total branch length of the tree. The principle of maximum parsimony searches for a tree that requires the smallest number of changes to explain the differences observed among the taxa under study.

A maximum-likelihood approach to phylogenetic inference evaluates the probability that the chosen evolutionary model has generated the observed data. The evolutionary model could simply mean that one assumes that changes between all nucleotides (or amino acids) are equally probable. The program will then assign all possible nucleotides to the internal nodes of the tree in turn and calculate the probability that each such sequence would have generated the data (if two sister taxa have the nucleotide "A," a reconstruction that assumes derivation from a "C" would be assigned a low probability compared with a derivation that assumes there already was an "A"). The probabilities for all possible reconstructions (not just the more probable one) are summed up to yield the likelihood for one particular site. The likelihood for the tree is the product of the likelihoods for all alignment positions in the data set.

Searching for Trees

The number of unique phylogenetic trees increases exponentially with the number of taxa, becoming astronomical even for, say, 50 sequences (Swofford et al., 1996; Li, 1997). In most cases, computational limitations permit exploration of only a small fraction of possible trees. The exact number will depend mainly on the number of taxa, the optimality criterion (e.g., MP is much faster than ML), the parameters (e.g., unweighted MP is much faster than weighted; ML with fewer preset parameters is much faster than with more and/or simultaneously optimized parameters), computer hardware, and computer software (some algorithms are faster than others; some software allows multiprocessing; some software limits the number and kind of trees that can be stored in memory). The search procedure is also affected by data structure: poorly resolvable data produce more "nearly optimal" trees that must be evaluated to find the most optimal.

Branch-swapping algorithms successively modify existing trees built by an initial step (Swofford et al., 1996). The algorithms range from those that generate all possible unique trees (exhaustive algorithms) to those that evaluate only minor modifications.

Quartet puzzling is a relatively rapid tree-searching algorithm available for ML tree building (Strimmer and von Haeseler, 1996) and is available in PUZZLE.

One of the best ways to economize the search effort is to prune the data set. For example, it might be apparent from the data alone or from preliminary searching that a particular cluster of five terminals is unresolvable, that the arrangement of these terminals does not impact the remainder of the topology, and/or that resolution of these terminals is not the objective of the analysis. Removing four of the terminals from the analysis simplifies the search by several orders of magnitude.

Every analysis is unique. The elements that influence the choice of optimal search strategy (amount of data, structure of data, amount of time, hardware, objective of analysis) are too variable to suggest a foolproof recipe. Thus, researchers must be familiar with their data; they must also have specific objectives in mind, understanding the various search procedures as well as the capabilities of their hardware and software.

Rooting Trees

The methods described above produce unrooted trees (i.e., trees having no evolutionary polarity). To evaluate evolutionary hypotheses, it is often necessary to locate the root of the tree. Rooting phylogenetic trees is not a trivial problem (Nixon and Carpenter, 1993).

If one accepts a molecular clock, then the root will always be at the midpoint of the longest span across the tree (Weston, 1994). Whether molecular evolution is indeed clocklike generally remains a contentious issue (Li, 1997), but most gene trees exhibit unclocklike behavior regardless of where the root is placed. Thus, rooting is generally evaluated by extrinsic evidence, that is, by means of determining where the tree would attach to an "outgroup," which can be any organism/sequence not descended from the nearest common ancestor of the organisms/sequences analyzed (for example, a bird sequence could be used to root an analysis of mammals). Outgroup rooting, however, creates a dilemma: an outgroup that is closely related to the ingroup might be simply an erroneously excluded member of the ingroup. A clearly distant outgroup (e.g., a fungus for an analysis of plants) can have a sequence so diverged that its attachment to the ingroup is subject to the long-branch attraction problem mentioned above. It is wise to examine the results obtained for trees both with and without an outgroup.

Another means of rooting involves analysis of a duplicated gene or gene with an internal duplication (Lawson et al., 1996). If all the paralogs from most or all of the organisms are included in the analysis, then one can logically root the tree exactly where the paralog gene trees converge, assuming that there are not long branch problems in all trees.

TREE EVALUATION

Several procedures are available that evaluate the phylogenetic signal in the data and the robustness of trees (Swofford et al., 1996; Li, 1997). The most popular of the former class are tests of data signal versus randomized data (skewness and permutation tests). The latter class includes tests of tree support from resampling of observed data (nonparametric bootstrap). The likelihood ratio test provides a means of evaluating both the substitution model and the tree.

Randomized Trees (Skewness Test)

Simulation studies indicate that the distribution of random MP tree lengths generated using random data sets will be symmetrical, whereas those using data sets with phylogenetic signal will be skewed. The critical value of the g_1 statistic of skewness will vary with the number of taxa and variable sites in the sequence. The test does not estimate the reliability of a particular topology, and it is sensitive to even very small amounts of signal present in an otherwise random data set. If taxa from groups that are obviously well supported by the data are selectively deleted, the test can be used to determine whether a phylogenetic signal remains, provided at least 10 variable characters and 5 taxa are examined. The procedure is implemented in PAUP.

Randomized Character Data (Permutation Tests)

The randomized data approach determines whether an MP tree or portion of it derived from the actual data could have arisen by chance. The data are not truly randomized but permuted within each aligned column, so that covariation in the initial data is removed. The result is an alignment of sequences that are not random sequences; rather, the base at each site in these sequences is randomly drawn from the population of bases occupying that site in the overall alignment. The permutation tail probability test (PTP) compares the score for the MP tree with trees generated by numerous permutations of the data at each site, determining only whether there is a phylogenetic signal in the original data. A topology-dependent test (T-PTP) compares the scores for specific trees to determine whether the difference can be attributed to chance. This method does not evaluate whether the tree or any portion of it is correct (Swofford et al., 1996). In particular, the T-PTP test will appear to corroborate groups that are in trees close to the MP tree but not in it. This is because the method detects the collective signal that places a taxon even approximately, if not actually, in its correct position. The results can be fine-tuned, however, by additional applications using relevant subsets of the data (Faith and Trueman, 1996). The procedure is implemented in PAUP.

Bootstrap

Bootstrapping is a resampling tree evaluation method that works with distance, parsimony, likelihood, and just about any other tree derivation method. It was invented in 1979 (Efron, 1979) and introduced as a tree evaluation method in phylogenetic analysis by Felsenstein (1985). The result of bootstrap analysis is typically a number associated with a particular branch in the phylogenetic tree that gives the proportion of bootstrap replicates that supports the monophyly of the clade.

How is this done practically? Bootstrapping can be considered a two-step process comprising the generation of (many) new data sets from the original set and the computation of a number that gives the proportion of times that a particular branch (e.g., a taxon) appeared in the tree. That number is commonly referred to as the bootstrap value. New data sets are created from the original data set by sampling columns of characters at random from the original data set with replacement. "With replacement" means that each site can be sampled again with the same probability as any of the other sites. As a consequence, each of the newly created data sets has the same number of total positions as the original data set, but some positions are duplicated or triplicated and others are missing. It is therefore possible that some of the newly created data sets are completely identical to the original set—or, on the other extreme, that only one of the sites is replicated, say, 500 times, whereas the remaining 499 positions in the original data set are dropped.

Although it has become common practice to include bootstrapping as part of a thorough phylogenetic analysis, there is some discussion on what exactly is measured by this method. It was originally suggested that the bootstrap value is a measure of repeatability (Felsenstein, 1985). In more recent interpretations, it has been considered to be a measure of accuracy—a biologically more relevant parameter that gives the probability that the true phylogeny has been recovered. On the basis of simulation studies, it has been suggested that, under favorable conditions (roughly equal rates of change, symmetric branches), bootstrap values greater than 70% correspond to a probability of greater than 95% that the true phylogeny has been found (Hillis and Bull, 1993). By the same token, under less favorable conditions, bootstrap values greater than 50% will be overestimates of accuracy (Hillis and Bull, 1993). Simply put, under certain conditions, high bootstrap values can make the wrong phylogeny look good; therefore, the conditions of the analysis must be considered. Bootstrapping can be used in experiments in which trees are recomputed after internal branches are deleted one at a time. The results provide information on branching orders that are ambiguous in the full data set (cf. Leipe et al., 1994).

Parametric Bootstrap

The parametric bootstrap differs from the nonparametric in that it uses simulated (yet actual) replicates rather than pseudoreplicates. In the case of phylogenetic sequence analysis, replicate data sets of the same size as the original data set are generated according to a specified model of sequence evolution, including the optimal tree topology determined according to that model (Huelsenbeck et al., 1996). Each data set is then analyzed according to the method of interest. Support for the branches in the test tree can be determined in much the same way as in the nonparametric bootstrap.

Likelihood Ratio Tests

As the name implies, likelihood ratio tests are applicable to ML analyses. A suboptimal likelihood value is evaluated for significance against a normal distribution of the error in the optimal model. In ideal applications, the error curve is presumed to be a χ^2 distribution. Thus, the test statistic is twice the difference between the optimal and test values, and the degrees of freedom is the number of parameter differences.

Application of the χ^2 test to alternative phylogenetic trees is problematic, especially because of the "irregularity of [the] parameter space" (Yang et al., 1995), but its use has been advocated for evaluating optimality of the substitution model when the number of parameters between models is known.

PHYLOGENETICS SOFTWARE

PHYLIP and PAUP compete as the most widely used phylogenetic analysis software, although other newer applications such as PUZZLE are beginning to compete. Here, PHYLIP and PAUP will be described in the most detail, with references made to other available packages that have useful features. However, the number of programs available is now so numerous, many each having their own useful features, that the reader is referred to the list of Internet resources at the end of this chapter for further information.

PHYLIP

PHYLIP (for **phyl**ogeny **i**nference **p**ackage) is a package now consisting of about 30 programs that cover most aspects of phylogenetic analysis. PHYLIP is free and available for a wide variety of computer platforms (Mac, DOS, UNIX, VAX/VMS, and others). According to its author, PHYLIP is currently the most widely used phylogeny program.

PHYLIP is a command-line program and does not have a point-and-click interface, as programs like PAUP do. The documentation is well written and very comprehensive, and the interface is straightforward. A program within PHYLIP is invoked by typing its name, which automatically causes the data to be read from a file called "infile" or a file name you specify if no infile exists. This infile must be in PHYLIP format; this format is clearly described in the documentation, and most sequence analysis programs offer the ability to export sequences in this format. For example, if an alignment is produced using CLUSTAL W or edited using GeneDoc, the alignment may be saved in PHYLIP format and then used in PHYLIP programs directly. Once the user activates a given PHYLIP program and loads the infile, the user can then choose from an option menu or accept the default values. The program will write its output to a file called "outfile" (and "treefile" where applicable). If the output is to be read by another program, "outfile" or "treefile" must be renamed before execution of the next program, as all files named outfile/tree file in the current directory are overwritten at the beginning of any program execution. The tree file generated is a widely used format that can be imported into a variety of tree-drawing programs, including DRAWGRAM and DRAWTREE that come with this package. However, these PHYLIP tree-drawing programs produce low-resolution graphics, so a program such as TreeView (described below) is instead recommended. Particulars of some of the PHYLIP tree-inference programs are discussed below.

PROTDIST is a program that computes a distance matrix for an alignment of protein sequences. It allows the user to choose between one of three evolutionary models of amino acid replacements. The simplest, fastest (and least realistic) model assumes that each amino acid has an equal chance of turning into 1 of the other 19 amino acids. The second is a category model in which the amino acids are redistributed among different groups; transitions in this model are evaluated differently depending on whether the change would result in an amino acid in the same or in a different group. The third (default) method, which is recommended, uses a table of empirically observed transitions between amino acids, the Dayhoff PAM 001 matrix (Dayhoff, 1979). More details can be found in the PHYLIP documentation and in a publication (Felsenstein, 1996).

NEIGHBOR is a tree-generating program that utilizes the distance matrix data generated from a program such as PROTDIST and generates a tree using the neighbor-joining method. This is one of the more popular methods, due to its speed of computation.

FITCH is another tree-generating program similar to NEIGHBOR but much more robust. It also uses distance matrix data, such as that described in PROTDIST, and generates a tree using the method of Fitch-Margoliash. This method, while more robust than NEIGHBOR, tends to produce a similar final answer, yet takes longer to compute. Although computational times are often significantly longer, the quality of the results produced by the method often makes this method the method of choice in these types of analyses.

PROTPARS is a parsimony program for protein sequences that generates trees without utilizing a distance matrix. The evolutionary model is different from the ones used in the PROTDIST program in that it considers the underlying changes in the nucleotide sequence to evaluate the probabilities of the observed amino acid changes. Specifically, it makes the (biologically meaningful) assumption that synonymous changes [e.g., GCA (alanine) → GCC (alanine)] occur more often than nonsynonymous changes. As a consequence, a transition between two amino acids that would require, for example, three nonsynonymous changes in the underlying nucleotide sequences, is assigned a lower probability than an amino acid change calling for two nonsynonymous changes and one synonymous change. PROTPARS does not have an option that uses empirical values for amino acid changes (e.g., PAM matrices).

DNADIST computes a distance matrix from nucleotide sequences. Trees are generated by running the output through NEIGHBOR or other distance matrix programs in the PHYLIP package. DNADIST allows the user to choose between three models of nucleotide substitution. The older Jukes and Cantor model is similar to the simple model in the PROTDIST program in that it assumes equal probabilities for all changes. The more recent Kimura two-parameter model is very similar but allows the user to weigh transversion more heavily than transitions. PHYLIP also comprises DNAML, a maximum-likelihood program for nucleotide data. Because the program is fairly slow, the use of its faster "sibling," the fastDNAml program (Olsen et al., 1994) described below, is recommended.

SEQBOOT and CONSENSE are required for bootstrap analysis. SEQBOOT is used to generate any number of replicates of the data; these replicates are then used in programs within the PHYLIP suite for analysis. The resulting tree file contains as many trees as there are replicates of the data, so this file needs to be run through CONSENSE to generate the consensus tree from the analysis. As an example, the steps involved in building a bootstrapped neighbor-joining tree for protein sequences are outlined in Figure 14.9.

→

Figure 14.9. Work flow for bootstrap analysis with the PHYLIP program. SEQBOOT accepts a file in PHYLIP format as input and multiplies it a user-specified number of times (e.g., 1,000). The resulting outfile can be used to calculate 1,000 distance matrices for DNA (DNADIST) or protein (PROTDIST) data. In this step, the actual data (nucleotides, amino acids) are discarded and replaced by a figure that is a measure for the amount of divergence between two sequences. The NEIGHBOR program will create 1,000 trees from these matrices. The CONSENSE program reduces the 1,000 trees to a single one and indicates the bootstrap values as numbers on the branches. The topology of the CONSENSE tree can be viewed with any text editor in the "outfile," whereas the "treefile" can be further processed for publication purposes. Treetool and TreeView allow the user to manipulate the tree (rerooting, branch rearrangements, conversions from dendrograms to phylograms, and so forth) and to save the file in commonly used graphic formats. Although these are not part of the PHYLIP package (indicated by boxes with dashed lines), they are freely available (see end-of-chapter list). Different file formats used during date processing through the stages of bootstrap analysis are also shown. Periods to the right and at the bottom of a box indicate files that were truncated to save space.

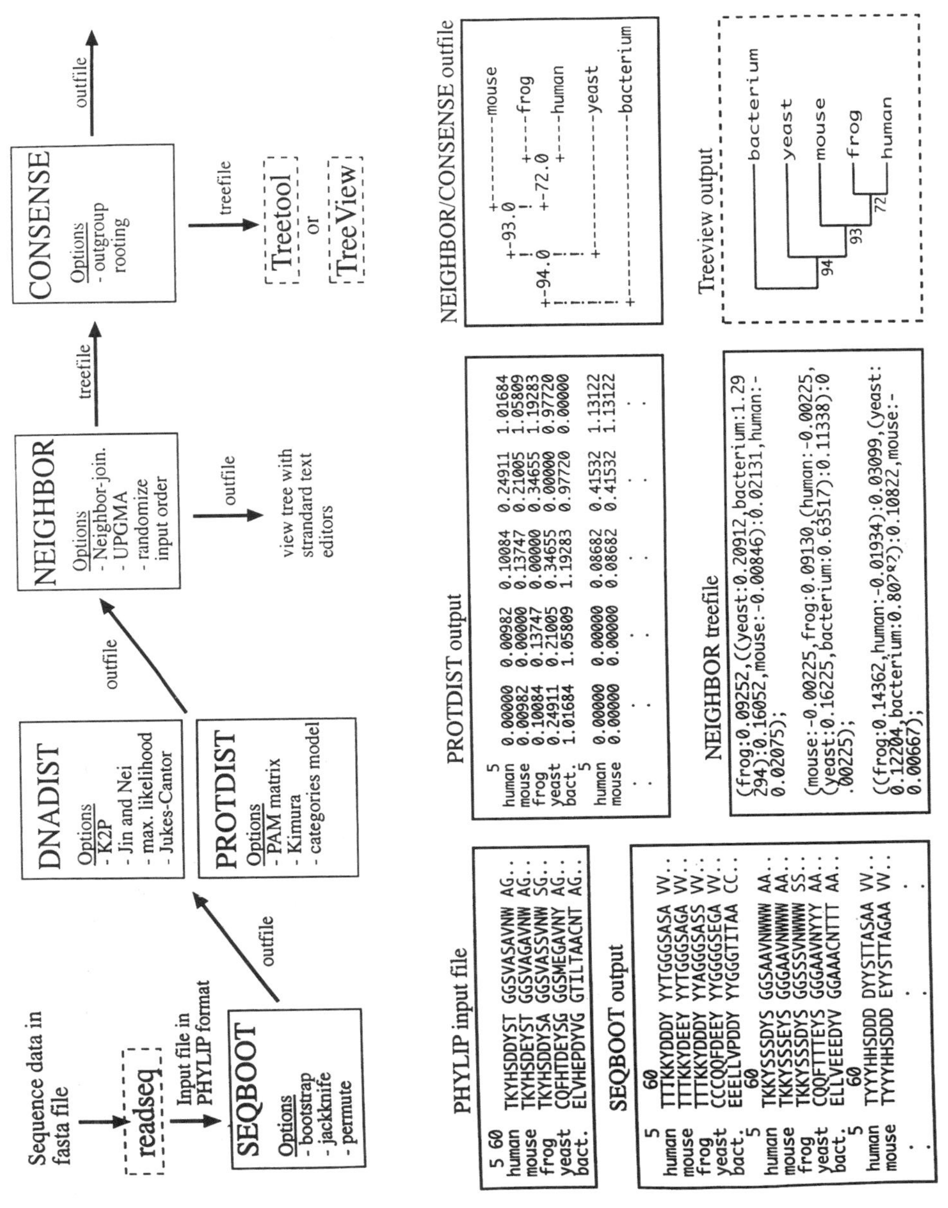
Sequence data in fasta file
readseq
Input file in PHYLIP format
SEQBOOT
Options
- bootstrap
- jackknife
- permute
outfile
DNADIST
Options
- K2P
- Jin and Nei
- max. likelihood
- Jukes-Cantor
PROTDIST
Options
- PAM matrix
- Kimura
- categories model
outfile
NEIGHBOR
Options
- Neighbor-join.
- UPGMA
- randomize input order
outfile
view tree with strandard text editors
treefile
CONSENSE
Options
- outgroup rooting
outfile
treefile
Treetool
or
TreeView
PHYLIP input file
5 60
human TKYHSDDYST GGSVASAVNW AG..
mouse TKYHSDEYST GGSVAGAVNW AG..
frog TKYHSDDYSA GGSVASSVNW SG..
yeast CQFHTDEYSG GGSMEGAVNY AG..
bact. ELVHEPDYVG GTILTAACNT AG..
SEQBOOT output
PROTDIST output
NEIGHBOR/CONSENSE outfile
NEIGHBOR treefile
(frog:0.09252,((yeast:0.20912,bacterium:1.29294):0.16052,mouse:-0.00846):0.02131,human:-0.02075);
Treeview output
bacterium
yeast
mouse
frog
human
94
93
72

PAUP

The objective of the development of PAUP is to provide a phylogenetics program that includes as many functions (including tree graphics) as possible in a single, platform-independent program with a menu interface. PAUP stands for **p**hylogenetic **a**nalysis **u**sing **p**arsimony and still contains one of the most sophisticated parsimony programs available. Version 3 performed only MP-associated tree-building and analytical functions. PAUP version 4 also includes distance and ML functions for nucleotide data and other new features.

Current tree-building functions in PAUP include MP, and, for nucleotide data, distance and ML using the fastDNAml algorithm. In addition, PAUP performs Lake's method of invariants (Swofford et al., 1996; Li, 1997). Each tree-building program permits a variety of options. The MP options include specification of any character-weighting scheme. Distance options include choice of NJ, ME, FM (see PAUP release notes regarding PHYLIP), and UPGMA procedures. The full range of options and their current values can be examined using the menu and/or by typing `pse[ttings] ?`, `dse[ttings] ?`, and `lse[ttings] ?` for parsimony, distance, and likelihood, respectively. Both distance and ML allow detailed specification of the substitution model (values of substitution, gamma, and invariant-sites parameters, assuming equal, specified, or empirical base frequencies), and these can be estimated for any tree by setting the parameter values to `est[imate]` and applying the `des[cribe tree]` command with a desired tree in memory.

According to the release notes accompanying PAUP test version 4, PAUP* *usually* finds trees with likelihoods as high or higher [i.e., better] than PHYLIP (both because PAUP*'s tree rearrangements are more extensive and because its convergence criterion for branch-length iteration is stricter).

With any tree-building method, PAUP allows a variety of tree search options. These include algorithm specification for generating the initial tree (starting tree): NJ, stepwise addition, or input tree(s). The stepwise-addition algorithm allows numerous options, including addition of taxa "as is" (taxa added in file order): closest, furthest, or random with any number of replicates. All the stepwise options allow for any maximal number of partial trees to be retained and built on during taxon addition. Increasing this number to, say, 100, is another means of increasing the diversity of starting topologies, although these are not random.

A random addition strategy provides a useful complement to the default search strategy (closest addition, TBR swapping, saving all best trees). In the random search, a large number of replicates can be combined with the faster NNI swapping algorithm. For MP analysis, in which a large number of equal-length trees might exist, the search should specify saving from each replicate only a few trees that match or are better than the score of the slower search. In addition, the number of suboptimal trees (the trees that will be swapped on to find better trees) should be limited by setting MAXTREES to a low number (e.g., 10). By using this strategy to explore areas of "tree space" possibly missed in the slow search, one sometimes finds better trees and/or additional unique optimal trees.

PAUP performs the nonparametric bootstrap for distance, MP, and ML, using all options available for tree building with these methods. When a bootstrap or jackknife with MP is under way, MAXTREES should be set between 10 and no more than 100. This is because poorly resolvable portions of an MP tree will usually be even less resolvable with resampled data; hence, a replicate could find astronomical numbers of equal-length trees. Because tree branches weakly supported by the full data set will not have high bootstrap or jackknife values, limiting MAXTREES will have little, if any, bearing on the results, especially if the number of replicates is increased to, say, 1,000.

In addition, PAUP performs the Kishino-Hasegawa test to compare MP or ML trees, computes four types of consensus of multiple trees (usually used for multiple equal-length MP trees), computes stepwise differences between MP trees, and evaluates signal conflicts between specified partitions of sites (e.g., nuclear and organellar sequence data in a combined analysis).

Other Programs

In addition to PAUP and PHYLIP, there are phylogenetics programs that have some unique capabilities but are generally more limited in their procedures and portability. These include FastDNAml, PUZZLE, MACCLADE, and MOLPHY.

FastDNAml. FastDNAml (Olsen et al., 1994) is a freestanding maximum-likelihood, tree-building program. Although it is currently not part of the PHYLIP package, it uses largely the same input and output conventions, and the results of fastDNAml and PHYLIP's DNAML should be very similar or identical. FastDNAml can be run on parallel processors, and it comes with a number of useful scripts (in particular for bootstrapping and jumbling the sequence input order). To take full advantage of the program, knowledge of UNIX is beneficial. The source code for UNIX systems is publicly available from the RDP Web site, and a Power Macintosh version is available by FTP.

PUZZLE or TREE-PUZZLE

PUZZLE or TREE-PUZZLE (Strimmer and von Haeseler, 1996), as it is now called, is a maximum likelihood-based program that implements a fast tree search algorithm (quartet puzzling) that allows analysis of large data sets and automatically assigns estimations of support to each internal branch. PUZZLE also computes pairwise maximum-likelihood distances as well as branch lengths for user-specified trees. PUZZLE also offers a novel method, likelihood mapping, to investigate the support of a hypothesized internal branch without computing an overall tree and to visualize the phylogenetic content of a sequence alignment. It conducts a number of statistical tests (χ^2 test for homogeneity of base composition, likelihood ratio clock test, Kishino-Hasegawa test) and includes a large range of models of substitution. Rate heterogeneity is modeled by a discrete gamma distribution and by allowing invariable sites.

MACCLADE. MACCLADE is an interactive Macintosh program for manipulating trees and data and studying the phylogenetic behavior of characters (Maddison and Maddison, 1992). It uses the NEXUS file format and will read PAUP data and tree files. Some information in PAUP files will be ignored in MACCLADE (e.g., gap mode), but information in a PAUP "assumptions" block will be imported, including character weightings and character and taxon sets. Several subtle differences exist between PAUP and MACCLADE files. Thus, PAUP files edited with MACCLADE and vice versa should be saved under new names and the unedited file maintained separately. PHYLIP, NBRF-PIR, and text files are also readable by MACCLADE. Any method can be used to generate the trees, but MACCLADE's functions are based strictly on parsimony. For example, the program allows one to trace the evolution of each individual character on any tree. The MP and ML reconstruction functions differ, however, and the ML function is considered more realistic (Swofford et al., 1996). Tree topologies can be manipulated by dragging branches, and flipping branches can produce aesthetic modifications in tree symmetry.

MACCLADE includes additional features relevant to sequence analysis, including a chart of character number versus number of changes in a tree, which is useful for visualizing among-site rate heterogeneity, and a chart of the overall numbers of changes from one base to another over an MP tree ("state changes and stasis" chart: the values are sometimes erroneously reported in the literature as substitution "rates," but there is no correction for branch lengths or among-site rate heterogeneity).

MOLPHY. MOLPHY is a shareware package of programs and utilities for ML analysis and statistics of nucleotide or amino acid sequences (Adachi and Hasegawa, 1996). It has been tested on Sun OS and HP9000/700 systems. Practical application requires some knowledge of UNIX file management. The ML procedures are similar to those in PHYLIP, but there is a wider range of amino acid substitution models and options for faster, heuristic searches, including an option to use "local bootstrap" analyses (i.e., a bootstrap on subtrees under the assumption that the remainder of the tree is correct) to search for better ML trees. The output includes branch-length estimates and standard error. Analysis of separate codon positions is possible. MOLPHY uses a subset of the nucleotide substitution models available in PAUP, although it allows user-specified parameter values. The current MOLPHY lacks a bootstrap option and also has no accommodation for among-site rate heterogeneity.

Tree Drawing. There are a number of tree-drawing programs available now, such as TreeTool (X-windows), TreeDraw (Macintosh), PHYLODENDRON (Macintosh), TreeView (Macintosh, Microsoft Windows), or the tree-drawing tool in PAUP, and all handle standard tree files. These programs facilitate not only the generation of trees suitable for publication or other presentation but also facilitate viewing of the data in general. For example, programs such as the freely available TreeView enable the user to manipulate the view of branching order, root the tree, and other graphical manipulations that aid the user.

INTERNET-ACCESSIBLE PHYLOGENETIC ANALYSIS SOFTWARE

Currently, there are few Web-based applications that will permit an investigator to perform phylogenetic analyses over the Web. However, these kinds of resources are appearing in increasing numbers, and, presumably as the Internet bandwidth increases and servers have faster CPUs, this may become even more common. Highlighted here are three Internet-based applications that provide phylogenetic analysis capabilities: WEBPHYLIP, PhyloBLAST, and the "BLAST2 & Orthologue Search Server." These illustrate the variety of applications currently available. Although all use PHYLIP programs, the latter two combine phylogenetic analysis with BLAST to aid the user in retrieving sequences for analysis.

WEBPHYLIP

WEBPHYLIP uses CGI/Perl scripts to produce a Web-based cut-and-paste interface to the PHYLIP programs. Unfortunately, the programs are not linked together; therefore, to generate a neighbor-joining tree, for example, the user must run multiple analyses (PROTDIST and NEIGHBOR in this case). Analyses may time-out if too intensive of an analysis is requested. Also, the trees cannot be easily viewed; for example, if the user has a PC, they must have ghostview or another such PostScript viewer installed to actually view the results. However, the Web site provides excellent flowcharts and other helpful documentation about running the programs, and an extensive collection of the PHYLIP programs is available.

PhyloBLAST

PhyloBLAST is also based on CGI/Perl scripts. PhyloBLAST compares a user's protein sequence to the SWISS-PROT/TREMBL databases using BLAST2 and then allows user-defined phylogenetic analyses to be performed on selected sequences from the BLAST output. Neighbor-joining and parsimony analyses may be performed, either with or without bootstrapping, using PHYLIP programs. Flexible features, such as the ability to input premade multiple sequence alignments and use all options found in the PHYLIP programs, provide additional functionality that goes beyond the simple analysis of a BLAST result. Because PHYLIP programs need to generate trees that are linked, there is less input required by the user than for WEBPHYLIP; however, DNA analysis and analysis using some programs (for example, FITCH) are not currently available. However, PhyloBLAST's ability to generate trees-containing hyperlinks to further protein sequence information or generate JPEG graphics of trees is a considerable advantage. Also, the program is set up to handle Web page time-outs so long analyses are not a problem (and can be E-mailed to the user if preferred).

BLAST2 & Orthologue Search Server

This is a fairly specialized application of phylogenetic analysis of a BLAST output for the identification of orthologs verses paralogs. It is based on the use of CLUSTALW, WU-BLAST2, and the tree-reconciling algorithm of Page (1994). This tool first performs a BLAST analysis and then performs a phylogenetic analysis on user-selected sequences based on a CLUSTAL W alignment and PHYLIP's neighbor-joining methods. This resulting neighbor-joining tree ("gene tree") is ultimately compared with a predicted species tree and the reconciled tree viewed for analysis. The philosophy here is that, whenever the phylogeny of the species matches the phylogeny of the gene tree, these genes will be deemed orthologous.

Although this is a useful tool, users should be cautioned that this does not represent a comprehensive phylogenetic analysis, due to the automated nature of the application. Its use should be primarily as intended: to gain insight into what homologous sequences are orthologous in an automated fashion. Further analysis should be performed for any particularly in-depth investigation using less automated alignment and phylogenetic analysis that suits the sequences being investigated.

SOME SIMPLE PRACTICAL CONSIDERATIONS

1. Paradoxical as it may sound, by far the most important factor in inferring phylogenies is not the method of phylogenetic inference but the quality of the input data. The importance of data selection and in particular of the alignment process cannot be overestimated. Even the most sophisticated phylogenetic inference methods are not able to correct for erroneous input data.
2. Look at the data from as many angles as possible. Use each of the three main methods (distance, maximum parsimony, maximum likelihood) and compare the resulting trees for consistency. At the same time, be aware that one cannot rely on having arrived at a good estimate for the true phylogeny just because all three methods produce the same tree. Unfortunately, consistency among results obtained by different methods does not necessarily mean that the result is statistically significant (or represents the true phylogeny), since there can be several reasons for such correspondence.
3. The choice of outgroup taxa can have as much influence on the analysis as the choice of ingroup taxa. Complication will occur in particular when the outgroup shares an unusual property (e.g., composition bias or clock rate) with one or several ingroup taxa. It is therefore advisable to compute every analysis with several outgroups and check for congruency of the ingroup topologies.
4. Be aware that programs can give different answers (trees) depending on the order in which the sequences appear in the input file. PHYLIP, PAUP and other phylogenetic software provide a "jumble" option that reruns the analysis with different (jumbled) input orders. If for whatever reason the tree must be computed in a single run, sequences that are suspected of being "problematic" should be placed toward the end of the input file, to lower the probability that tree rearrangement methods will be negatively influenced by a poor initial topology stemming from any problematic sequences.

INTERNET RESOURCES FOR TOPICS PRESENTED IN CHAPTER 14

Compilation of available phylogeny programs	*http://evolution.genetics.washington.edu/phylip/software.html*
BLAST2 & Orthologue Search	*http://www.Bork.EMBL-Heidelberg.DE/Blast2e/*
CLUSTAL W	*http://www-igbmc.u-strasbg.fr/BioInfo/*
GeneDoc	*http://www.psc.edu/biomed/genedoc/*
GeneTree	*http://taxonomy.zoology.gla.ac.uk/rod/genetree/genetree.html*
PHYLIP	*http://evolution.genetics.washington.edu/phylip.html*
PhyloBLAST	*http://www.pathogenomics.bc.ca/phyloBLAST/*
Phylogenetic Resources	*http://www.ucmp.berkeley.edu/subway/phylogen.html*
PUZZLEBOOT	*http://www.tree-puzzle.de*
ReadSeq	*http://dot.imgen.bcm.tmc.edu:9331/seq-util/Options/readseq.html*
RDP Tree	*http://rdp.life.uiuc.edu/RDP/commands/sgtree.html*

TreeView	*http://taxonomy.zoology.gla.ac.uk/rod/treeview.html*
WebPHYLIP	*http://sdmc.krdl.org.sg:8080/~lxzhang/phylip/*
WHS	*http://www.cladistics.org/education.html*

REFERENCES

Adachi, J., and Hasegawa, M. (1996). *MOLPHY Version 2.3. Programs for Molecular phylogenetics based on maximum likelihood* (Tokyo: Institute of Statistical Mathematics).

Efron, B. (1979). Bootstrapping methods: Another look at the jackknife. *Ann. Stat.* 7, 1–26.

Dayhoff, M. O., Schwartz, R. M., and Orcutt, B. C. (1978). A model of evolutionary change in proteins. In *Atlas of Protein Sequence and Structure* M. O. Dayhoff, Ed. (Washington, DC: National Biomedical Research Foundation), p. 345–362.

Dayhoff, M. O. (1979). *Atlas of Protein Sequence and Structure*, Volume 5, Supplement 3, 1978. National Biomedical Research Foundation, Washington, D.C.Faith, D. P., and Trueman, J. W. H. (1996). When the topology-dependent permutation test (T-PTP) for monophyly returns significant support for monophyly, should that be equated with (a) rejecting the null hypothesis of nonmonophyly, (b) rejecting the null hypothesis of "no structure," (c) failing to falsify a hypothesis of monophyly, or (d) none of the above? *Syst. Biol.* 45, 580–586.

Felsenstein, J. (1985). Confidence intervals on phylogenies: An approach using the bootstrap. *Evolution* 39, 783–791.

Felsenstein, J. (1996). Inferring phylogenies from protein sequences by parsimony, distance, and likelihood methods. *Methods Enzymol.* 266, 418–427.

Felsenstein, J. (1997). An alternative least-squares approach to inferring phylogenies from pairwise distances. *Syst. Biol.* 46, 101–111.

Feng, D. F., and Doolittle, R. F. (1996). Progressive alignment of amino acid sequences and construction of phylogenetic trees from them. *Methods Enzymol.* 266, 368–382.

Gatesy, J., DeSalle, R., and Wheeler, W. (1993). Alignment-ambiguous nucleotide sites and the exclusion of systematic data. *Mol. Phylogenet. Evol.* 2, 152–157.

Hershkovitz, M.A., and Lewis, L.A. (1996). Deep-level diagnostic value of the rDNA ITS region. *Mol. Biol. Evol.* 13, 1276–1295.

Hillis, D. M., Allard, M. W., and Miyamoto, M. M. (1993). Analysis of DNA sequence data: Phylogenetic inference. *Methods Enzymol.* 224, 456–487.

Hillis, D. M., and Bull, J. J. (1993). An empirical test of bootstrapping as a method for assessing confidence in phylogenetic analysis. *Syst. Biol.* 42, 182–192.

Hillis, D. M., Huelsenbeck, J. P., and Cunningham, C. W. (1994). Application and accuracy of molecular phylogenies. *Science* 264, 671–677.

Huelsenbeck, J. P. (1995). Performance of phylogenetic methods in simulation. *Syst. Biol.* 44, 17–48.

Huelsenbeck, J. P., Hillis, D. M., and Jones, R. (1996). Parametric bootstrapping in molecular phylogenetics. In *Molecular Zoology: Advances, Strategies, and Protocols*, J. D. Ferraris and S. R. Palumbi, Eds. (New York: Wiley-Liss), p. 19–45.

Hughey, R., Krogh, A., Barrett, C., and Grate, L. (1996). SAM: Sequence alignment and modelling software. University of California, Baskin Center for Computer Engineering and Information Sciences. (*http://www.cse.ucsc.edu/research/compbio/papers/sam_doc/sam_doc.html*)

Kimura, M. (1983). *The Neutral Theory of Molecular Evolution*. Cambridge University Press, Cambridge.

Kumar, S., Tamura, K., and Nei, M. (1994). MEGA: Molecular Evolutionary Genetics Analysis software for microcomputers. *Comput. Appl. Biosci.* 10, 189–191.

Lake, J. A. (1994). Reconstructing evolutionary trees from DNA and protein sequences: Paralinear distances. *Proc. Natl. Acad. Sci. U.S.A.* 91, 1455–1459.

Lawrence, C. E., Altschul, S. F., Boguski, M. S., Liu, J. S., Neuwald, A. F., and Wootton, J. C. (1993). Detecting subtle sequence signals: A Gibbs sampling strategy for multiple alignments. *Science* 262, 208–214.

Lawson, F. S., Charlebois, R. L., and Dillon, J. A. (1996). Phylogenetic analysis of carbamoylphosphate synthetase genes: Complex evolutionary history includes an internal duplication within a gene which can root the tree of life. *Mol. Biol. Evol.* 13, 970–977.

Leipe, D. D., Wainright, P. O., Gunderson, J. H., Porter, D., Patterson, D. J., Valois, F., Himmerich, S., and Sogin, M. L. (1994). The Stramenopiles from a molecular perspective: 16S-like rRNA sequences from *Labyrinthuloides minutum* and *Cafeteria roenbergensis. Phycologia* 33, 369–377.

Li, W.-H. (1997). *Molecular Evolution* (Sunderland, MA: Sinauer Associates).

Maddison, W. P., and Maddison, D. R. (1992). *MacClade: Analysis of Phylogeny and Character Evolution. Version 3.0* (Sunderland, MA: Sinauer Associates).

Nixon, K. C., and Carpenter, J. M. (1993). On outgroups. *Cladistics* 9, 413–426.

Olsen, G. J., Matsuda, H., Hagstrom, R., and Overbeek, R. (1994). fastDNAml: A tool for construction of phylogenetic trees of DNA sequences using maximum likelihood. *Comput. Appl. Biosci.* 10, 41.

Page, R.D.M. (1994). Maps between trees and cladistic analysis of historical associations among genes, organisms, and areas. *Systematic Biol.* 43, 58–77.

Rzhetsky, A., and Nei, M. (1992). A simple method for estimating and testing minimum evolution trees. *Mol. Biol. Evol.* 9, 945–967.

Rzhetsky, A., and Nei, M. (1994). METREE: A program package for inferring and testing minimum-evolution trees. *Comput. Appl. Biosci.* 10, 409–412.

Saitou, N. (1996). Reconstruction of gene trees from sequence data. *Methods Enzymol.* 266, 427–449.

Strimmer, K., and von Haeseler, A. (1996). Quartet puzzling: A quartet maximum likelihood method for reconstructing tree topologies. *Mol. Biol. Evol.* 13, 964–969.

Swofford, D. L., Olsen, G. J., Waddell, P. J., and Hillis, D. M. (1996). Phylogenetic inference. In *Molecular Systematics*, D. M. Hillis, C. Moritz, and B. K. Mable, Eds. (Sunderland, MA: Sinauer Associates), p. 407–514.

Thompson, J. D., Higgins, D. G., and Gibson, T. J. (1994). Clustal W: Improving the sensitivity of progressive multiple alignment through sequence weighting. *Nucleic Acids Res.* 22, 4673–4680.

Thorne, J. L., and Kishino, H. (1992). Freeing phylogenies from artifacts of alignment. *Mol. Biol. Evol.* 9, 1148–1162.

Weston, P. H. (1994). Methods for rooting cladistic trees. In *Models in Phylogeny Reconstruction*, R. W. Scotland, D. J. Siebert, and D. M. Williams, Eds. (Oxford: Systematics Association), p. 125–155.

Yang, W. M., Inouye, C. J., Zeng, Y., Bearss, D., and Seto, E. (1996). Transcriptional repression by YY1 is mediated by interaction with mammalian homolog of the yeast global regulator RPD3. *Proc. Natl. Acad. Sci. U.S.A.* 93, 12845–12850.

Yang, Z., Goldman, N., and Friday, A. (1995). Maximum likelihood trees from DNA sequences: A peculiar statistical problem. *Syst. Biol.* 44, 384–399.

15

COMPARATIVE GENOME ANALYSIS

Michael Y. Galperin and Eugene V. Koonin

National Center for Biotechnology Information
National Library of Medicine
National Institutes of Health
Bethesda, Maryland

The first complete genome sequences of cellular life forms have become available in just the last several years. In 1995, the genomes of the first two bacteria, *Haemophilus influenzae* and *Mycoplasma genitalium*, were reported (Fleischmann et al., 1995; Fraser et al., 1995). One year later, the first archaeal (*Methanococcus jannaschii*) and the first eukaryotic (yeast *Saccharomyces cerevisiae*) genomes were completed (Bult et al., 1996; Goffeau et al., 1996). 1997 was marked by a landmark achievement—the sequencing of the genomes of the two best-studied bacteria, *Escherichia coli* (Blattner et al., 1997) and *Bacillus subtilis* (Kunst et al., 1997). Many more bacterial and archaeal genomes, as well as the first genome of a multicellular eukaryote, the nematode *Caenorhabiditis elegans*, have since been sequenced (see details below), providing ample material for comparative analysis.

A notable (and perhaps disappointing to many biologists) outcome of these first genome projects is that at least one-third of the genes encoded in each genome had no known or predictable function; for many of the remaining genes, only a general functional prediction appeared possible. The depth of our ignorance becomes particularly obvious on examination of the genome of *Escherichia coli* K12, arguably the most extensively studied organism among both prokaryotes and eukaryotes. Even in this all-time favorite model organism of molecular biologists, at least 40% of the genes have unknown function (Koonin, 1997). On the other hand, it turned out that the level of evolutionary conservation of microbial proteins is rather uniform, with ~70% of gene products from each of the sequenced genomes having homologs in distant genomes (Koonin et al., 1997). Thus, the functions of many of these genes

Bioinformatics: A Practical Guide to the Analysis of Genes and Proteins
Edited by A. D. Baxevanis and B. F. F. Ouellette
ISBN 0-471-38390-2 (cloth), ISBN 0-471-383910 (paper)

can be predicted simply by comparing different genomes and by transferring functional annotation of proteins from better-studied organisms to their orthologs from lesser-studied organisms. This makes comparative genomics a powerful approach for achieving a better understanding of the genomes and, subsequently, of the biology of the respective organisms. Here, we describe databases that store genomic information and bioinformatics tools that are used in the computational analysis of complete genomes. The subject of comparative genomics includes a number of distinct aspects, and it is unrealistic to cover them all in a brief chapter. We limit the discussion to the analysis of protein sets from completely sequenced genomes. Because most of the latter are from prokaryotes, there is an inevitable focus on prokaryotic biology in the presentation. Furthermore, in our choice of the genome analysis tools to discuss in detail, we decided to concentrate largely on Web-based ones that are readily accessible to any user, as opposed to stand-alone software that has more limited applicability.

PROGRESS IN GENOME SEQUENCING

By the beginning of 2000, genomes of 23 different unicellular organisms (5 archaeal, 17 bacterial, and 1 eukaryotic) had been completely sequenced. At least 70 more microbial genomes were in different stages of completion with respect to sequencing. Periodically updated lists of both finished and unfinished publicly funded genome sequencing projects are available in the GenBank Entrez Genomes division and at the site maintained by The Institute for Genome Research (TIGR) and at Integrated Genomics. A complete list of sequencing centers world-wide can be found at the NHGRI Web site. One can retrieve the actual sequence data from the NCBI FTP site or from the FTP sites of each individual sequencing center. A convenient sequence retrieval system is maintained also at the DNA Data Bank of Japan. In the framework of the Reference Sequences (RefSeq) project, NCBI has recently started to supplement the lists of gene products with some valuable sequence analysis information, such as the lists of best hits in different taxa, predicted functions for uncharacterized gene products, frame-shifted proteins, and the like. On the other hand, sequencing centers like TIGR have been updating their sequence data, correcting some of the sequencingerrors and, accordingly, their sites may contain more recent data on unfinished genome sequences.

General-Purpose Databases for Comparative Genomics

Because the World Wide Web makes genome sequences available to anyone with Internet access, there exists a variety of databases that offer more or less convenient access to basically the same sequence data. However, several research groups, specializing in genome analysis, maintain databases that provide important additional information, such as operon organization, functional predictions, three-dimensional structure, and metabolic reconstructions.

PEDANT. This useful Web resource provides answers to most standard questions in genome comparison (Frishman and Mewes, 1997). PEDANT provides an easy way to ask simple questions, such as finding out how many proteins in *H. pylori* have known (or confidently predicted) three-dimensional structures or how many

NAD^+-dependent alcohol dehydrogenases (EC 1.1.1.1) are encoded in the *C. elegans* genome (Fig. 15.1). The list of standard PEDANT queries includes EC numbers, PROSITE patterns, Pfam domains, BLOCKS, and SCOP domains, as well as PIR keywords and PIR superfamilies. Although PEDANT does not allow the users to enter their own queries, the variety of data available at this Web site makes it a convenient entry point into the field of comparative genome analysis.

COGs. The Clusters of Orthologous Groups (COGs) database has been designed to simplify evolutionary studies of complete genomes and improve functional assignments of individual proteins (Tatusov et al., 1997, 2000). It consists of ~2,800 conserved families of proteins (COGs) from each of the completely sequenced genomes. Each COG contains orthologous sets of proteins from at least three phylogenetic lineages, which are assumed to have evolved from an individual ancestral protein. By definition, *orthologs* are genes that are connected by vertical evolutionary descent (the "same" gene in different species) as opposed to *paralogs*—genes related by duplication *within* a genome (Fitch, 1970; Henikoff et al., 1997). Because orthologs typically perform the same function in all organisms, delineation of orthologous families from diverse species allows the transfer of functional annotation from better-studied organisms to the lesser-studied ones. The protein families in the COG database are separated into 17 functional groups that include a group of uncharacterized yet conserved proteins, as well as a group of proteins for which only a general functional assignment appeared appropriate (Fig. 15.2). This site is particularly useful for functional predictions in borderline cases, where protein similarity levels are fairly low. Due to the diversity of proteins in COGs, sequence similarity searches against the COG database can often suggest a possible function for a protein that otherwise has no clear database hits. This database also offers some convenient tools for a comparative analysis of complete genomes as will be described below.

KEGG. The Kyoto Encyclopedia of Genes and Genomes (KEGG) centers around cellular metabolism (Kanehisa and Goto, 2000). This Web site presents a comprehensive set of metabolic pathway charts, both general and specific, for each of the completely-sequenced genomes, as well as for *Schizosaccharomyces pombe, Arabidopsis thaliana, Drosophila melanogaster*, mouse, and human. Enzymes that are already identified in a particular organism are color-coded, so that one can easily trace the pathways that are likely to be present or absent in a given organism (Fig. 15.3). For the metabolic pathways covered in KEGG, lists of orthologous genes that code for the enzymes participating in these pathways are also provided. It is also indicated whenever these genes are adjacent, forming likely operons. A very convenient search tool allows the user to compare two complete genomes and identify all cases in which conserved genes in both organisms are adjacent or located relatively close (within 5 genes) to each other. The KEGG site is continuously updated and serves as an ultimate source of data for the analysis of metabolism in various organisms.

MBGD. The Microbial Genome Database (MBGD) at the University of Tokyo offers another convenient tool for searching for likely homologs among all sequenced microbial genomes. In contrast to COGs, MBGD assigns homology relationships based solely on sequence similarity (BLASTP values of 10^{-2} or less). MBGD allows the user to submit several sequences at once (up to 2,000 residues) for searching

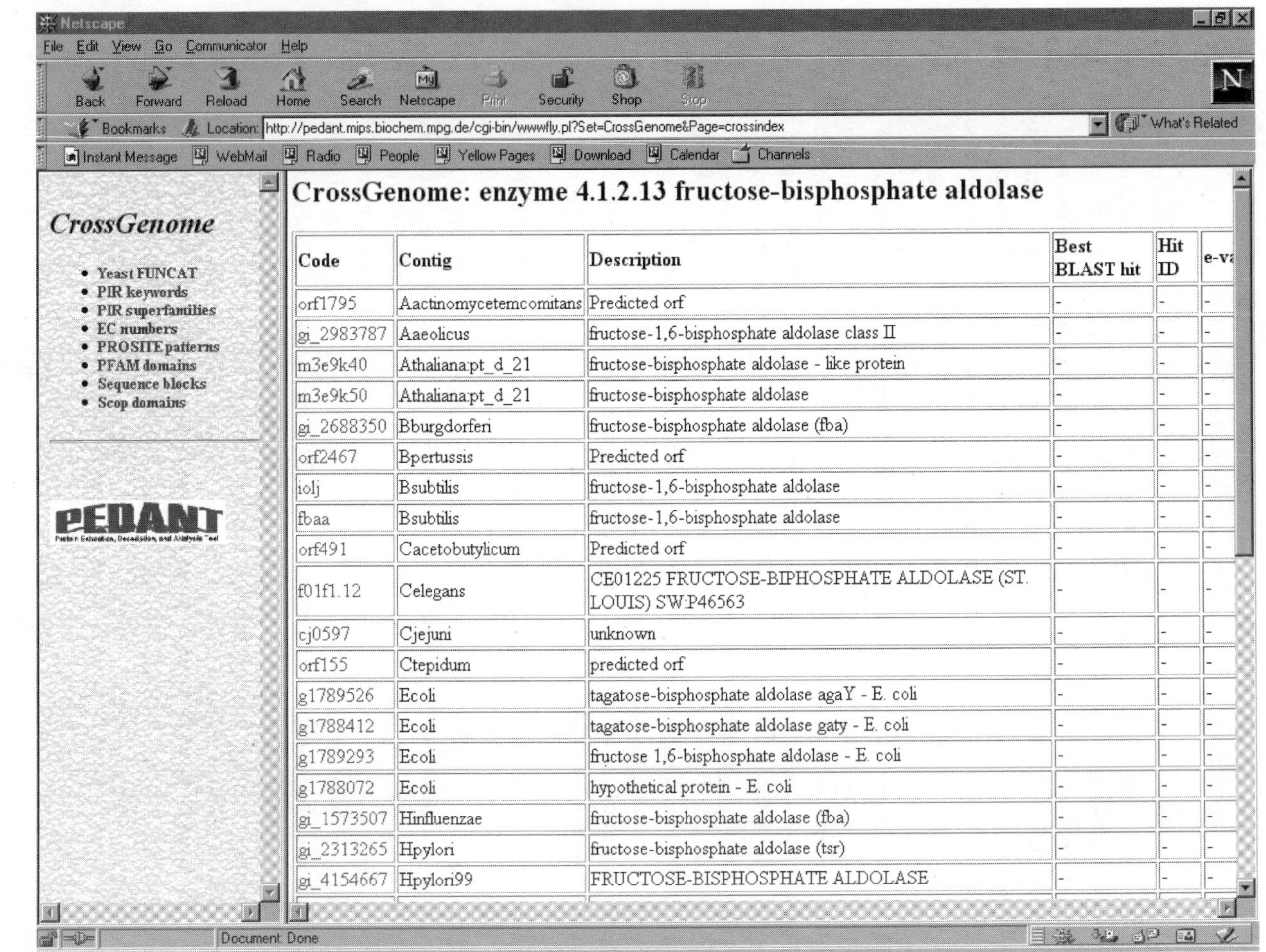

Code	Contig	Description	Best BLAST hit	Hit ID	e-va
orf1795	Aactinomycetemcomitans	Predicted orf	-	-	-
gi_2983787	Aaeolicus	fructose-1,6-bisphosphate aldolase class II	-	-	-
m3e9k40	Athaliana:pt_d_21	fructose-bisphosphate aldolase - like protein	-	-	-
m3e9k50	Athaliana:pt_d_21	fructose-bisphosphate aldolase	-	-	-
gi_2688350	Bburgdorferi	fructose-bisphosphate aldolase (fba)	-	-	-
orf2467	Bpertussis	Predicted orf	-	-	-
iolj	Bsubtilis	fructose-1,6-bisphosphate aldolase	-	-	-
fbaa	Bsubtilis	fructose-1,6-bisphosphate aldolase	-	-	-
orf491	Cacetobutylicum	Predicted orf	-	-	-
f01f1.12	Celegans	CE01225 FRUCTOSE-BIPHOSPHATE ALDOLASE (ST. LOUIS) SW:P46563	-	-	-
cj0597	Cjejuni	unknown	-	-	-
orf155	Ctepidum	predicted orf	-	-	-
g1789526	Ecoli	tagatose-bisphosphate aldolase agaY - E. coli	-	-	-
g1788412	Ecoli	tagatose-bisphosphate aldolase gaty - E. coli	-	-	-
g1789293	Ecoli	fructose 1,6-bisphosphate aldolase - E. coli	-	-	-
g1788072	Ecoli	hypothetical protein - E. coli	-	-	-
gi_1573507	Hinfluenzae	fructose-bisphosphate aldolase (fba)	-	-	-
gi_2313265	Hpylori	fructose-bisphosphate aldolase (tsr)	-	-	-
gi_4154667	Hpylori99	FRUCTOSE-BISPHOSPHATE ALDOLASE	-	-	-

Figure 15.1. The list of fructose-1,6-bisphosphate aldolases in different genomes, as generated by PEDANT. The left frame lists standard PEDANT queries.

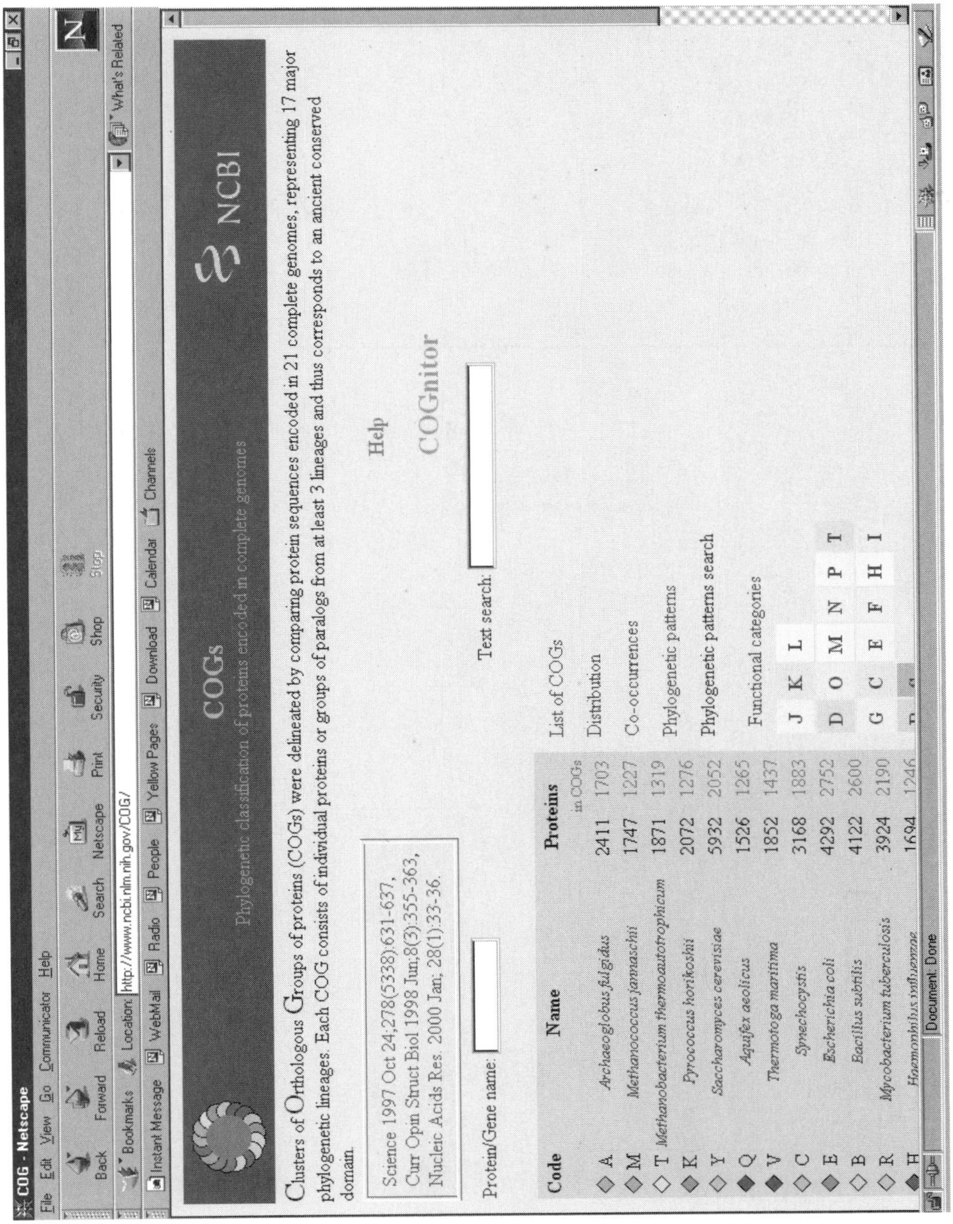

Figure 15.2. The home page of the Clusters of Orthologous Groups of proteins (COGs) database. See text for details.

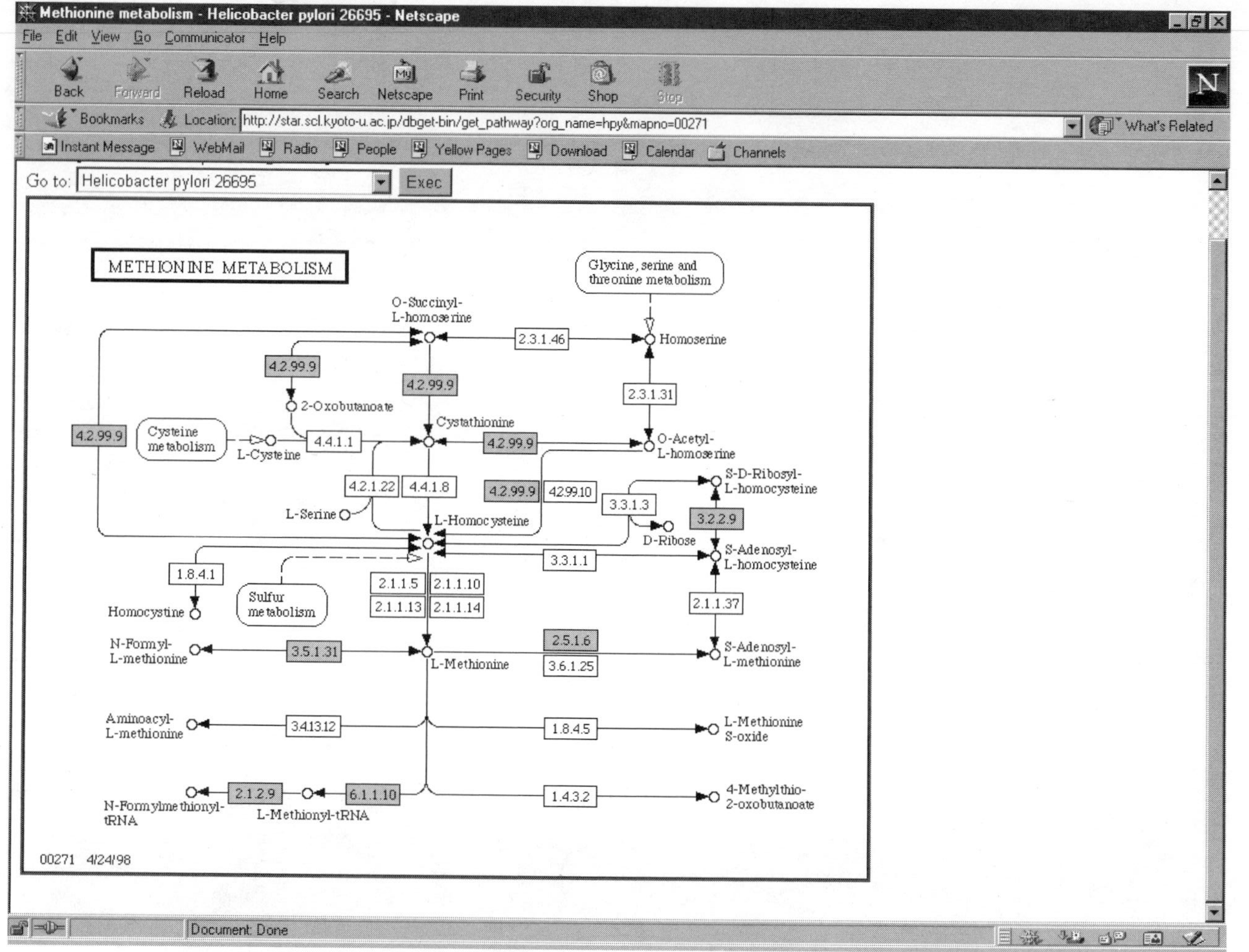

Figure 15.3. The KEGG chart of methionine metabolism in *Helicobacter pylori*. The chart shows the general organization of methionine metabolism in various species. Boxes represent enzymes with the EC numbers shown, and circles represent the intermediates of the pathway. The enzymes for which a gene has been identified in the *H. pylori* genome (according to the KEGG authors) are indicated by shaded boxes.

against all of the complete genomes available, displays color-coded functions of the detected homologs, and shows their location on a circular genome map. The output of MBGD's BLAST search also shows the degree of overlap between the query and target sequences, which could help in discerning multidomain proteins. For each sequenced genome, MBGD provides convenient lists of all recognized genes that are involved in a particular function, e.g., the biosynthesis of branched-chain amino acids or the degradation of aromatic hydrocarbons.

WIT. The WIT ("What Is There?") database, like KEGG, aims at metabolic reconstruction for completely sequenced genomes (Overbeek et al., 2000). The distinguishing features of the WIT approach are that it (1) considers as a pathway any sequence of reactions between two bifurcations and (2) includes proteins from many partially-sequenced genomes. These features allow WIT to offer many more sequences of the same enzymes from different organisms than any other database, which significantly facilitates the recognition of additional members of these enzyme families. On the other hand, complex pathways like glycolysis or the TCA cycle have been split into many separate reactions, which sometimes makes pathway analysis unnecessarily complicated. However, anyone who overcomes the initial difficulties in using the WIT system will be rewarded by the ability to easily predict metabolic pathways in many organisms with complete and still unfinished genomes.

Organism-Specific Databases

In addition to general genomics databases, there exists a variety of databases that center around a particular organism or a group of organisms. Although all of them are useful for specific purposes, those devoted to *E. coli, B. subtilis*, and yeast are probably the ones most widely used for functional assignments in other, less studied organisms. Following are short descriptions of the most frequently used of these databases.

Escherichia coli. The importance of *E. coli* for molecular biology is reflected in the large number of databases dedicated to this organism. Two of these are maintained at the University of Wisconsin-Madison and at the Nara Institute of Technology, the research groups that carried out the actual sequencing of the *E. coli* genome. Because the Wisconsin group is now involved in sequencing the enteropathogenic *E. coli* O157:H7 and other enterobacteria, their database is most useful for analysis of enteric pathogens. The group at the Nara Institute of Technology is primarily interested in resolving the functions of still-unannotated *E. coli* genes and strives to create an ultimate resource for further studies of *E. coli*. Their site provides a convenient link of genomic data to the Kohara restriction map of *E. coli* and allows one to search for Kohara clones that cover the region of interest. Another useful database on *E. coli*, EcoCyc, lists all experimentallystudied *E. coli* genes; it also provides exhaustive coverage of the metabolic pathways identified in *E. coli*.

The goal of another *E. coli* database, EcoGene, is to provide curated sequences of the *E. coli* proteins. This is a good source for frame-shifted and potentially mistranslated proteins. Finally, Colibri and RegulonDB are the databases of choice for those interested in regulatory networks of *E. coli*. The *E. coli* Genetic Stock Center (CGSC) Web site also provides gene linkage and function information; it also lists the mutations available at the CGSC.

Mycoplasma genitalium. *Mycoplasma* has the smallest genome of all known cellular life forms, which offers some clues as to what is the lower limit of genes necessary to sustain life (the "minimal genome"). Its comparison to the second smallest known genome, that of *Mycoplasma pneumoniae*, is available online. Recent data from TIGR provides insight into the range of *M. pneumoniae* and *M. genitalium* genes that can be mutated without loss of viability. From both computational analysis and mutagenesis studies, it appears that 250–300 genes are absolutely essential for the survival of mycoplasmas.

Bacillus subtilis. The *B. subtilis* genome also attracts considerable attention from biologists and, like that of *E. coli*, is being actively studied from the functional perspective. The Subtilist Web site, maintained at the Institute Pasteur, is constantly updated to include the most recent results on functions of new *B. subtilis* genes. In addition, a convenient index of *B. subtilis* sporulation genes is maintained at the Royal Holloway University of London.

Saccharomyces cerevisiae. The major databases specifically devoted to the functional analysis of yeast *S. cerevisiae* genome are the Saccharomyces Genome Database (SGD) at Stanford University, the Yeast Database at Munich Institute for Protein Sequences (MIPS), and Yeast Protein Database (YPD) at Proteome, Inc. All three databases provide periodically updated lists of yeast proteins with known or predicted functions, appropriate references, and mutant phenotypes and reflect the ongoing efforts aimed at complete characterization of all yeast proteins. SGD is probably the largest and most comprehensive source of information on the current status of the yeast genome analysis and includes the *Saccharomyces* Gene Registry. The MIPS database provides most of the same data and serves as a resource for new results coming from the multinational EUROFAN project. YPD is a curated database that is an useful resource for current information on the function of yeast proteins. YPD now allows free access for academic researchers using the database for non-commercial purposes.

Other useful sites for yeast genome analysis include *Saccharomyces cerevisiae* Promoter Database, listing known regulatory elements and transcriptional factors in yeast; TRansposon-Insertion Phenotypes, Localization, and Expression in *Saccharomyces* (TRIPLES) database, which tracks the expression of transposon-induced mutants and the cellular localization of transposon-tagged proteins, and the *Saccharomyces* Cell Cycle Expression Database, presenting the first results on changes in mRNA transcript levels during the yeast cell cycle.

GENOME ANALYSIS AND ANNOTATION

With recent progress in rapid, genome-scale sequencing, sequence analysis and annotation of complete genomes have become the limiting steps in most genome projects. This task is particularly daunting given the paucity of functional information for a large fraction of genes even in the best-understood model organisms, let alone poorly-studied ones such as those from Archaeal species. The standard steps involved in the structural-functional annotation of uncharacterized proteins includes (1) sequence similarity searches using programs such as BLAST, FASTA, or the Smith-Waterman algorithm; (2) identifying functional motifs and structural domains by comparing the protein sequence against PROSITE, BLOCKS, SMART, or Pfam; (3) predicting structural features of the protein, such as likely signal peptides, transmembrane segments, coiled-coil regions, and other regions of low sequence complexity; and (4) generating a secondary (and, if possible, tertiary) structure prediction. All these steps have been automated in several software packages, such as GeneQuiz, MAGPIE, PEDANT, Imagene, and others. Of these, however, MAGPIE and PEDANT do not allow outside users to submit their own sequences for analysis and display only the authors' own results. GeneQuiz offers a limited number of searches (up to 100 a day) to general users but is still a good entry point for comparative genome analysis (Andrade et al., 1999; Hoersch et al., 2000). However, GeneQuiz relies on unrealistically high cutoff scores to infer homology, which inevitably results in relatively low sensitivity. In some cases, the user may be better off by simply using the same tools that are packaged in the aforementioned programs separately. To perform sequence analysis on a large scale, it is frequently desirable to run the requisite software locally, in batch mode. One such package that is currently available for free downloading is SEALS, developed at NCBI. It consists of a number of UNIX-based tools for retrieving sequences from GenBank, running database search programs such as BLAST and MoST, viewing and parsing search outputs, searching for sequence motifs, and predicting protein structural features (Walker and Koonin, 1997). A similar package, Imagene, has been developed at Université Paris VI (Medigue et al., 1999).

Using Genome Comparison for Prediction of Protein Functions

Analysis of the first several bacterial, archaeal, and eukaryotic genomes to be sequenced showed that the sequence comparison methods mentioned above failed to predict protein function for at least one-third of gene products in any given genome. In these cases, other approaches can be used that take into consideration all other available data, putting them into "genome context" (Huynen and Snel, 2000). By taking advantage of the availability of multiple complete genomes, these approaches offer new opportunities for predicting gene functions in each of these genomes. All these approaches rely on the same basic premise, that the organization of the genetic information in each particular genome reflects a long history of mutations, gene duplications, gene rearrangements, gene function divergence, and gene acquisition and loss that has produced organisms uniquely adapted to their environment and capable of regulating their metabolism in accordance with the environmental conditions. This means that cross-genome similarities can be viewed as meaningful in the *evolutionary* sense and thus are potentially useful for functional analysis. The most promising comparative methods—specifically employ information derived from multiple genomes to achieve robustness and sensitivity that are not easily attainable with standard tools. It seems that they are indeed the tools for the "new genomics," whose impact will grow with the increase in the amount and diversity of genome information available. Here, some of these new approaches are briefly reviewed using for illustration, whenever possible, examples provided by currently available Web-based tools. A disproportionate number of these examples are from the COG system. This should not be construed as a claim that this is, in any sense, the best tool for genome annotation; rather, it reflects a degree of flexibility in formulating queries that is provided by the COGs as well as the subjective factor of the authors' familiarity with the organization of this system.

Transfer of Functional Information. The simplest and by far the most common way to utilize the information embedded in multiple genomes (at least at this time) is the transfer of functional information from well-characterized genomes to poorly-studied ones. Implicitly, this is done whenever a prediction is made for a newly sequenced gene on the basis of a database hit(s). There are, however, many pitfalls that tend to hamper accurate functional prediction on the basis of such hits. Perhaps the most important ones relate to the lack of sufficient sensitivity, error propagation because of reliance on incorrect or imprecise annotations already present in the general-purpose databases, and the difficulty in distinguishing orthologs from paralogs. The issue of orthology vs. paralogy is critical because transfer of functional information is likely to be reliable for orthologs (direct evolutionary counterparts) but may be quite misleading if paralogs (products of gene duplications) are involved. All these problems are, in part, obviated in the COG system, which consists of carefully annotated sets of likely orthologs and does not rely on arbitrary cutoffs for assigning new proteins to them.

The COGs can be employed for annotation of newly-sequenced genomes using the COGNITOR program. This program assigns new proteins to COGs by comparing them to protein sequences from all genomes included in the COG database and detecting genome-specific best hits (BeTs). When three or more BeTs fall into the same COG, the query protein is considered a likely new COG member. The reasoning is that it is extremely unlikely that such coherence occurs by chance, even if the observed sequence similarity *per se* is not statistically significant. The requirement of multiple BeTs for a protein to be assigned to a COG serves, to some extent, as a safeguard against the propagation of errors that might be present in the COG database itself. Indeed, if a COG contains one or even two false-positives, this will not result in a false assignment by COGNITOR under the three-BeT cutoff rule. Figure 15.4 shows two examples of the COGNITOR application to proteins from the bacterium *Deinococcus radiodurans* and the archaeon *Aeropyrum pernix* that have not been assigned a function in the original genome annotation.

Phylogenetic Patterns (Profiles). The COG-type analysis applied to multiple genomes provides for the derivation of ***phylogenetic patterns***, which are potentially useful in many aspects of genome analysis and annotation (Tatusov et al., 1997). Similar concepts have been introduced by others in the form of phylogenetic profiles (Gaasterland and Ragan, 1998; Pellegrini et al., 1999). The phylogenetic pattern for each protein family (COG) is defined as the set of genomes in which the family is represented. The COG database is accompanied by a pattern search tool that allows the user to select COGs with a particular pattern (Fig. 15.5A). Predictably, genes that are functionally related (e.g., those that encode different subunits of the same enzyme or participate in consecutive steps of the same metabolic pathway) tend to have the same phylogenetic pattern (Fig. 15.5B). In a complementary fashion, closely related species tend to co-occur in COGs. Because of these features, phylogenetic patterns can be used to improve functional predictions in complete genomes. When a particular genome is represented in the COGs for a subset of components of a particular complex or pathway but is missing in the COGs for other components, a focused search for the latter is justified. The same applies to cases in which a gene is found in one of two closely related genomes, but not the other, particularly if it is conserved in a broad range of other genomes (Fig. 15.5C). There are several reasons why unexpectedly incomplete phylogenetic patterns may be observed.

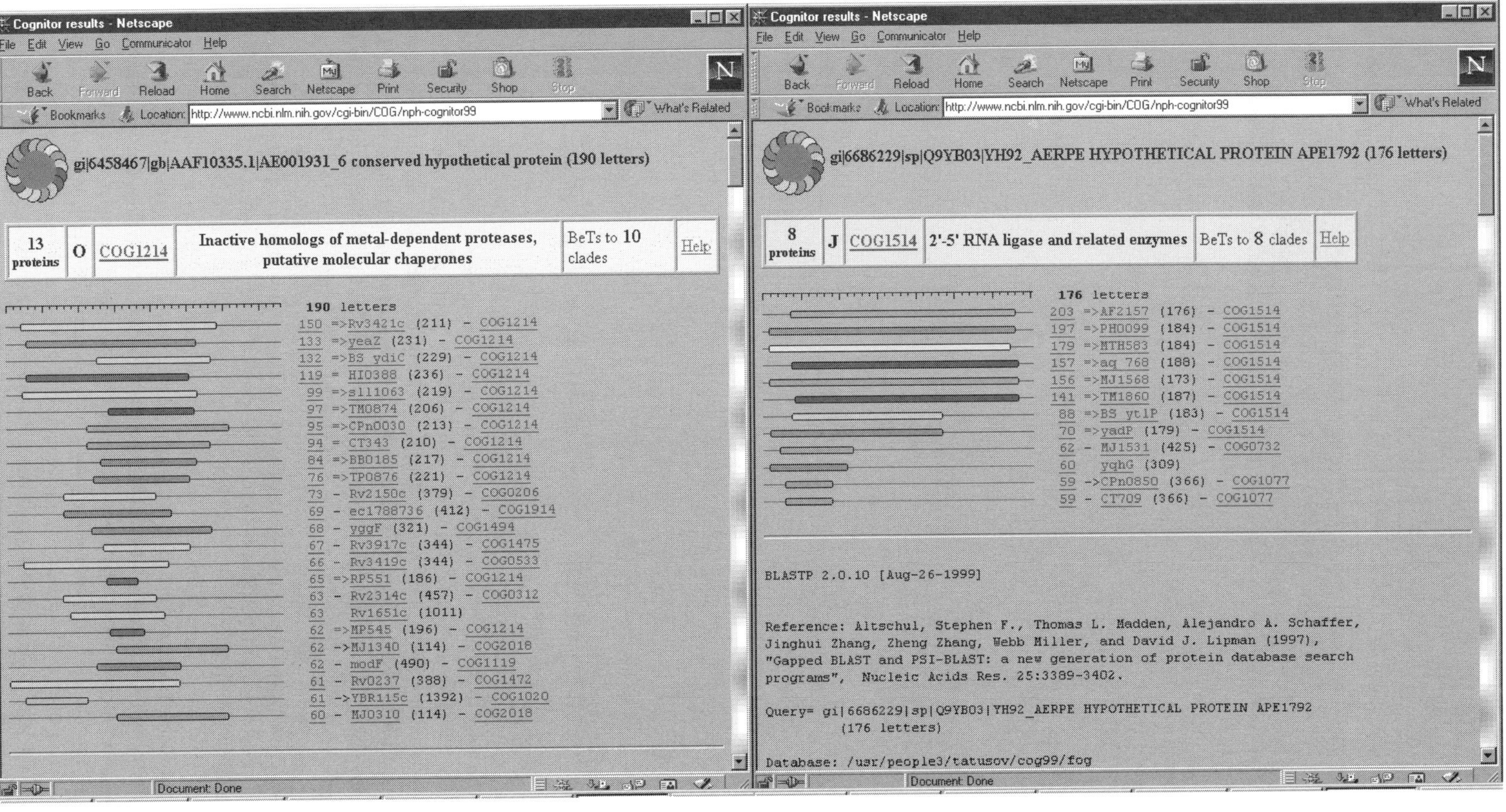

Figure 15.4. Assigning proteins to COGs using COGNITOR. (A) Uncharacterized protein from the bacterium *Deinococcus radiodurans*. (B) Uncharacterized protein from the archaeon *Aeropyrum pernix*. COGNITOR compares the query sequence to the database of protein sequences from complete genomes, registers genome-specific best hits (shown by arrows), and, in case three or more of these fall within the same COG, assigns the new protein to this COG. If this happens, the COG name should be perceived as a general functional prediction for the query. For more details, see the online COG Help pages.

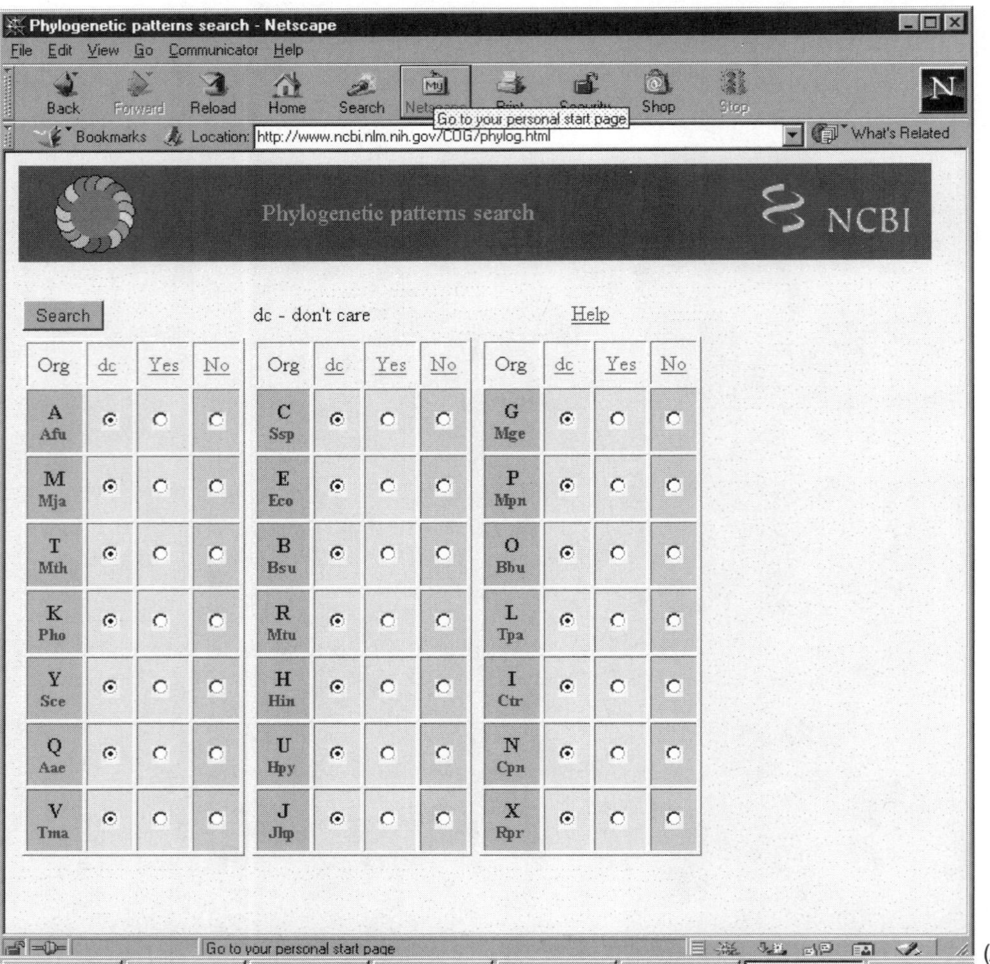

Figure 15.5. Applications of phylogenetic patterns. (A) The COG phylogenetic pattern search tool. Species name abbreviations: A (Afu), *Archaeoglobus fulgidus*; M (Mja), *Methanococcus jannaschii*; T (Mth), *Methanobacterium thermoautotrophicum*; K (Pho), *Pyrococcus horikoshii*; Y (Sce), yeast *Saccharomyces cerevisiae*; Q (Aae), *Aquifex aeolicus*; V (Tma), *Thermotoga maritima*; C (Ssp), *Synechocystis sp.*; E (Eco), *Escherichia coli*; B (Bsu), *Bacillus subtilis*; R (Mtu), *Mycobacterium tubeculosis*; H (Hin), *Haemophilus influenzae*; U (Hpy), *Helicobacter pylori*; J (Jhp), *Helicobacter pylori* J strain; G (Mge), *Mycoplasma genitalium*; P (Mpn), *Mycoplasma pneumoniae*; O (Bbu), *Borrelia burgdorferi*; L (Tpa), *Treponema pallidum*; I (Ctr), *Chlamydia trachomatis*; N (Cpn), *Chlamydia pneumoniae*; X (Rpr), *Rickettsia prowazekii*. (B) Phylogenetic pattern conservation in the threonine biosynthesis pathway. The patterns are shown using the one-letter designations for species as in (A); a letter indicates that the given species is represented in the respective COG and a dash means that it is not. Note that aspartokinase and aspartate-semialdehyde dehydrogenase are found in Chlamydiae and Rickettsiae, even though these parasitic bacteria do not encode the entire pathway of threonine biosynthesis or those for methionine and lysine biosynthesis that share the same first stages. The functions of these enzymes in these cases remain unclear. The absence of detectable homoserine kinase in *A. fulgidus* and *M. thermoautotrophicum* is probably due to nonorthologous gene displacement with a distinct kinase(s). (C) Differential genome display—COGs represented in *C. trachomatis*, but not in *C. pneumoniae*. In the generalized representation of the phylogenetic pattern above the list of COGs, asterisks indicate that the respective species may be either present or absent.

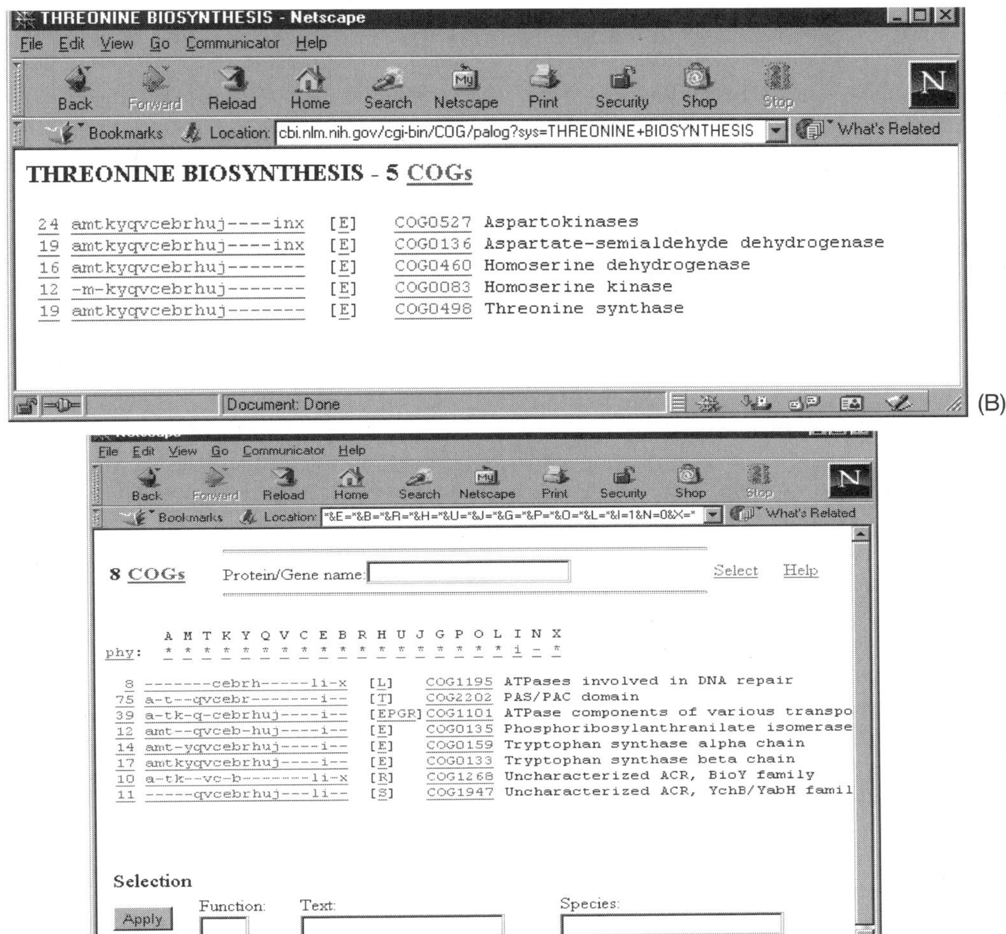

Figure 15.5. Continued

In the simplest case, certain proteins, typically small ones, could have been missed in genome translation (Natale et al., 2000). Thus, the apparent absence of *secE* genes in the genomes of *Aquifex aeolicus* and *Helicobacter pylori* that encoded all the other components of the Sec protein translocation machinery suggests that the *secE* genes could have been missed in the original genome annotation for these two bacteria. Indeed, these genes are easily recognized by searching the six-frame translation of the respective genomes using TBLASTN. Examination of the COGs that miss one representative from a group of close species similarly may result in the identification of otherwise undetected genes. For example, only one COG (COG1546) contained an *M. genitalium* protein but not an *M. pneumoniae* protein. A search for a possible missing *M. pneumoniae* counterpart identified a candidate, whose inclusion into this COG was subsequently verified by COGNITOR and sequence alignment (Natale et al., 2000).

An unexpected absence of a species in a phylogenetic pattern also may indicate that the given species encodes a highly diverged member of the respective orthologous family. For example, the presence of easily-recognizable A, B, D, and I subunits of the archaeal type H^+-ATPase in *Borrelia burgdorferi, Treponema pallidum*, and both chlamydia (COGs 1155, 1156, 1394, and 1269) immediately suggests that membrane-bound subunits of this enzyme should also be encoded in these genomes. Indeed, genes for the E and K subunits of the H^+-ATPase could be recognized in these genomes (COGs 1390 and 0636) despite their low sequence similarity to the corresponding subunits of the archaeal enzymes. The gene for the F subunit, however, has been identified so far only in *T. pallidum* but not in the three other bacterial species (see COG1436), whereas the gene for the C subunit (COG1527) has not been recognized in any of them.

Finally, unexpected "holes" in phylogenetic patterns and differences between components of the same complex or pathway may be the manifestation of a phenomenon termed *nonorthologous gene displacement*, in which unrelated or distantly related proteins are responsible for the same function in different organisms. When essential functions are involved, this tends to result in phylogenetic patterns that are perfectly or partially complementary, together spanning the entire range of genomes. Figure 15.6 shows two examples of COGs with such complementary phylogenetic patterns. Note that, in each case, the complementarity is not perfect because certain genomes encode both forms of the respective enzyme. In the case of lysyl-tRNA synthetases (Fig. 15.6A), the two forms of the enzyme are completely unrelated, whereas the two fructose-biphosphate aldolases (Fig. 15.6B) are distantly related, but are not orthologs. During the analysis of new genomes, it is possible to focus on families with complementary phylogenetic patterns to identify candidates for missing components of complexes and pathways.

Use of Phylogenetic Patterns for Differential Genome Display. The phylogenetic pattern approach and, specifically, the pattern search tool associated with the COGs can be used in a systematic fashion to perform formal logical operations (AND, OR, NOT) on gene sets—an approach suitably dubbed "differential genome display" (Huynen et al., 1997). Figure 15.7 shows examples of such analyses. This type of genome comparison allows a researcher to delineate subsets of gene products that are likely to contribute to the specific lifestyles of the respective organisms, for example, thermophily (Fig. 15.7A). The use of this approach to identify candidate drug targets in pathogenic bacteria is perhaps of special interest. It seems logical to look for such targets among those genes that are shared by several pathogenic organisms, but are missing in eukaryotes. Simple exercises in this direction show, however, that this is not a straightforward strategy. It is tempting to suggest that the best targets for new broad-spectrum antimicrobial agents would be genes that are shared by all pathogenic microbes, but not by any other organisms. The trouble is that such genes do not seem to exist, even if one allows for those that are missing in mycoplasmas, which have by far the smallest genomes (Fig. 15.7B). Furthermore, even when the conditions are relaxed so that it is only required that the genes be present in all pathogenic bacteria (except possibly mycoplasmas) and absent in yeast and *E. coli* (the dominant component of the normal gut microbial population), the net comes back empty (Fig. 15.7C). It seems therefore that the best one can do to search for such potentially universal antimicrobial agents is to isolate the genes that are present in all pathogens, possibly in other bacteria and archaea but not in eukaryotes. This results in a list of 35 families, most of which are, in fact, represented in all bacteria (Fig. 15.7D); it seems likely that some of these proteins are indeed good candidates for drug targets. More specifically directed searches can be easily set up; for example, searching for families that are represented in two species of chlamydia, possibly other pathogenic bacteria, but not any other genomes produces just two COGs (Fig. 15.7E). These could be of interest for a detailed experimental study aimed at the development of new agents that could be active against both chlamydia and spirochetes.

Examination of Gene (Domain) Fusions. Another recently developed comparative genomic approach involves systematic analysis of protein and domain fusion (and fission) (Enright et al., 1999; Marcotte et al., 1999; Snel et al., 2000). The basic assumption is that fusion would be maintained by selection only when it facilitates functional interaction between proteins, for example, kinetic coupling of consecutive enzymes in a pathway. Thus, proteins that are fused in some species can be expected to interact, perhaps physically or at least functionally, in other organisms. A straightforward example of functional inferences that can be drawn from domain fusion is seen in the histidine biosynthesis pathway, which in *E. coli* and *H. influenzae* includes two two-domain proteins, HisI and HisB (Fig. 15.8). The two domains of HisI catalyze two sequential steps of histidine biosynthesis and thus represent subunits that are likely to physically interact even when produced as separate proteins; this correlates with the predominance of the domain fusion among these enzymes (Fig. 15.8A). In contrast, the two domains of HisB catalyze the seventh and ninth steps of the pathway and hence are not likely to physically interact, which is compatible with the relative rarity of the fusion (Fig. 15.8B). The COG database includes about 700 distinct multidomain architectures that have stand-alone counterparts. Thus, using domain fusion for functional prediction has considerable heuristic potential although this approach will not work for "promiscuous" domains such as, for example, the DNA-binding helix-turn-helix domain, which can be found in combination with a wide variety of other domains (Marcotte et al., 1999).

In addition, several databases (with accompanying search tools) have recently been developed for detecting domains and exploring architectures of multidomain proteins: Pfam (Bateman et al., 2000), ProDom (Corpet et al., 2000), and SMART (Schultz et al., 1998, 2000).

Although not comprehensive as of this writing, SMART seems to be the most advanced of these systems, combining high sensitivity of domain detection with accuracy, high speed, and extremely informative presentation of domain architectures. Rapid searches for protein domains, based on a modification of the PSI-BLAST program is now available through the Conserved Domains Database (CDD) at NCBI (cf. Chapter 11).

It seems worth considering an example of a complex multidomain protein analysis in some detail, to see how assigning functions to various domains of a multidomain protein helps one understand its likely cellular role(s). The *M. tuberculosis* protein Rv1364c consists of 653 amino acid residues. Its annotation in GenBank correctly indicates that it has statistically significant similarity to the *B. subtilis* sigma factor regulation protein, RSBU_BACSU, which, however, is only 335 amino acids long. The region of similarity between these two proteins is said to be even shorter, 244 amino acid residues. Thus, in addition to the portion apparently homologous to RsbU, Rv1364c probably contains other domains. Submitting Rv1364c for a Pfam

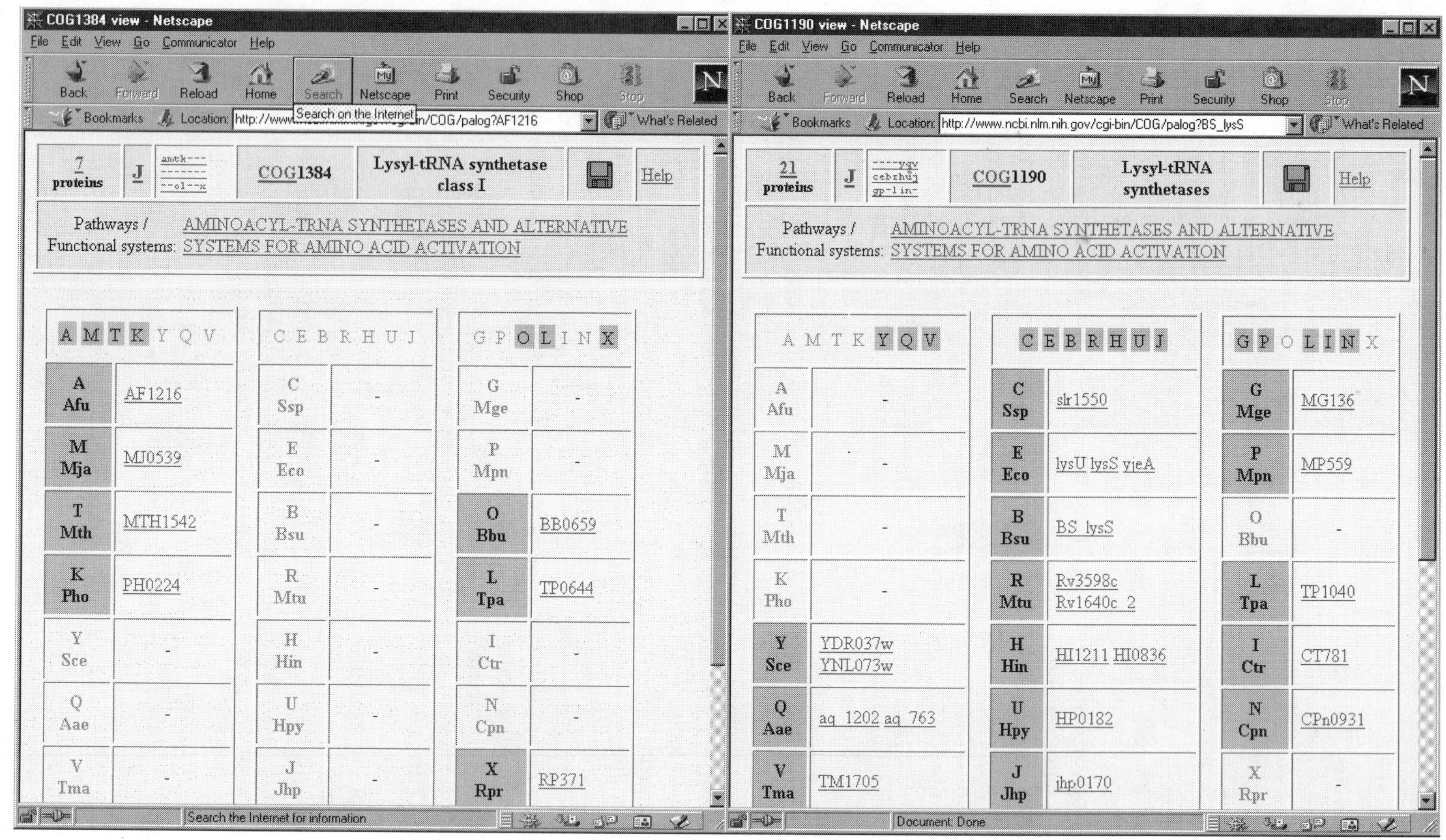

(A)

Figure 15.6. Complementary phylogenetic patterns. (A) The two classes of lysyl-tRNA synthetases. Note that both COGs include *T. pallidum*; it encodes an unusual class II enzyme that might not be involved in translation. (B) Two classes of fructose-1,6-bisphosphate aldolase. Note that *E. coli* and *A. aeolicus* encode both types of aldolases.

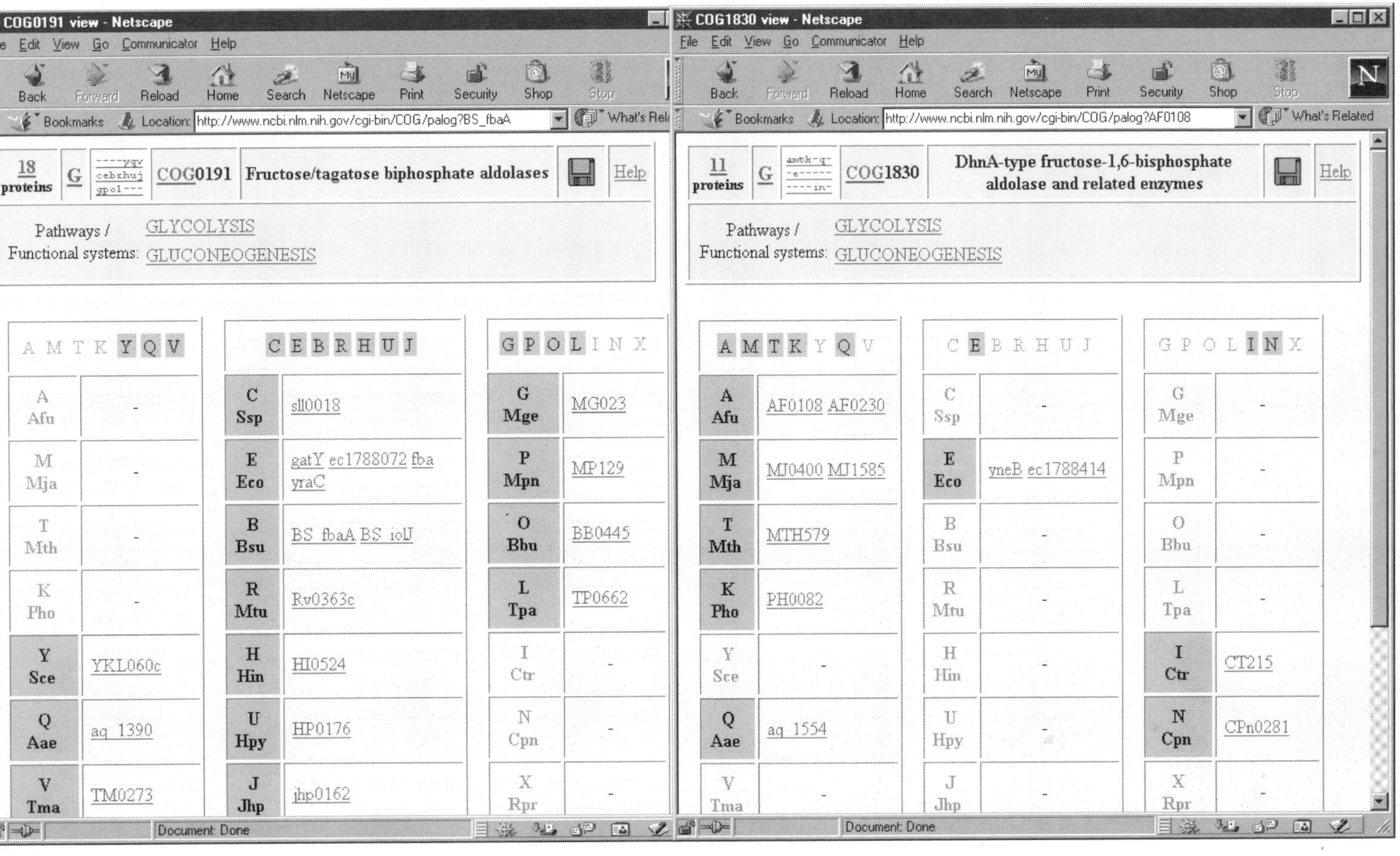

(B)

Figure 15.6. Continued

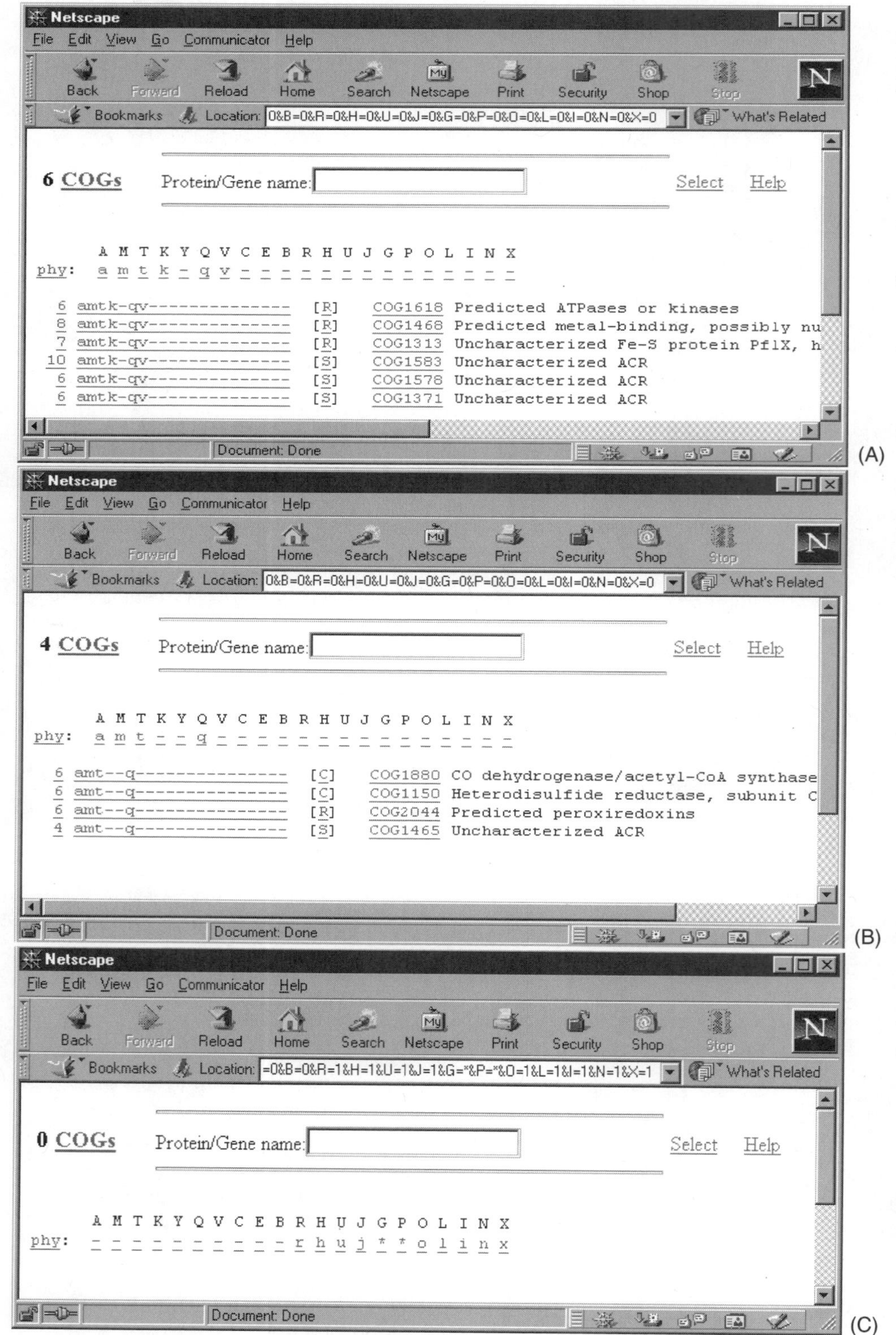

Netscape
File Edit View Go Communicator Help
Back Forward Reload Home Search Netscape Print Security Shop Stop
Bookmarks Location: 0&B=0&R=0&H=0&U=0&J=0&G=0&P=0&O=0&L=0&I=0&N=0&X=0
What's Related
6 COGs
Protein/Gene name:
Select Help
A M T K Y Q V C E B R H U J G P O L I N X
phy: a m t k - q v - - - - - - - - - - - - - -
6 amtk-qv-------------- [R] COG1618 Predicted ATPases or kinases
8 amtk-qv-------------- [R] COG1468 Predicted metal-binding, possibly nu
7 amtk-qv-------------- [R] COG1313 Uncharacterized Fe-S protein PflX, h
10 amtk-qv-------------- [S] COG1583 Uncharacterized ACR
6 amtk-qv-------------- [S] COG1578 Uncharacterized ACR
6 amtk-qv-------------- [S] COG1371 Uncharacterized ACR
Document: Done
(A)
Netscape
File Edit View Go Communicator Help
Back Forward Reload Home Search Netscape Print Security Shop Stop
Bookmarks Location: 0&B=0&R=0&H=0&U=0&J=0&G=0&P=0&O=0&L=0&I=0&N=0&X=0
What's Related
4 COGs
Protein/Gene name:
Select Help
A M T K Y Q V C E B R H U J G P O L I N X
phy: a m t - - q - - - - - - - - - - - - - - -
6 amt--q--------------- [C] COG1880 CO dehydrogenase/acetyl-CoA synthase
6 amt--q--------------- [C] COG1150 Heterodisulfide reductase, subunit C
6 amt--q--------------- [R] COG2044 Predicted peroxiredoxins
4 amt--q--------------- [S] COG1465 Uncharacterized ACR
Document: Done
(B)
Netscape
File Edit View Go Communicator Help
Back Forward Reload Home Search Netscape Print Security Shop Stop
Bookmarks Location: =0&B=0&R=1&H=1&U=1&J=1&G=*&P=*&O=1&L=1&I=1&N=1&X=1
What's Related
0 COGs
Protein/Gene name:
Select Help
A M T K Y Q V C E B R H U J G P O L I N X
phy: - - - - - - - - - - r h u j * * o l i n x
Document: Done
(C)

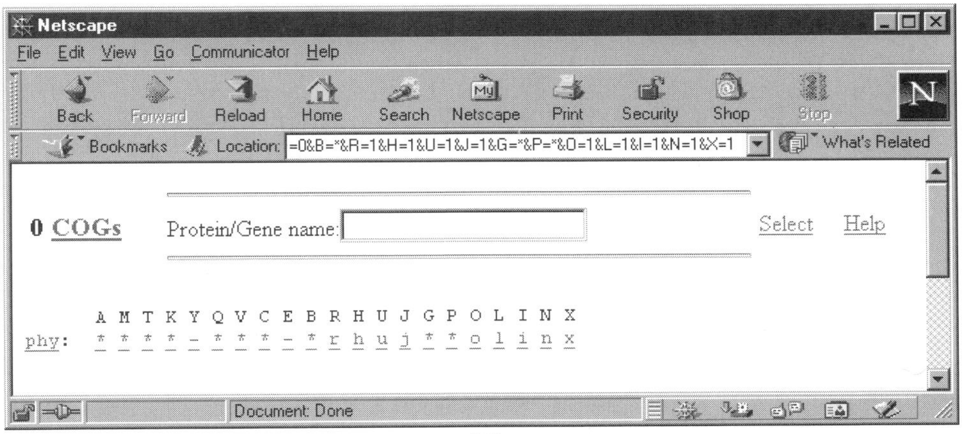

Netscape

File Edit View Go Communicator Help

Back Forward Reload Home Search Netscape Print Security Shop Stop

Bookmarks Location: =0&B=*&R=1&H=1&U=1&J=1&G=*&P=*&O=1&L=1&I=1&N=1&X=1 What's Related

0 COGs Protein/Gene name: Select Help

```
     A M T K Y Q V C E B R H U J G P O L I N X
phy: * * * * - * * * - * r h u j * * o l i n x
```

Document: Done

(D)

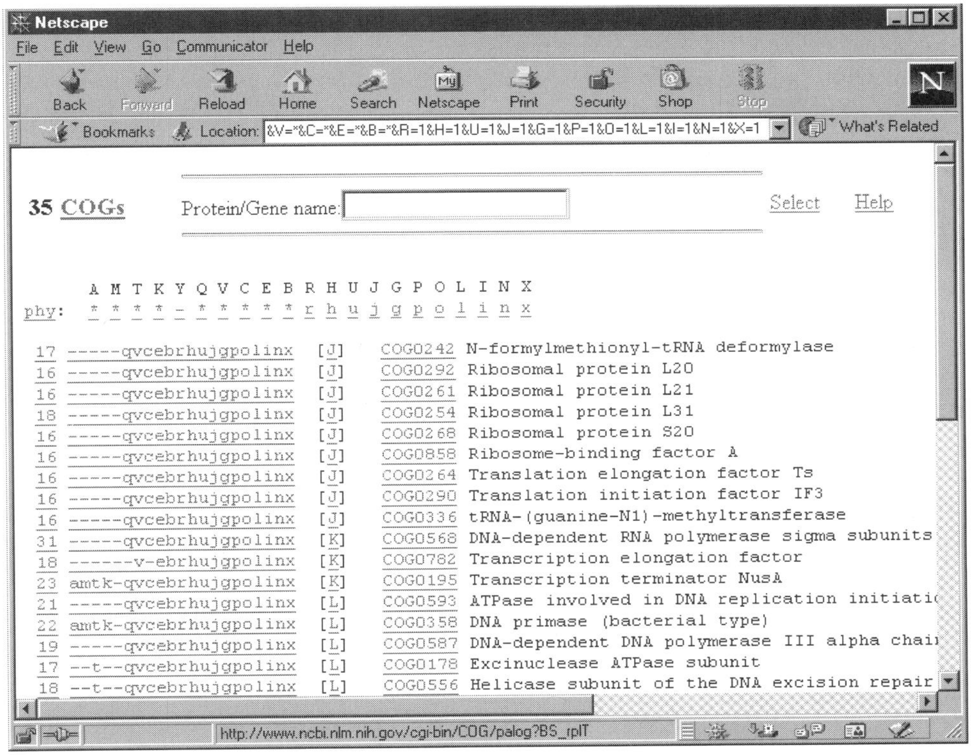

Netscape

File Edit View Go Communicator Help

Back Forward Reload Home Search Netscape Print Security Shop Stop

Bookmarks Location: &V=*&C=*&E=*&B=*&R=1&H=1&U=1&J=1&G=1&P=1&O=1&L=1&I=1&N=1&X=1 What's Related

35 COGs Protein/Gene name: Select Help

```
     A M T K Y Q V C E B R H U J G P O L I N X
phy: * * * * - * * * * * r h u j g p o l i n x

17 -----qvcebrhujgpolinx  [J]  COG0242 N-formylmethionyl-tRNA deformylase
16 -----qvcebrhujgpolinx  [J]  COG0292 Ribosomal protein L20
16 -----qvcebrhujgpolinx  [J]  COG0261 Ribosomal protein L21
18 -----qvcebrhujgpolinx  [J]  COG0254 Ribosomal protein L31
16 -----qvcebrhujgpolinx  [J]  COG0268 Ribosomal protein S20
16 -----qvcebrhujgpolinx  [J]  COG0858 Ribosome-binding factor A
16 -----qvcebrhujgpolinx  [J]  COG0264 Translation elongation factor Ts
16 -----qvcebrhujgpolinx  [J]  COG0290 Translation initiation factor IF3
16 -----qvcebrhujgpolinx  [J]  COG0336 tRNA-(guanine-N1)-methyltransferase
31 -----qvcebrhujgpolinx  [K]  COG0568 DNA-dependent RNA polymerase sigma subunits
18 ------v-ebrhujgpolinx  [K]  COG0782 Transcription elongation factor
23 amtk-qvcebrhujgpolinx  [K]  COG0195 Transcription terminator NusA
21 -----qvcebrhujgpolinx  [L]  COG0593 ATPase involved in DNA replication initiati
22 amtk-qvcebrhujgpolinx  [L]  COG0358 DNA primase (bacterial type)
19 -----qvcebrhujgpolinx  [L]  COG0587 DNA-dependent DNA polymerase III alpha chai
17 --t--qvcebrhujgpolinx  [L]  COG0178 Excinuclease ATPase subunit
18 --t--qvcebrhujgpolinx  [L]  COG0556 Helicase subunit of the DNA excision repair
```

http://www.ncbi.nlm.nih.gov/cgi-bin/COG/palog?BS_rplT

(E)

Figure 15.7. Delineation of distinct subsets of COGs using the phylogenetic pattern search tool. (A) COGs represented exclusively in thermophiles. (B) Search for COGs represented exclusively in pathogens (with a possible exception of the mycoplasmas). (C) Search for COGs represented in pathogens (with a possible exception of the mycoplasmas), but not in yeast or *E. coli*. (D) COGs represented in all pathogens, but not in yeast. (E) COGs represented in the two Chlamydia and possibly in other pathogens, but not in free-living organisms.

search gives an unexpected result: Pfam search identifies a SpoIIAA(RsbV)-like domain in the 550–652 region of the protein. The confidence level is not particularly high (E = 0.0049), but examination of the alignment shows conservation of the phosphorylatable serine residue and the surrounding motif, as well as conservation of the secondary structure elements (not shown). This suggests that Rv1364c actually contains *four* domains, the second and the fourth of which correspond to *B. subtilis* RsbU and RsbV proteins. The structures and functions of the first (residues 1–155) and the third (334–550) domains remain to be analyzed. This could be done by PSI-BLAST analysis of individual segments encompassing the presumptive domains, but this route would involve careful examination of the complex search outputs. In contrast, SMART provides a one-step solution (Fig. 15.9). The SMART output indicates that the protein under analysis contains N-terminal PAS and PAC domains, which are ligand-binding sensor domains present in many histidine kinases and other signal-transduction proteins, a PP2C-like phosphatase domain (the actual biochemical activity of the RsbU domain) and a histidine kinase-type ATPase domain (residues 433–527). For the latter domain, however, the statistical significance of the hit is low (E = 0.16) and the assignment needs further verification. A PSI-BLAST search started with the segment of Rv1364c identified by SMART as the ATPase domain reveals similarity to the RsbW proteins, a distinct group of serine kinases of the histidine kinase fold involved in the anti-sigma regulatory systems (hence, the low-significance hit to the general profile for this domain in SMART). SMART does not detect the C-terminal SpoIIAA domain of Rv1364c, which has been identified by Pfam, emphasizing the importance of complementary methods for complete assignments of domains and functions. The domain organization of the protein can also be probed using the COG database. Entering `Rv1364c` as a COGNITOR query assigns its domains to four COGs (where the protein already belongs since it originates from a completely sequenced genome): (1) Rv1364c_1-COG2202 "PAS/PAC domain," (2) Rv1364c_2-COG2208 "Serine phosphatase RsbU, regulator of sigma subunit," (3) Rv1364c_3-COG2172 "Anti-sigma regulatory factor (Ser/Thr protein kinase)," and (4) Rv1364c_4-COG1366 "Anti-anti-sigma regulatory factor (antagonist of anti-sigma factor)." Thus, taken together, the results obtained using different methods converge on an unprecedented four-domain architecture for Rv1364c that juxtaposes the sensor PAS/PAC domain with all three components of the anti-sigma regulatory system fused within a single protein (Fig. 15.9). The PAS/PAC domain is most likely involved in sensing the energetic state of the cell, similarly to the recently characterized Aer protein of *E. coli* (Taylor and Zhulin, 1999), whereas the phosphatase, kinase, and phosphorylatable adapter domains are expected to efficiently transmit this information to the downstream signal response machinery. Thus, we can tentatively annotate Rv1364c protein as a complex regulator of sigma factor activity; the exact implications of the unusual domain fusion remain to be investigated experimentally.

Analysis of Conserved Gene Strings (Operons). An approach that is conceptually similar to the analysis of gene fusions, but is more general, if less definitive, involves systematic analysis of gene "neighborhoods" in genomes (Overbeek et al., 1999). Because functionally linked genes frequently form operons in bacteria and archaea, gene adjacency may provide important functional hints. Of course, many functionally related genes never form operons, and, in many instances, adjacent genes are not connected in any way. However, due to the lack of overall conservation of

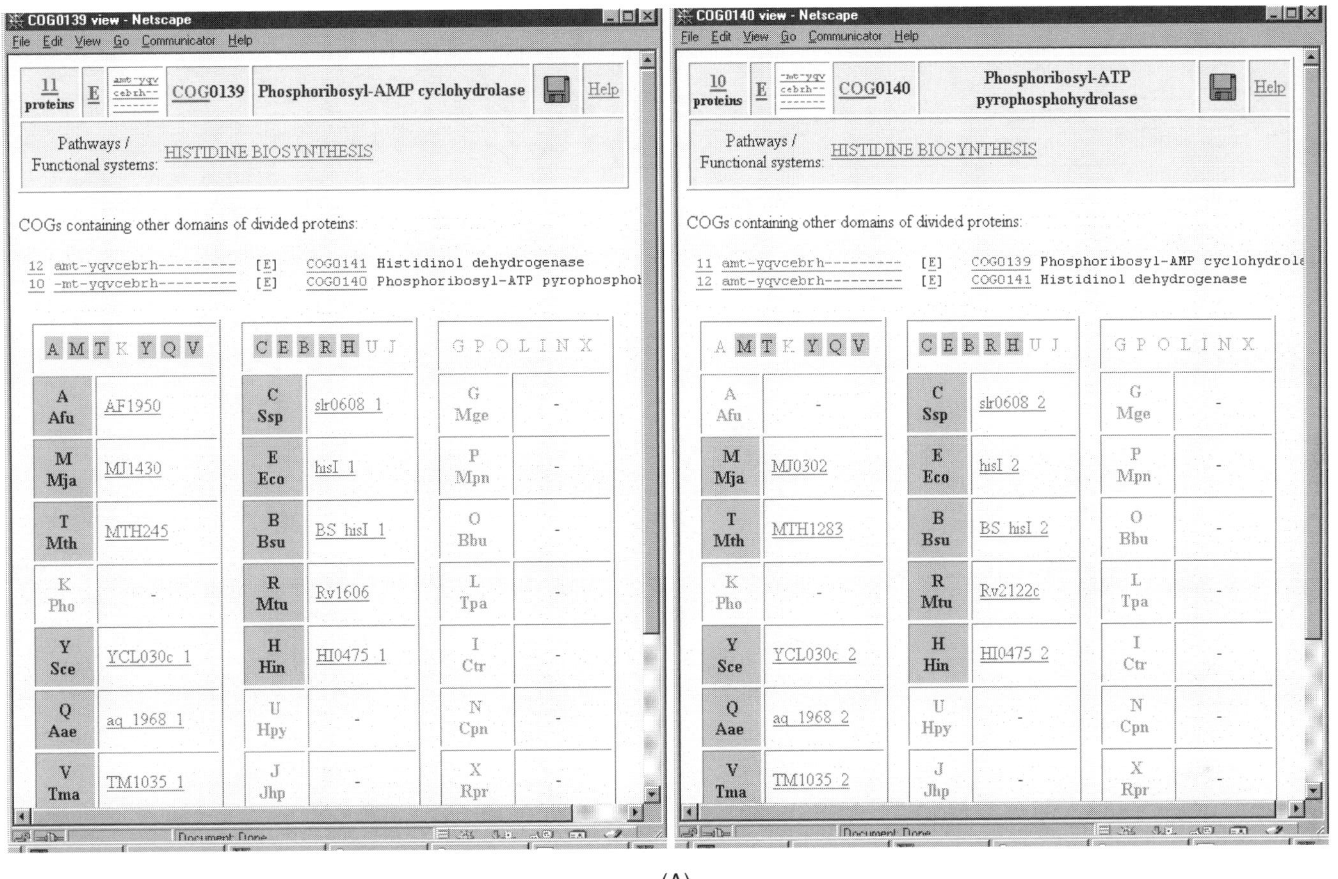

(A)

Figure 15.8. Multidomain proteins in the COGs. (A) HisI proteins. Note the fusion of the two enzymes in all bacteria that possess histidine biosynthesis as opposed to the stand-alone proteins in archaea. (B) HisB protein. Note the limited phylogenetic distribution of the fusion.

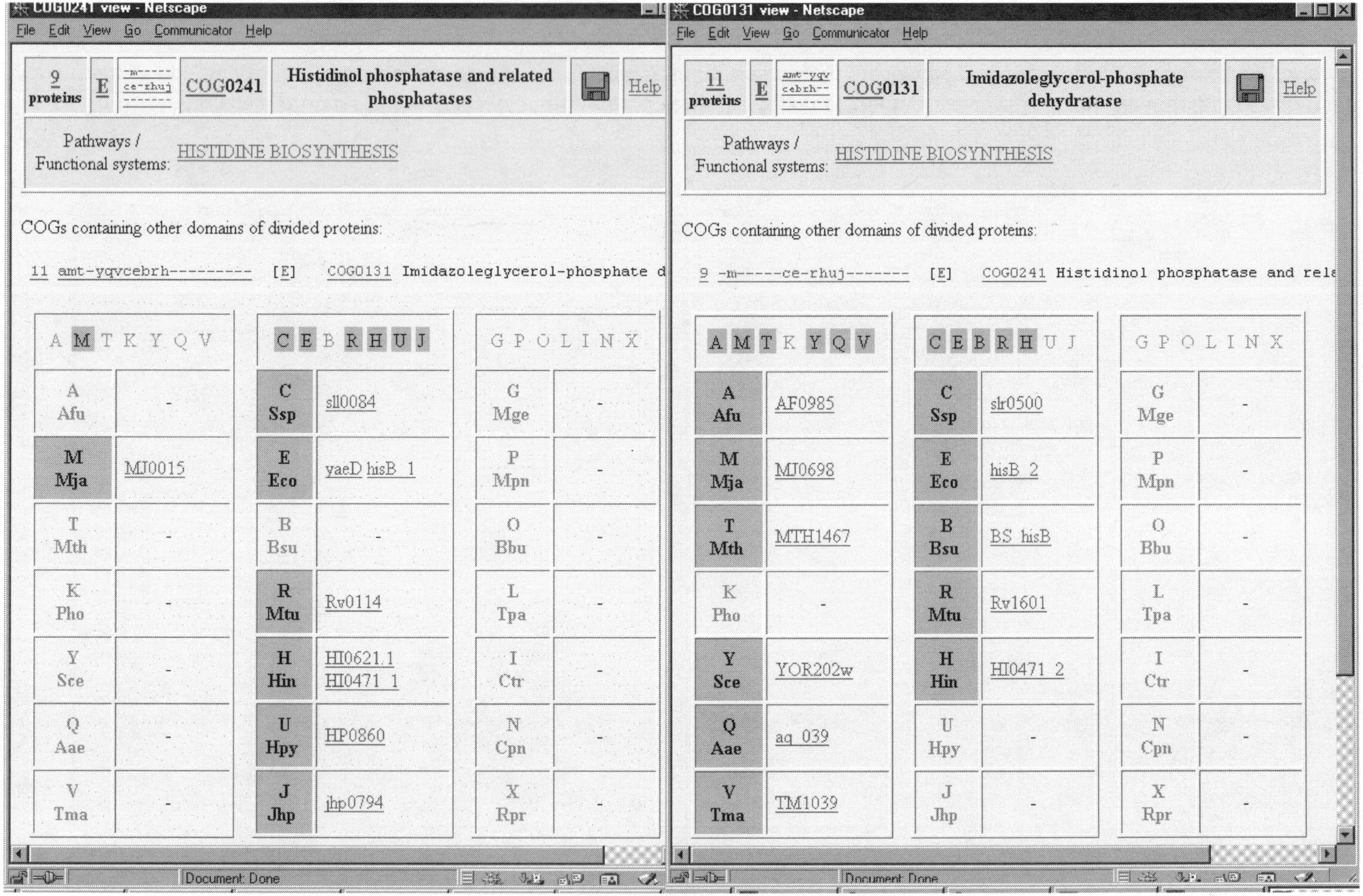

COG0241 view - Netscape

File Edit View Go Communicator Help

9 proteins | E | -m----- ce-rhuj | COG0241 | Histidinol phosphatase and related phosphatases | Help

Pathways / Functional systems: HISTIDINE BIOSYNTHESIS

COGs containing other domains of divided proteins:

11 amt-yqvcebrh--------- [E] COG0131 Imidazoleglycerol-phosphate d

A M T K Y Q V		C E B R H U J		G P O L I N X	
A Afu	-	C Ssp	sll0084	G Mge	-
M Mja	MJ0015	E Eco	yaeD hisB_1	P Mpn	-
T Mth	-	B Bsu	-	O Bbu	-
K Pho	-	R Mtu	Rv0114	L Tpa	-
Y Sce	-	H Hin	HI0621.1 HI0471_1	I Ctr	-
Q Aae	-	U Hpy	HP0860	N Cpn	-
V Tma	-	J Jhp	jhp0794	X Rpr	-

Document: Done

COG0131 view - Netscape

File Edit View Go Communicator Help

11 proteins | E | amt-yqv cebrh-- ------- | COG0131 | Imidazoleglycerol-phosphate dehydratase | Help

Pathways / Functional systems: HISTIDINE BIOSYNTHESIS

COGs containing other domains of divided proteins:

9 -m-----ce-rhuj------- [E] COG0241 Histidinol phosphatase and rela

A M T K Y Q V		C E B R H U J		G P O L I N X	
A Afu	AF0985	C Ssp	slr0500	G Mge	-
M Mja	MJ0698	E Eco	hisB_2	P Mpn	-
T Mth	MTH1467	B Bsu	BS_hisB	O Bbu	-
K Pho	-	R Mtu	Rv1601	L Tpa	-
Y Sce	YOR202w	H Hin	HI0471_2	I Ctr	-
Q Aae	aq_039	U Hpy	-	N Cpn	-
V Tma	TM1039	J Jhp	-	X Rpr	-

Document: Done

(B)

Figure 15.8. Continued

Domains within the query sequence of 653 residues

Mouse over domain / undefined region to see the limits; click on it to go to further annotation
Signal peptides determined by the *SignalP* program (), transmembrane segments predicted by the *TopPred* program (very likely () and possible ()), coiled coil regions determined by the *Coils2* program () and Segments of low compositional complexity, determined by the *SEG* program () are shown in the line below the SMART domains.

Figure 15.9. Elucidation of a protein's domain architecture using SMART. The SMART output for the Rv1364c protein is shown. The additional bar above the line shows the location of the SpoIIAA(RbsV) domain that is not recognized by SMART.

gene order in prokaryotes, the presence of a pair of adjacent orthologous genes in three or more genomes or the presence of three orthologs in a row in two genomes can be considered a statistically meaningful event and can be used to infer potential functional interaction for the products of these genes. The simplest current tool for identification of conserved gene strings in any two genomes is available as part of KEGG. It allows the user to select any two complete genomes (e.g., *B. burgdorferi* and *R. prowazekii*) and look for all genes whose products are similar to each other (e.g., have BLAST scores greater than 100) and are located within a certain distance from each other (that is, separated by 0–5 genes). The results are displayed in a graphical format illustrating the gene order and the presumed functions of gene products. In the example shown in Fig. 15.10, the uncharacterized conserved protein BB0788 from *B. burgdorferi* is similar to the RP042 protein of *R. prowazekii*, and BB0789 is similar to RP043. These pairs of genes indeed have been identified as orthologs in the COGs (COG0037 and COG0465, respectively), and examination of the relative genomic locations of other members of these COGs shows that orthologous gene strings are present in the genomes of *C. trachomatis* (CT840-CT841), *C. pneumoniae* (CPn0997-CPn0998), and *T. maritima* (TM0579-TM0580), whereas in *B. subtilis* the corresponding genes (*yacA* and *ftsH*) are one gene apart. This conservation of gene juxtaposition in phylogenetically distant bacteria is suggestive of a functional connection. The functions of one of the genes in all these pairs are well known, as they are clear orthologs of the *E. coli ftsH* (*hflB*) gene. This gene encodes an ATP-dependent metalloprotease, which is responsible for the degradation of short-lived cytoplasmic proteins and a distinct class of membrane proteins. By association, BB0788 and its orthologs may be expected to also play some, perhaps regulatory, role in the degradation of specific protein classes. The *E. coli* ortholog of BB0788, MesJ, is annotated in the SWISS-PROT database as a putative cell cycle protein. We have been unable to identify the source of this information. Nevertheless, it is compatible with a functional (and possibly also physical) interaction between MesJ and FtsH, which also has been implicated in cell division on the basis of genetic data. In the COG database, MesJ and its orthologs (COG0037) are annotated as "Predicted ATPases of the PP-loop superfamily." All these enzymes share a diagnostic sequence motif, which has been discovered previously in a number of ATP pyrophosphatases (Bork and Koonin, 1994). This motif is clearly seen in the multiple alignment accompanying the COG and in the ProDom entry PD000352. Therefore,

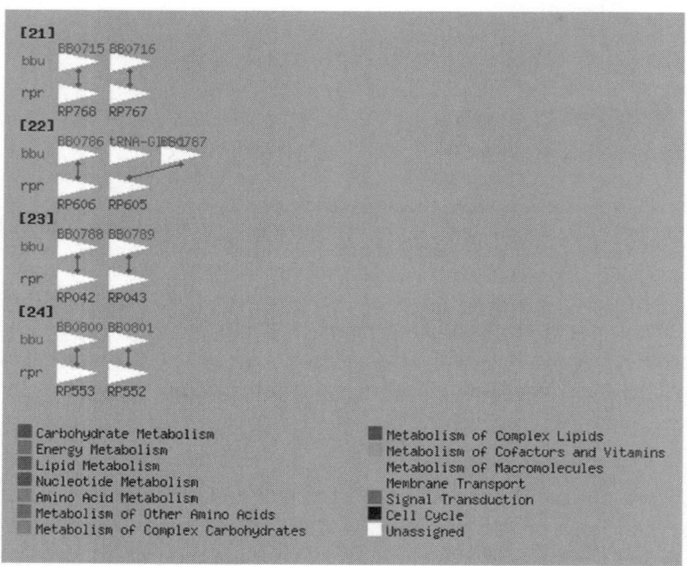

Figure 15.10. The results of a search for conserved gene strings (operons) in *Borrelia burgdorferi* (upper arrows) and *Rickettsia prowazekii* (lower arrows) using the KEGG gene cluster tool. The arrows are color-coded according to the gene function. Genes without an assigned function (all of the genes in this page) are shown in white.

by combining operon information with sequence-based prediction of the biochemical activity, we hypothesize that MesJ and its orthologs are ATP-pyrophosphatases involved in the regulation of FtsH-mediated proteolysis of specific bacterial proteins, which may be important for cell division. Because a PP-loop superfamily ATPase would comprise a novel class of cell division regulators, experimental verification of this hypothesis will be of considerable interest.

APPLICATION OF COMPARATIVE GENOMICS—RECONSTRUCTION OF METABOLIC PATHWAYS

To succinctly recap the genome analysis tools discussed above, we present here a reconstruction of the glycolytic pathway in the archaeon *Methanococcus jannaschii*. Metabolic reconstruction is one of the indispensable final steps of all genome analyses and a natural convergence point for the data produced by different methods. Glycolysis is perhaps *the* central pathway of cellular biochemistry as it becomes obvious from a cursory exploration of the general scheme of biochemical pathways, available in the interactive form on the ExPASy Web site. The names of all the enzymes and metabolites on this map are hyperlinked and searchable. Entering "glycolysis" as the search term finds three fields, B5, F5 and U6, the first two of which indicate the border areas of glycolysis, which actually stretches from C5 to E5. The enzyme names are hyperlinked to the ENZYME database, which is now the official site of the Enzyme Commission (IUPAC-IUBMB Joint Commission for Biochemical Nomenclature). The ENZYME database lists names and catalyzed reactions for all the enzymes that have been assigned official Enzyme Commission (EC) numbers, whether or not their protein sequences are known. Thus, clicking on the name "phosphoglucomutase" will bring up the corresponding page in the ENZYME database, which will also indicate that the official name of this enzyme is glucose-6-phosphate isomerase (EC 5.3.1.9).

Glycolysis Step-by-Step

1. Glucose-6-Phosphate Isomerase. Each page in the ENZYME database lists all enzymes with the corresponding EC number that are included in the current release of SWISS-PROT. There is, however, no entry for *M. jannaschii* under glucose-6-phosphate isomerase; therefore, we will instead use the KEGG database. On entering KEGG, one can go to Open KEGG, then to Metabolic Pathways, then to Glycolysis. This takes one to an easy-to-navigate chart of compounds and enzymes that participate in glycolysis and gluconeogenesis. The pull-down menu in the upper left corner allows one to select the organism of choice. When *Methanococcus jannaschii* is selected, the boxes that indicate enzymes already recognized in this organism become shaded green. One can see that the box 5.3.1.9 is highlighted. Clicking on this box shows *M. jannaschii* protein MJ1605, which indeed can be confidently identified as phosphoglucomutase (glucose-6-phosphate isomerase). In the SWISS-PROT/TREMBL database, this protein (Q59000) is currently annotated as "similar to prokariota glucose-6-phosphate isomerase." To verify that MJ1605 is indeed the ortholog of known glucose-6-phosphate isomerases, one can use the COGs, WIT, or MBGD systems. For example, in the COG database, MJ1605 is the only member of COG0166 from *M. jannaschii*; since this COG includes glucose-6-phosphate isomerases from a number of species, identification of the *M. jannaschii* protein appears to be reliable. This can be confirmed by examination of the multiple sequence alignment associated with this COG, which shows a particularly high similarity between MJ1605 and phosphoglucoisomerases from *Thermotoga maritima* and *B. subtilis*. The WIT database will show that MJ1605 is even more similar to the (predicted) glucose-6-phosphate isomerases of *Campylobacter jejuni* and *Streptococcus pyogenes*. Collectively, this evidence leaves no doubt that we have identified the correct *M. jannaschii* protein.

2. Phosphofructokinase. The next glycolytic enzyme, phosphofructokinase (EC 2.7.1.11), illustrates the opportunities and limitations of metabolic reconstruction based on comparative genomics. The most common version of this enzyme, PfkA, uses either ATP (in bacteria and many eukaryotes) or pyrophosphate (primarily in plants) as the phosphate donor. In addition, *E. coli* encodes a second version of this enzyme, PfkB, which is unrelated to PfkA and instead belongs to the ribokinase family of carbohydrate kinases. All the databases we can use agree that there is no readily identifiable candidate for this activity in *M. jannaschii*. Indeed, the KEGG chart for *M. jannaschii* does not show this enzyme as predicted, WIT does not suggest any candidates for this function, and the COG database does not show any archaeal members in its COG0205 (6-phosphofructokinase) or COG1105 [fructose-1-phosphate kinase and related fructose-6-phosphate kinase (PfkB)] entries. Thermophilic archaea possess a distinct, ADP-dependent phosphofructokinase, the gene(s) for which has been recently identified in *Pyrococcus furiosus* (Tuininga et al., 1999).

An ortholog of this protein is readily identifiable in other Pyrococci and in *M. jannaschii* (MJ1604) but not in any other archaeal or bacterial species; its sequence shows no detectable similarity to other kinases. This is a clear case of nonorthologous gene displacement; due to the limited phylogenetic distribution of this novel Pfk, the candidate protein in *M. jannaschii* could not have been detected by computational means until its ortholog had been experimentally characterized in *P. furiosus.*

3. Aldolase. The next glycolytic enzyme, fructose-1,6-bisphosphate aldolase (EC 4.1.2.13), is found in two substantially different variants, namely, metal-independent (class I) and metal-dependent (class II) aldolases in bacteria and eukaryotes. There is no ortholog of either of these in *M. jannaschii.* Instead, predicted archaeal aldolases comprise COG1830 ("DhnA-type fructose-1,6-bisphosphate aldolase and related enzymes"), which includes two *M. jannaschii* proteins, MJ0400 and MJ1585. This COG includes orthologs of the recently described class I aldolase of *E. coli*, which is only very distantly related to the typical class I enzymes (Galperin et al., 2000). As indicated above, the phylogenetic patterns for typical bacterial class II aldolase (COG0191) and this DhnA-type aldolase (COG1830) complement each other, with the exception that both types of aldolases are encoded by *E. coli* and *A. aeolicus*. This complementarity allows one to predict that the DhnA-type enzyme functions as the only fructose-1,6-bisphosphate aldolase in chlamydiae and archaea, including *M. jannaschii.*

4–6. Triosephosphate Isomerase, Glyceraldehyde-3-Phosphate Dehydrogenase, and Phosphoglycerate Kinase. These enzymes catalyze the next three steps of glycolysis and are nearly-uniformly represented in all organisms. The *M. jannaschii* candidates for these activities (MJ1528, MJ1146, and MJ0641, respectively) can be easily identified by a BLAST search. Accordingly, all major databases converge on these functional assignments.

7. Phosphoglycerate Mutase. The activity of phosphoglycerate mutase (EC 5.4.2.1) has been experimentally demonstrated in a close relative of *M. jannaschii*, but there are no obvious candidate proteins to carry out this function. As a result, KEGG does not show this enzyme as encoded in the *M. jannaschii* genome. WIT does suggest a candidate protein (MJ1612 or RMJ05975 in WIT) but annotates it as "phosphonopyruvate decarboxylase." Indeed, WIT shows only limited sequence similarity of this protein to the phosphoglycerate mutases from the mycoplasmas and *H. pylori*, whereas all close homologs of MJ1612 seem to be phosphonopyruvate decarboxylases. Finally, a search of the COG database for "phosphoglycerate mutase" retrieves three COGs, only one of which, COG1015 (phosphopentomutase/predicted phosphoglycerate mutase and related enzymes), contains archaeal members, including two *M. jannaschii* proteins, MJ1612 and MJ0010. A detailed analysis shows that both of them could possess phosphoglycerate mutase activity (Galperin et al., 1998). The definitive identification of the phosphoglycerate mutase in *M. jannaschii* awaits direct biochemical studies.

8–9. Enolase and Pyruvate Kinase. In *M. jannaschii*, these enzymes are readily identified through strong sequence similarity to the corresponding bacterial orthologs. As a result, there is general consensus on assigning these functions to MJ0232 and MJ0108, respectively. Thus, although no glycolytic enzymes has been experimentally characterized in *M. jannaschii*, computational analysis provides for the identification of all of them, with uncertainty remaining with regard to just one step.

Glycolysis in* H. pylori? *A Cautionary Note. Metabolic reconstruction based on genome analysis requires considerable caution to be exerted with regard to the plausibility of functional predictions in the general context of the biology of the organism in question. So as to not depart from glycolysis, consider the case of phosphofructokinase in the gastric ulcer-causing bacterium, *Helicobacter pylori*. This organism lacks a homolog of PfkA but does encode a close homolog of PfkB. Accordingly, WIT suggests this PfkB homolog, HP0858, as a candidate for the phosphofructokinase function in *H. pylori*. Perusal of the COG database, however, suggests a different solution, since HP0858 consists of two distinct domains, one of which indeed belongs with PfkB and other sugar kinases (COG0524), whereas the other one is predicted to be a nucleotidylyltransferase (COG0615). Given this domain architecture, it appears most likely that this protein is an ADP-heptose synthetase, an enzyme of peptidoglycan biosynthesis. A simple analysis of the biology of *H. pylori* as an acid-tolerant bacterium shows that it is likely to use gluconeogenesis, but not glycolysis, which makes phosphofructokinase unnecessary for this organism. Thus, the assignment of this activity to HP0858, which appeared to be statistically supported, is, in all likelihood, biologically irrelevant. Examination of this protein's domain architecture could be the first indication of its role in a process other than glycolysis, with biological considerations further supporting this interpretation.

AVOIDING COMMON PROBLEMS IN GENOME ANNOTATION

Due to its intrinsic complexity, genome annotation defies full automation and is inherently errorprone. Accidental error rate can be minimized only through further development of the semiautomated annotation systems and the appropriate training of annotators. There are, however, several sources of systematic error that plague genome analysis. Awareness of these could help improve the quality of genome annotation (Brenner, 1999; Galperin and Koonin, 1998).

Error Propagation and Incomplete Information in Databases

Sequence databases are prone to error propagation, whereby erroneous annotation of one protein causes multiple errors as it is used for annotation of new genomes. Furthermore, database searches have the potential for noise amplification, so that the original annotation could have involved a minor inaccuracy or incompleteness, but its transfer on the basis of sequence similarity aggravates the problem and eventually results in outright false functional assignments (Bhatia et al., 1997). These aspects of current sequence databases make the common practice of assigning gene function on the basis of the annotation of the best database hit (or even a group of hits with compatible annotations) highly error-prone. Time- and labor-consuming as this may be, adequate genome annotation requires that each gene be considered in the context of both its phylogenetic relationships and the biology of the respective organism, hence the rather disappointing performance of automated systems for genome annotation. There are numerous reasons why functional annotation may be wrong in the first place, but two main groups of problems have to do with database search methods and with the complexity and diversity of the genomes themselves.

False Positives and False Negatives in Database Searches

It is customary in genome annotation to use a cutoff for "statistically significant" database hits. It can be expressed in terms of the false-positive expectation (E) value for the BLAST searches and is set routinely at values such as $E = 0.001$ or $E = 10^{-5}$. The problem with this approach is that the distribution of similarity scores for evolutionarily and functionally relevant sequence alignments is very broad and that a considerable fraction of them fail the E-value cutoff, resulting in undetected relationships and missed opportunities for functional prediction (false negatives). Conversely, spurious hits may have E-values lower than the cutoff, resulting in false positives. The latter is most frequently caused by compositional bias (low-complexity regions) in the query sequence and in the database sequences. Clearly, there is a trade-off between *sensitivity* (false-negative rate) and *selectivity* (false-positive rate) in all database searches, and it is particularly difficult to optimize the process in genome-wide analyses. There is no single recipe to circumvent these problems. To minimize the false-positive rate, appropriate procedures for filtering low-complexity sequences are critical (Wootton and Federhen, 1996). Filtering using the SEG program is the default for Web-based BLAST searches, but additional filtering is justified for certain types of proteins. For example, filtering of predicted nonglobular domains using SEG with specifically adjusted parameters and filtering for coiled-coil domains using the COILS2 program is one way to minimize the false positive rate. Minimizing the false-negative rate (that is, maximizing sensitivity) is an open-ended problem. It should be kept in mind that a standard database search (e.g., using BLAST) with the protein sequences encoded in the given genome as queries is insufficient for an adequate annotation. To increase the sensitivity of genome analysis, it should be supplemented by other, more powerful methods such as screening the set of protein sequences from the given genome with preformed profile libraries (see above).

Genome, Protein, and Organismal Context as a Source of Errors

As discussed above, protein domain architecture, genomic context and an organism's biology may serve as sources of important, even if indirect, functional information. However, those same context features, if misinterpreted, may become one of the major sources of error and confusion in genome annotation. Standard database search programs are not equipped with the means to explicitly address the implications of the multidomain organization of proteins. Therefore, unless specialized tools such as SMART or COGs are employed and/or the search output is carefully examined, assignment of the function of a single-domain protein to a multidomain homolog and vice versa becomes frequent in genome annotation. Promiscuous, mobile domains are particularly likely to wreak havoc in the annotation process, as demonstrated, for example, by the proliferation of "IMP-dehydrogenase-related proteins" in several genomes. In reality, most or all of these proteins (depending on the genome) share with IMP dehydrogenase the mobile CBS domain but not the enzymatic part (Galperin and Koonin, 1998).

As discussed above, it is also critical for reliable genome annotation that the biological context of the given organism is taken into account. In a simplistic example, it is undesirable to annotate archaeal gene products as nucleolar proteins, even if their eukaryotic homologs are correctly described as such. As a general guide to functional annotation, it should be kept in mind that current methods for genome analysis, even the most powerful and sophisticated of them, facilitate, but do not supplant the work of a human expert.

CONCLUSIONS

With an increasing number of complete genome sequences becoming available and specialized tools for genome comparison being developed, the comparative approach is becoming the most powerful strategy for genome analysis. It seems that the future should belong to databases and tools that consistently organize the genomic data according to phylogenetic, functional, or structural principles and explicitly take advantage of the diversity of genomes to increase the resolution power and robustness of the analysis. Many tasks in genome analysis can be automated, and, given the rapidly growing amount of data, automation is critical for the progress of genomics. This being said, the ultimate success of comparative genome analysis and annotation critically depends on complex decisions based on a variety of inputs, including the unique biology of each organism. Therefore, the process of genome analysis and annotation taken as a whole is, at least at this time, not automatable, and human expertise is necessary for avoiding errors and extracting the maximum possible information from the genome sequences.

INTERNET RESOURCES FOR TOPICS PRESENTED IN CHAPTER 15

General	
NCBI	*http://www.ncbi.nlm.nih.gov/*
EBI	*http://www.ebi.ac.uk/*
DDBJ	*http://www.ddbj.nig.ac.jp/*
ExPASy	*http://www.expasy.ch/*
Genome projects	
GenBank Entrez Genomes division	*http://www.ncbi.nlm.nih.gov/PMGifs/Genomes/bact.html*
The Institute for Genome Research (TIGR) Microbial Database	*http://www.tigr.org/tdb/mdb/mdb.html*
Integrated Genomics Inc.	*http://wit.integratedgenomics.com/GOLD*
NHGRI List of Genetic and Genomic Resources	*http://www.nhgri.nih.gov/Data*
The Sanger Centre	*http://www.sanger.ac.uk*
Washington University-St. Louis	*http://genome.wustl.edu*
Ohlahoma University	*http://www.genome.ou.edu/*
Microbial Genome Database	*http://mbgd.genome.ad.jp*

GENOME ANALYSIS SYSTEMS	
MAGPIE	*http://genomes.rockefeller.edu/magpie*
GeneQuiz	*http://jura.ebi.ac.uk:8765/ext-genequiz/*
PEDANT	*http://pedant.mips.biochem.mpg.de*
Clusters of Orthologous Groups of Proteins (COGs)	*http://www.ncbi.nlm.nih.gov/COG*
Kyoto Encyclopedia of Genes and Genomes (KEGG)	*http://www.genome.ad.jp/kegg*
What Is There (WIT)	*http://wit.integratedgenomics.com/IGwit/*
DATABASES AND TOOLS FOR ANALYSIS OF PROTEIN DOMAINS	
ProDom	*http://protein.toulouse.inra.fr/prodom.html*
Pfam	*http://pfam.wustl.edu*
SMART	*http://smart.embl-heidelberg.de*
Protein modules	*http://www.bork.embl-heidelberg.de/Modules/sinput.shtml*
CDD search	*http://www.ncbi.nlm.nih.gov/Structure/cdd/wrpsb.cgi*
INDIVIDUAL MICROBIAL GENOME DATABASES	
Escherichia coli	
University of Wisconsin-Madison	*http://www.genetics.wisc.edu*
Nara Institute of Technology	*http://ecoli.aist-nara.ac.jp*
EcoCyc	*http://ecocyc.doubletwist.com*
EcoGene	*http://bmb.med.miami.edu/ecogene*
Colibri	*http://bioweb.pasteur.fr/GenoList/Colibri*
RegulonDB	*http://www.cifn.unam.mx/Computational_Biology/regulondb*
E. coli Genetic Stock Center (CGSC)	*http://cgsc.biology.yale.edu*
Mycoplasma genitalium	
Mycoplasma genome	*http://www.zmbh.uni-heidelberg.de/M_pneumoniae/*
Essential genes	*http://www.sciencemag.org/feature/data/1042937.shl*
Bacillus subtilis	
Subtilist	*http://bioweb.pasteur.fr/GenoList/SubtiList*
Sporulation genes	*http://www1.rhbnc.ac.uk/biological-sciences/cutting*
Yeast	
Saccharomyces Genome Database (SGD)	*http://genome-www.stanford.edu/Saccharomyces*
MIPS Yeast Database	*http://www.mips.biochem.mpg.de/proj/yeast*
Yeast Protein Database (YPD)	*http://www.proteome.com/databases*
Promoter Database	*http://cgsigma.cshl.org/jian*
TRIPLES database	*http://ygac.med.yale.edu*
Cell Cycle Expression Database	*http://genomics.stanford.edu*

Metabolic reconstruction

Biochemical pathways map	*http://www.expasy.ch/cgi-bin/search-biochem-index*
ENZYME	*http://www.expasy.ch/enzyme/*

PROBLEMS FOR ADDITIONAL STUDY

The reader who decides to address these problems is expected to use the tools described in this chapter including general-purpose ones such as different versions of BLAST. In most cases, it is advisable to apply more than one method when trying to solve a problem. The reader should keep in mind that some of the problems may not have a single "correct" solution but rather one or perhaps even two or three most likely solutions.

1. Three archaeal species, *M. jannaschii, M. thermoautotrophicum*, and *A. fulgidus*, typically are found in COGs together. The fourth species, *P. horikoshii*, is frequently missing from these COGs (see the table of co-occurrence of genomes in COGs, which is available on the COG Web site). How do you explain this? What kind of genes are absent from *P. horikoshii*?

2. Which bacterial species share the greatest number of genes with Archaea, if measured as the ratio of the shared genes (COGs) to the total number of genes in the bacterial genome? Use the table of co-occurrence of genomes in COGs to identify the trend. Once it is clear, discuss different explanations for the observations.

3. What is the function of the *E. coli* HemK protein and its orthologs? Explain the basis for your conclusion and possible alternatives.

4. How many IMP dehydrogenases are there in *A. fulgidus*? In *A. aeolicus*? What are the domain organizations and functions of "IMP dehydrogenase-related" proteins?

5. Methanobacterial protein MTH1425 is annotated in GenBank as *O*-sialoglycoprotein endopeptidase. Describe the domain organization of this protein. Should you detect an unexpected domain fusion, could it be due to a sequencing error? What are the possible functional implications?

6. What are the functions of the following proteins:

   a) MJ1612
   b) MJ1001
   c) *E. coli* NagD

7. Describe the domain architectures and functions and of the following proteins:

   a) *E. coli* YfiQ
   b) *B. subtilis* YtvA
   c) slr1759

8. Suggest a comparative-genomic approach to search for new targets for anti-ulcer drugs; which of the identified proteins could be also potential targets for anti-tuberculosis drugs?

9. The Glu-tRNA^{Gln} amidotransferase consists of A, B, and C subunits. Compare the phylogenetic patterns for these. How can you explain the differences? Test some of the explanations. What is unusual about the Glu-tRNA^{Gln} amidotransferase complex of the mycoplasmas?

10. Families of orthologous proteins involved in translation frequently contain one representative of each of the bacterial and archaeal genomes, but two members from yeast. How would you explain this pattern? Use at least two lines of evidence in support of your explanation.

REFERENCES

Andrade, M. A., Brown, N. P., Leroy, C., et al. (1999). Automated genome sequence analysis and annotation. *Bioinformatics*. 15, 391–412.

Bateman, A., Birney, E., Durbin, R., Eddy, S. R., Howe, K. L., and Sonnhammer, E. L. (2000). The pfam protein families database. *Nucleic Acids Res.* 28, 263–266.

Bhatia, U., Robison, K., and Gilbert, W. (1997). Dealing with database explosion: a cautionary note. *Science*. 276, 1724–1725.

Blattner, F. R., Plunkett, G., 3rd, Bloch, C. A., et al. (1997). The complete genome sequence of *Escherichia coli* K-12. *Science*. 277, 1453–1474.

Bork, P., and Koonin, E. V. (1994). A P-loop-like motif in a widespread ATP pyrophosphatase domain: implications for the evolution of sequence motifs and enzyme activity. *Proteins*. 20, 347–355.

Brenner, S. E. (1999). Errors in genome annotation. *Trends Genet.* 15, 132–133.

Bult, C. J., White, O., Olsen, G. J., et al. (1996). Complete genome sequence of the methanogenic archaeon, *Methanococcus jannaschii. Science*. 273, 1058–1073.

Corpet, F., Servant, F., Gouzy, J., and Kahn, D. (2000). ProDom and ProDom-CG: tools for protein domain analysis and whole genome comparisons. *Nucleic Acids Res.* 28, 267–269.

Enright, A. J., Illopoulos, I., Kyrpides, N. C., and Ouzounis, C. A. (1999). Protein interaction maps for complete genomes based on gene fusion events. *Nature*. 402, 86–90.

Fitch, W. M. (1970). Distinguishing homologous from analogous proteins. *Syst. Zool.* 19, 99–113.

Fleischmann, R. D., Adams, M. D., White, O., et al. (1995). Whole-genome random sequencing and assembly of *Haemophilus influenzae* Rd. *Science*. 269, 496–512.

Fraser, C. M., Gocayne, J. D., White, O., et al. (1995). The minimal gene complement of *Mycoplasma genitalium. Science*. 270, 397–403.

Frishman, D., and Mewes, H.-W. (1997). PEDANTic genome analysis. *Trends Genet.* 13, 415–416.

Gaasterland, T., and Ragan, M. A. (1998). Microbial genescapes: phyletic and functional patterns of ORF distribution among prokaryotes. *Microb Comp Genomics*. 3, 199–217.

Galperin, M. Y., Aravind, L., and Koonin, E. V. (2000). Aldolases of the DhnA family: a possible solution to the problem of pentose and hexose biosynthesis in archaea. *FEMS Microbiol. Lett.* 183, 259–264.

Galperin, M. Y., Bairoch, A., and Koonin, E. V. (1998). A superfamily of metalloenzymes unifies phosphopentomutase and cofactor-independent phosphoglycerate mutase with alkaline phosphatases and sulfatases. *Protein Sci.* 7, 1829–1835.

Galperin, M. Y., and Koonin, E. V. (1998). Sources of systematic error in functional annotation of genomes: domain rearrangement, non-orthologous gene displacement and operon disruption. *InSilico Biol.* 1, 55–67.

Goffeau, A., Barrell, B. G., Bussey, H., et al. (1996). Life with 6000 genes. *Science.* 274, 546–567.

Henikoff, S., Greene, E. A., Pietrokovski, S., Bork, P., Attwood, T. K., and Hood, L. (1997). Gene families: the taxonomy of protein paralogs and chimeras. *Science.* 278, 609–614.

Hoersch, I., Leroy, I., Brown, N. P., Andrade, M. A., and Sander, I. (2000). The GeneQuiz web server: protein functional analysis through the Web. *Trends Biochem Sci.* 25, 33–35.

Huynen, M. A., Diaz-Lazcoz, Y., and Bork, P. (1997). Differential genome display. *Trends Genet.* 13, 389–390.

Huynen, M. A., and Snel, B. (2000). Gene and context: integrative approaches to genome analysis. *Adv. Protein Chem.* 54, 345–379.

Kanehisa, M., and Goto, S. (2000). KEGG: kyoto encyclopedia of genes and genomes. *Nucleic Acids Res.* 28, 27–30.

Koonin, E. V. (1997). Genome sequences: genome sequence of a model prokaryote. *Curr. Biol.* 7, R656–R659.

Koonin, E. V., Mushegian, A. R., Galperin, M. Y., and Walker, D. R. (1997). Comparison of archaeal and bacterial genomes: computer analysis of protein sequences predicts novel functions and suggests a chimeric origin for the archaea. *Mol. Microbiol.* 25, 619–637.

Kunst, F., Ogasawara, N., Moszer, I., et al. (1997). The complete genome sequence of the gram-positive bacterium *Bacillus subtilis. Nature.* 390, 249–256.

Marcotte, E. M., Pellegrini, M., Ng, H. L., Rice, D. W., Yeates, T. O., and Eisenberg, D. (1999). Detecting protein function and protein-protein interactions from genome sequences. *Science.* 285, 751–753.

Medigue, C., Rechenmann, F., Danchin, A., and Viari, A. (1999). Imagene: an integrated computer environment for sequence annotation and analysis. *Bioinformatics.* 15, 2–15.

Natale, D. A., Galperin, M. Y., Tatusov, R. L., and Koonin, E. V. (2000). Using the COG database to improve gene recognition in complete genomes. *Genetica.* 108, 9–17.

Overbeek, R., Fonstein, M., D'Souza, M., Pusch, G. D., and Maltsev, N. (1999). The use of gene clusters to infer functional coupling. *Proc. Natl. Acad. Sci. U. S.A.* 96, 2896–2901.

Overbeek, R., Larsen, N., Pusch, G. D., D'Souza, M., Jr, E. S., Kyrpides, N., Fonstein, M., Maltsev, N., and Selkov, E. (2000). WIT: integrated system for high-throughput genome sequence analysis and metabolic reconstruction. *Nucleic Acids Res.* 28, 123–125.

Pellegrini, M., Marcotte, E. M., Thompson, M. J., Eisenberg, D., and Yeates, T. O. (1999). Assigning protein functions by comparative genome analysis: protein phylogenetic profiles. *Proc. Natl. Acad. Sci. U.S.A.* 96, 4285–4288.

Schultz, J., Copley, R. R., Doerks, T., Ponting, C. P. and Bork, P. (2000). SMART: a web-based tool for the study of genetically mobile domains. *Nucleic Acids Res.* 28, 231–234.

Schultz, J., Milpetz, F., Bork, P. and Ponting, C. P. (1998). SMART, a simple modular architecture research tool: identification of signaling domains. *Proc. Natl. Acad. Sci. U.S.A.* 95, 5857–5864.

Snel, B., Bork, P., and Huynen, M. (2000). Genome evolution. Gene fusion versus gene fission. *Trends Genet.* 16, 9–11.

Tatusov, R. L., Galperin, M. Y., Natale, D. A., and Koonin, E. V. (2000). The COG database: a tool for genome-scale analysis of protein functions and evolution. *Nucleic Acids Res.* 28, 33–36.

Tatusov, R. L., Koonin, E. V., and Lipman, D. J. (1997). A genomic perspective on protein families. *Science*. 278, 631–637.

Taylor, B. L. and Zhulin, I. B. (1999). PAS domains: internal sensors of oxygen, redox potential, and light. *Microbiol. Mol. Biol. Rev.* 63, 479–506.

Tuininga, J. E., Verhees, C. H., van der Oost, J., Kengen, S. W., Stams, A. J., and de Vos, W. M. (1999). Molecular and biochemical characterization of the ADP-dependent phosphofructokinase from the hyperthermophilic archaeon Pyrococcus furiosus. *J. Biol. Chem.* 274, 21023–21028.

Walker, D. R., and Koonin, E. V. (1997). SEALS: a system for easy analysis of lots of sequences. *ISMB*. 5, 333–339.

Wootton, J. C., and Federhen, S. (1996). Analysis of compositionally biased regions in sequence databases. *Methods Enzymol.* 266, 554–71.

16

LARGE-SCALE GENOME ANALYSIS

Paul S. Meltzer

Cancer Genetics Branch
National Human Genome Research Institute
National Institutes of Health
Bethesda, Maryland

INTRODUCTION

The availability of complete or near-complete catalogs of genes for organisms of increasing complexity has created opportunities for studying numerous aspects of gene function at the genomic level. Gene expression, defined by steady-state levels of cellular mRNA, has emerged as the first aspect of gene function amenable to genome-scale measurement with readily-available technology. It is now possible to carry out massively parallel analysis of gene expression on tens of thousands of genes from a given sample. For model organisms such as *S. cerevisiae*, total genome expression analysis is now routine (Lashkari et al., 1997; Wodicka et al., 1997). For higher eukaryotes, expression measurements that cover a significant proportion of the genome are currently possible, and complete genome expression analysis appears to be an achievable goal. Moreover, this analysis can be repeated on multiple samples, allowing for the statistical analysis of the behavior of genes across a large number of conditions. The massive quantities of data generated by this type of analysis have created new bioinformatics challenges, but these obstacles are well worth overcoming as this type of information has already begun to provide new insights into genome function. So far, this promise has been best realized in *S. cerevisiae*, in which whole-genome measurements have been used to examine fundamental processes such as the cell cycle and the roles of specific transcription factors (DeRisi et al., 1997; Cho et al., 1998; Spellman et al., 1998; Holstege et al., 1998; Chu et al., 1998; Roberts et al., 2000). Although significantly more difficult, similar prob-

Bioinformatics: A Practical Guide to the Analysis of Genes and Proteins
Edited by A. D. Baxevanis and B. F. F. Ouellette
ISBN 0-471-38390-2 (cloth), ISBN 0-471-383910 (paper)

lems are now being productively approached in mammalian systems (DeRisi et al., 1996; Amundson et al., 1999; Khan et al., 1999; Iyer et al., 1999; Feng et al., 2000). In addition to their impact on fundamental questions in biology, these technologies are being used to probe the complexity of human diseases, particularly cancer (Khan et al., 1998; Golub et al., 1999; Alon et al., 1999; Alizadeh et al., 2000; Ross et al., 2000; Scherf et al., 2000; Bittner et al., 2000).

Access to large numbers of measurements allows for the statistical analysis of gene expression across multiple samples. In the broadest sense, this opens the possibility of identifying patterns of coregulation among genes, which, in turn, reflects underlying regulatory mechanisms and functional interrelationships. Because mammalian genomes are mainly populated by genes of unknown function, it is theoretically possible to assign anonymous genes to pathways or at least to generate hypotheses as to function by identifying the circumstances that alter their expression. Computational techniques for processing gene expression data are rapidly evolving as many investigators attack the problem from diverse perspectives. The following discussion will present an overview of the technologies for generating and processing expression data, with an emphasis on those techniques and databases that are publicly available.

TECHNOLOGIES FOR LARGE-SCALE GENE EXPRESSION

Measurements

Two broad categories of technology have emerged that can provide large-scale expression data. The first is sequence-based, as exemplified by the serial analysis of gene expression (SAGE) (Velculescu et al., 1995). An alternative hybridization approach is generically termed microarray hybridization (Schena et al., 1995; Lockhart et al., 1996). These two technologies have distinct advantages and disadvantages. In SAGE, a short unique sequence tag is generated from each gene by a PCR-based strategy (Fig. 16.1). Concatemerized tags are sequenced, and the abundance of these tags provides a measurement of the level of gene expression in the starting material. As illustrated below, SAGE tags can be linked to a specific transcript designation in an appropriate database of unique transcripts, such as UniGene. Thus, SAGE is essentially an accelerated technique for cDNA library sequencing. Because it is sequence intensive, SAGE is not well suited to the analysis of large numbers of samples. However, because SAGE does not require a priori knowledge of the pattern of gene expression in a given mRNA source, the same biochemical and bioinformatics procedure can be applied to any sample, given the availability of the appropriate reference database of SAGE tags.

By contrast, in the microarray strategy, labeled cDNAs (the "target") are hybridized to an array of DNA elements (the "probes") affixed to a solid support (Fig. 16.2). The array elements can either be synthetic oligonucleotides or larger DNA fragments. High densities are achievable and enable the measurement of over 10,000 genes with either type of microarray. In contrast to SAGE, microarray analysis can only measure the expression of genes that correspond to the sequences included in the array fabrication process. Therefore, complete genome expression analysis by microarrays requires the construction of complete microarrays. This difficulty is counterbalanced by the ease of individual experiments that enable the analysis of

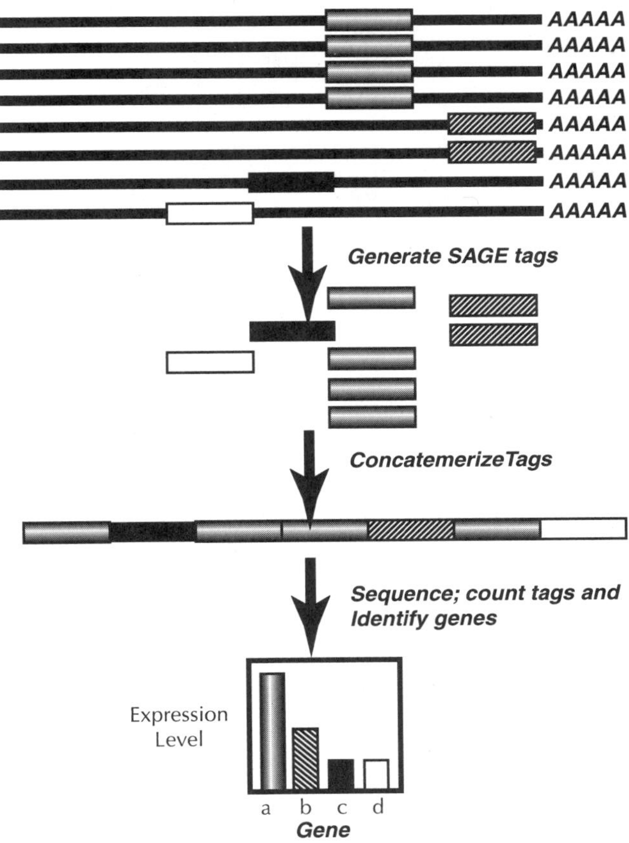

Figure 16.1. Serial Analysis of Gene Expression (SAGE) depends on the generation of a tag from the 3′ end of an mRNA. Tags are concatemerized and sequenced. These data are compared with a database of tags linked to individual transcripts to generate the frequency of each tag in the library, a measure of the expression level for that gene. (See color plate.)

numerous samples. Microarray expression databases are being developed that contain information derived from hundreds or even thousands of samples.

Informatics Aspects of Microarray Production

For organisms with completely sequenced and well-annotated genomes such as *S. cerevisiae*, the production of arrays is relatively straightforward. By selecting a primer pair that amplifies each ORF from genomic DNA, a set of PCR products encompassing the complete genome can be produced. Primer pairs for this purpose are commercially available. Likewise, a series of oligonucleotide array elements can be selected from each ORF. More complex genomes, which may not yet be fully annotated, are more problematic. The consensus at this juncture is to utilize a catalog of genes such as UniGene, generated by reduction of EST sequencing data to unique, clustered transcripts. Clustered ESTs can then be used to predict oligonucleotide sequences for array fabrication or to select a representative EST clone for each

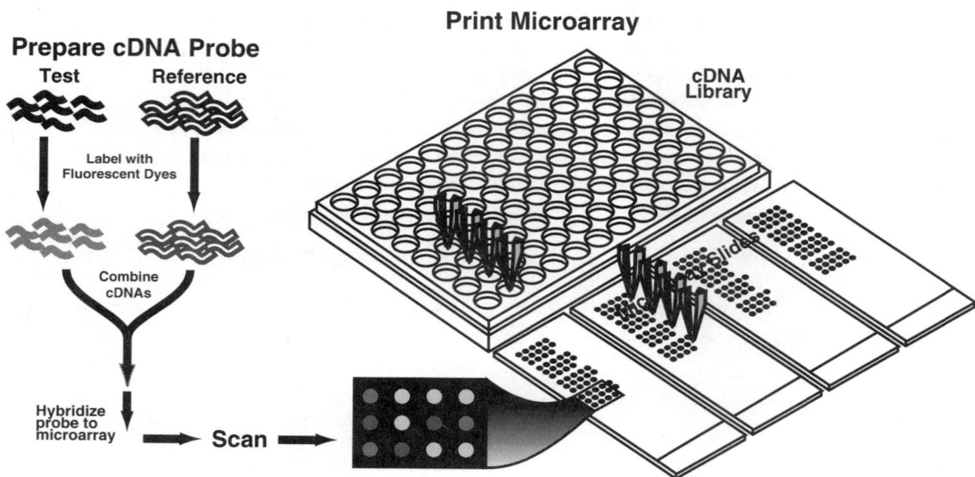

Figure 16.2. The process of microarray hybridization using printed DNA probes. A robotic printer deposits DNA in a regular array on a series of glass slides. After they are processed, the slides are hybridized to a mixture of two cDNA pools derived from test and reference samples that have been labeled with spectrally distinct fluorochromes. After stringency washes, the microarray is scanned in a laser-scanning device, and the image is processed to generate numerical data. (See color plate.)

cluster; in turn, this is then used to generate a cDNA fragment by PCR for deposition on a microarray. This approach is intrinsically limited by the quality and completeness of the EST database, as well as the intrinsic instability of EST cluster designations that evolve along with the EST sequencing projects. A further practical difficulty unique to the cDNA microarray strategy is brought about by the requirement of this system for a physical clone. Recovery of low-redundancy clones may prove difficult due to problems intrinsic to the handling of arrayed libraries. Ideally, EST-based approaches will be supplemented (or even supplanted) by gene catalogs based on genomic sequence that will ultimately provide stable databases for array construction.

In order for the information generated by expression analysis to be useful, it is necessary for each EST to be linked to as much biological information as possible. The accessibility and quality of this information depend on the organism. Because the interpretation of results often relies on the expertise of an individual investigator, issues surrounding gene nomenclature and annotation are significant. For example, an investigator may be familiar with a gene by its common literature alias, but it is cataloged under a more cryptic "official" name. ESTs from known genes may fail to be annotated as genes altogether if the available sequence is of insufficient quality to pass the filters required for cluster assembly with known mRNAs. Fortunately, mammalian gene catalogs such as NCBI's LocusLink are being developed that attempt to address these problems.

What is Actually Measured?

A detailed review of array technologies is beyond the scope of this chapter, but the user of expression data must have an understanding of the types of measurement reported by various expression platforms and incorporated in databases. To understand the options available to mine information in expression data, it is essential to understand the measurements generated by the various technologies presently available. Ideally, an expression measurement would be converted into copies of a given mRNA per cell. Unfortunately, no technology measures this value directly. The variety of expression measurement platforms in use creates problems in the cross-comparison of data generated by technologies and in some cases even within a given technology. For example, SAGE measures the abundance of a particular tag. The validity of this number will depend on the number of tags sequenced. A sample of 1,000 tags will be much less reliable than a sample of 20,000 tags. Hybridization-based systems measure the abundance of cDNAs in a synthetic population of nucleic acids, which is a representation of the mRNA composition of the cell. Variations in biochemical manipulations, hybridization efficiency across array elements, and cross-hybridization may distort the accuracy of measurements. This calls for caution when comparing data generated by different techniques. For example, most synthetic oligonucleotide array assays are based on hybridization of a representation synthesized by amplification from a cDNA template that has been engineered to contain a phage promoter. In contrast, most fluorescent cDNA microarray assays use direct incorporation of tagged nucleotides in the cDNA prepared by reverse transcription of mRNA from the source of interest.

There are few data that directly compare the results of these three assays on the same samples. Moreover, abundance measurements from oligonucleotide array assays are reported as expression levels on a continuous scale. In contrast, printed fluorescent cDNA arrays require the use of a two-color system incorporating a reference mRNA to compensate for variations in the performance of individual array elements and array slides. Although the intensities of each channel are measured and reported, the most robust measurement is expressed as the normalized ratio of intensities for each gene. Radioactive hybridization to cDNA microarrays printed on nylon membranes presents additional problems because only a single channel is measured in a given hybridization, and normalization must rely on cross-comparison of experiments. In addition, expression scales vary in an individual way from linearity and have distinct thresholds and saturation levels depending on the technology and instrumentation utilized. In principle, measurements from these various systems could be converted to a common format, but as yet there are no standards for such a conversion. To be certain that computational techniques are appropriately applied, users of array expression databases and experimentalists venturing into microarray research need to be aware of the characteristics of the data generated by the particular experimental platform in use.

Aspects of the primary analysis of array data are illustrated for printed cDNA microarrays in the following example. This methodology achieves accurate ratio measurements of the relative abundance of each mRNA in the test and reference sources by combining the test cDNA pool with a reference cDNA labeled with a spectrally distinct fluorophore (Fig. 16.2). This is accomplished by obtaining images of the hybridized microarray with a two-channel laser-scanning device for analysis with software (such as DeArray, ScanAlyze, CrazyQuant, or proprietary software bundled with scanner instruments) that measures the signal intensity over background, normalizes the two channels, calculates ratios, generates overall array statistics, and outputs a pseudocolor image and data spreadsheet. This process is illustrated here using DeArray, a module of Array Suite. Examining the scatter plot of one channel versus the other (Fig. 16.3) is extremely useful. In most hybridizations, the majority of genes will show minimal differences, so the scatter plot would then be centered on the diagonal line. Deviation from this behavior, as illustrated in the example at lower intensities, indicates the signal intensity level at which the data should be filtered. In other cases, the scatter plot might simply tend to widen as intensities drop, again indicating that confidence in ratio measurements declines with intensity. A similar scatter plot constructed from single-channel data from two one-

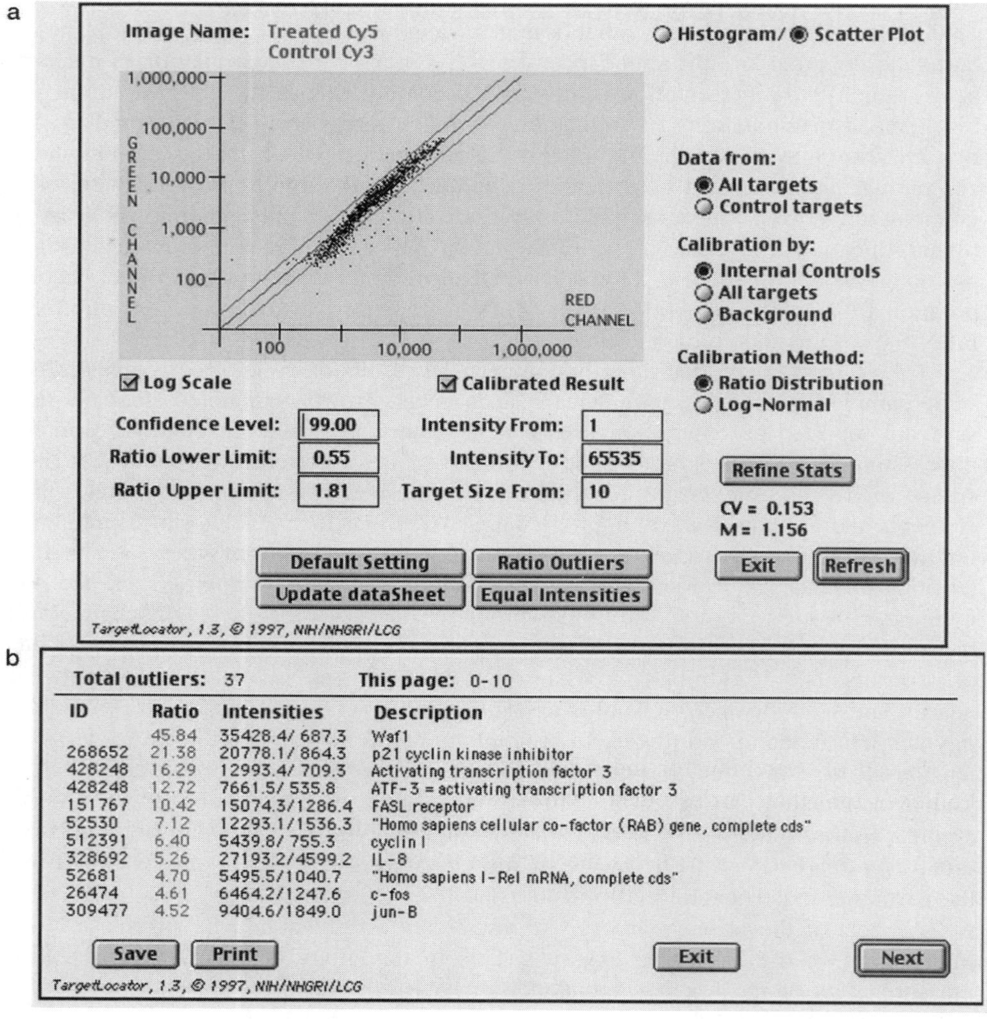

Figure 16.3. (a) Scatter plot illustrating the fluorescent intensity measurements in two channels plotted against each other. Note that most measurements fall on the diagonal. Deviation from this pattern at low intensity indicates that the data are skewed due to nonspecific fluorescence in one channel. For analysis, the data should be filtered to exclude these points. (b) After data are filtered for intensity and a minimum spot size (to exclude spurious measurements), an outlier list can be displayed. A spreadsheet containing the data from the entire microarray can also be saved.

color or radioactive hybridizations is a useful adjunct to the analysis of data generated from these platforms. The analysis of a single microarray experiment is straightforward but not usually very illuminating. The real power of high-throughput gene expression technologies emerges when multiple experiments are subjected to comparison and statistical analysis. To accomplish this goal, expression data must be stored in an appropriate database.

COMPUTATIONAL TOOLS FOR EXPRESSION ANALYSIS

Public Databases

At the present time, the only centralized, publicly maintained repositories of high-throughput gene expression data, aside from EST sequencing databases, contain SAGE data. A single, central repository for all expression data comparable to GenBank is presently not available. This situation may change if standards for the uniform reporting of expression data are adopted. Many useful databases containing the results of various studies are maintained by individual laboratories listed at the end of this chapter.

NCBI maintains a SAGE database called SAGEmap, currently containing data from 69 libraries (Lal et al., 1999). These data can be searched either for data on individual genes or for lists of genes differentially expressed between libraries or pools of libraries. Given a cDNA sequence (which must include the 3′ end), the user can search for its representation in a number of SAGE libraries. SAGE library data are downloadable, and a submission tool for SAGE laboratories is also available. To search the SAGEmap database for data on a given gene using the "virtual Northern tool," the target cDNA sequence is submitted and possible tags are returned (Fig. 16.4). As in the example, one gene may contain multiple tags. Linked to this tag is a display indicating the relative abundance of this tag in a SAGE library with a proportional gray scale image providing a "virtual Northern" (Fig. 16.5). In addition, the data are linked to the UniGene cluster(s) corresponding to this tag. The UniGene designation provides a link to an abundance table of all tags for this particular UniGene entry (Fig. 16.6).

Instead of searching for individual genes, SAGEmap can be queried using x-Profiler, a tool that carries out a "virtual subtraction" to develop lists of tags that are present in one library or group of libraries at a differing frequency from a reference set. After the user selects the libraries to compare (Fig. 16.7a) and the ratio cutoff between the libraries, xProfiler returns a list (Fig. 16.7b) of differentially expressed genes in the two sources that can be downloaded for further analysis. It should be noted that SAGE data are subject to error, primarily related to sequencing error in the SAGE libraries and in the sequence databases that are used to define gene clusters. The impact of this error is inversely proportional to the number of times a given tag is counted. Tags that occur only once in the library, therefore, represent the least reliable data. xProfiler can be applied in essentially the same fashion to provide a virtual subtraction of EST sequence data compiled as part of the Cancer Genome Anatomy Project (CGAP). Another tool available for screening CGAP and dbEST library data, digital differential display (DDD), provides a gray-scale output indicating the relative abundance of statistically significant differentially expressed genes.

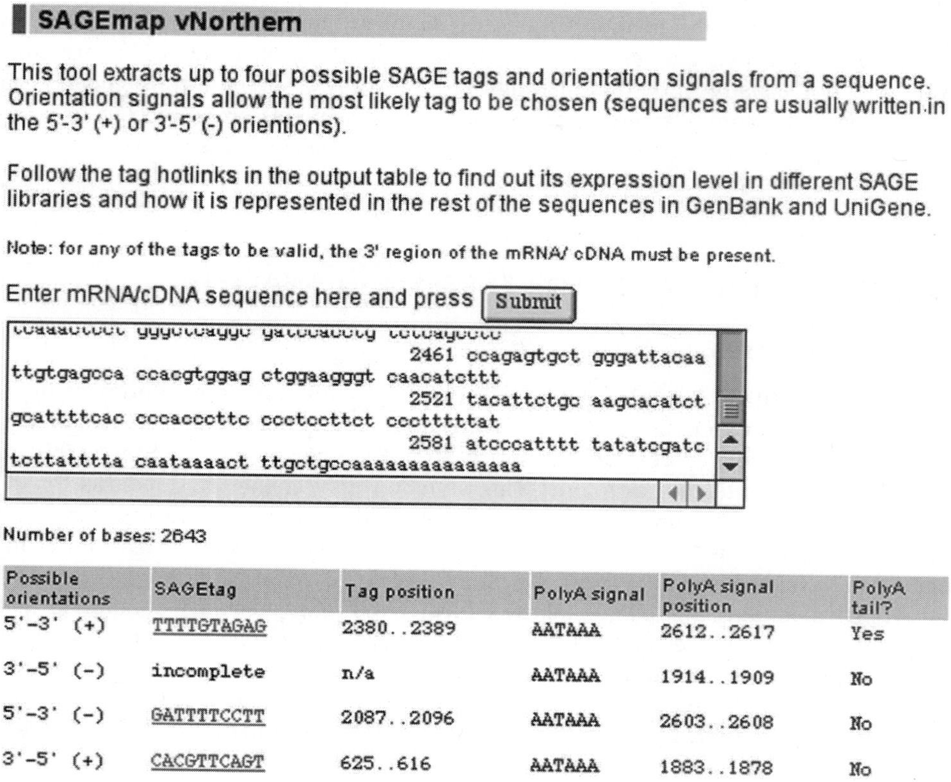

SAGEmap vNorthern

This tool extracts up to four possible SAGE tags and orientation signals from a sequence. Orientation signals allow the most likely tag to be chosen (sequences are usually written in the 5'-3' (+) or 3'-5' (-) orientions).

Follow the tag hotlinks in the output table to find out its expression level in different SAGE libraries and how it is represented in the rest of the sequences in GenBank and UniGene.

Note: for any of the tags to be valid, the 3' region of the mRNA/ cDNA must be present.

Enter mRNA/cDNA sequence here and press Submit

```
                                       2461 ccagagtgct gggattacaa
ttgtgagcca ccacgtggag ctggaagggt caacatcttt
                                       2521 tacattctgc aagcacatct
gcattttcac cccacccttc ccctccttct ccctttttat
                                       2581 atcccatttt tatatcgatc
tcttatttta caataaaact ttgctgccaaaaaaaaaaaaaa
```

Number of bases: 2643

Possible orientations	SAGEtag	Tag position	PolyA signal	PolyA signal position	PolyA tail?
5'-3' (+)	TTTTGTAGAG	2380..2389	AATAAA	2612..2617	Yes
3'-5' (-)	incomplete	n/a	AATAAA	1914..1909	No
5'-3' (-)	GATTTTCCTT	2087..2096	AATAAA	2603..2608	No
3'-5' (+)	CACGTTCAGT	625..616	AATAAA	1883..1878	No

Figure 16.4. Searching SAGEmap for tags representing a given gene ("virtual Northern"). The example illustrates the cDNA sequence of human p53 (*TP53*). Potential tags and their positions in the p53 mRNA are returned.

Most public microarray data are accessible only through individual laboratory Web sites, many of which are listed at the end of this chapter. It can be anticipated that unified public expression databases will be developed once issues as to data format and cross-platform comparison are resolved. One of the preliminary efforts currently under development and aimed at the public use and dissemination of gene expression data is the NCBI Gene Expression Omnibus (GEO). This database is intended to house different types of expression data, including oligonucleotide and cDNA microarray data, hybridization filter data, and SAGE data. Although this platform will undoubtedly evolve as more and more gene expression information becomes available, GEO is currently envisioned as having four primary entities:

1. Submitter, which contains contact and login information on the submitter,
2. Platform, which contains information on the physical reagents used in the actual experiment,
3. Sample, which deals with the mRNA samples in question and the data generated from the actual experiment, and

SAGE Tag to Gene Mapping

SAGEtag (10 bases): TTTTGTAGAG

Reliable UniGene clusters matched to this tag:

Hs.1846 : **tumor protein p53 (Li-Fraumeni syndrome)**

SAGE library data for this tag:

Library name	*Tags per million*		*Tag counts*	*Total tags*
SAGE Duke BB542 normal cerebellum	24		1	41450
SAGE Chen LNCaP	31		2	62681
SAGE Chen LNCaP no-DHT	15		1	65206
SAGE Chen Normal Pr	29		2	66687
SAGE Duke GBM H1110	42		3	71138
SAGE SW837	32		2	61290
SAGE RKO	19		1	52423
SAGE CPDR LNCaP-C	95		4	41848
SAGE CPDR LNCaP-T	45		2	44370
SAGE pooled GBM	15		1	63078
SAGE BB542 whitematter	20		2	96344
SAGE Panc 91-16113	58		2	34159
SAGE OVCA432-2	345		1	2894
SAGE OV1063-3	50		2	39466
SAGE Duke H341	44		2	44983
SAGE OVP-5	359		4	11118
SAGE OVT-6	69		3	43074
SAGE OVT-7	18		1	55476
SAGE SKBR3	119		1	8379
SAGE Duke HMVEC	37		2	53089
SAGE Duke HMVEC+VEGF	17		1	58499
SAGE DCIS	24		1	41540
SAGE Duke 757	50		1	19674
SAGE OVT-8	29		1	34096
SAGE A2780-9	88		2	22551
SAGE ML10-10	122		7	57326
SAGE Duke H247 normal	16		1	61096
SAGE A+	47		1	21231
SAGE IOSE29-11	122		6	48876

Number of SAGE libraries: 69
Total tags in all SAGE libraries: 2877861

Figure 16.5. Following the link from a given tag in the virtual Northern results (Fig. 16.4) provides a display of the abundance of the tag in the database with a gray scale "virtual Northern" to facilitate scanning the long list.

SAGE Gene to Tag Mapping

UniGene cluster: Hs. 1846 Submit

Hs.1846 : **tumor protein p53 (Li-Fraumeni syndrome)**

SAGE library data and reliable tag summary:

Reliable tags found in SAGE libraries:

GGGTCTAGAA	Library name	Tags per million		Tag counts	Total tags
	SAGE Duke H1020	18		1	53554
	SAGE 293-CTRL	44		2	44667
	SAGE Chen LNCaP	47		3	62681
	SAGE Chen Normal Pr	44		3	66687
	SAGE Chen Tumor Pr	14		1	69202
	SAGE Duke H392	34		2	58099
	SAGE Duke GBM H1110	14		1	71138
	SAGE NHA(5th)	18		1	53219
	SAGE Tu102	17		1	58190
	SAGE Tu98	20		1	49527
	SAGE Duke H341	22		1	44983
	SAGE ES2-1	31		1	31763
	SAGE LNCaP	43		1	22935
	SAGE Duke HMVEC+VEGF	17		1	58499
	SAGE A2780-9	44		1	22551
	SAGE ML10-10	17		1	57326
	SAGE Br N	26		1	38274
	SAGE IOSE29-11	40		2	48876
TGCATTTTCA	**Library name**	**Tags per million**		**Tag counts**	**Total tags**
	SAGE 293-CTRL	22		1	44667
	SAGE Duke mhh-1	20		1	48959
	SAGE SciencePark MCF7 control 3h	168		1	5924
TTCAAGACAG	**Library name**	**Tags per million**		**Tag counts**	**Total tags**
	SAGE IOSE29-11	20		1	48876
TTTTGTAGAG	**Library name**	**Tags per million**		**Tag counts**	**Total tags**
	SAGE Duke BB542 normal cerebellum	24		1	41450
	SAGE Chen LNCaP	31		2	62681
	SAGE Chen LNCaP no-DHT	15		1	65206

Figure 16.6. The complete list of tags from the virtual Northern results (Figs. 16.4 and 16.5) can also be used to query the database, finding the occurrence of their tags in all SAGE libraries.

4. Series, which houses information on collections of samples and the relationship between the samples. The series entity will also contain the results of any data clustering.

Even when public databases for microarray expression data are established, investigators who are setting up microarray laboratories soon learn that data storage is an essential requirement for their research. Sources for freely available software for expression databases are listed at the end of this chapter. One example of such a

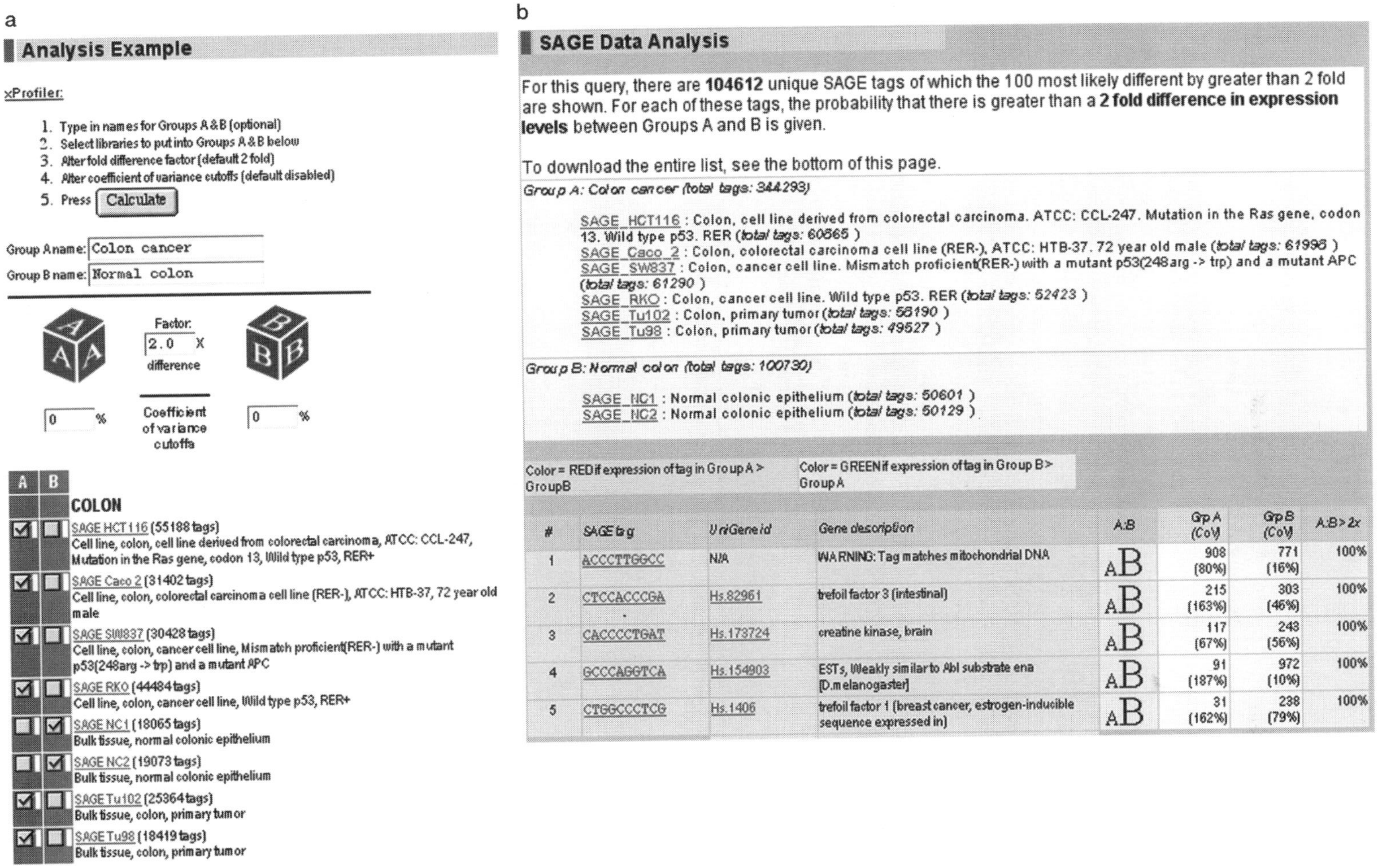

a

Analysis Example

xProfiler:

1. Type in names for Groups A & B (optional)
2. Select libraries to put into Groups A & B below
3. Alter fold difference factor (default 2 fold)
4. Alter coefficient of variance cutoffs (default disabled)
5. Press [Calculate]

Group A name: Colon cancer
Group B name: Normal colon

Factor: 2.0 X difference

0 % Coefficient of variance cutoffs 0 %

A	B	COLON
☑	☐	SAGE HCT116 (55188 tags) Cell line, colon, cell line derived from colorectal carcinoma, ATCC: CCL-247, Mutation in the Ras gene, codon 13, Wild type p53, RER+
☑	☐	SAGE Caco 2 (31402 tags) Cell line, colon, colorectal carcinoma cell line (RER-), ATCC: HTB-37, 72 year old male
☑	☐	SAGE SW837 (30428 tags) Cell line, colon, cancer cell line, Mismatch proficient(RER-) with a mutant p53(248arg -> trp) and a mutant APC
☑	☐	SAGE RKO (44484 tags) Cell line, colon, cancer cell line, Wild type p53, RER+
☐	☑	SAGE NC1 (18065 tags) Bulk tissue, normal colonic epithelium
☐	☑	SAGE NC2 (19073 tags) Bulk tissue, normal colonic epithelium
☑	☐	SAGE Tu102 (25364 tags) Bulk tissue, colon, primary tumor
☑	☐	SAGE Tu98 (18419 tags) Bulk tissue, colon, primary tumor

b

SAGE Data Analysis

For this query, there are **104612** unique SAGE tags of which the 100 most likely different by greater than 2 fold are shown. For each of these tags, the probability that there is greater than a **2 fold difference in expression levels** between Groups A and B is given.

To download the entire list, see the bottom of this page.

Group A: Colon cancer (total tags: 344293)

SAGE_HCT116 : Colon, cell line derived from colorectal carcinoma. ATCC: CCL-247. Mutation in the Ras gene, codon 13. Wild type p53. RER (*total tags: 60865*)
SAGE_Caco_2 : Colon, colorectal carcinoma cell line (RER-), ATCC: HTB-37. 72 year old male (*total tags: 61998*)
SAGE_SW837 : Colon, cancer cell line. Mismatch proficient(RER-) with a mutant p53(248arg -> trp) and a mutant APC (*total tags: 61290*)
SAGE_RKO : Colon, cancer cell line. Wild type p53. RER (*total tags: 52423*)
SAGE_Tu102 : Colon, primary tumor (*total tags: 58190*)
SAGE_Tu98 : Colon, primary tumor (*total tags: 49527*)

Group B: Normal colon (total tags: 100730)

SAGE_NC1 : Normal colonic epithelium (*total tags: 50601*)
SAGE_NC2 : Normal colonic epithelium (*total tags: 50129*)

Color = RED if expression of tag in Group A > Group B | Color = GREEN if expression of tag in Group B > Group A

#	SAGE tag	UniGene id	Gene description	A:B	Grp A (CoV)	Grp B (CoV)	A:B > 2x
1	ACCCTTGGCC	N/A	WARNING: Tag matches mitochondrial DNA	AB	908 (80%)	771 (16%)	100%
2	CTCCACCCGA	Hs.82961	trefoil factor 3 (intestinal)	AB	215 (163%)	303 (46%)	100%
3	CACCCCTGAT	Hs.173724	creatine kinase, brain	AB	117 (67%)	243 (56%)	100%
4	GCCCAGGTCA	Hs.154903	ESTs, Weakly similar to Abl substrate ena [D.melanogaster]	AB	91 (187%)	972 (10%)	100%
5	CTGGCCCTCG	Hs.1406	trefoil factor 1 (breast cancer, estrogen-inducible sequence expressed in)	AB	31 (162%)	238 (79%)	100%

Figure 16.7. (a) The xProfiler tool allows the user to perform an electronic subtraction between sets of SAGE libraries by selecting libraries in this window. In this example, normal colonic epithelium is subtracted from colon cancer. (b) The results of the electronic subtraction are displayed in tabular form.

database is ArrayDB (Masiello et al., 1999). ArrayDB, which requires either a Sybase or Oracle client, was designed for printed microarray data and allows the storage of experimental data and simple data queries. ArrayDB is designed to store a database of clones used for array fabrication and the data output from individual hybridization experiments. Links to the UniGene database are maintained. Data can be downloaded or viewed through a Web-based Java applet. On entry to the database, one selects the experiment to view from a pull-down menu and, on loading, a histogram of ratios appears (Fig. 16.8). This window contains several selectable options for viewing portions of the data. Importantly, the data can be filtered according to intensity and spot size to remove insignificant measurements from subsequent analyses. On submitting a query, a new window opens providing an image of the array and a table of genes that match the query results (Fig. 16.9). The image is useful because it allows confirmation that a given value is not the consequence of hybridization artifact such as a scratch or dye precipitate. The table contains the relevant intensity, spot size, and ratio data for each gene meeting the search criteria. Clicking on the Clone ID field in this view opens a window containing data on the clone printed at that point on the array with links to relevant databases including UniGene, OMIM, GenBank, and GeneCards, which are useful for acquiring functional information about a given gene. Additional features of ArrayDB include the ability to see a list

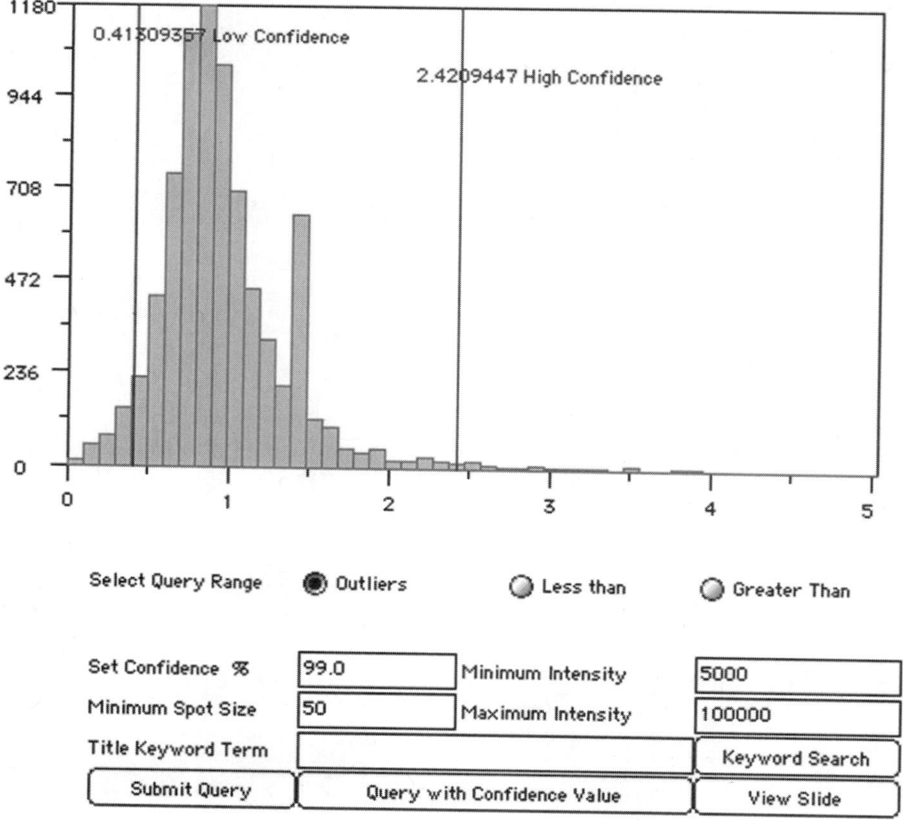

Figure 16.8. Ratio histogram of a microarray hybridization retrieved from ArrayDB.

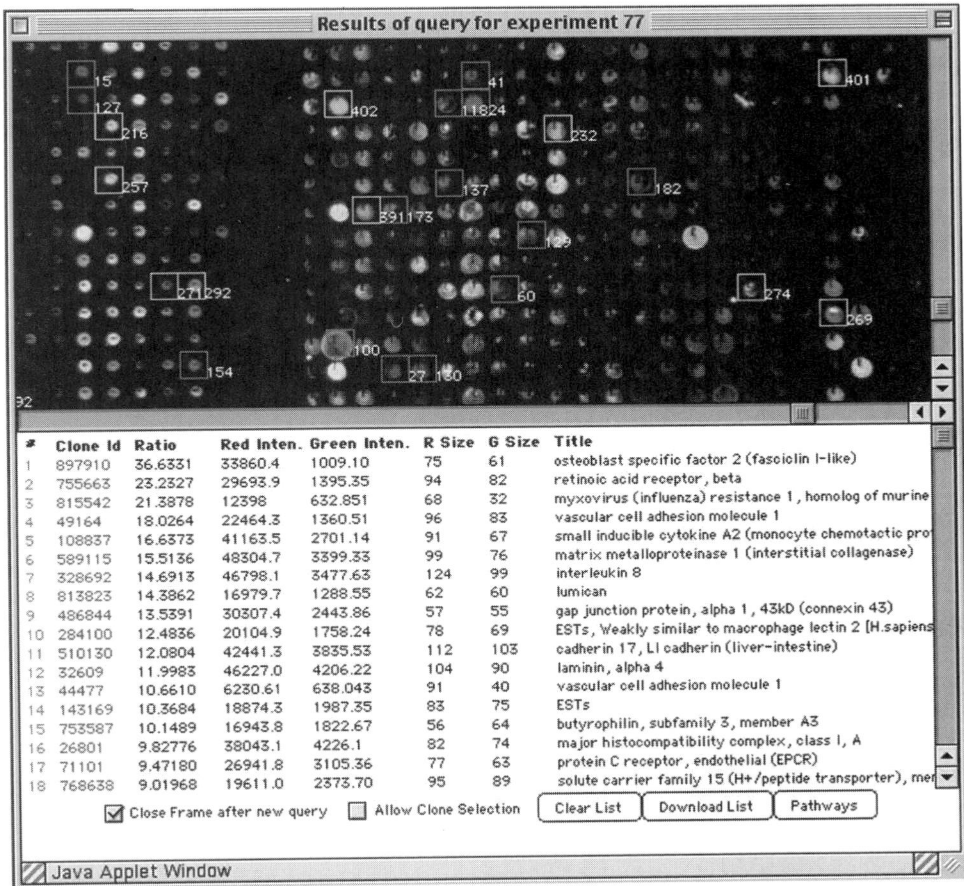

#	Clone Id	Ratio	Red Inten.	Green Inten.	R Size	G Size	Title
1	897910	36.6331	33860.4	1009.10	75	61	osteoblast specific factor 2 (fasciclin I-like)
2	755663	23.2327	29693.9	1395.35	94	82	retinoic acid receptor, beta
3	815542	21.3878	12398	632.851	68	32	myxovirus (influenza) resistance 1, homolog of murine
4	49164	18.0264	22464.3	1360.51	96	83	vascular cell adhesion molecule 1
5	108837	16.6373	41163.5	2701.14	91	67	small inducible cytokine A2 (monocyte chemotactic pro
6	589115	15.5136	48304.7	3399.33	99	76	matrix metalloproteinase 1 (interstitial collagenase)
7	328692	14.6913	46798.1	3477.63	124	99	interleukin 8
8	813823	14.3862	16979.7	1288.55	62	60	lumican
9	486844	13.5391	30307.4	2443.86	57	55	gap junction protein, alpha 1, 43kD (connexin 43)
10	284100	12.4836	20104.9	1758.24	78	69	ESTs, Weakly similar to macrophage lectin 2 [H.sapiens
11	510130	12.0804	42441.3	3835.53	112	103	cadherin 17, LI cadherin (liver-intestine)
12	32609	11.9983	46227.0	4206.22	104	90	laminin, alpha 4
13	44477	10.6610	6230.61	638.043	91	40	vascular cell adhesion molecule 1
14	143169	10.3684	18874.3	1987.35	83	75	ESTs
15	753587	10.1489	16943.8	1822.67	56	64	butyrophilin, subfamily 3, member A3
16	26801	9.82776	38043.1	4226.1	82	74	major histocompatibility complex, class I, A
17	71101	9.47180	26941.8	3105.36	77	63	protein C receptor, endothelial (EPCR)
18	768638	9.01968	19611.0	2373.70	95	89	solute carrier family 15 (H+/peptide transporter), me

Figure 16.9. In ArrayDB, query for outliers returns an image of the microarray, with outlying genes highlighted in the image and listed below the image, along with intensity data and clone identifiers. (See color plate.)

of biological pathways that gene products of interest are involved in; the positions of these gene products within known biological pathways are also shown graphically using image maps from the Kyoto Encyclopedia of Genes and Genomes (KEGG). Investigators involved in microarray research soon learn the critical importance of rapid access to descriptive information on gene function. Most of the genes of interest in individual experiments are unfamiliar, and an efficient mechanism for obtaining a synopsis of gene function is essential to the interpretation of microarray data.

Central storage of microarray data with investigator access via individual work stations is clearly the preferred mode for information management for centers generating significant quantities of microarray data and ultimately will be required for public distribution of expression data. However, it is important to note that projects of significant size can be accommodated in low-cost databases maintained on desktop computers. As an example of this approach, FileMaker Pro templates for cDNA microarray data can be downloaded from the NHGRI microarray site. Projects of up to 50 experiments can easily be accommodated. Although, as discussed below, the statistical analysis of expression data with specialized tools is important, the value of simple relational databases should not be underestimated. Even in a database such as FileMaker Pro, it is possible to define patterns of genes expressed consistently across experiments and possible to generate useful graphs displaying numerical data as color blocks (Fig. 16.10). This type of graphic display is much easier to scan by eye than a large table of numbers. Reanalyzing a data set using various filters is greatly facilitated by placing the entire set of experiments in a searchable database. Various settings can be tested until the output is optimized.

Initial processing of data requires the application of significance filters to the raw data set so that only meaningful measurements enter into downstream analysis. This requires the application of a sensitivity threshold to remove genes that have not been measured accurately. The importance of this step was illustrated earlier in this chapter. Individual laboratories will need to establish thresholds applicable to their own data, and investigators should exercise caution when using publicly accessible databases to be certain whether the data has been filtered for a detection threshold. Of course, there is no fixed cutoff that must be applied to all expression data, but, in general, as the threshold is lowered toward the background noise of an assay, the quality of the data will diminish. To maximize the yield of information from an experiment and to simplify data displays by removing uninformative genes, in most cases it is also useful to apply a filter that removes genes that, although measured accurately, do not fluctuate across a series of experiments.

Once an appropriately filtered data set has been extracted, the data are ready for analysis. In some instances, simple searches may be sufficient to generate lists of genes that are expressed under a given condition. For example, from a set of experiments in which cells have been treated with a drug or transfected with a gene, simple searching and sorting of the filtered data will yield a list of genes induced or

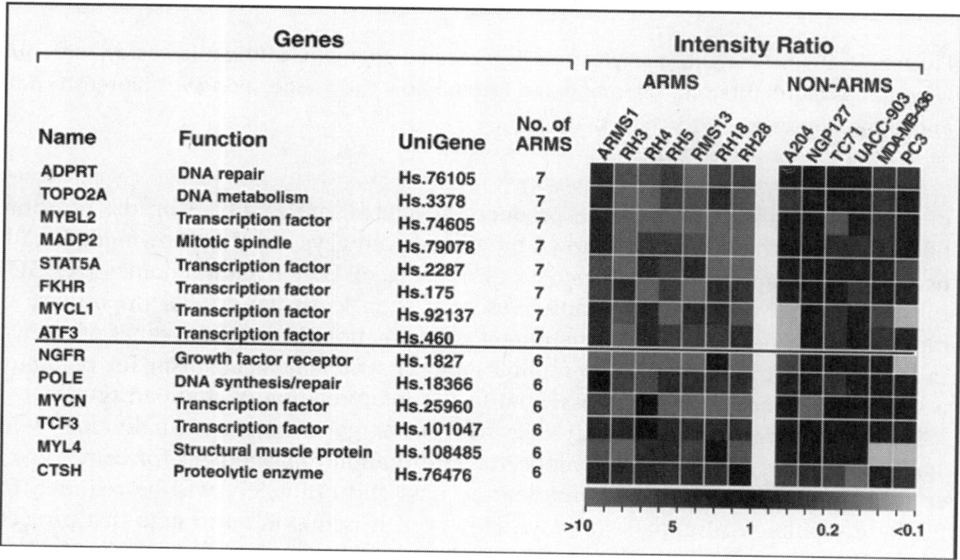

Figure 16.10. Display of microarray results retrieved from FileMaker Pro. This example illustrates the results of a query for genes upregulated in a series of cancer cell lines with ratios coded by a red to green color map (Khan et al., 1998). (See color plate.)

repressed by the treatment. However, the limitations of simple searches are quickly reached with larger data sets involving multiple samples or conditions and computational techniques for data organization are essential.

Cluster Analysis

The goal of cluster analysis is to reveal underlying patterns in data sets that contain hundreds of thousands of measurements and to present this data in a user-friendly manner. Ideally, patterns of similarities and difference among samples are identified and genes are coregulated in distinct patterns. A number of tools for clustering have been developed, which are freely available. In general, these are based on conventional statistical techniques widely used in other contexts and are largely dependent on linear correlation analysis. It is certain that future research efforts will be directed at the development of increasingly sophisticated tools designed for mining expression data. By applying statistical analysis, definition of subsets in apparently homogeneous tissue samples, development of classifiers for various disease entities, and identification of groups of coregulated genes may be possible. These possibilities are especially intriguing because they may provide a way to assign the large number of anonymous genes in higher eukaryotes to functional groups.

HIERARCHICAL CLUSTERING

Agglomerative hierarchical clustering has established itself as the most frequently applied technique for processing array data for inspection (Weinstein et al., 1997; Khan et al., 1998; Eisen et al., 1998). All pairwise comparisons of expression levels are made between experiments (Fig. 16.11). The resulting matrix of scatter plots can be reduced to a matrix of Pearson correlation coefficients. This is readily displayed in two dimensions as a hierarchical dendrogram (Fig. 16.12). Both genes and samples can be clustered in this fashion. By color coding expression levels, a large numerical table can be replaced by a much more compact and easily inspected color plot

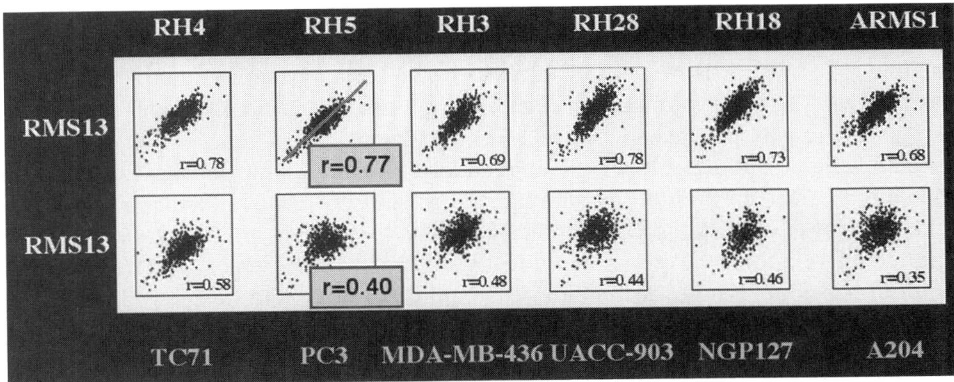

Figure 16.11. Portion of a scatter plot matrix illustrating pairwise plots of log ratios across a series of microarray experiments. Each scatter plot is used to calculate a Pearson correlation coefficient between the two sets of measurements (Khan et al., 1998).

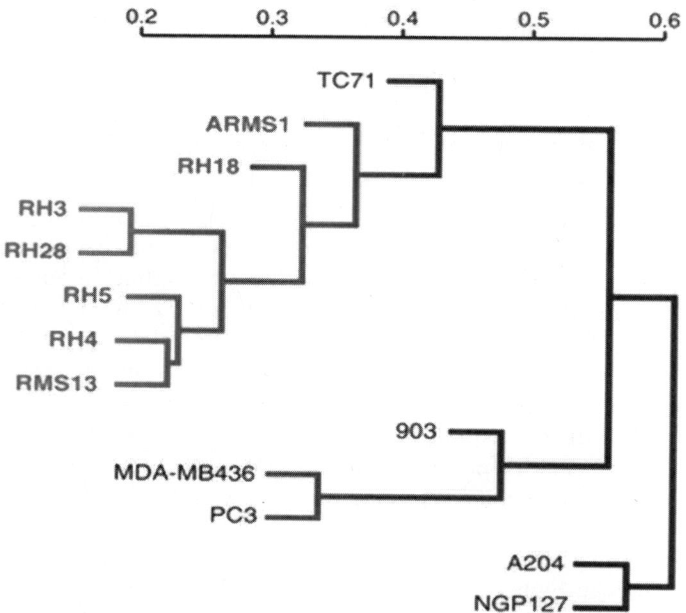

Figure 16.12. Hierarchical clustering dendrogram plotted from the Pearson correlation coefficients calculated across a series of experiments (Khan et al., 1998).

(Weinstein et al., 1997; Eisen et al., 1998). Inspection of the dendrogram and color plot will reveal samples with closely related patterns of gene expression as well as clusters of genes with similar expression pattern. Software for this type of display is freely available. This approach has been productively applied to yeast and human expression data. It is important to note that there is an arbitrary character to the manner in which a dendrogram is drawn. Clusters can be rotated about the point of bifurcation affecting the apparent proximity of the edges of a cluster with adjacent clusters. The important information is contained in the cluster contents and their similarity.

An alternative approach for displaying the same information is multidimensional scaling (MDS) (Fig. 16.13). MDS software is available in standard statistical software such as MATLAB. The distance measure for plotting individual samples is based on $1 - r$ where r is the Pearson correlation coefficient. Thus samples that are closely related will plot closely together, and widely differing samples plot farther apart. Although MDS does not allow simultaneous display of gene and sample clusters, it has the virtue of retaining a graphical display that places each sample (or gene) plotted in relation to every other. Ideally, an MDS plot, which is a reduction of multidimensional data to a three-dimensional display, should be viewed on a computer screen that allows rotation of the data so that the viewer is not deceived by any single two-dimensional projection.

Several other statistical tools based on linear and nonlinear methods have been used to analyze array data, including self-organizing maps, k-means clustering, and principal component analysis (Tamayo et al., 1999; Ben-Dor et al., 1999). Each of these tools has merit, and software for some of these is publicly available. However,

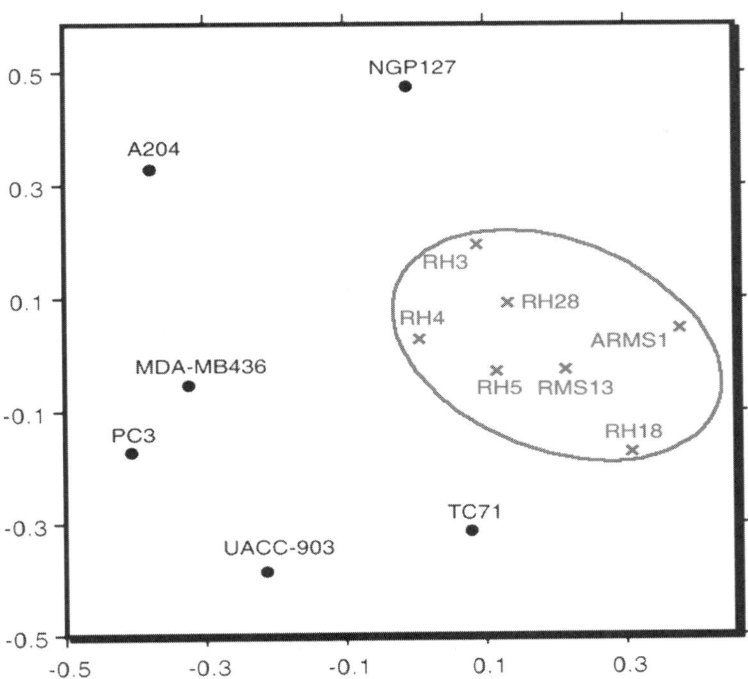

Figure 16.13. Multidimensional scaling plot of the same data represented in Figure 16.12. Note that the samples that fall closely together in the dendrogram also plot near to one another in the MDS plot. Ideally, this plot should be viewed in three dimensions on a computer screen so that it can be rotated to allow better inspection of the distances between the data points.

the precise selection of statistical methods for the analysis of array data is still the subject of active investigation, and the limits of these techniques are not well defined (Bittner et al., 1999). No single method can be recommended at this time to the exclusion of others, and most investigators will base their choice on the availability of software tools and Web-based resources. The range and quality of these tools will certainly improve rapidly in the coming years.

PROSPECTS FOR THE FUTURE

Existing technologies for high-throughput gene expression analysis and the informatics approaches to analysis of data arising from these methods are already producing an extremely interesting and novel view of genome function. However, it is apparent that there are limitations to current approaches that present the opportunity for significant improvements. At the level of technology, it can be anticipated that microarrays will move progressively closer to whole genome analysis with improved sensitivity and reduced sample requirements. At the level of informatics, several areas of progress can be foreseen. Tools for providing precomputed informative summaries of function for named genes are likely to be developed. Methods of linking gene expression data to genomic sequence would be of great value in dissecting networks of coregulated genes by identifying regulatory motifs shared by clustered genes. Most far-reaching would be methods of defining gene networks in terms of codetermination through the development of computational tools that predict the expression of a gene based on the expression of other genes (Seungchan et al., 2000). It is hoped that the recognition of gene networks will facilitate the assignment of function to anonymous genes and the definition of pathways. Ultimately, models may be developed that accurately reflect the mesh of self-compensating regulatory networks that maintain ordered genome function.

INTERNET RESOURCES FOR TOPICS PRESENTED IN CHAPTER 16

NCBI Databases and Tools

DDD	*http://www.ncbi.nlm.nih.gov/CGAP/info/ddd.cgi*
NCBI Gene Expression Omnibus	*http://www.ncbi.nlm.nih.gov/geo*
SAGEmap	*http://www.ncbi.nlm.nih.gov/SAGE/*
xProfiler	*http://www.ncbi.nlm.nih.gov/CGAP/hTGI/xprof/cgapxpsetup.cgi*

Software

ArrayDB	*http://genome.nhgri.nih.gov/arraydb/*
Cluster, TreeView, and ScanAlyze	*http://rana.Stanford.EDU/software/*
CrazyQuant	*http://chroma.mbt.washington.edu/mod_www/tools/index.html*
GeneCluster	*http://www.genome.wi.mit.edu/MPR/*
GeneX	*http://www.ncgr.org/research/genex/*
P-SCAN	*http://abs.cit.nih.gov/main/pscan.html*
ScanAlyze; AMAD	*http://www.microarrays.org/*

An up-to-date list of the links to laboratories involved in microarray technology development can be found at the NHGRI Web site (*http://www.nhgri.nih.gov/DIR/LCG/15K/HTML/*).

REFERENCES

Alizadeh, A. A., Eisen, M. B., Davis, R. E., Ma, C., Lossos, I. S., Rosenwald, A., Boldrick, J. C., Sabet, H., Tran, T., Yu, X., Powell, J. I., Yang, L., Marti, G. E., Moore, T., Hudson, J., Jr., Lu, L., Lewis, D. B., Tibshirani, R., Sherlock, G., Chan, W. C., Greiner, T. C., Weisenburger, D. D., Armitage, J. O., Warnke, R., Staudt, L. M., et al. (2000). Distinct types of diffuse large B-cell lymphoma identified by gene expression profiling. *Nature* 403, 503–511.

Alon, U., Barkai, N., Notterman, D. A., Gish, K., Ybarra, S., Mack, D., and Levine, A. J. (1999). Broad patterns of gene expression revealed by clustering analysis of tumor and normal colon tissues probed by oligonucleotide arrays. *Proc. Natl. Acad. Sci. USA* 96, 6745–6750.

Amundson, S. A., Bittner, M., Chen, Y., Trent, J., Meltzer, P., and Fornace, A. J., Jr. (1999). Fluorescent cDNA microarray hybridization reveals complexity and heterogeneity of cellular genotoxic stress responses. *Oncogene* 18, 3666–3672.

Ben-Dor, A., Shamir, R., and Yakhini, Z. (1999). Clustering gene expression patterns. *J. Comput. Biol.* 6, 281–97.

Bittner, M., Meltzer, P., Chen, Y., Jiang, Y., Seftor, E., Hendrix, M., Radmacher, M., Simon, R., Ben-Dor, A., Sampas, N., Dougherty, E., Wang, E., Marincola, F., Gooden, C., Lueders, C., Glatfelter, A., Pollock, P., Carpten, J., Gillanders, E., Leja, D., Dietrich, K., Beaudry, C., Berens, M., Alberts, D., Sondak, V., Hayward, N., and Trent, J. M. (2000). Molecular classification of cutaneous malignant melanoma by gene expression profiling. *Nature* 406, 536–540.

Bittner, M., Meltzer, P., and Trent, J. (1999). Data analysis and integration: of steps and arrows. *Nat. Genet.* 22, 213–215.

Cho, R. J., Campbell, M. J., Winzeler, E. A., Steinmetz, L., Conway, A., Wodicka, L., Wolfsberg, T. G., Gabrielian, A. E., Landsman, D., Lockhart, D. J., and Davis, R. W. (1998). A genome-wide transcriptional analysis of the mitotic cell cycle. *Mol. Cell.* 2, 65–73.

Chu, S., DeRisi, J., Eisen, M., Mulholland, J., Botstein, D., Brown, P. O., and Herskowitz, I. (1998). The transcriptional program of sporulation in budding yeast [published erratum appears in Science 1998 Nov 20; 282(5393):1421]. *Science* 282, 699–705.

DeRisi, J., Penland, L., Brown, P. O., Bittner, M. L., Meltzer, P. S., Ray, M., Chen, Y., Su, Y. A., and Trent, J. M. (1996). Use of a cDNA microarray to analyse gene expression patterns in human cancer. *Nat. Genet.* 14, 457–460.

DeRisi, J. L., Iyer, V. R., and Brown, P. O. (1997). Exploring the metabolic and genetic control of gene expression on a genomic scale. *Science* 278, 680–686.

Eisen, M. B., Spellman, P. T., Brown, P. O., and Botstein, D. (1998). Cluster analysis and display of genome-wide expression patterns. *Proc. Natl. Acad. Sci. USA* 95, 14863–14868.

Feng, X., Jiang, Y., Meltzer, P., and Yen, P. M. (2000). Thyroid hormone regulation of hepatic genes in vivo detected by complementary DNA microarray. *Mol. Endocrinol.* 14, 947–955.

Golub, T. R., Slonim, D. K., Tamayo, P., Huard, C., Gaasenbeek, M., Mesirov, J. P., Coller, H., Loh, M. L., Downing, J. R., Caligiuri, M. A., Bloomfield, C. D., and Lander, E. S. (1999). Molecular classification of cancer: class discovery and class prediction by gene expression monitoring. *Science* 286, 531–537.

Holstege, F. C., Jennings, E. G., Wyrick, J. J., Lee, T. I., Hengartner, C. J., Green, M. R., Golub, T. R., Lander, E. S., and Young, R. A. (1998). Dissecting the regulatory circuitry of a eukaryotic genome. *Cell* 95, 717–728.

Iyer, V. R., Eisen, M. B., Ross, D. T., Schuler, G., Moore, T., Lee, J. C. F., Trent, J. M., Staudt, L. M., Hudson, J., Jr., Boguski, M. S., Lashkari, D., Shalon, D., Botstein, D., and Brown, P. O. (1999). The transcriptional program in the response of human fibroblasts to serum. *Science* 283, 83–87.

Khan, J., Bittner, M. L., Saal, L. H., Teichmann, U., Azorsa, D. O., Gooden, G. C., Pavan, W. J., Trent, J. M., and Meltzer, P. S. (1999). cDNA microarrays detect activation of a myogenic transcription program by the PAX3-FKHR fusion oncogene. *Proc. Natl. Acad. Sci. USA* 96, 13264–13269.

Khan, J., Simon, R., Bittner, M., Chen, Y., Leighton, S. B., Pohida, T., Smith, P. D., Jiang, Y., Gooden, G. C., Trent, J. M., and Meltzer, P. S. (1998). Gene expression profiling of alveolar rhabdomyosarcoma with cDNA microarrays. *Cancer Res.* 58, 5009–5013.

Lal, A., Lash, A. E., Altschul, S. F., Velculescu, V., Zhang, L., McLendon, R. E., Marra, M. A., Prange, C., Morin, P. J., Polyak, K., Papadopoulos, N., Vogelstein, B., Kinzler, K. W., Strausberg, R. L., and Riggins, G. J. (1999). A public database for gene expression in human cancers. *Cancer Res.* 59, 5403–5407.

Lashkari, D. A., DeRisi, J. L., McCusker, J. H., Namath, A. F., Gentile, C., Hwang, S. Y., Brown, P. O., and Davis, R. W. (1997). Yeast microarrays for genome wide parallel genetic and gene expression analysis. *Proc. Natl. Acad. Sci. USA* 94, 13057–13062.

Lockhart, D. J., Dong, H., Byrne, M. C., Follettie, M. T., Gallo, M. V., Chee, M. S., Mittmann, M., Wang, C., Kobayashi, M., Horton, H., and Brown, E. L. (1996). Expression monitoring by hybridization to high-density oligonucleotide arrays. *Nat. Biotechnol.* 14, 1675–1680.

Masiello, A. J., Chen, Y., Bittner, M. L., Meltzer, P. S., Trent, J. M., and Baxevanis, A. D. (1999). ArrayDB 2.0: Software for the Exploration and Analysis of Microarray Gene Expression Data. Cold Spring Harbor Meeting on Genome Sequencing and Biology, Cold Spring Harbor, New York, p. 151.

Roberts, C. J., Nelson, B., Marton, M. J., Stoughton, R., Meyer, M. R., Bennett, H. A., He, Y. D., Dai, H., Walker, W. L., Hughes, T. R., Tyers, M., Boone, C., and Friend, S. H. (2000). Signaling and circuitry of multiple MAPK pathways revealed by a matrix of global gene expression profiles. *Science* 287, 873–880.

Ross, D. T., Scherf, U., Eisen, M. B., Perou, C. M., Rees, C., Spellman, P., Iyer, V., Jeffrey, S. S., Van de Rijn, M., Waltham, M., Pergamenschikov, A., Lee, J. C., Lashkari, D., Shalon, D., Myers, T. G., Weinstein, J. N., Botstein, D., and Brown, P. O. (2000). Systematic variation in gene expression patterns in human cancer cell lines. *Nat. Genet.* 24, 227–235.

Schena, M., Shalon, D., Davis, R. W., and Brown, P. O. (1995). Quantitative monitoring of gene expression patterns with a complementary DNA microarray. *Science* 270, 467–470.

Scherf, U., Ross, D. T., Waltham, M., Smith, L. H., Lee, J. K., Tanabe, L., Kohn, K. W., Reinhold, W. C., Myers, T. G., Andrews, D. T., Scudiero, D. A., Eisen, M. B., Sausville, E. A., Pommier, Y., Botstein, D., Brown, P. O., and Weinstein, J. N. (2000). A gene expression database for the molecular pharmacology of cancer. *Nat. Genet.* 24, 236–244.

Seungchan, K., Dougherty, E. K., Chen, Y., Krishnamoorthy, S., Meltzer, P., Trent, J. M., and Bittner, M. (2000). *Multivariate measurement of gene expression relationships. Genomics* 67, 201–209.

Spellman, P. T., Sherlock, G., Zhang, M. Q., Iyer, V. R., Anders, K., Eisen, M. B., Brown, P. O., Botstein, D., and Futcher, B. (1998). Comprehensive identification of cell cycle-regulated genes of the yeast *Saccharomyces cerevisiae* by microarray hybridization. *Mol. Biol. Cell* 9, 3273–3297.

Tamayo, P., Slonim, D., Mesirov, J., Zhu, Q., Kitareewan, S., Dmitrovsky, E., Lander, E. S., and Golub, T. R. (1999). Interpreting patterns of gene expression with self-organizing maps: methods and application to hematopoietic differentiation. *Proc. Natl. Acad. Sci. USA* 96, 2907–2912.

Velculescu, V. E., Zhang, L., Vogelstein, B., and Kinzler, K. W. (1995). Serial analysis of gene expression. *Science* 270, 484–487.

Weinstein, J. N., Myers, T. G., O'Connor, P. M., Friend, S. H., Fornace, A. J., Jr., Kohn, K. W., Fojo, T., Bates, S. E., Rubinstein, L. V., Anderson, N. L., Buolamwini, J. K., van Osdol, W. W., Monks, A. P., Scudiero, D. A., Sausville, E. A., Zaharevitz, D. W., Bunow, B., Viswanadhan, V. N., Johnson, G. S., Wittes, R. E., and Paull, K. D. (1997). An information-intensive approach to the molecular pharmacology of cancer. *Science* 275, 343–349.

Wodicka, L., Dong, H., Mittmann, M., Ho, M. H., and Lockhart, D. J. (1997). Genome-wide expression monitoring in *Saccharomyces cerevisiae. Nat. Biotechnol.* 15, 1359–1367.

17

USING PERL TO FACILITATE BIOLOGICAL ANALYSIS

Lincoln D. Stein

The Cold Spring Harbor Laboratory
Cold Spring Harbor, New York

Consider a situation in which an investigator is studying genes that affect neuronal signaling in *C. elegans*, with a primary interest in identifying those with gene products that may be secreted. A Web site that reports the results of a large systematic study of predicted *C. elegans* genes using the RNA-induced inhibition of gene expression (RNAi) technique is available, and the investigator can download a summary file based on a few thousand experiments. The latest release of WormPep, which reports peptide sequences of over 19,000 known and predicted worm genes, is also available. What is now needed is the ability to search the RNAi results file for those genes that affect worm movement in some way (e.g., a common phenotype for genes affecting neuronal activity), and extract the sequence of those genes from WormPep. The ultimate plan is to submit these sequences of interest to SignalP, an E-mail-based signal peptide cleavage site predictor based at the Technical University of Denmark (discussed in Chapter 11).

How would one accomplish this? One way is to do the job by hand. First, one would need to read the RNAi summary file into a word processing program, cull it for experiments that affected locomotion in some way, and then assemble a list of all of the genes that produce relevant phenotypes. Next, one would open WormPep, search for the corresponding sequence for each of these genes, and cut-and-paste these sequences into another file. The last step would be to reformat the entries into the format required by the SignalP server, pasting all of the entries into an E-mail message.

Bioinformatics: A Practical Guide to the Analysis of Genes and Proteins
Edited by A. D. Baxevanis and B. F. F. Ouellette
ISBN 0-471-38390-2 (cloth), ISBN 0-471-383910 (paper)

It should be immediately apparent that there are some problems with this approach. If there are more than just a few genes of interest to analyze, the job quickly becomes rather tedious, if not overly time consuming. Worse yet, the next time that new RNAi results are released, the whole process will need to be repeated, determining which RNAi entries are new. The step involving loading WormPep into a word processor might not even be tenable due to the sheer size of the database, well over 10 megabases.

This is the type of problem that a Perl script can help with. The Perl programming language excels at slicing, dicing, and integrating data files and is the language of choice for the many bioinformatics researchers. This chapter will provide a gentle introduction to Perl, with examples designed to illustrate the usefulness of learning this language.

GETTING STARTED

During the course of this chapter, a solution will be developed for the data integration problem introduced in the first paragraph of this chapter. Before this problem is attacked, however, some very simple scripts that illustrate the basics of Perl programming will be discussed. In considering these examples, the reader is strongly encouraged to follow along by typing the examples into a Perl interpreter to get a better sense for what these short scripts actually do. Modifying and experimenting with the scripts for individual use is also encouraged.

Perl interpreters are available for the Macintosh, Windows, and UNIX operating systems and more often than not is made available as freeware. A number of download sites are listed at the end of this chapter. Because Perl program files are usually compressed to speed downloads, Macintosh users will need to use a utility like UnStuffIt to uncompress the package, whereas Windows users will need WinZip or PKZip. Perl should come as part of the standard software installation on most UNIX-based machines. Typing `perl -v` at the UNIX prompt will indicate whether the program is indeed installed and, if so, the version number that is available on that machine.

Perl consists of two essential parts:

the *interpreter*, called perl on UNIX systems, MacPerl on Macintoshes, and perl.exe on Windows machines; and

scripts, text files that are written by the user describing a discrete set of steps to be performed by the interpreter. The scripts are actually computer programs, and the words script and program can be used interchangeably.

The process of writing and running a Perl script is similar on Windows and UNIX systems but slightly different for Macintosh users. The basic steps are described below.

On a Windows or UNIX system, create a text-only file containing the following lines:

```
#!/usr/bin/perl
print "My first Perl script.";
```

Any word processing program may be used, as long as the file is saved as `text only`. The Windows Notepad program is good for this task, as it saves its files in text-only format by default. Name the file `first.pl`.

This newly created file contains two lines. The first line is a comment that identifies the file as a Perl script (indicated by the `#!` at the beginning of the line). The second line is a *print statement* that tells Perl to print out the text

```
My first Perl script
```

The name of the file ends in the extension `.pl`, which is a standard naming convention for Perl scripts.

To run this command, open a command-line window (the DOS window on Windows systems, a shell window on UNIX), change to the directory that contains the file, and type the command

```
% perl first.pl
My first Perl script.
```

where `%` represents the command-line prompts for both the Windows and UNIX systems, and boldface type represents input typed by the user at the keyboard.

What this particular command does is invoke the Perl command interpreter, passing to it the name of the file that should be run. The interpreter dutifully processes the script line-by-line, sees the single print command, and executes it. The output of the script appears in the command window.

If Perl is installed in the standard way on the computer, the `perl` command does not have to be explicitly typed. On UNIX systems, the script file can be marked as being directly executable using the `chmod` command:

```
% chmod +x first.pl
```

On Windows systems, the file does not have to be explicitly marked as executable because Perl is usually installed in such a way that any file ending in the `.pl` extension is associated with the Perl interpreter. Simply typing in the name of the program will run it:

```
% first.pl
My first Perl script.
```

Mac-based Perl does not have a command window, so the process of creating and running a Perl script is somewhat different. On Macintoshes, the MacPerl application is launched by double clicking on its icon (a pyramid with a camel). This will launch a special-purpose text editor from which the user can create, run, and debug Perl scripts. Select New from the File menu to bring up a text-editing window. Type the script shown above, beginning with the `#!/usr/bin/perl` line and choose `Save As` from the `File` menu. Give the script a name, such as `first.pl`, and from the list of Type options in the Save dialog, choose Droplet. This will save the script as a miniature application called a *droplet*, which has a distinctive pyramid-shaped icon (containing another camel). Double clicking on this new icon will produce a text window containing the output of the script, which simply reads `My first`

`Perl script`. The script can also be run directly through the MacPerl application by choosing Run Script from the Script pull-down menu.

HOW SCRIPTS WORK

A script consists of a series of commands, more formally called *statements*, that are meaningful to the Perl interpreter. Unless told otherwise, the interpreter starts at the top of the script file and works its way down to the bottom, executing each statement in turn.

Consider this new script:

```
#!/usr/bin/perl
# preamble...
print "I can do math!\n";
# do some calculations
$sum = 3 + 4;
# print the result
print "The sum of 3+4 is ",$sum,".\n";
```

This script consists of three statements; the first one (`print "I can do math!\n"`) tells the interpreter to print out the indicated text. As we shall discuss in more detail later, the special character sequence `\n` is not interpreted literally but instead prints out as a *newline* character. The second statement, `$sum = 3 + 4`, adds the numbers three and four together and stores the result in a variable named `$sum`. The last statement prints the text `The sum of 3+4 is`, followed by the contents of the variable `$sum`, followed by a period and a newline.

Notice that each statement ends with a semicolon. The semicolon tells Perl where one statement ends and another begins. Blank lines and other white space can help make the script more readable but are ignored by the interpreter. Any line that begins with a pound sign (#) is a comment. When Perl sees a pound sign it simply ignores everything between it and the end of the line. The use of comments is strongly encouraged, since it allows other users to better understand what the programmer was trying to accomplish in a particular block of code, as well as reminding the programmer themselves of the same when reexamining code written long before.

The topmost line is also a Perl comment, but on UNIX systems it serves double-duty as a directive to the UNIX shell to tell it to execute the command `/usr/bin/perl` when the script file is executed. On Macintosh and Windows systems, this line is extraneous, but it is better to include it to maintain portability with UNIX machines and as a matter of good form.

Unless otherwise instructed, the Perl interpreter starts at the first statement and works its way to the last. When this script is run, the following output is produced:

```
I can do math!
The sum of 3+4 is 7.
```

Notice that the user only sees the output from the two print statements. The statement that performs the addition acts silently, behind the scenes.

STRINGS, NUMBERS, AND VARIABLES

Perl can deal with an astonishing number of data types, including but not limited to text, integers, floating point numbers, complex numbers, and binary numbers.

Following a long computer science tradition of using obscure terms for simple concepts, the Perl term for text is *string*. Strings are surrounded by single or double quotation marks:

```
'I am a string.'
"I am another string."
```

Having two types of quotation marks available makes it easier to create strings that contain embedded quotation marks:

```
'"Anna," she wailed "come quickly! The tiara is gone!"'
```

There are also some more substantial differences between the two types of quotes that will be discussed later in this chapter under *Variable Interpolation.*

Numbers are written just as one would expect:

```
1
49
28.2
-109
6.04E23
```

The last example shows how one represents scientific notation. The `E` means exponent, and the number should be interpreted as $6.02\ 10^{23}$.

If you want your strings to contain special characters, such as tabs or new lines, Perl provides special *escape sequences* to represent them. These escape sequences consist of a backslash followed by a single character. The two most commonly used are `\n`, which begins a new line, and `\t`, which inserts a tab. For example,

```
print "There is a newline\nhere and a tab\tthere.\n";
```

produces the following output:

```
There is a newline
here and a tab        there.
```

Perl only interprets escape sequences when they occur in double-quoted strings. In single-quoted strings, the backslash and the character that follows it are interpreted literally.

Variables provide temporary storage for strings, numbers, and other values. In Perl, variables are arbitrary names preceded by a dollar sign. Examples of valid variable names are shown below.

```
$x
$X
```

```
$i_am_a_variable
$LongVariableName
```

Perl variables are case sensitive. In the list above, `$x` is one variable, and `$X` is a different one entirely.

When first created, variables are empty or *undefined.* Values are assigned to variables using the = sign, also known as the assignment operator:

```
$x = 42;
print 'The value of $x is ',$x,"\n";
```

Assignment works from right to left. In the example above, the number 42 is assigned to the variable `$x`. Once assigned, the variable can be used in the place of a value, as the print statement above shows. The same variable can be used multiple times, using the assignment operator to change its contents.

```
print 'The value of $x is still ',$x,"\n";
$x = 'Mary had a little lamb';
print 'But now the value of $x is ',$x,"\n";
```

This code fragment will now print out the following:

```
The value of $x is still 42
But now the value of $x is Mary had a little lamb
```

Note that there is no restriction on the type of data a variable can hold. In this example, `$x` initially contained an integer and then held a string. Unlike some programming languages, Perl does not require the user to *declare* (formally describe) variables before using them, although this type of checking can be activated if desired. Perl actually has several types of variables. In addition to variables that hold a single value, which are technically called *scalar* variables, there are *arrays* and *hashes,* two types of variables that are capable of holding multiple values. These will be discussed in an upcoming example.

ARITHMETIC

Perl knows basic arithmetic. Symbols known as *operators* are responsible for the various arithmetic operations:

```
+  addition
-  subtraction
*  multiplication
/  division
** exponentiation
() grouping
```

The following example does a little math and prints out the result:

```
$x = 4;
$y = 2;
$z = 3 + $x * $y;
print $z,"\n";
```

The result that gets printed out is 11. However, a more succinct way to express this would be to combine the first three lines into a single expression passed to print:

```
print 3+4*2,"\n";
```

Note that the arithmetic expression is processed as 3+(4*2) rather than (3+4)*2. When evaluating numeric expressions, Perl uses the standard rules of precedence. The precedence can be changed by explicitly using parentheses:

```
print (3+4)*2,"\n";
```

VARIABLE INTERPOLATION

Another interesting difference between double- and single-quoted strings is what happens when a variable is embedded inside a string. In double-quoted strings, the variable is expanded to its contents, a process known as *string interpolation*. This can aid readability considerably:

```
print "The value of $x is $x\n";
```

Assuming that `$x` again contains "Mary had a little lamb," the above statement outputs

```
The value of Mary had a little lamb is Mary had a little lamb
```

Single-quoted strings do not work in this fashion. If the print statement used a single-quoted string instead, it would print

```
The value of $x is $x\n
```

The user can precisely control whether variable interpolation occurs in double-quoted strings by placing a backslash in front of variables that should not be interpolated.

```
print "The value of \$x is $x\n";
```

In this statement, the first occurrence of the `$x` variable is protected against interpolation because of the backslash, but the second is not.

```
The value of $x is Mary had a little lamb
```

The backslash character can also be used to embed a double-quote character inside a double-quoted string. How is the following different from the outputs shown above? Run it through the Perl interpreter to check your conclusion:

```
print "The value of \$x is \"$x\"\n";
```

Variable interpolation only extends to the contents of the variable itself. Perl will not try to evaluate arithmetic expressions or other programming statements that are embedded in double-quoted strings. For example, the statements

```
$y = 19;
print "The result is $y+3\n";
```

produces the output

```
The result is 19+3
```

To evaluate the expression, put it outside the double quotes. Perl will do the arithmetic, and the print statement will output the result.

```
print "The result is ",$y+3,"\n";
```

BASIC INPUT AND OUTPUT

Input, in programming parlance, is how data "get into" a script. Output is, of course, what comes out of the script. Most scripts will do both, inputting data from one source and outputting it to another.

The main way of producing output is to use the **print** function. Print takes a list of one or more arguments separated by commas and sends them to the current output device, which by default is the computer's screen. As already shown, print can deal equally well with text, numbers, and variables:

```
$sidekick = 100;
print "Maxwell Smart's sidekick is ",$sidekick-1,".\n";
print "If she had a twin, her twin might be called"
2*($sidekick-1),".\n";
```

The result of this script is

```
Maxwell Smart's sidekick is 99.
If she had a twin, her twin might be called 198.
```

The main way to read input data is to use the angle bracket operator (<>), which reads a line of input from the current input device. You will usually call this operator in conjunction with the assignment operator to save the returned information into a variable:

```
$line = <>;
print "Got $line";
```

With these two operations, one can now write a fully interactive program named "dog years," which converts your age in human years to your age in dog years:

```
#!/usr/bin/perl
print "Enter your age: ";
$age = <>;
print "Your age in dog years is ",$age/7,"\n";
```

When this program is run, the result will look like this:

```
% dog_years.pl
Enter your age: 42
Your age in dog years is 6
```

where the 42 is typed in at the keyboard. Another tiny program illustrates one of the idiosyncrasies of the <> operator:

```
#!/usr/bin/perl
print "Enter your name: ";
$name = <>;
print "Hello $name, happy to meet you!\n";
```

Running this program produces output that might not be what is expected:

```
% hello.pl
Enter your name: Lincoln
Hello Lincoln
, happy to meet you!
```

What's going on? In fact, when the <> operator reads a line of input, the newline character at the end of the typed data is still there! More often than not, the newline character must be removed. Obligingly, Perl provides a function named **chomp** that will do exactly that, removing the terminal newline from a string. The rewritten program looks like this:

```
#!/usr/bin/perl
print "Enter your name: ";
$name = <>;
chomp $name;
print "Hello $name, happy to meet you!\n";
```

With the newline removed, the output looks the way it should:

```
% hello.pl
Enter your name: Lincoln
Hello Lincoln, happy to meet you!
```

The program can still be made a bit shorter by combining the input and chomp statements into a single statement, at the risk of making the program slightly harder to understand:

```
#!/usr/bin/perl
print "Enter your name: ";
chomp($name = <>);
print "Hello $name, happy to meet you!\n";
```

The parentheses control the precedence of the operation so that Perl does the input first and then passes `$name` to the chomp function.

FILEHANDLES

When lines of input are read, the data come from the keyboard by default. When the script writes lines of output, the output goes to the screen by default. What if one wants to change these defaults so that input comes from a file or output goes into one?

There are several ways to do this, but the most straightforward is to use *filehandles*. A filehandle is the connection between a script and a file. Scripts can read from a filehandle to get the contents of the file a line at a time and print to a filehandle to add data to a file.

To open a file for reading, use the **open** function:

```
open MYFILE,'data.txt';
```

The open function expects exactly two arguments. The first is a name for the filehandle (`MYFILE`). This is an arbitrary name that the user chooses, with the convention being to use all uppercase letters. The second argument is the name of the file to open. If just the file name is provided, as shown in the example, Perl will attempt to open a file by that name in the current directory (on Windows and UNIX systems, the directory in which the command to run the script was given; on Macintoshes, the folder that the script is located in). The full path, explicitly giving the location of the file, can also be given.

Unfortunately, the way one specifies a file path is different on the three different operating systems. On UNIX systems, a path begins with a forward slash and each directory is separated by additional slashes. On Windows systems, a path begins with the drive letter (e.g., `C:`) and uses backslashes to separate directories. On Macintoshes, the path begins with the name of the hard disk, and colons separate the name of each successive folder. Examples of fully qualified path names on UNIX, Windows, and Macintosh systems are shown below.

```
UNIX        /usr/local/blast/data/cosmids.txt
Windows     C:\Documents\Blast\Data\Cosmids.txt
Macintosh HD:Sequence Data:Blast:cosmids
```

The open command may not be able to open a file for reading if, for instance, the file does not actually exist. Ways to detect and handle this kind of error will be discussed in the section *Filehandle Errors*.

Once a filehandle is open, it can be read from using the <> operator:

```
$line = <MYFILE>;
chomp $line;
$next_line = <MYFILE>;
chomp $next_line;
```

The only difference between reading from the keyboard and reading from a file is that, instead of using an empty pair of angle brackets (<>), the open filehandle is placed between them (<MYFILE>).

Each time you call <MYFILE>, a new line of data will be read from the file. When the last line has been read, the operation will return the undefined value; detecting this type of error will be discussed later. When a filehandle is no longer needed, it can be closed using the **close** function:

```
close MYFILE;
```

Writing to a file works in much the same way. The main difference is that, when the file is opened, Perl is instructed to write to the file by placing a > sign before the filename:

```
open MYFILE,'>data.txt';
```

If the file does not already exist, Perl will create it and open it for writing. If the file already exists, then Perl will empty out its existing contents before opening it. This ensures that the new data written to the file will replace anything that was already there. To add data to the end of the file without disrupting its current contents, the file can be opened for appending using the >> sign:

```
open MYFILE,'>>data.txt';
```

Data written to the file will now be appended to the end of its current contents rather than writing over the file. If the file does not already exist, then an empty one is automatically created.

Once a filehandle is opened for writing, data can be sent to it, using the filehandle as the print command's first argument:

```
print MYFILE "Your age in dog years is ",$age/7,"\n";
```

This will write the indicated line of text to the file associated with MYFILE. Notice that there is no comma between the filehandle and the list of data arguments! The full, generalized syntax for print is

```
print [FILEHANDLE] $data1 [,$data2 [,$data3 ....]]
```

The square brackets mean an argument is optional. One or more spaces is used to separate the optional filehandle from the first data argument, and commas are used to separate the individual items to be written to the filehandle.

When finished writing to a filehandle, use the close function to `close` it as before. If the program ends without explicitly closing the filehandle, Perl will close the file automatically.

Nothing prevents a script from having multiple filehandles open at the same time. This odd little program will write the first two odd-numbered lines of input to the file `odd.txt` and even-numbered lines to the file `even.txt`:

```
#!/usr/bin/perl
open ODD, ">odd.txt";
open EVEN, ">even.txt";
$line = <>;
print ODD $line;
$line = <>;
print EVEN $line;
$line = <>
print ODD $line;
$line = <>
print EVEN $line;
close ODD;
close EVEN;
```

There is a certain amount of repetition in this script, for sake of clarity. This script will be revisited shortly, using a more elegant loop to perform the same function.

Each time <> is called, a new line of input is obtained either from the keyboard or the file. What happens when the end of the file is reached? Because there aren't any more lines to read, <> simply returns the undefined value.

MAKING DECISIONS

So far, the programs that have been discussed have been very linear (see Fig. 17.1). The Perl interpreter starts at the first statement and works its way to the last, executing each of them along the way.

Life is full of decisions, however, and so are Perl scripts. Often, there should be different paths followed if a particular condition is true, another if the condition is false. Considering the dog years calculator shown above, what would happen if the user had entered a negative number for the age or a number that's unreasonably large? It would be desirable to reject the input outright, rather than produce and output a preposterous answer. This simple modification to the original script achieves this goal.

```
#!/usr/bin/perl
print "Enter your age: ";
$age = <>;
die "Preposterous age" if $age <= 0 or $age >= 100;
print "Your age in dog years is ",$age/7,"\n";
```

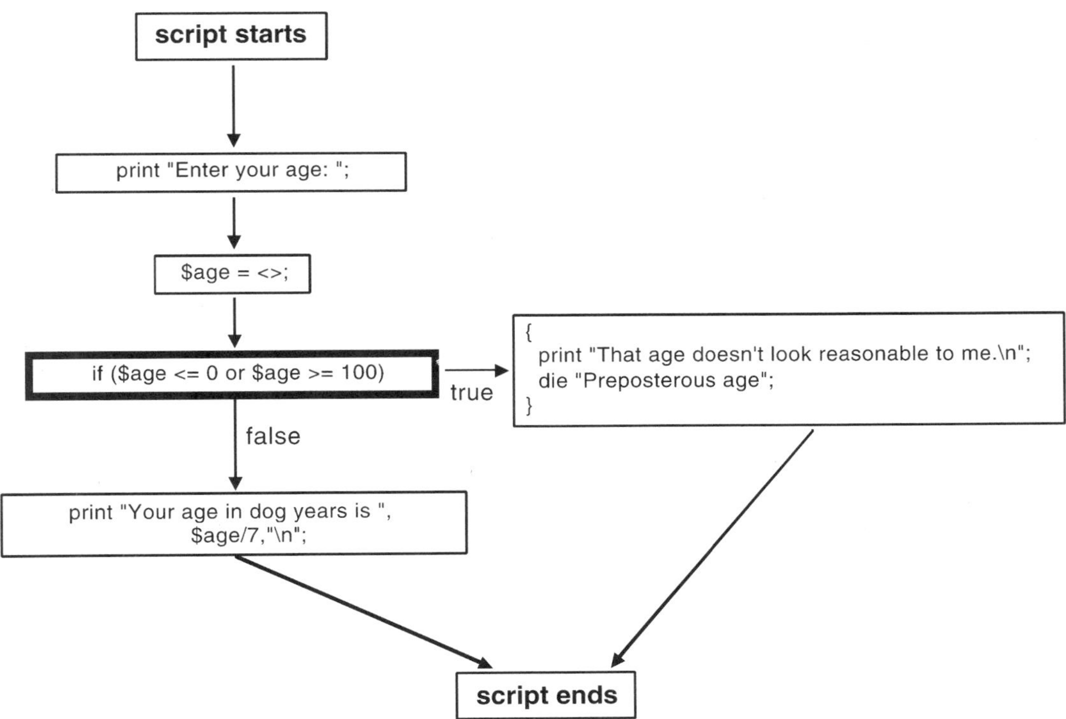

Figure 17.1. The "dog years" calculator provides an example of entirely linear program flow.

The fourth line is the new one, consisting of two parts. The first part (`die "Preposterous age"`) introduces the **die** function, which is a lethal form of the `print` command. It prints out the error message given to it, along with information indicating the current line number. It then immediately terminates the program. It is great for handling those occasions when a fatal, unrecoverable error has occurred. The second part of the line (`if $age <= 0 or $age >= 100`) tests the `$age` variable's numeric value. If `$age` is either less than or equal to zero or greater than or equal to 100, then the test is true and the die function is executed. If `$age` satisfies neither test, then the `die` function is skipped and the program goes on as before. In plain English, the statement can be read as, "Die if the age is either less than or equal to zero or the age is greater than or equal to 100." If the age is invalid, the program will terminate at the die statement, and the last print statement will never be executed. Testing the potential scenarios of the now-modified program, we find

```
% dog_years.pl
Enter your age: -20
Preposterous age at dog_years.pl line 4, <> chunk 1.
% dog_years.pl
Enter your age: as old as Methuselah
Preposterous age at dog_years.pl line 4, <> chunk 1.
% dog_years.pl
Enter your age: 42
Your age in dog years is 6
```

TABLE 17.1. Numeric Comparison Operators

Operator	Description	Example
==	Equality	`$a == $b`
!=	Not equal	`$a != $b`
<	Less than	`$a < $b`
>	Greater than	`$a > $b`
<=	Less than or equal to	`$a <= $b`
>=	Greater than or equal to	`$a >= $b`
!	Logical not	`$ = !$b`

Perl has a complete set of comparison operators that work on numbers and strings. To compare numbers, use any of the operators shown in Table 17.1. To compare strings, use any of the operators shown in Table 17.2. The numeric comparison operators are straightforward because they look, for the most part, like conventional expressions used in algebra. The big trap is the == operator that is used to test the equality of two numbers. Two equal signs are used instead of just one.

If you forget and accidentally use the assignment operator (=), then you will not get the result you expect:

```
$a == $b; # compare $a to $b, return true if equal
$a = $b;  # assign contents of $b to $a
```

Read the ! operator as *not*. Unlike the other operators, it takes a single argument and reverses its truth. True expressions become false and vice versa. For example

```
print "the number is not greater than 0" if !($a > 0);
```

This statement first compares the current value of `$a` to zero and returns a true value if `$a` is greater than zero. The `!` operator then reverses the test so that the expression as a whole is true only if `$a` is less than or equal to zero.

The string comparison operators are funny two-letter commands. The one used most frequently is `eq`, for testing whether two strings are the same. Consider this new version of the `hello.pl` script.

TABLE 17.2. String Comparison Operators

Operator	Description	Example
`eq`	Equality	`$a eq $b`
`ne`	Not equal	`$a ne $b`
`lt`	Less than	`$a lt $b`
`gt`	Greater than	`$a gt $b`
`le`	Less than or equal to	`$a le $b`
`ge`	Greater than or equal to	`$a ge $b`
`=~`	Pattern match	`$a =~ /gattc/`

```
#!/usr/bin/perl
print "Enter your name: ";
chomp($name = <>);
print "Hello $name, happy to meet you!\n";
print "Hail great leader!\n" if $name eq 'Lincoln';
```

The output of the program uses `eq` to give Lincoln a special greeting:

```
% hello.pl
Enter your name: George
Hello George, happy to meet you!
% hello.pl
Enter your name: Lincoln
Hello Lincoln, happy to meet you!
Hail great leader!
```

Other handy string comparison operators are `lt`, which is true if the first string is less than the second string, and `gt` if the first is greater than the second. Perl compares strings alphabetically but uses criteria different from the telephone book. Among other subtle (and not-so-subtle) differences, the set of uppercase letters is less than the set of lowercase letters; in Perl, `Z` is less than `a`. Be careful not to use == to compare strings or `eq` to compare numbers. The handiest string comparison operator of them all is =~, the pattern matching operator. This is the most powerful of Perl operators, deserving its own discussion later in this chapter.

Two or more comparison operations can be combined using the operators `and` and `or`. An expression involving `and` is true only if *both* the right and left sides are true, whereas `or` expressions are true if *either* side is true. There is also a `not` operator, which reverses the sense of whatever comes to the right, making true expressions false and false ones true, just like Big Brother did in George Orwell's *1984*.

The use of the `if` statement has already been demonstrated, executing a statement only when the condition that follows is true. As always with Perl, the opposite operator exists, called `unless`, executing a statement only when the condition following is false. Returning to the program testing for preposterous ages, the test could be rewritten this way:

```
die "Preposterous age" unless 0 < $age and $age < 100;
```

Read the statement this way: "Die unless the age is greater than zero and less than 100." The effect is the same. Sometimes it will seem more natural to write the conditional with an `if` sometimes with an `unless`.

CONDITIONAL BLOCKS

How would one approach executing several statements conditionally? In this case, the statements can be grouped into a *block*, using curly braces. The grouped statements can then be executed altogether, using the block form of `if`. Returning again to the dog-age calculator

```
#!/usr/bin/perl
print "Enter your age: ";
$age = <>;
if ($age <= 0 or $age >= 100) {
  print "That age doesn't look reasonable to me.\n";
  die "Preposterous age";
}
print "Your age in dog years is ",$age/7,"\n"';
```

In this example, `if` controls a block of two statements surrounded by the curly braces. The first statement prints out a warning, and the second one terminates the program with die. If the age does not fall within the range of 0–100, then the statements within the block *will* execute and the program will end before reaching the last line of code. If the age does fall within the specified range, the two statements in the `if` block are ignored and the program goes on to print out the calculated result, as before. The effect is to create two alternative paths in the program, one of which leads to termination with an error statement (Fig. 17.2).

`If` blocks have this general form:

```
if (TEST) {
                              STATEMENT 1;
                              STATEMENT 2;
                              STATEMENT 3;
                                ...
}
```

The test itself must be enclosed by parentheses in the manner shown, but any comparison operation involving numbers or strings is allowed. The indentation is a matter of style—although Perl does not depend on the indentation to interpret the

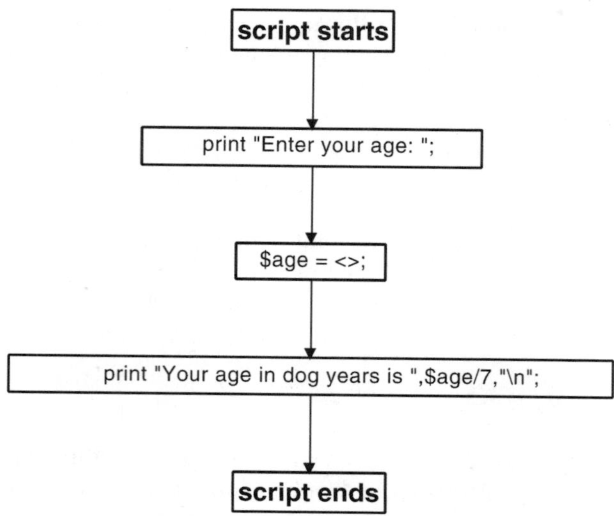

Figure 17.2. If blocks can change the flow of execution.

code, indenting code in this fashion is invaluable when debugging a program, as it is easier to see where a given block of code begins and ends.

Perl can also handle situations in which one set of statements should be executed if the test condition is true, and another set of statements if the test condition is false. To do so, an `else` block is added to the if structure, changing the construct from an *if* statement to an *if—else* statement.

```
#!/usr/bin/perl
print "Enter your age: ";
$age = <>;
if ($age <= 0 or $age >= 100) {
  print "That age doesn't look reasonable to me.\n";
} else {
  print "Your age in dog years is ",$age/7,"\n";
}
```

The if-else statement shown above has two blocks, each surrounded by curly braces. The contents of the first block is executed when the test is true. Otherwise the second block is executed. The result is that, if the entered age falls outside the acceptable range for the program, an error message is printed; if not, it prints the calculated results instead.

```
% dog_years.pl
Enter your age: eighteen
That age doesn't look reasonable to me.
% dog_years.pl
Enter your age: 21
Your age in dog years is 3
```

There happens to be a single statement in each block in this example, but there is no limit to the number of statements that can be enclosed within a block. Finally, for sake of completeness, code that specifies what is wrong with bad input data can be included in the program. A handy `elsif` block can test one string after another.

```
#!/usr/bin/perl
print "Enter your age: ";
$age = <>;
if ($age <= 0) {
  print "You are way too young to be using a computer.\n";
} elsif ($age >= 100) {
  print "Not in a dog's life!\n";
} else {
  print "Your age in dog years is ",$age/7,"\n";
}
```

There are now two `$age` checks in this program. The first test compares the age with zero and prints out a warning message if it is less than or equal to zero. If the test is false, then the program proceeds to the `elsif` block and tries the second test, which compares `$age` with 100. If *this* test is true, then the second error message is printed out. Otherwise, if both tests are false, the program falls through to the `else` block. The possible outcomes are now as follows:

```
% dog_years.pl
Enter your age: -20
You are way too young to be using a computer.
% dog_years.pl
Enter your age: 999
Not in a dog's life!
% dog_years.pl
Enter your age: 28
Your age in dog years is 4
```

WHAT IS TRUTH?

The words *true* and *false* have been tossed around rather blithely in this chapter. It is now appropriate to define these terms more pricisely. Regardless of the meaning of truth in the broader, philosophical sense, truth in Perl boils down to four very simple rules:

Zero (0) is *false*.
The empty string (" ") is *false*.
The undefined value is *false*.
Everything else is *true*.

The various numeric and string comparison tests that were illustrated in the previous section evaluate to 1 when true and to the undefined value when false.

LOOPS

Conditional statements allow the flow of a program to be modified so that sections of code can be executed or skipped over as needed. They cannot, however, make the program execute a particular section of code more than once. For this, Perl (and most programming languages) utilizes what are known as *loops*. Perl has a lot of different types of loops available to the user, but the most useful one is the **while** loop. The while loop looks very much like an **if** block, but instead of executing the contents of the block once if the test is true, it executes the statements repeatedly so long as the test is true. An example of this type of loop is illustrated by this simple counting program:

```
#!/usr/bin/perl
$count = 1;
while ($count <= 5) {
  print "$count potato\n";
  $count = $count + 1;
}
```

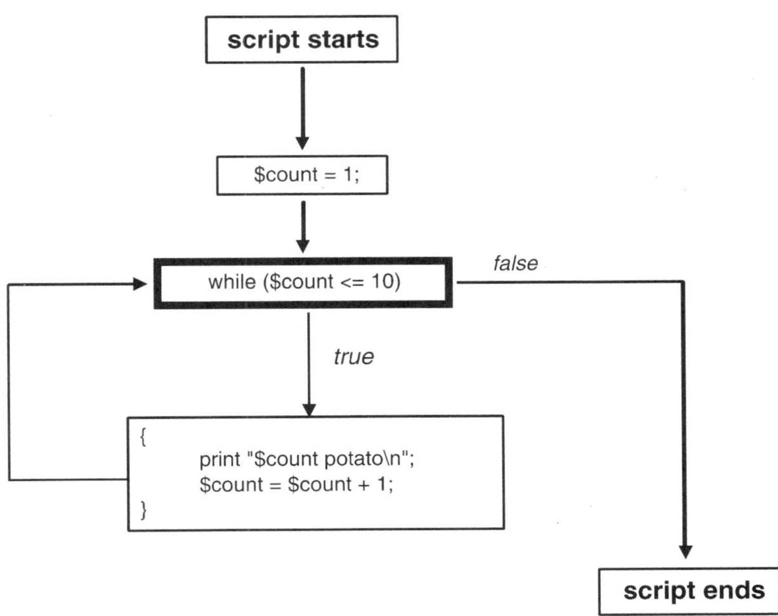

Figure 17.3. While loops execute the same block repeatedly until a defined condition is satisfied.

Before the loop begins, the program creates a variable named `$count` and sets its value to 1. The `while` loop test checks whether `$count` is less than or equal to 10 and executes the two statements contained in the curly braces as long as this condition holds true. The first statement prints out the current value of `$count`, and the second increments the variable by 1. The first time the while statement is encountered, `$count` was set to 1, so the block is executed. The second time, `$count` is 2, and the block is executed again. This continues until `$count` is 6, at which point the test is no longer true (because `$count` is greater than 5) and the loop terminates. The output for the program is quite simply

```
% count.pl
1 potato
2 potato
3 potato
4 potato
5 potato
```

In the same way that `unless` reverses the sense of the `if` statement, `until` can be used as an alternative to `while`. `Until` loops will execute the contents of a block until a certain test becomes true. The counting program could then be rewritten as follows, to yield the same output:

```
#!/usr/bin/perl
$count = 1;
```

```
until ($count > 5) {
  print "$count potato\n";
  $count = $count + 1;
}
```

COMBINING LOOPS WITH INPUT

Loops become very powerful when combined with input statements. Consider this simple example:

```
#!/usr/bin/perl
print "Type something> ";
while ( defined($line = <> ) ) {
  chomp $line;
  print "You typed '$line'\n\n";
  print "Type something> ";
  }
```

This script prints out the prompt `Type something>` and immediately enters a `while` loop. The `while` loop here is a little different from the ones shown before. Instead of doing a comparison, the while loop's test contains the expression `defined($line = <>)`. This expression looks a bit bizarre but can be explained very simply. The first thing that happens is that a line of input is read and assigned to the variable `$line`. Then, `$line` is then passed to a new function named `defined`, which returns true if the contents of a variable are defined, false otherwise. Recall that <> reads lines from a file (or keyboard) until it reaches the end of the file, at which point it returns undefined. This test is telling the while loop to read input lines one at a time until the end of the file is reached. The parentheses ensure that defined is called after the line is read into `$line`, not before.

For each line read in this way, the while loop executes three statements. The first statement calls chomp to remove the newline character from the end of `$line`. The second statement uses variable interpolation to insert the input line into the string `"You typed '$line'\n\n"` and then prints the resulting string out. The last statement displays the prompt again. The result looks like this:

```
% echo.pl
Type something> hi there
You typed 'hi there'
Type something> this is something
You typed 'this is something'
Type something> ^D
%
```

As you can see, this program echoes back everything typed in, like a particularly annoying child. Even if nothing were typed in, the program would simply echo back an empty string; this is because the script would read a single newline character, which it would then remove with chomp, yielding an empty string. To stop the program, an *end-of-file* character has to be sent. On UNIX and Macintosh systems, this is done by typing control-D (written ^D for short). On Microsoft Windows systems, this is done by typing control-Z (^Z). An alternative is to kill the program by typing control-C (^C). This stops the program dead in its tracks.

Because it can be inconvenient to remember obscure control characters, the program can be made a bit friendlier by allowing the user to type `quit` to exit the loop:

```
#!/usr/bin/perl
print "Type something. 'quit' to finish> ";
while ( defined($line = <>) ) {
  chomp $line;
  last if $line eq 'quit';
  print "You typed '$line'\n\n";
  print "Type something> ";
  }
print "goodbye!\n";
```

The main change is the line that immediately follows the `chomp`. This introduces a new function named `last`, which acts as a loop modifier. It is only allowed to occur within the body of a loop. When executed, `last` causes the script to exit the loop immediately, even though the loop test may still be true. This statement compares the contents of `$line` to the string `quit`. If they match, the `last` function is executed and the loop finishes. Running the program now, entering the word **quit** exits the loop so that the last line (which prints "goodbye!") can be executed.

Because the process of reading and processing incoming data a line at a time is so common, Perl provides a handy shortcut. If <> appears all alone in the test part of a `while` statement, Perl will read a line into an automatic variable with the odd-looking name `$_` (which is read as "dollar sign underscore") and then call `defined` on your behalf. Furthermore, many text-processing functions operate on `$_` by default, including `chomp`. Taking advantage of this shortcut, one can rewrite the previous example in this way:

```
#!/usr/bin/perl
print "Type something. 'quit' to finish> ";
while (<>) {
  chomp;
  last if $_ eq 'quit';
  print "You typed '$_'\n\n";
  print "Type something> ";
  }
print "goodbye!\n";
```

STANDARD INPUT AND OUTPUT

When a Perl script reads a line from <> and prints without using a specified filehandle, it instead uses two automatic filehandles called `STDIN` and `STDOUT`. The statements

```
print "The answer to life is...";
print STDOUT "The answer to life is...";
```

are exactly equivalent. In the same vein, the statements `$line = <>;` and `$line = <STDIN>;` are almost equivalent. The subtle differences between `<>` and `<STDIN>` will be discussed below.

The names `STDIN` and `STDOUT` are derived from *standard input* and *standard output*, an idea popularized by the UNIX operating system. Standard input and standard output are abstract files from which a script can accept input and send output, respectively. When a script is first launched, standard input corresponds to the keyboard and standard output corresponds to the computer screen. On Windows and UNIX systems, standard output appears in the command interpreter window. On Macs, the output appears in a small scrolling window that MacPerl creates specifically for this purpose. When a Perl script is launched, the user has the option of changing where standard input and output come from and go to. The user can also arrange for the standard output of one script to be sent to the standard input of another script in assembly-line fashion. This is actually a very powerful facility, but one that is beyond the scope of this chapter.

To redirect standard *input* from a file in either Microsoft Windows or UNIX systems, use the less-than (<) symbol to indicate the file:

```
% count_words.pl <C:\My Documents\cosmids\11_22_00cosmids.txt
```

To redirect standard *output* to a file on Windows or UNIX systems, use the greater-than (>) symbol:

```
% reverse_translate.pl > dna.txt
```

Standard input and output can be redirected simultaneously by using both symbols on the same command line:

```
% reverse_translate.pl < protein.txt > dna.txt
```

There is actually a third automatic filehandle called `STDERR`, for *standard error*. Perl sends its error messages and other diagnostics to standard error rather than standard output. When die is used to display error messages, the messages go to the `STDERR` filehandle automatically. `STDERR` is initially attached to the screen, just like `STDOUT`. On UNIX systems, `STDERR` can be redirected using >&. The following command will send the standard output of reverse_translate.pl to a file named `dna.txt`, and send any warnings or errors to a file named `errors.out`.

```
% reverse_translate.pl > dna.txt >& errors.out
```

Perl in Windows and Macintosh systems do not use the concept of a separate, standard error, so this type of redirection will not work on those systems.

Returning to the idea that `<STDIN>` and `<>` are almost—but not exactly—the same, the difference lies in the fact that `<>` contains some additional magic that makes it easy to process command line arguments. Given a file named unsorted.txt that is to be processed using a file named odd_even.pl, one way to process this file would be to redirect standard input, like so:

```
% odd_even.pl < unsorted.txt
```

However, because odd_even.pl contains a line using <> operator to read from standard input, the command can also be written in the following way:

```
% odd_even.pl unsorted.txt
```

The file name unsorted.txt is now being passed to the script as an *argument* and not as standard input. When <> is used, it looks for any filenames on the command line, opens them, and reads from them a line at a time. If no files are mentioned on the command line, <> reads from standard input. This feature can be used to process multiple files at once as well, reading from each of the files mentioned on the command line as is they were all a single, continuous file. Macintosh users, who do not have a command line available to them, can still take advantage of this <> feature. When a Perl script is saved as a droplet, text files can be dragged and dropped on top of its icon. Each of the dropped text files will be passed to <> for reading.

FINDING THE LENGTH OF A SEQUENCE FILE

Moving to a biologically based example, consider a text file containing a large DNA sequence in single-letter format. The sequence is of unknown length, and it would be desirable to quickly determine the number of bases in the sequence. Using the file size alone would not be appropriate, since the presence of end-of-line characters at the end of each line would artificially inflate the number. The Perl script that will be developed in this section will answer this question as well as count up the number of lines in the original text file. This short script will read in the file, one line at a time, removing each line's terminal newline character and determining the length of what is left. The length for that line is added to a running total, with a counter tracking the number of lines in the file.

```
#!/usr/bin/perl
# file_size.pl
$length = 0;  # set length counter to zero
$lines = 0;   # set number of lines to zero
while (<>) {  # read file one line at a time
  chomp;  # remove terminal newline
   $length = $length + length $_;
  $lines = $lines + 1;
}
print "LENGTH = $length\n";
print "LINES = $lines\n";
```

A built-in function named `length` is invoked to determine the length of the line once the terminal newline is removed. When the script is done, it prints out the ultimate values of `$length` and `$lines`:

```
% file_size.pl dna.txt
LENGTH = 50649
LINES = 1387
```

PATTERN MATCHING

The script from the previous section was useful only for calculating DNA lengths in the unusual case of having a file that contains nothing but raw DNA sequence. More frequently, however, sequence data come in FASTA format. To tally up the length of all the sequences in a FASTA file, the lines that begin with > must be ignored. Perl's pattern-matching operations make this easy to do.

A pattern match is a special type of text comparison. It is something like `eq`, but, instead of testing for an exact match between two strings, it tests a string against a pattern, using a pattern description language known as a *regular expression*. A simple example of a pattern match would be

```
print "EcoRI site found!" if $dna =~ /GAATTC/;
```

This if statement compares the contents of the variable `$dna` against the pattern `GAATTC`. The funny-looking =~ symbol is the pattern match comparison operator; think of it as an "approximately equal" comparison. If the string to the left of the pattern match operator contains the indicated pattern, it will return true. In the script fragment above, the program will print out `EcoRI site found!` if the string contains `GAATTC` anywhere along its length.

Regular expression patterns are delimited by forward slashes. The simplest ones contain a sequence of normal characters that must match somewhere within the body of a string. The *EcoR* I-site detector is one such example. Regular expressions are much more powerful than this, however. For example, square brackets can be used to specify a set of alternative characters in the manner shown here:

```
$dna =~ /GGG[GATC]CCC/
```

This pattern matches a sequence of characters beginning with GGG, followed by any of the characters G, A, T, or C, followed by the sequence CCC. In other words, this pattern searches for GGGNCCC.

To search for a series of alternative patterns, you can use the | symbol to separate the alternatives. For example, this will search for either *EcoR* I sites or *Hind* III sites:

```
$dna =~ /GAATTC|AAGCTT/;
```

This facility is greatly enhanced by *metacharacters* and *quantifiers*. A metacharacter represents a whole class of characters. For example, a single dot (.) will match any character except the end of a line, whereas `\d` signifies any digit. There

TABLE 17.3. Regular Expression Metacharacters

Metacharacter	Description
`.`	Any character except newline
`^`	The beginning of a line
`$`	The end of a line
`\w`	Any word character (non-punctuation, non-white space)
`\W`	Any non-word character
`\s`	White space (spaces, tabs, carriage returns)
`\S`	Non-white space
`\d`	Any digit
`\D`	Any non-digit

are also metacharacters that will match the beginning and ending of lines and match the boundaries between one word and the next. Table 17.3 lists some of the more common metacharacters. Notice that there are many cases in which a metacharacter representing a character set is paired with its complement. For example, `\s` matches white space, and `\S` matches nonwhite space (in other words, printing characters). Another frequently used pair, `^` and `$`, match the beginning and end of lines, respectively.

For example, to match a "ZIP+4" format ZIP code, the required regular expression would be written as follows:

```
$address =~ /\d\d\d\d\d-\d\d\d\d/;
```

For a regular expression to match a metacharacter *literally*, it must be preceded by a backslash. For example, to match DNA sequence IDs of the form "M58200.2," where a dot is used literally, the regular expression should be written

```
$sequence_id =~ /\w+\.\d+/;
```

By default, any character or metacharacter in a regular expression matches exactly once. By placing a quantifier after the character, Perl can match a character a specific number of times or a range of times. The simplest type of quantifier is `{M}`, which tells Perl to match the pattern exactly M times. Using this notation, the ZIP+4 regular expression could be rewritten as

```
$address =~ /\d{5}-\d{4}/;
```

Similar quantifiers include the form `{M,N}`, which will match at least *M* times but no more than *N*; `{M,}`, which matches at least *M* times; and `{,N}`, which will match no more than *N* times. To bring the examples back into a biological context, the following expression will match the plant mcrBC methylation site Pu-C-X(40–80)-Pu-C, in which the center of the recognition site can be anywhere from 40 to 80 nucleotides long.

```
$sequence =~ /[AG]C[GATC]{40,80}[AG]C/;
```

Although the curly braces can be used to describe any quantification, there are some short-cut metacharacters that are used for the frequent cases (Table 17.4).

Parentheses can be used to group parts of a regular expression, and then a quantifier can be applied to the entire group. For example, this regular expression will match normal five-digit ZIP codes as well as the ZIP+4 form:

```
$address =~ /\d{5}(-\d{4})?/;
```

The `\d{5}` part matches a digit repeated exactly five times. This is followed by an optional section containing `-\d{4}`, a hyphen followed by four digits. The optional section is completely surrounded by parentheses to group it, and the group is followed by a `?` symbol, meaning that it can match at most once and possibly not at all. Parenthesized groups can also be used to extract portions of regular expressions.

With regular expressions, the DNA length calculator can be rewritten so that it correctly ignores lines that begin with the > sign. The modified program looks like this:

```
#!/usr/bin/perl
# file_size2.pl
$length = 0;
$lines = 0;
while (<>) {
chomp;
$length = $length + length $_ if $_ =~ /^[GATCNgatcn]+$/;
$lines = $lines + 1;
}
print "LENGTH = $length\n";
print "LINES = $lines\n";
```

The key modification to the original script is the insertion of a conditional test on the statement that tallies the length of each line of text. The script tests `$_` for a pattern match with the regular expression `/^[GATCNgatcn]+$/`, which matches lines containing DNA sequence. The initial ^ character matches the beginning of the line, `[GATCNgatcn]+` matches one or more of the characters GATCN or their lower-case equivalents, and the $ matches the end of the line. Other lines in the FASTA file, including blank lines and the description lines, are ignored. The file length that is printed out corresponds now to the sequence only.

TABLE 17.4. Regular Expression Quantifiers

Quantifier	Description
?	0 or 1 occurrence
+	1 or more occurrences
*	0 or more occurrences
{N,M}	Between N and M occurrences
{N, }	At least N occurrences
{ ,M}	No more than M ocurrences

Before leaving this script, some syntactical tricks can be applied to make the script more concise. If a regular expression appears alone without a =~ operator, then Perl assumes that the variable to be tested for a pattern match is `$_`. The string length function also behaves this way, returning the length of `$_` if no variable is explicitly specified. So, the length-tallying line can be rewritten as

```
$length = $length + length if /^[GATCNgatcn]+$/;
```

A second shortcut is to append an `i` flag at the end of the regular expression right after the second slash. This puts the regular expression into *case-insensitive* mode, and allows the statement to be written as

```
$length = $length + length if /^[GATCN]+$/i;
```

Finally, because adding a value to a variable and storing the sum back into the same variable is such a common operation, Perl provides the shortcut operator += (read as "plus equals"). This operator takes a numeric value on its right side, adds it to the contents of the variable on its left side, and stores the result back into the same variable, all in one graceful step. Taking advantage of this feature gives statements of the form

```
$length += length if /^[GATCN]+$/i;
```

Similar assignment shortcuts are summarized in Table 17.5. Putting all these shortcuts together gives the final version of the DNA length-tallying script:

```
#!/usr/bin/perl
# file_size3.pl
$length = 0;
$lines = 0;
while (<>) {
  chomp;
  $length += length if /^[GATCN]+$/i;
  $lines += 1;
}
print "LENGTH = $length\n";
print "LINES = $lines\n";
```

TABLE 17.5. Assignment Shortcut Operators

Operator	Example	Description
`+=`	`$a += 3`	Add a number to a variable
`-=`	`$a -= 3`	Subtract a number from a variable
`*=`	`$a *= 10`	Multiply variable by a number
`/=`	`$a /= 2`	Divide a variable by a number
`.=`	`$txt .= "abc"`	Append a string to a variable

EXTRACTING PATTERNS

Not only are regular expressions good for detecting patterns in text, but they can be used to extract matching portions of the text as well. To see how this might be useful, consider a FASTA description line like this one:

```
>M18580 Clone 305A4, complete sequence
```

In addition to the initial >, the description line contains two different fields that we might like to capture. The first is "M18580," the mandatory sequence ID. The second is "Clone 305A4, complete sequence," an optional human-readable comment. In regular expression terms, the description line looks like this:

```
/^>\S+\s*.*$/
```

Reading from left to right is the beginning of the line (^), a > sign, one or more non-white-space characters corresponding to the sequence ID, zero or more spaces or other white space, zero or more of any character corresponding to the optional description, and the end of the line ($). This regular expression match can be made to extract the ID and description lines simply by putting parentheses around the parts that should be captured:

```
/^>(\S+)\s*(.*)$/;
$id = $1;
$description = $2;
```

When a string successfully matches a regular expression, any portions of the expression that are contained within parentheses are extracted and placed into the automatic variables `$1, $2, $3`, and so forth. The extraction works from left to right and only happens if the entire regular expression matches.

This trick can be used to fix a deficiency in the DNA length calculator from the previous section. Previous versions of this script naively treated the entire FASTA file as a single DNA sequence. However, a FASTA file usually contains entries for multiple sequences. With pattern matching, the description lines can be identified and the sequence IDs extracted, allowing for the length of each sequence to be printed out separately.

```
#!/usr/bin/perl
$id = '';                 # holds sequence ID of current sequence
$length = 0;              # holds length of current sequence
$total_length = 0;        # tallies aggregate length of all seqs
while (<>) {
  chomp;
  if (/^>(\S+)$/) { # found a new description line
    print "$id: $length\n" if $length > 0;
    $id = $1;
    $length = 0;
  } else {
    $length += length;
```

```
    $total_length += length;
  }
}
print "$id: $length\n" if $length > 0; # last entry
print "TOTAL LENGTH = $total_length\n";
```

After initializing the three variables at the top of the program, the script enters a while loop. As before, it reads a line at a time into $_ and removes the terminating newline. The new feature is an if-else block. The block performs a pattern match on the line, looking for a FASTA description line. If one is found, it signals the beginning of a new sequence. The following then occurs:

- If `$length` is non-zero, the ID and length of the previous sequence are printed. The check on `$length` prevents the program from printing out an empty line when it hits the very first sequence in the file.
- The matched sequence ID is copied into `$id`.
- `$length` is set to zero.

Otherwise, the program is in the middle of a sequence, in which case the length of the current line is added to the appropriate counters. After the last line is read, the ID and length of the last sequence in the file are printed, as well as the total length of all the sequences:

```
% fasta_length.pl ests.fasta
D28205: 1105
BCD207F: 402
BCD207R: 332
BCD386F: 192
BCD386R: 362
CDO98F: 374
TOTAL LENGTH = 2767
```

ARRAYS

Previous examples have worked with single-valued *scalar* variables only. However, Perl has the ability to work with multivalued variables as well. There are two basic multivalued variables, named *arrays* and *hashes*. An array is a list of data values indexed by number. A hash is a list of data values indexed by string. Both are very easy to use and incredibly handy. To understand arrays, consider how one might keep track of a large number of identifiers, such as clone names. With scalar variables, one approach could be to assign each clone name to a different variable:

```
$clone1 = '192a8';
$clone2 = '18c10';
$clone3 = '327h1';
...
```

The problem with this approach, besides being tedious, is that it does not offer any way to step through the entire list of clones one by one, performing the same operation on each one. Arrays circumvent this problem. An array can be defined as follows:

```
@clones = ('192a8','18c10','327h1','201e4');
```

This new array, named `@clones`, contains four strings. Array variables begin with an `@` sign to distinguish them from scalar variables, which begin with a `$`. Scalars and array variables are completely separate. In fact, a scalar variable can be named `$clones` and an array variable can be named `@clones` within the same program. They will not interact in any way. As alluded to above, operations can be applied to arrays as a whole. For example, the = operator can be used to copy one array into another:

```
@old_clones = @clones;
```

Items can be added to the end of an array using the **push** function:

```
push @clones,'281e3';
```

After this statement executes, `@clones` will contain five items. The opposite of push is **pop**, which removes the last item from the array, reducing it in size by one, and returns the removed item as its result. This statement will reduce `@clones` back to a length of four , assigning `281e3` to the scalar variable `$last_clone`:

```
$last_clone = pop @clones;
```

Two array operations are particularly common: accessing an arbitrary array element by its positions in the array using *indexing* and looping over each element of the array in order using the **foreach** loop. Considering indexing first, to copy the third element of `@clones` into a variable named `$third_clone`, the statement would be written as

```
$third_clone = $clones[2];
```

This will—and should—look strange at first. The numeral in the square brackets, `[2]`, is the *index*. Perl numbers its arrays starting with *zero*, so the first item is actually index 0, the second item is index 1, and the third item is index 2. Any expression can be placed within the square brackets, as long as that expression ultimately evaluates to an integer. As an example, if a scalar variable `$i` contains a number, to address the next element in the series, it could be referred to as `$clones[$i+1]`. More mystifying, however, is the `$` at the beginning of the array name. What happened to `@clones`?

When one indexes into an array, the symbol at the front refers to the individual array element, *not* the array as a whole. Because the element itself is a scalar, the symbol at the front should be a `$`. To clarify, look at the following two examples.

```
@old_clones = @clones;
$first_clone = $clones[0];
```

The first line would copy the *entire* array, whereas the second line copies just a single element within the array.

Arrays can be extremely long; ones with thousands of elements are not unusual. A common operation is to loop through each member of an array and do something with it. The `foreach` loop makes this possible. For instance, say that the array `@dna` contains a list of DNA sequences and that a printout of the length of each element would be helpful. The following small loop would accomplish this.

```
foreach $dna (@dna) {
  print length $dna,"\n";
}
```

The `foreach` loop has three parts: the name of a scalar variable known as the *loop variable*, an array name enclosed in parentheses, and a block containing the statements to be executed. `Foreach` steps through the array one element at a time, placing each element in the loop variable and executing the statements within the block. The statements may examine the contents of the loop variable and act on it or even change the loop variable to change the corresponding array element. After the loop is finished, the loop variable will again contain whatever it had before the loop began or be undefined if this is the only time it was used.

To illustrate how assigning to the loop variable changes the contents of the array, below is a fragment of code that will treat every element of an array as a DNA sequence, replacing it with its reverse complement:

```
foreach $dna (@dna) {
  $dna = reverse $dna; # reverse it
  $dna =~ tr/gatcGATC/ctagCTAG/; # complement it
}
```

The two statements in the block show off a pair of Perl functions that have not yet been discussed. The first of these, **reverse**, returns the reverse of a scalar variable, turning `GGGGTTTT` into `TTTTGGGG`. The reversed sequence is assigned back into the loop variable. The next statement uses the **tr** function (for *translate*) to substitute one set of characters with another. `tr` has an unusual syntax that is rooted in Perl's historical origins. It uses the slash as a delimiter, replacing the list of characters between the first set of delimiters with the characters in the second set. The replacement occurs on whatever variable `tr` is bound to using the =~ operator (the syntax should be reminiscent of pattern matching). Characters not mentioned in the list are left unchanged. In the example above, the list gatcGATC is replaced with ctagCTAG. What happens is that "g" is replaced with "c," "a" with "t", and so forth. Thus TTTTGGGG becomes AAAACCCC, which is the reverse complement of the original element, `GGGGTTTT`.

ARRAYS AND LISTS

Perl lists are closely related to arrays. Lists are a set of constants or variables enclosed in parentheses. An example of a list of strings would be (`"one", "two", "buckle my shoe"`), whereas an example of a list of variables would be (`$a, $b, $c`). A list that combines variables, constant strings, and constant numbers might be something like (`$a, "the Roman empire", 3.1415926, $ipath`). Lists can be thought of as being related to array variables in the same way that the constant 123.4 might be related to the scalar variable `$total`.

Lists are useful for performing operations in parallel. For example, lists can be assigned to array variables to make the array identical to the list. In fact, one example of this was shown earlier:

```
@clones = ('192a8','18c10','327h1','201e4');
```

Arrays can be assigned to lists, provided that each element of the list is a variable and not a constant. For example, to extract the first three elements of an array, one could write

```
($first,$second,$third) = @clones;
```

After this operation, `$first` will contain `18c10`, and so on. Naturally enough, lists can be assigned to lists as well, again provided that the list on the left contains variables only:

```
($one,$two,$three) = (1,2,3);
```

SPLIT AND JOIN

It is often very useful to transform strings into arrays and to join the elements of arrays together into strings. The **split** and **join** functions allow for these operations. Split takes two arguments: a delimiter and a string. It splits the string at delimiter boundaries, returning an array consisting of the split elements. The delimiters themselves are discarded. To illustrate this, consider a case requiring the manipulation of a long file containing comma-delimited files, such as the following:

```
192a8,The Sanger Centre,GGGTTCCGATTTCCAA,CCTTAGGCCAAATTAAGGCC
```

Split makes it easy to convert the long string into a more manageable array. To split on the comma, the comma is used as the delimiter:

```
chomp($line = <>);           # read the line into $line
@fields = split ',',$line;
```

`@fields` will now contain the five individual elements, which can now be indexed or looped over. Split is often used with a list on the left side instead of an array, allowing one to go directly to assigning to the list. For example, rather than creating an array named `@fields`, the result of a split command can assign values to a list of named, scalar variables:

```
($clone,$laboratory,$left_oligo,$right_oligo) = split ',',$line;
```

The `join` function has exactly the opposite effect of split, taking a delimiter and an array (or list) and returning a scalar containing each of the elements joined together by the delimiter. Thus, continuing the earlier example, the `@fields` array can be turned into a tab-delimited string by joining on the tab character (whose escape symbol is `\t`):

```
$tab_line = join "\t",@fields;
```

After this operation, `$tab_line` will look like this:

```
192a8 The Sanger Centre GGGTTCCGATTTCCAACCTTAGGCCAAATTAAGGCC
```

HASHES

The last Perl data type that will be considered is the **hash**. Hashes are similar to arrays in many respects. They hold multiple values, they can be indexed, and they can be looped over, one element at a time. What distinguishes hashes from arrays is that the elements of a hash are unordered, and the indexes are not numbers but strings. A few examples will clarify this:

```
%oligos = ();
$oligos{'192a8'} = 'GGGTTCCGATTTCCAA';
$oligos{'18c10'} = 'CTCTCTCTAGAGAGAGCCCC';
$oligos{'327h1'} = 'GGACCTAACCTATTGGC';
```

In this example, an empty hash named `%oligos` is created, and three elements are then added to it. Each element has an index named after the clone from which it was derived, and a value containing the sequence of the oligo itself. After the assignments, the values can be accessed by indexing into the hash with curly braces.

```
$s = $oligos{'192a8'};
print "oligo 192a8 is $\n";
print "oligo 192a8 is ",length $oligos{'192a8'}," base pairs long\n";
print "oligo 18c10 is $oligos{'18c10'}\n";
```

This will print out

```
oligo 192a8 is GGGTTCCGATTTCCAA
oligo 192a8 is 16 base pairs long
oligo 18c10 is CTCTCTCTAGAGAGAGCCCC
```

Just as a variable containing an integer can be used as an index into an array, a variable containing a string can be used as the index into a hash. The following example uses a loop to print out the sequence of each of the three oligos previously defined by `%oligos`:

```
foreach $clone ('327h1','192a8','18c10') {
          print "$clone: $oligos{$clone}\n";
}
```

As with arrays, there is a distinction between the hash as a whole and individual elements of a hash. When referring to an element of a hash using its index surrounded by curly braces, one is referring to the *scalar* value contained within the hash, so the $ symbol must be used as a prefix. To refer to the hash variable as a whole, use the % symbol as the prefix. This allows for one hash to be assigned to another, as well as the ability to perform other whole-hash operations.

For historical reasons, the indexes of a hash are called its *keys*. Calling the **keys** function produces a list of all the keys in the hash. Using a command of the form `@clones = keys %clones;` will assign to the array three string elements (`'327h1','192a8','18c10'`), *but in no predictable order*. The elements of a hash are unordered, and the order in which they are put into a hash has no effect on the order in which they are returned.

To get all the values of a hash, use the **values** function:

```
@oligos = values %clones;
```

The `@oligos` array will now contain a three-element list consisting of each of the oligo sequences that were placed into the hash. As with keys, the values are returned in an unpredictable order. However, the order of elements retrieved by keys will match the order retrieved by values. Hence, the position of clone 192a8 in the `@clones` array will match the position of its corresponding oligo in `@oligos`.

A REAL-WORLD EXAMPLE

At this point in the discussion, all the tools needed to solve the problem posed at the very beginning of this chapter are now at hand. The problem will be approached in steps. The first task is to scan through a file of RNAi results, collecting the genes that have anything to do with locomotion. Assume that the input file is named `rnai.txt` and contains lines in the following format; each field is separated by a tab.

Gene	Date	Status	Phenotype Summary
B0310.2	2/18/2000	complete	larval arrest
B0379.4	2/18/2000	complete	none
B0496.8	2/19/2000	incomplete	
ZK899.6	2/19/2000	complete	uncoordinated, coils and kinks
ZK945.6	2/19/2000	complete	hermaphrodites sterile
M6.1	2/19/2000	complete	flaccid paralysis
...			

```
#!/usr/bin/perl

# STEP 1: extract RNA inhibition data
open RNAi,"rnai.txt" or die "Couldn't open rnai.txt: $!";
%interesting_genes = ();
while (<RNAi>) {
      chomp;
      ($gene,$date,$status,$phenotype) = split "\t";
      $interesting_genes{$gene} = $phenotype
            if $phenotype =~ /uncoordinated|paraly|coil|movement|kink|jerk/;
}
close RNAi;

# STEP 2: extract protein sequence data
open WP,"WormPep.fasta" or die "Can't open WormPep: $!";
%sequences = ();
while (<WP>) {
      if (/^>(\S+)$/) { # found a new description line
            $id = $1;
      } else {
            $sequences{$id} .= $_ if $interesting_genes{$id};
      }
}
close WP;

# STEP 3: reformat for submission to SignalP
open SIGNALP,">signalP.txt" or die "Couldn't open signalP.txt: $!";
print SIGNALP "euk\n";
print SIGNALP "graphics\n";
foreach $gene (keys %sequences) {
      print SIGNALP ">$gene $interesting_genes{$gene}\n";
      print SIGNALP $sequences{$gene};
}
close SIGNALP;
```

Figure 17.4. This script reformats a set of neuron-related *C. elegans* genes for submission to the SignalP signal peptide prediction program.

The main challenge here is to search the Phenotype Summary field for results having to do with locomotion. This is a fuzzy sort of problem, best solved using Perl's pattern-matching facility. After scanning through the results for a while, we decide to search for any of the following keywords: uncoordinated, paralysis, paralyzed, coils, coiler, movement, kinky, and jerky.

With this decision made, the first part of the script can be written (Fig. 17.4, Step 1). The script attempts to open the file `rnai.txt`. If unsuccessful, it dies with an error message. If successful, it initializes the hash `%interesting_genes` to empty. This hash will be used to hold the list of locomotion-related genes and will be set up so that the keys are gene names and the values are the corresponding phenotypes. The program then steps through the input file, one line at a time. It chomps off the newline from the end of each line and then uses the `split` function to split each line into fields, using the tab character as the delimiter. This results in a four-element list, which is assigned to an array containing the variables `$gene`, `$date`, `$status`, and `$phenotype`.

Once this is done, the contents of `$phenotype` are compared to a regular expression. The regular expression is a series of alternatives, separated by the | character. Any phenotype that contains any of the strings listed will produce a match. Notice that some of the keywords have been shortened to reduce the length of the expression, as well as pull in some terms that may not have been anticipated. For example, `paraly` will match both `paralysis` and `paralyzed`, without much risk of matching something unintended. If the pattern matches, then the corresponding gene is recorded. By the end of the loop, `%interesting_genes` will be populated with all of the genes whose phenotypes matched the target list. Because `rnai.txt` will no longer be used, its filehandle is closed.

In Step 2 of Figure 17.4, the WormPep set of predicted *C. elegans* proteins will be stepped through, pulling out all the ones matching the collection of genes identified in Step 1. WormPep's format is similar to the standard FASTA format:

```
>2L52.1 CE20433 Zinc finger, C2H2 type (CAMBRIDGE) protein id:CAA21776.1
MSMVRNVSNQSEKLEILSCKWVGCLKSTEVFKTVEKLLDHVTADHIPEVIVNDDGSEEVV
CQWDCCEMGASRGNLQKKKEWMENHFKTRHVRKAKIFKCLIEDCPVVKSSSQEIETH...
```

The portion of the definition line that immediately follows the > is the name of the gene. The program needs to step through all of these entries, extracting the gene names and saving the sequences of those contained in the collection of interesting genes. The script begins Step 2 by opening the WormPep file using a filehandle named `WP`. If successful, it initializes a hash named `%sequences`. This hash will have keys corresponding to the names of the interesting genes, with values consisting of their peptide sequences. The script then enters a loop in which it retrieves each line of the WormPep file, pattern-matching it against a regular expression that examines the def lines. If the program detects a definition line, meaning the beginning of a new sequence, it puts the gene name into a scalar variable named `$id`, taking advantage of Perl's ability to extract parenthesized portions of regular expressions.

If the current line does *not* match the regular expression, then the program has hit a sequence line, with the name of the current gene being held in `$id`. The `if` statement tests whether the current gene is an element of `%interesting_genes`, and, if so, the sequence will be read in, growing one line at a time, until the next definition line is reached. At the end of the loop, `%interesting_genes` will be fully populated with the sequences of interest. Note that the newline has not been removed from the end of each line of input sequence data; in this case, it is desirable to keep the newlines, to facilitate later parts of the program.

In Step 3 of Figure 17.4, the gene sequences are formatted into an E-mail message for submission to the SignalP server. The server expects E-mail submissions in the following format:

```
euk
graphics
>ID1 Comments (ignored)
MLETLCYNYLPLCEQLEPVLNVRDKEDLATSLVRVMYKHNLAKEFLCDLIMKEVEKL...
>ID2 More comments (ignored)
MPARRHLSQPAREGSLRACRSHESLLSSAHSTHMIELNEDNRLHPVHPSIFEVPNCF...
.
```

The first two lines contain information required by SignalP to properly process the sequences; here, the server is being instructed that the sequences are from a eukaryote and that graphics of the predictions should be returned. Following these flags are the sequences, in FASTA format. The sequence ID is required, and anything following it is ignored by the server. A dot follows the last sequence in the file.

The script begins Step 3 by attempting to open the file `signalP.txt` for writing. If successful, it writes the top two lines of the outgoing E-mail message to the file. The program then enters a `foreach` loop, calling the `keys` function to recover all keys from the `%sequences` hash. This retrieves all of the names of all of the genes for which sequence information has been assembled. For each gene, a description line containing the gene name and its phenotype is printed to the filehandle. Although the phenotype will be ignored by SignalP, the information is retained for future reference. After this, the sequence of the gene is printed, newlines and all. At the end of the loop, the `signalP.txt` filehandle is closed. The final step in the analysis is for the researcher to take the newly created `signalP.txt` file and to e-mail it to the SignalP server. The most useful part of this script is that it can be automatically rerun each time WormPep is updated to repeat the analysis.

WHERE TO GO FROM HERE

Perl has many features that cannot be covered in a single chapter. For example, *subroutines* allow one to define customized functions that take arguments and return a result. *References* allow sophisticated data structures such as lists of lists to be created. *Objects* allow large, complex programs to be written so that code can be reused in different contexts. *Pipes and processes* allow for the control of external programs, perhaps to create an automated pipeline invoking commonly used programs.

Last, but not nearly least, there are *modules*, which are libraries of useful code routines put together by Perl programmers around the world and made available for public use. For example, the `Mail::Sendmail` module would enable the SignalP-processing script to E-mail its formatted message directly to the SignalP server without creating an intermediate file. Other modules allow for the creation of interactive, graphical front ends for programs or the creation of dynamic Web pages. Most relevant for biologists are the modules that form Bioperl, an extremely powerful collection of tools for searching biological databases, manipulating and processing sequences, and analyzing nucleotides and proteins.

INTERNET RESOURCES FOR TOPICS PRESENTED IN CHAPTER 17

Perl home page	*http://www.perl.com/*
Comprehensive Perl Archive Network	*http://www.cpan.org/*
BioPerl	*http://www.bioperl.org/*

SUGGESTED READING

Schwartz, R. L. (1998). Learning Perl, 2nd Ed. (O'Reilly & Associates).

Christiansen, T., and Torkington, N. (1999) Perl Cookbook (O'Reilly & Associates).

GLOSSARY

An extensive glossary of genetic terms can be found at the Web site for the National Human Genome Research Institute (*http://www.nhgri.nih.gov/DIR/VIP/Glossary/*). The entries in this glossary provide a brief written definition of the term; the user can also listen to an informative explanation of the term using RealAudio.

algorithm Any sequence of actions (e.g., computational steps) that perform a particular task.

analogous In phylogenetics, characters that have descended in a convergent fashion from unrelated ancestors.

browser Program used to access sites on the World Wide Web. By using hypertext markup language (HTML), browsers are capable of representing a Web page the same way regardless of computer platform.

candidate gene A gene that is implicated in the causation of a gene. Candidate genes lie in a region that has been identified through genetic mapping. The protein product of a candidate gene may implicate the candidate gene as being the actual disease gene being sought.

cDNA library A collection of double-stranded DNA sequences that are generated by copying mRNA molecules. Because these sequences are derived from mRNAs, they contain only protein-coding DNA.

characters and character states In phylogenetics, characters are homologous features in different organisms. The exact condition of that feature in a particular individual is the character state. As an example, the character "hair color" can have the character states "gold," "red," and "yellow." In molecular biology, the character states can be one of the four nucleotides (A, C, T, or G) or one of the 20 amino acids. Please note that some authors define character to mean the character state as defined here.

client A computer, or the software running on a computer, that interacts with another computer at a remote site (server). Note the difference between client and user.

contig Short for "contiguous." Refers to a contiguous set of overlapping DNA sequences.

cytogenetic map The representation of a chromosome on staining and examination by microscopy. Visually distinct light and dark bands give each chromosome a unique morphological appearance and allow for the visual tracking of cytogenetic abnormalities such as deletions or inversions.

descriptor Information about a sequence or set of sequences whose scope depends on its placement in a record. It is placed on a set of sequences to reduce the need to save multiple redundant copies of information.

domain name Refers to one of the levels of organization of the Internet and used to both classify and identify host machines. Top-level domain names usually indicate the type of site or the country in which the host is located.

download The act of transferring a file from a remote host to a local machine via FTP.

E-mail Electronic mail. Refers to messages that can be composed on the computer and transmitted via the Internet to a remote location within seconds. [*Ant*: snail mail, postal mail.]

EST Expressed Sequence Tags. These are usually short (300–500 bp) single reads from mRNA (cDNA) that are usually produced in large numbers. They represent a snapshot of what is expressed in a given tissue or at a given developmental stage. They represent tags (some coding, others not) of expression for a given cDNA library.

exon Within a gene, a region that codes for part of the gene's protein product; the "**ex**pressed regi**on**" of a gene.

FAQ Frequently asked questions. Exactly what it sounds like: a compiled list of questions and answers intended for new users of any computer-based resource, such as mailing lists or newsgroups.

feature Annotation on a specific location on a given sequence.

firewall A computer separating a company or organization's internal network from the public part, if any, of the same network. Intended to prevent unauthorized access to private computer systems.

FTP File transfer protocol. The method by which files are transferred between hosts.

genetic map Gives the relative positions of known genes and or markers. Markers must have two or more alleles that can be readily distinguished.

genome All of the DNA found within each of the cells of an organism. Eukaryotic genomes can be subdivided into their nuclear genome (chromosomes found within the nucleus) and their mitochondrial genome.

Gopher A document delivery system allowing the retrieval and display of text-based files.

GSS Genome Survey Sequences. This DDBJ/EMBL/GenBank division is similar in nature to the EST division, except that its sequences are genomic in origin, rather than being cDNA (mRNA). The GSS division contains (but will not be limited to) the following types of data: random "single pass read" genome survey sequences, single pass reads from cosmid/BAC/YAC ends (these could be chromosome specific, but need not be), exon-trapped genomic sequences, and Alu PCR sequences.

GUI Graphical user interface. Refers to software front ends that rely on pictures and icons to direct the interaction of users with the application.

heuristic algorithm An economical strategy for deriving a solution to a problem for which an exact solution is computationally impractical or intractible. Consequently, a heuristic approach is not guaranteed to find the optimal or "true" solution.

homologs/homologous In phylogenetics, particular features in different individuals that are genetically descended from the same feature in a common ancestor are termed homologous. In molecular biology, homologous is often used simply to mean similar, regardless of genetic relationship.

homoplasy Similarity that has evolved independently and is not indicative of common phylogenetic origin.

host Any computer on the Internet that can be addressed directly through a unique IP address.

HTG/HTGS High Throughput Genome Sequences. Various genome sequencing centers worldwide are performing large-scale, systematic sequencing of human and other genomes of interest. The databases have deemed it beneficial to put the unfinished sequences resulting from such sequencing efforts in a separate division. HTG sequence entries undergo a maturation process. In Phase 0, the entry contains a single-to-few pass read of a single clone. In Phase 1, the entry contains unfinished sequence, which may be unordered, contain unoriented contigs, or a large number of gaps. In Phase 2, the entry still contains unfinished sequence but is ordered, with oriented contigs that may or may not contain gaps. In Phase 3, the entry contains finished sequence, with no gaps; at this point, the entry is moved into the appropriate primary GenBank division.

HTML Hypertext markup language. The standard, text-based language used to specify the format of World Wide Web documents. HTML files are translated and rendered through the use of Web browsers.

haplotype Sets of alleles that are usually inherited together.

hyperlink A graphic or text within a World Wide Web document that can be selected using a mouse. Clicking on a hyperlink transports the user to another part of the same Web page or to another Web page, regardless of location.

hypertext Within a Web page, text that is differentiated either by color or by underlining, which functions as a hyperlink.

indel Acronym for "**in**sertion or **del**etion." Applied to length-variable regions of a multiple alignment when it is not specified whether sequence length differences have been created by insertions or deletions.

Internet A system of linked computer networks used for the transmission of files and messages between hosts.

IP address The unique, numeric address of a computer host on the Internet.

Intranet A computer network internal to a company or organization. Intranets are often not connected to the Internet or are protected by a firewall.

Java A programming language developed by Sun Microsystems that allows small programs (applets) to be run on any computer. Java applets are typically invoked when a user clicks on a hyperlink.

LAN Local Area Network. A network that connects computers in a small, defined area, such as the offices in a single wing or a group of buildings.

LOD score For "**l**og **od**ds," a statistical estimate of the linkage between two loci on the same chromosome.

molecular clock The hypothesis that nucleotide or amino acid substitutions occur at more or less a fixed rate over evolutionary time, like the slow ticking of a clock. It has been proposed that given a calibration date and a constant molecular clock, the amount of sequence divergence can be used to calculate the time that has elapsed since two molecules diverged.

mutation studies In Sequin, a set of sequences for the same gene in the same species, perhaps the same individual, in which several different induced mutations are isolated and sequenced.

oligo For oligonucleotide, a short, single-stranded DNA or RNA. Most often used as probes for the detection of complementary DNA or RNA.

orthologs/orthologous Homologous sequences are said to be orthologous when they are direct descendants of a sequence in the common ancestor, i.e., without having undergone a gene duplication event. See also homologs and paralogs.

PAM matrix PAM (percent accepted mutation) and BLOSUM (blocks substitution matrix) are matrices that define scores for each of the 210 possible amino acid substitutions. The scores are based on empirical substitution frequencies observed in alignments of database sequences and in general reflect similar physicochemical properties (e.g., a substitiution of leucine for isoleucine, two amino acids of similar hydrophobicity and size, will score higher than a substitution of leucine for glutamate.)

paralogs/paralogous Homologous sequences in two organisms A and B that are descendants of two different copies of a sequence created by a duplication event in the genome of the common ancestor. See also homologous and orthologs.

pedigree A tree representation of a family (cohort) showing the relationships between members and the pattern of inheritance of a given trait.

phylogenetic studies In Sequin, a set of sequences for the same gene in individuals of different species. The presumption is that the individuals cannot interbreed. Sequin does not allow a single organism name but expects the organism to be encoded in the definition line. It does, however, present a control for setting the proper genetic code.

physical map A genome map showing the exact location of genes and markers. The highest-resolution physical map is the DNA sequence itself.

platform Properly, the operating system running software on a computer, e.g., UNIX or Windows 95. More often used to refer to the type of computer, such as a Macintosh or PC-compatible.

polymorphism A gene or locus that can have one or more alleles or haplotypes.

population studies In Sequin, a set of sequences for the same gene in individuals of a single species. The presumption is that the individuals can interbreed. Sequin allows entry of a single organism name, although it would expect that some distinguishing source information, such as strain, clone, or isolate, be entered for each sequence.

positional cloning Relies on the identification of a gene through pedigree analysis, genetic and physical mapping, and mutation analysis. Does not require extensive knowledge of the biochemistry of the disease to determine the gene responsible for the disease. [*Ant*: functional cloning.]

protein name In a sequence record, the preferred field for a protein feature.

protein description In a sequence record, used if the protein name is not known.

server A computer that processes requests issued from remote locations by client machines.

site An individual column of residues in an amino acid or nucleotide alignment. The residues at a site are presumed to be homologous.

spam Postings to newsgroups or mail broadcast to a large number of E-mail accounts that usually are wholly irrelevant or not of interest to the recipients. Analogous to postal junk mail.

STS Sequenced Tagged Sites. An operationally unique sequence that identifies the combination of primer pairs used in a PCR assay, generating a reagent that maps to a single position within the genome. STS sequences are usually on the order of 200–500 bases in length. dbSTS is a division of GenBank devoted to STS sequences; it is intended to facilitate cross-comparison of STSs with sequences in other divisions for the purpose of correlating map positions of anonymous sequences with known genes.

Telnet An Internet protocol or application that allows users to connect to computers at remote locations and use these computers as if they were physically sitting at that computer.

URL Uniform resource locator. Used within Web browsers, URLs specify both the type of site being accessed (FTP, Gopher, or Web) and the address of the Web site.

user The person using client-server or other types of software.

World Wide Web A document delivery system capable of handling various types of non-text-based media.

INDEX